Mastering Fireground Command: Calm the Chaos!

MASTERING FIREGROUND COMMAND

CALM THE CHAOS!

ANTHONY KASTROS & BRIAN BRUSH

FIRE ENGINEERING™
• BOOKS •

Disclaimer

The recommendations, advice, descriptions, and methods in this book are presented solely for educational purposes. Photos are for instructional purposes only. Always wear the proper level of approved PPE when conducting training drills and operating at incidents. The author and publisher assume no liability whatsoever for any loss or damage that results from the use of any of the material in this book. Use of the material in this book is solely at the risk of the user.

Fire Engineering Books
110 S. Hartford Ave., Suite 200
Tulsa, Oklahoma 74120 USA

800.752.9764
+1.918.831.9421
info@fireengineeringbooks.com
www.FireEngineeringBooks.com

Executive Vice President: Eric Schlett
Vice President, Group Publishing: Amanda Champion
Acquisitions: David Rhodes, Diane Rothschild
Vice President of Content Operations: Starlet Franz
Production Manager: Tony Quinn
Customer Success Manager: Lane Nash
Book Designer: Robert Kern, TIPS Publishing Services, Carrboro, NC
Cover Designer: Jared Hood

Library of Congress Cataloging-in-Publication Data

Names: Kastros, Anthony, author. | Brush, Brian, author.
Title: Mastering fireground command : calm the chaos! / Anthony Kastros & Brian Brush.
Description: Tulsa, Oklahoma, USA : Fire Engineering Books, [2024] | Includes index.
Identifiers: LCCN 2024016059 (print) | LCCN 2024016060 (ebook) | ISBN 9781593705992 (hardback) | ISBN 9781593705992 (epub)
Subjects: LCSH: Command and control at fires. | Fire extinction. | Fire risk assessment.
Classification: LCC TH9310.8 .K37 2024 (print) | LCC TH9310.8 (ebook) | DDC 628.9/25--dc23/eng/20240422
LC record available at https://lccn.loc.gov/2024016059
LC ebook record available at https://lccn.loc.gov/2024016060

Printed in the United States of America

5 27 26

The IC

by Anthony Kastros

Upon arrival on this chaotic scene,
You swear you'd already seen it in a horrible dream.
A radio blares in muffled voices and tones,
You feel a sudden chill in your bones.
"The fire is growing!" the companies yell,
You call dispatch for an additional bell.
Before you assume command on the call,
Everyone thinks your eye's on the ball.
Hoping to pull something out of your sleeve,
When a mother runs up screaming, "My baby won't breathe!"
With soot-stained lips the baby lies in her arms,
It's time to call another alarm.
A jake from the rescue suddenly runs by,
Swiftly taking the baby from Mom without a goodbye.
She looks to you as if to say, "What do I do?"
When the radio blares with "Command, we've lost Engine 2!"
You take a deep breath, remain calm, collected, no fear,
"Hold all the traffic, keep this frequency clear!"
The world on your shoulders, you pray to God for some grace,
When one of the captains at the right time and right place…
"Command, this is Engine 5 in the rear…"
"We've found Engine 2, there's nothing to fear."
You acknowledge the message and take a deep breath,
Thanking God for the gift to have just cheated death.
You set up divisions, two groups, and tell the next crew,
"Come to the command post, I've got a job for you."
"What is it, chief?" the new captain inquires,
"See if you can do something about those downed wires."
Medic 10 transmits a message to you,
"The baby is breathing, we're taking the jake from Truck 2."
The mother has gone so you ask a cop to find her,
He runs off to get her in a blink and a blur.
Just then Division says the fire is knocked,
You think, "Thank God we won't burn down the block!"
Canceling the third alarm with a sigh,
Hoping no one sees the tear in your eye.
Once it's all done and everyone's home,
You settle into bed all alone.
And thank God one more time for the favor He's shown,
And for the skippers you trained, and you've grown.
For they made the fight, won the battle, saved lives,
Now you let the tears flow from your eyes.

Contents

Section I
Fundamentals of Incident Command

Section II
Commanding Structure Fires

Section III
Commanding All-Risk Incidents

Section IV
Appendixes

Foreword

Command of firefighters in potentially dangerous, even deadly conditions is one of the most vital tasks a firefighter can aspire to. It is not for the faint of heart or the novice. *Mastering Fireground Command—Calm the Chaos* is a vital guide for anyone who aspires to that role.

This text is *not just* an Incident Command System training manual, teaching division and group definitions or titles for the various roles that will be applied, although through the various chapters there is plenty of opportunity to familiarize oneself with each of these items. The authors, two highly seasoned chief officers, have chosen the practical approach to incident command, describing not just the various positions, but what each are trying to accomplish at a wide variety of real-world incidents, ranging from simple private dwelling fires to much more complex issues like strip mall and high-rise fires and even vegetation (wildland) fires that can grow to astounding proportions.

A manageable span of control has been proven to be vital to firefighter safety. Dividing the fireground into manageable pieces has been going on for decades, but these two chiefs highlight the ways that each of the various players can enhance their effectiveness in each of their roles and improve communications, command, and control in the process.

Chief Kastros and Chief Brush illustrate the lessons learned at real world incidents in their own lives and share what went well, as well as how applying the lessons they share in this text would have improved the outcome. This book will be a valuable resource for fire officers in any size fire department and will help them to approach incidents that are starting to spiral out of control around them and hopefully be able to *Calm the Chaos*.

I wish I had read it 30 years ago.

—John Norman

Preface

Welcome to *Mastering Fireground Command—Calm the Chaos*!

The genesis for this curriculum was a terrible apartment fire in 2006 where two young children and their father died. I was the incident commander (IC) and a one-year battalion chief (BC). Early one weekday morning, we were dispatched to an apartment fire with a baby trapped. Prior to arrival, we were updated that another child was trapped, and the father went in to rescue his kids and was trapped as well. All three died and all of us who responded were impacted for years to come.

I wept that day. I had lost children and babies in fires before. Each was tragic. This time, I was a new father, and this was the first time I lost a child as a dad. And to top it off, I was the IC.

While I didn't have the horrible experience of removing the babies or their dad, like our firefighters did, I was the IC. The buck stopped with me. I was made painfully aware of my blind spots that day. And, while we tried to console ourselves with comments like, "there was nothing we could have done," or "they were gone before we arrived on scene," nothing comforted us.

I vowed that day to do it better—everything better: incident command and how to use the Incident Command System (ICS) more effectively; strategic and tactical decision-making; training, officer development, hands-on training, multi-company drills, and incident reviews; standard operating guidelines (SOGs); and whatever else necessary to prevent that kind of kick in the gut from ever happening to any of us ever again.

One of my mentors was Chief Alan Brunacini, and he had always been there for me to lend an ear and offer words of wisdom. This was no different. He also had a tragic loss when Brett Tarver died at the Southwest Supermarket Fire in Phoenix, AZ. He built a command training center (CTC) that I had visited back when I was a captain in the training division.

We built a CTC at Metro Fire. We revised our SOGs. And we trained and trained and trained. I started recording my fires with a dash camera, way before GoPro cameras. We reviewed every fire and had an unquenchable thirst to do it better—everything better.

Then, 4 years later, on August 28, 2010, we had an extremely rare second chance. We had a second chance at redemption. Incident #0102713 was dispatched. Again, we were dispatched to a working apartment fire. Again, we were updated that a child was trapped. And again, it was a two-story garden apartment complex with fire on the second floor.

This time, we had 4 years of intense training, incident reviews, and dash camera videos behind us. We had done simulations, hands-on drills, vent-enter-search (VES) drills (before it was VEIS), and table-top scenarios, and I had trained all my captains to be incident commanders, division/groups supervisors, tacticians, decision-makers, and combat leaders. And they trained me. They helped me understand how to better support them from the command post, better train them, and get my act together.

When the apartment fire was dispatched, we were already on the road, turned out and responding to another fire that was clearly extinguished. It was a sign fire at a Goodwill store. We had to pass the apartment fire to get to the extinguished fire. We diverted and responded to the apartment fire with the child trapped. Since the Computer Aided Dispatch (CAD) system showed our alarm responding to the Goodwill store, it dispatched the next closest alarm to the apartment fire. The net result between our diverted alarm and the second alarm from CAD was 9 engines, 6 trucks, 4 BCs, and 3 fire medics responding to the apartment fire. I had a 21:1 span-of-control with multiple apartments burning, the attic, and a child trapped. I needed to organize and calm the chaos RIGHT NOW!

We had practiced and implemented the curriculum in which you are about to embark for 4 years, continuously revising as we learned. The 21:1 span-of-control was reduced to 4:1: fire attack group, rescue group, roof division and medical group. Other than medical, supervised by a paramedic, each group/division was supervised by a company officer. Within 4 ½ minutes of arrival, the boy was on the gurney. He was not breathing, and his carbon monoxide (CO) level was 23%—a lethal dose.

The supervising company officers were empowered to make tactical decisions, account for crews, and act as the safety officers embedded into the incident. Through training and time, we developed immense trust with each other. They didn't need to ask me for additional hose-lines or ask if they could do VES, they just did it. Why? Because training and time equals trust. I trusted them to do the right thing at the right time. Scan the QR code below to watch the dashcam video of the El Camino Fire.

After overhaul, I went to the emergency department to check on the boy. He was in an induced coma, on a ventilator. The doctor said something to me that I will never forget: "He had no respiratory burns. I think your crews may have saved his life." I wept again. This time, they were tears of joy, redemption, and deliverance. I knew in my heart that he was going to make it. He was transported to the Shriner's Children Hospital in downtown Sacramento where they kept him in an induced coma on a ventilator for another week. We prayed with his family for his recovery. One week later, they took him off the meds, off the vent, and he sat up. He had no deficits, just second degree burns to his leg and arm. He walked out of the hospital (fig. P–1).

The firefighters on Engine and Truck 106 were heroes that day. They made it happen. Since then, I have had the honor of telling their story. This was a collective team effort. **Incident command on today's fires and other events is a team sport.** Fires grow way too fast and the window to make a successful rescue closes even faster. The margin of error is zero.

Therefore, we must re-think incident command and evolve beyond the challenges of modern fire behavior, building construction, low mass synthetics, staffing levels, attrition rates, and lack of experience. We must re-think how we think. So, please be ready to challenge yourself and your own assumptions. I was fortunate to be a line BC for 13 years in a busy, diverse, and dynamic metropolitan area. Every day was a learning day, and every day was a training day. My hope is that you will learn from my experiences, both good and bad. I don't want you to get the same occupational road rash that I did.

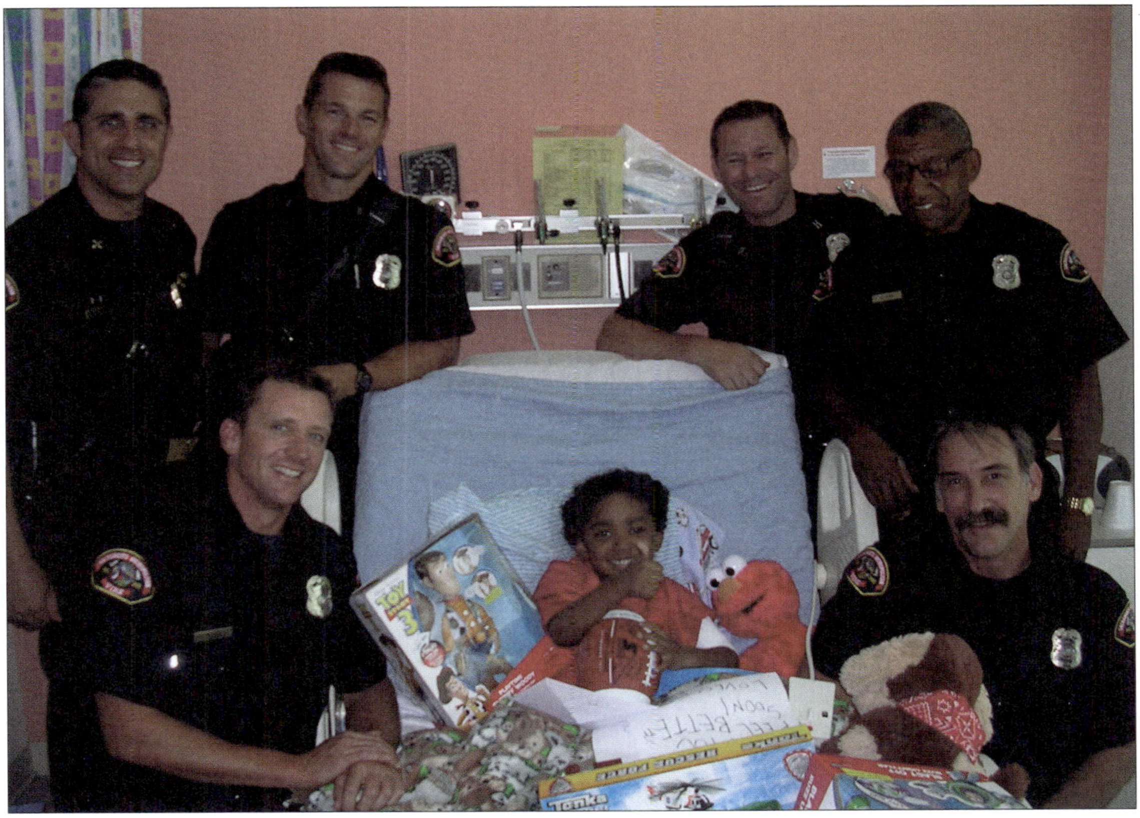

Fig. P–1. This is why we exist—to save lives!

By divine intervention, I reconnected with Brian Brush after reading his article about Firefighter Rescue Survey. I was blown away. The statistics, methods and mission were just what I was looking for. I often asked classes, "If we have the golden hour for trauma, what is the 'golden' timeframe to save fire victims?" Brian and his research answered the question. I called him immediately. Turns out, we had met years before. Long story short, we are kindred spirits and have been training together in the mission to save more civilians ever since.

His contributions, hard work, research, experience and dedication are second to none. I am thankful for him and his friendship. He suggested that we write a book based upon our collective work. Our team at TrainFirefighters.com had already launched our 32-hour online Mastering *Fireground Command—Calm the Chaos* workshop and had taught the hands-on train-the-trainer workshops for years. And Brian's master's thesis and the Firefighter Rescue Survey data and surveys seemed like a natural fit to meet a common goal. So, the book was born.

This curriculum has been taught throughout America since 2011. We have developed a 32-hour online course entitled *Mastering Fireground Command—Calm the Chaos.* The self-paced course is a study guide for this text, and full of incident dashcam footage from over a decade as a line battalion chief. You can see countless examples of the principles from this text applied on real fires from the command post. There are also 115 simulations for you to practice what you learn in this book.

Through our team at TrainFirefighters.com and the hands-on train-the-trainer workshop, fire departments throughout the United States and beyond are training their own firefighters and we could not be more humbled and thankful. We have been blessed to teach this curriculum to thousands of firefighters and they have since saved lives! There is nothing more gratifying, more rewarding, or more important than that.

If you have any questions, please contact our team at Info@TrainFirefighters.com, or scan the QR code to link to our website and learn more about this training.

Keep the faith and keep training! Thank you.

—Anthony Kastros

It is an honor to be a part of this fireground command textbook and to get the opportunity to work with Chief Kastros and so many other very experienced and talented legends in the discipline of command, as I personally am not one of them. I did not come to be involved in this book to share my experience in the seat of command; instead, I came to be involved in this book as a result of identifying a need and being challenged to be a part of the solution.

I began my fire service career in 1996, and through the years I have had a front row seat to witness the influence and impact on our profession of both 9/11 and the technology/information age that arrived with the 21st century. The professional standards and expectations for the fire service, internal and external command and leadership models, and access to evidence-based information for training and decision making have changed drastically over the course of just one generation. At the same time, because of technology changes accelerated by the COVID-19 pandemic, more and more firefighters are accessing programs, training, and education through online delivery methods that improve access. For the student this allows for the resources to respond faster to updates in information, current events, and fire service trends.

In 2017, after over 20 years in the field, I took a position in the training division and I now serve as a training chief. While I had been observing many changes take place around me from a seat on a rig, it wasn't until then that I saw how they all came together. I was in a place that was responsible for the preparation of our members from recruit to chief officer, supporting and guiding the direction of the organization and finding and maintaining resources to do it. Very quickly I realized that many of these things were working against each other and this was not unique to just our department. Over the course of the next few years in training and through the completion of the Executive Fire Officer Program at the National Fire Academy and the Master of Science degree in Fire and Emergency Management Administration at Oklahoma State University, I dedicated much of my time to academics and researching our past, present, and potential future. In that work, it became very clear that the gap between the leadership and command models of our peers in the military and emergency management and that of the fire service was growing, as were the differences in educational content and delivery models the various generations inside the fire service were receiving.

Through traveling the country conducting hands-on training programs and at the department coordinating recruit fire academies, I found our newest firefighters were drawn to and hungry for more programs rooted in concepts such as Extreme Ownership by Jocko Willink and Leif Babin, or those that blended the science and detailed data from United Laboratories Fire Safety Research Institute (ULFSRI) studies with hands-on application in high-quality drills. In conferences across the nation, this need was being met by third-party trainers for task-level firefighters and growing. Unfortunately, the lack of similar training and education for chief and company level officers only continued to grow the gap in comfort and understanding between these groups, and in many departments this was causing friction.

When I attended my first *Calm the Chaos* workshop from Chief Kastros, I saw the foundations of what the fire service needed. Looking further into his success with Mastering the Fire Service Assessment Center with regard to professional development, taking the hands-on portion of *Calm the Chaos* and seeing the decentralized command models in real time practice, and working with him on the deployment of his online curriculum for greater access, I knew this was the method and what was currently missing.

"If you are going to take the time to bring me a problem, you had better also take the time to present a solution." In the fall of 2021, we met with the late Chief Bobby Halton to present our plan. While the training, classes, articles, and online programs were successful, they were each limited in their own way. For the fire service to move from interest to institutionalizing, it would take a new textbook, written specifically to professional qualifications, standards, and higher education expectations, that could serve as an in-hand reference for personnel and policy development, promotional material, and an educational course resource. The fire service has had incredible long-standing resources like this for decades and they were doing their best to meet needs through edition updates, but now was the time to start fresh, codify the present state of the fire service as a blend of experience and evidence, and begin to support our future in a new way. We humbly asked Bobby for his trust in supporting us to take on this challenge as a team, along with so many of our peers who have contributed to this project, and he obliged. What is contained in this book is our first bold step towards our goal of bridging not only the tactical gaps on the fire ground, but also the professional gaps in our service. It is our hope that it leads to further progress.

Thank you for joining us.

—Brian Brush

Acknowledgments

Both Anthony Kastros and Brian Brush would first like to thank God for our wives and children who have always remained steadfast to our united mission to train the fire service and save civilian lives. You are the greatest blessings in our lives.

We also thank God for being firefighters, having been brought together, and for the gift of teaching. Another dear friend in this project is Chief Bobby Halton, for whom we both owe tremendous debts of gratitude for his giving each of us a voice and believing in this project from the beginning.

Thank you to Editor-In-Chief David Rhodes for continual support of this book and Diane Rothschild for keeping us in line. Thank you also to Chris Barton, Tony Quinn, Holly Fournier, Mark Haugh, and the rest of the Fire Engineering/Clarion family for your tremendous support.

Thank you, John Norman, for the incredible honor of writing the foreword.

A very special thank you to our contributing authors from across America who shared your experiences and wisdom as incident commanders: Luis Aldana, Gregg Avery, Anthony Avillo, Robby Bergerson, Jamie Bowron, Mark Brunton, Greg Cassell, Joe Castro, Ken Cook, Dan Cotrell, Scott Dean, Dave Dodson, Ryan Fields-Spack, Mark Ghilarducci, Bill Gustin, Jeff Helvin, Kiley Keeley, Frank Leeb, Shawn Lemon, Brett Loomis, Stephen Marsar, Dave McGrail, Joel Mendoza, Doug Mitchell, Jack Murphy, Nick Papa, David Rhodes, Frank Ricci, Eric Saylors, Dan Shaw, Danny Sheridan, Chris Stavros, Rusty Van Vuren, and Mike Walker.

A very special thank you to the Los Angeles City Fire Department for access to amazing photographs. Thank you specifically to Henry Berkson, David Blair, Greg Doyle, Steve Gentry, James Hedlund-Kaiser, Zak Holman, Joe Lyneis, Rick McClure, Michael Meadows, David Ortiz, Jacob Salzman, Jacob Valin, Cody Weireter, and the Los Angeles City Public Services Bureau. Also, thank you to the Sacramento Metropolitan Fire District and specifically, Battalion Chief Parker Wilbourn.

A tremendous debt of gratitude to Sam Wallace at SawIllustrations.com for the incredible artwork!

Anthony Kastros

I would like to thank my brothers Mitch and Demetrius Kastros for their continued leadership, mentoring, and paving the way in the fire service. This would not be possible without you. Thank you to Brett Loomis, John Wagner, Kevin Wegener, Alda "Mama Bear" Hearne, and the rest of the TrainFirefighters.com team. You are saving lives. Thank you, Cynthia, for saving *my* life, being the love of my life, supporting all the teaching, writing, and doing all the proofreading. And thank you to my precious daughters Sophia and Aubrey for being loving and supportive. All my love, all my life. I adore you.

Brian Brush

I was raised in life by incredible parents who instilled strong values and were powerful examples of determination, service, and work ethic. I was raised in the fire service by so many from Forestville Fire to California Department of Forestry, Hilton Station, and West Metro Fire Rescue who mirrored the same in our profession.

I have been blessed to have my loving and supportive wife Sarah in my life longer than I was without her, and now married for more than 20 years together with our three children Keegan, Lila, and Luke. You all bring out the best in me through your love, faith, and trust.

To my great friends in and outside the fire service: I wish there were enough pages to personally thank you all with a specific memory of your support, but it would be impossible and that is a powerful thing to realize. From Portland, Oregon to Pensacola Beach and New York City to Orange County California, the brotherhood is strong!

Introduction

This book is the sum of over 1000 years of experience. In addition to the two authors, we asked 33 of our colleagues from around America to add their wisdom and insight. These "Wisdom from the Masters" and "Case Studies in Command" add tremendous value to this textbook, for experience must be passed on, internalized, filtered, honestly evaluated, and re-applied to be useful. Unlike any previous work, this book takes the best from all around the United States of America, from the Los Angeles City Fire Department (LAFD) to the Fire Department of New York (FDNY) and dozens of agencies in between, of all sizes and demographics. We sought out the most experienced officers who have commanded some of the most infamous incidents in the last decade. These first-hand accounts overflow with lessons learned to ensure you are successful on your next fire, multi-casualty incident (MCI), or any number of unforeseen events that continually redefine the new normal of incident command and the landscape of emergency operations.

In addition, science now plays a much more significant role in fireground operations. Thanks to the Underwriters Laboratories Fire Safety Research Institute (UL FSRI), Firefighter Rescue Survey, and thousands of hours of research and incident reviews, we can bring quantitative *and* qualitative data to tactical decision-making. This unprecedented addition of data-based decision-making now informs the modern incident commander, well before they ever assume command of their first incident.

We have also analyzed military doctrine and the principles of mission command and decentralization. We have spoken to Navy Seals and 4-star generals. The military has evolved over centuries in the arena of command and control to meet new threats and the ever-changing landscape of the modern battlefield. There is much to be learned and applied that seamlessly transcends the parallel worlds of the fire service and military.

The combination of experience and science, east and west coast philosophy, dozens of case studies, military doctrine, and thousands of hours of training with fire departments from all over America is what lies in the pages of this textbook. And the actual, practical, proven, rapid application of the Incident Command System (ICS) will be introduced in ways never seen in any previous textbook. You will learn how to quickly and effectively apply ICS to house fires, apartment fires, hotels, motels, commercial buildings, taxpayers, big boxes, strip malls, row houses, high rises, maydays, vegetation fires, wildland urban interface, multi-casualty incidents, hazardous materials incidents, and even unified command.

You will learn about how the National Institute of Occupational Safety and Health (NIOSH) 5 falls into the tactical gap, how to train and empower officers at all levels to meet today's and tomorrow's challenges, how to remain ahead of the Incident Power Curve, how to maintain incident alignment and the best tactics to save the most savable civilian lives while ensuring that your troops go home safe from each fire and each run. You will learn what the military teaches: that incident command is now a team sport and you will learn how to build the team for success.

We can never have too much training and wisdom from those who have gone before us. We can never read enough After-Action Reports, NIOSH line-of-duty death (LODD) reports, Green Sheets, Blue Sheets, articles, or books, nor go to enough classes or do enough hands-on training to become bulletproof, immune to danger, gravity, asphyxiation, thermodynamics, construction hazards, illegal remodels, and new unprecedented hazards that yet to be encountered by firefighters.

We must continue to voraciously devour knowledge, train constantly, stay humble, and pass on our wisdom. We will never achieve perfection nor mastery in this business; however, on the way to perfection, we will reach excellence. Mastery is a constant endeavor that is never quite attained.

The training and education you have received up to the moment of the alarm will serve you well. The topics, tactics, tasks, and philosophies you have most recently or most frequently trained upon will be in the front of your mind. Also, the principles you learned in the academy, no matter how many years ago, will also creep back into consciousness without your even realizing it.

New officers are rising through the ranks like never before. New chiefs are leading new company officers who are leading probationary firefighters. This curriculum is proven on a myriad of incident types and at thousands of responses. It has been trained upon and embraced by countless fire departments across America. This will help you promote, be a better fireground officer and incident commander, and move your fire department (FD) to the next level. You can get your entire region on the same page, all the while, using experience, science, east and west, ICS, Fire and Emergency Services Higher Education (FESHE), National Fire Protection Association (NFPA), and countless after-action reports.

1

History of Command

Fig. 1–0. The chief is at front and center with his bugle, the first portable radio, circa 1884.
Source. Photo courtesy of Redwood City (CA) Fire Department

Introduction

As with any journey, you must know where you have been if you want to know where you are going. This first chapter will provide a historical perspective to bring us up date. Before we jump into the nuts and bolts of this book, we thought it would be important to review the history of command and what the current landscape is across America. The *Mastering Fireground Command—Calm the Chaos* curriculum has been taught throughout the United States since 2011. In that time, we have welcomed hundreds of fire departments and thousands of firefighters from small, medium, large, volunteer, combination, and career fire departments. They span from California to New York and Washington to Florida, with everywhere in between. Each class, workshop, hands-on evolution, debriefing, after-action review, and the continuous exposure to incident command has given us the opportunity to work with the finest firefighters in America. We have learned from you and, along with our own first-hand experiences, those of others, and countless hours of research and study has yielded a perspective that is a crucial foundation to understanding incident command.

One thing is for sure: everything is different! Fire departments across North America have countless permutations of how they perform incident command and tactics. From terminology, interpretation, application, consistency, and training, to resources, staffing, frequencies, equipment, apparatus, construction, and population demographics, everything is different.

How did we get here? What can we learn from the past? What will the future look like? Those and other questions are what this chapter aims to answer. Only then can we get into the meat and potatoes of incident command and tactical decision-making.[1] So, enjoy this first chapter as we look into the past. Some names may be familiar, some not. All have contributed something and played a role in where we are today.

Modern Incident Command

The modern fire service demands a highly mature combination of independence and obedience.

—Alan Brunacini

Fire departments have long been described as paramilitary in rank, structure, and operation to highlight similarities between our work and the armed services. Over the last 50 years, fire departments have continued to expand well beyond fire responses alone to all risks, or all-hazard organizations. Each decade has had a new trend. In the 1970s it was Emergency Medical Services (EMS) with Johnny and Roy on *Emergency!* (Look it up if you don't know who Johnny and Roy are.) The 1980s was hazmat response. The 1990s was Urban Search and Rescue (USAR). The 2000s was terrorism and white powder. The 2010s was active shooter. The 2020s have been COVID-19. Nothing has been taken off the table; our mission has only increased. Buildings and contents have also changed.

Duties in the 21st century are moving fire departments more toward the emergency *management* field. With roots in both military and emergency management, municipal fire departments are a natural downstream result of those two tributaries.

Beyond the changes in service types, we also face a change in tempo. In the compartment of a residential structure fire, the wildland urban interface, or, unfortunately, an active shooter event, the pace at which hazards develop and expand is now measured in seconds. The fireground and any emergency scene has become more complex and moves exponentially faster. Our ability to catch it and remain ahead of the incident is vital to the lives of our troops and those we are sworn to save. These changes in duty and in society are not unique or temporary and adaptation is mandatory.

Chief Alan Brunacini and his book *Fire Command* were the first bridges for the fire service from traditional incident command and control models to developing incident management systems. His quote speaks directly to the gap that he sought to address and the way that he identified how our service, and the dynamic nature of our environments, "demands a highly mature combination" of independence to act, but obedience to order.[2] We have all benefited from his wisdom and teachings, and the fire service owes a tremendous debt of gratitude to Chief Brunacini. His mentoring and friendship will be forever cherished. This book would not be possible without him (fig. 1–1).

Over the decades that have followed Chief Brunacini's text, our mission has become more complex and the pace at which our incidents and firegrounds grow has changed. In addition, our military and the emergency management field have further identified systemic operational and organizational conflicts to our mission in the modern setting of personnel, technology, and environment that results from a strict adherence to the traditional command and control model.

The United States has also since experienced the attacks of September 11th, 2001, and Hurricane Katrina in August of 2005. Both events shifted paradigms in incident management theories and practices. The late Dr. Russel Dynes, founder of the Disaster Research Center and widely recognized as an expert in modern emergency management, explains this shift: "Emergencies are characterized by decentralized and pluralistic decision making, therefore the autonomy of decision making should be encouraged, not the centralization of authority."[3] This is not just for the large, once-in-a-career, national-level incidents. This is for house fires, apartment fires, commercial structure fires, vegetation fires, hazardous materials incidents, and multi-casualty incidents.

A chain can be used metaphorically to describe connections, links, and continuity. A chain can also be used physically to secure, prevent, or restrict movement. It is critical that organizations that subscribe to the chain of command actively evaluate their intent, interpretation, and balance of these two, both administratively and operationally. Don't let the chains restrict.

Fig. 1–1. Chief Brunacini (left) and Chief Kastros

Command and Control

Command and control in the fire service today is both unequivocally recognized and routinely misunderstood. The root of this term dates back over 250 years and is credited to Napoleon Bonaparte, where it originated as *Chaos, Command, and Control.* This methodology was developed at a time when armies were expanding empires and conquering lands and peoples. Where *chaos* was encountered due to resistance or language barriers, firm and strict *command* of subordinates had to be established to gain and maintain *control.* During the late 1700s, this model centralized and concentrated power in the French Army to the mind of General Bonaparte. This is the reason his name is so prevalent in history as a great military leader. He had the greatest control.

As empires expanded, armies needed to expand as well. Military service was far from the professional process we know today. People were largely forced into service with the few exceptions of the wealthy and educated taking roles in formal leadership. The success of the authoritarian application of the chaos, command, and control approach to expanding the empire quickly merged into the model used for personnel development and management within the army. To establish order among soldiers who were forced into service, firm and strict command of the subordinate had to be established to gain and maintain control. In modern terms, some may classify the militaries of this era as "low trust" organizations. The truth is that the original organized military models were built on distrust of the subordinate, and this includes that of the United States (fig. 1–2).[4]

Napoleon Bonaparte is recognized for his successes as a conqueror, commander, and leader in the professionalization of the French Army. The peak of the Bonaparte reign was also the infancy of the United States. As the new nation developed, our forefathers looked for models of success from which to build on and many of the Napoleonic ideals were adopted.

What is left out of most historical records is how quickly the model which led to his success contributed to his failure.[5] During the end of the Napoleonic era (1807 to 1815), as his subordinates became better trained, developed experience, and professionalized organizationally and strategically, the strict command and control model came to be viewed as micromanagement and restrictive. At the same time, the armies of his enemies such as Austria, Spain, and Russia also began to professionalize and elevate their armies, making them more capable adversaries. Historian Harold Parker had this to say of Napoleon: "The one-time god of battles [Napoleon] was overthrown by the dynamics of warfare that he had unleashed but failed to comprehend." Napoleon demonstrated success in shifting an approach but failed to continue to evolve. In less than a 20-year period (one generation), the failure to extend *trust and*

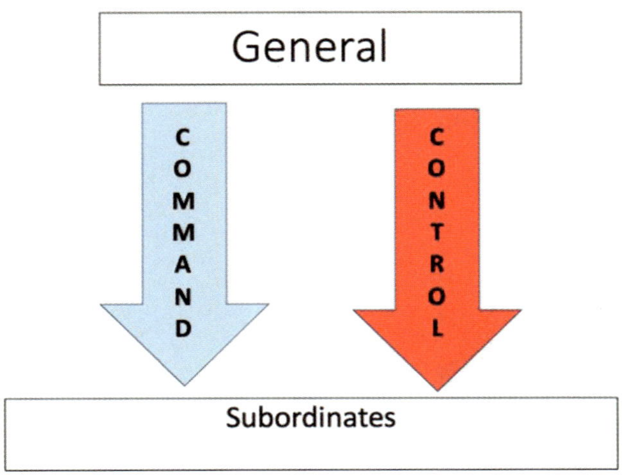

Fig. 1–2. Original military model of command and control

responsibility to developing subordinate leaders and adapt to the new pace of war led to his downfall. In short, he failed to change his command and control model from a centralized to a decentralized ethos, which was organically evolving as a byproduct of the development of his subordinates.

Meanwhile, the fire service of the late 1800s maintained a centralized system of command and control of the chaos where the fire chief would command through his bugle from in front of the fire. This was primarily due to the lack of true organization in the fire service at the time period. Countless pictures from that era show the fire chief with a bugle in posed photos like Figure 1–0 at the start of this chapter, or at the front of the fire in paintings from that era (fig. 1–3).

These bugles were far more than presentation pieces; they were the first communication devices and concentrated authority to a single source. Unfortunately, the fire service held on to this methodology long after changes in technology. Even as portable radios became commonplace and in turn connected everyone on the fireground, some still have the incident commander (IC) controlling all, standing in front with their modern 800 MHz bugle.

Early Officer Development

As the influence of Napoleon faded in Europe, the Prussian Army moved to the forefront of military leadership. When Frederick the Great died, the Prussians realized that power had become too centralized and they were too reliant on the genius of a single person. They, too, would suffer the fate of the French if this was not addressed. During the early 1800s, the Prussians were credited with formalizing the first decision-making training and development programs for staff officers.

Fig. 1–3. Louis Maurer, Rushing to the Conflict, circa 1858[18]

As the professional development and decision making models began to gain traction, the Prussian Great General Staff became the bar by which all other nations began to measure their military command and strategy against. It wasn't until 1901 that the United States Army was in a position to institute a similar program. The U.S Army General Service and Staff College was established in Fort Leavenworth, Kansas. The training curriculum from that school became U.S. Army doctrine in 1910 when "Estimating Tactical Situations and Publishing Field Orders" was published.[6]

Early officer decision and command development training was still concentrated at the highest levels of command. This was steady progress but not a complete paradigm shift. From the civil war into the early 1900s, the U.S. military transitioned from a state of distrust in its subordinates to a low-trust environment as there were some improvements in organization, professionalization, and cohesion. Unfortunately, this stalled as the great demand for mobilization and the cost of attrition associated with World War I limited opportunity. For over a decade, there was virtually no return on investment in any training and development beyond preparing for the battlefield and learning from the field of battle.

In the 1930s and early 40s prior to World War II, officer development began to emphasize communication further down the chain of command. In 1932 the U.S. Army published the first Military Decision-Making Process (MDMP). The direction of the model had five steps for field commanders to consider.[7]

1. Mission
2. Opposing forces
    a. Enemy forces
    b. Own forces
    c. Relative combat strength
3. Enemy situation
    a. Plans open to the enemy
    b. Analysis of enemy plans
    c. Enemy's probable intentions
4. Own situation
    a. Plans open to you
    b. Analysis of plans
5. Decision

Transition to the Fire Service

Reviewing the history of the MDMP is part of fire service history. World War II advanced the fire service further in a shorter period than at any other point in its history. Some of these catalysts included Army officer training that made significant improvements in tactical officer development. The Navy took the role as leaders in fire control techniques and training. Fire apparatus and equipment produced and dispatched around the world to bases and air strips advanced technologies at a pace that was never demanded by domestic municipal fire departments. When the war was over, service members returned to their cities and communities and took on domestic roles in the disciplines they had exposure to and developed experience in during the war. At the same time, the United States began a period of great expansion into the suburban era and the growth of new cities and towns presented more opportunities for positions in organized fire departments than ever before. All the while, the fire service continued to trail the professional development processes of the military by decades.

There were some advancements, however. The most identifiable demonstration of this was Chief Lloyd Layman. Chief Layman authored two books that were published by NFPA, *Attacking and Extinguishing Interior Fires* (1952)[8] and *Fire Fighting Tactics* (1953),[9] both of which provide guidance to U.S. fire departments on how to apply military models and experience to their organizations and operations. Through his work he introduced the fire service to the Facts, Probabilities, Own Situation, Decision, Plan of Operation (FPODP) decision-making process, which is still a foundational approach over 70 years later, as well as Rescue, Exposures, Confinement, Extinguishment, Overhaul, Ventilation, Salvage (RECEO-VS) (fig. 1–4).

1. Facts
    a. What you know for sure
2. Probabilities
    a. Educated guess to the extent of the threat
3. Own situation
    a. Current or responding resource level
4. Decision
    a. Strategy
5. Plan of operation
    a. Tactics

Fig. 1–4. FPODP and RECEO-VS were both introduced to the American Fire Service by Lloyd Layman in 1953.

Prior to World War II, organized career fire services were primarily found in larger cities. The rank structure and organization of these departments was considered paramilitary because of the similarities in terminology, but that was where most of the parallels stopped. Following World War II, fire departments, organization saw significant expansion, even in volunteer departments. As men returned from war and time in the armed forces to their home communities and jobs in public service, the influence of the military poured in. When the FPODP model is compared to the MDMP, that influence is black and white. While tactical and strategic decision making models transitioned, at the highest levels of departments, where command culture is determined, a centralized command and control or low trust model remained.

Early Emergency Management

While there are records of federal disaster relief efforts in the United States that date back to the early 1800s, the start of the field of emergency management was the Federal Civil Defense Administration and later the Federal Civil Defense Authority. Civil Defense was the agency for domestic disaster preparedness. During the 1950s and '60s Civil Defense expanded rapidly, as did our nation, and just as was demonstrated in the fire service, the postwar military influence was significant. During the 1960s the United States was also engaged in the Cold War, and for this reason the concentration of Civil Defense efforts was geared towards preparing the civilian population for a potential attack. Unfortunately, with the perception that the domestic civilian population would panic in the event of an attack and cause chaos, the approach to administration was a very strict and traditional command and control model.

In the mid- to late-1960s the U.S. experienced a series of significant natural disasters, and it became clear that there was a disconnect in the plans and needs for domestic preparedness. When President Nixon took office in 1969, he tasked the Office of Emergency Planning with evaluating Civil Defense, and it found that the nation's preparedness for natural disasters was minimal to non-existent. For the first time in history, federal funding dedicated to national preparation for defense against military attacks was made available to state and local governments for natural disaster preparedness. This period in the history of the emergency management field is identified as the point where the direction was first shifted from control to coordination.

Regionalization of Command Models

During the fall of 1970, a period of unseasonably hot weather combined with dry winds resulted in a wildland fire outbreak in Southern California that claimed 16 lives and more than 700 structures (fig. 1–5).

The timing of this disaster with the aforementioned direction of the president to the federal government to support disaster preparedness led to the creation of a task force to evaluate the wildfire problem.[10] Recommendations were taken to congress and in 1971 the House of Representatives appropriated "$900,000 to strengthen fire command and control systems research."[11] Engagement in the process was incredible and it involved organizations from the local, state, and federal level. Both the public and private sector contributed, from fire departments, county sheriffs, and forest rangers to aviation and communication industries. For years these groups worked across the western United States with the purpose of creating and testing a system that could effectively and efficiently coordinate interagency action and allocation of resources

The San Diego Union

HOME EDITION

SAN DIEGO, CALIFORNIA

Thousands Flee Raging County Fire

Losses Soar In L.A. Area

150,000 Acres, 223 Homes Burn; Worst In History

TOWN EVACUATED

Fig. 1–5. This 1970 headline was a new phenomenon.

in dynamic large-scale situations. In May of 1980 Firefighting Resources of California Organized for Potential Emergencies (FIRESCOPE), a new concept in multiagency fire suppression coordination, was published and the Incident Command System (ICS) was introduced.

Throughout the 1980s and the '90s ICS quickly became the standard practice in the Western U.S. as comfort with the model and terminology grew. Outside of just the application on the scene, many departments began to bring concepts into organizational structures and the titles of \ chiefs or branch director became positions in departments. While incident command and control was experiencing an evolution as a result of this influence in the western states, the eastern U.S. remained in the more traditional model.

As mentioned in the introduction to this chapter, Chief Brunacini recognized that, nationally, fire department incident command models, terminology, and practices were beginning to grow apart. With his text *Fire Command*, first published in 1985, he attempted to balance these differences and also introduce the national fire service audience to the core principles of the Incident *Management* System (IMS). FIRESCOPE/ICS and Chief Brunicini/IMS came of age in the 1970s. Chief Brunacini and IMS were more fireground-centric, whereas ICS was intended for any type of incident, including structure fires.

WISDOM FROM THE MASTERS

On-Scene and In Charge—The Evolution of ICS

CHIEF JOE CASTRO, LOS ANGELES CITY (CA) FIRE DEPARTMENT

Early in my career, the Los Angeles City Fire Department's legendary Deputy Chief, Commander of Emergency Operations Donald Anthony, would arrive on-scene at a major emergency incident (more than 15 fire companies working) and our dispatch would announce over the tactical radio frequency, "Chief Anthony is on-scene and in charge." Chief Anthony was an iron-fisted fireground commander who made decisions with a trip-hammer level of efficiency. Just hearing his name announced as the incident commander gave all responders an immediate sense of relief and confidence. If things were going wrong, he'd fix it. We all knew the incident was going to be run in a manner that was efficient, and objectives and tasks would be relevant and clearly communicated, staffed, and supported. The incident would be organized, radio discipline would be maintained, safety elements would be in place, and personal risk would be consistent with potential gain. All this because of the mention of his name. He was THE man!

Chief Anthony was one of the Incident Command System's (ICS) founding inspirations, and the first to move the concept past the wildland firefighting arena and into an all-risk application. In a sense, he was the nation's first ICS cop in that he required all of his command officers to actually use ICS, at every incident (from house fires to high rise fires) (fig. 1–6).

Every commander reading this should establish as an end-result goal for their individual dive into learning and practicing ICS this same effect when they arrive on-scene and take charge: a tangible reaction from their firefighters, an acknowledgment of confidence, that they will have a positive effect on ensuring an organized and well-run incident. That is why ICS is in existence and exactly what happened on the First Interstate Tower Fire on May 4, 1988 when Chief Anthony was the IC (fig. 1–7).

I was very fortunate to work for an organization that embraced ICS for my entire professional life. My 39-year career with the Los Angeles City Fire Department began in 1978, not too long after ICS was developed by the visionary participants of FIRESCOPE. History tells us that FIRESCOPE members, in reaction to a series of catastrophic wildfires in Southern California in the 1970s, analyzed why these fires resulted in the loss of so many structures. Their analysis revealed that the firefighters were well trained and equipped, and resources were on-scene and ready for action at each of the fires. However, structures were lost because they didn't have a system that could effectively determine relevant strategy and appropriately deploy available resources in a complex and rapidly changing wildland incident. FIRESCOPE's initial blush to addressing that dilemma was called ICS.

The initial rollout of ICS was rather insignificant outside of the California wildland firefighting arena. The Los Angeles City Fire Department then began to codify incident-specific iterations of the brush fire ICS model, high rise ICS, HazMat ICS, and medical, multi-casualty incident (MCI) ICS, with the significant changes and additions being in the types of functional groups beneath the operations section.

Yet the prevalent perception was that ICS was only for large incidents and it had little value at routine and smaller incidents (structure fires).

Fig. 1–6. Chief Don Anthony applied ICS to structure fires in the Los Angeles City Fire Department and expected all officers to do so.

Fig. 1–7. Deputy Chief Don Anthony was the IC of the First Interstate Fire and utilized ICS to bring it under control within 4 hours, utilizing 55 companies, 4 helicopters and 270 personnel.

Most of the rank-and-file members of various fire departments viewed ICS as a system for Chiefs, with little application or usefulness to the ground-pounding mission workers' efforts at the emergency. This contention was further authenticated by the way most people *learned* ICS in the 1970s and 1980s: by filling out a blank ICS organization chart. This, almost subconsciously, taught firefighters that the first step in implementing ICS was to fill in all of the boxes of the organizational chart. This inadvertently created a generation of box checkers who participated in a form over function exercise that was not driven by the needs of the specific and unique incident. Therefore, it's no wonder that firefighters saw little application of this complex and verbose organizational chart in the daily application of routine (smaller) incident management. What was not taught back in those early years and is one ICS's greatest strengths is its flexibility and scalability. In the 1980s the Los Angeles Fire Department began to require all officers (captain and above) to use ICS terminology at all incidents. This in itself was the beginning of modern all-risk ICS.

Of course, the turning point for ICS arose on September 11th, 2001. New York City was sucker-punched with an unimaginably complex and novel incident. 14 months later, the Department of Homeland Security was formed, and a couple of years later the 9/11 Commission report was first published. In 2004 the first edition of the National Incident Management System (NIMS) doctrine was published. A component of the NIMS doctrine was ICS. The foundational element of the NIMS-ICS component was, principally, the system built by FIRESCOPE in the mid-1970s. Local, state, territorial and tribal jurisdictions were required to adopt NIMS (and ICS) in order to receive federal preparedness grants, essentially establishing ICS as the law of the land for the entire country. The goal of the requirement was to establish one national system of incident management for all risk and hazards, and all response and recovery disciplines (not just firefighters).

This would allow the catastrophes of the 21st century to be managed with a single system that would enable large incidents to accept resources from different agencies and disciplines (local, state, and federal) into a commonly understood system of incident management. Suddenly, nationwide, firefighters, law enforcement officers, and non-traditional response entities were all (somewhat) forced to abandon whatever home-grown system of incident management they used previously and learn ICS.

For many, these lessons resulted in as many questions as there were answers. The federal ICS classes of 100, 200, 300, and 400 taught a system and a method of implementation, with emphasis on a myriad of rules and terms and focused heavily on the executive-level operational planning P, which codified the cyclical planning process for multiple operational periods. This learning modality further communicated ICS's value for a large incident with multiple operational periods. The obvious problem was that if a jurisdiction's firefighters and law enforcement officers learned the system, as directed, the frequency of large incidents on which to practice was far too low to remain proficient.

Like all incident management skillsets, your ability to block out the plethora of complicated inputs at an emergency and actually execute your skill is dependent upon how many times you've practiced. It's still all about sets and reps. To that end, those fire officers who used ICS at smaller/routine incidents found themselves much more effective at larger and complex incidents because they had trained themselves not only on implementation of the system, but also began to learn the true art of ICS.

Now let's drill back down to practical ICS use for firefighters and its value to a first-arriving fire captain at a house fire. *It's based upon the size, scope, complexity, novelty and duration of the incident*, meaning modern ICS is an objective-driven system of incident management and no two incidents are the same.

If you think about it, modern ICS has been designed utilizing the best practices of the most experienced officers in the country; their daily, routine experiences evolved the system into something you can (and should) use at every incident, big or small. You arrive on-scene, take charge, and gain situational awareness (determine what is happening). From that awareness you develop incident objectives (determine what needs to be done). Then you order the appropriate (or available) resources to accomplish those objectives. Next you develop a plan or method to communicate to everyone (radio channels). You then deploy those arriving resources to the most urgent objectives, and organize those resources into an ICS-compliant organizational (org) chart. As you can see, the organizational chart is the seventh step, not the first, and the org chart will be filled out to accomplish the incident objectives, not arbitrarily composed prior to the incident.

Fireground officers now use ICS as a predictable tool to maintain personnel accountability at an emergency, as mandated by NFPA 1561. Emergency managers use ICS for event action planning for gatherings of people at parades, football games and graduations. All of these applications have the same end-result goal: to address the specific threats of any incident (or event), and from those threats, determine what needs to be accomplished to abate or diminish them, and then use whatever tools you have in your ICS toolbox to get that done.

In conclusion, as you dig into this book and learn from the best ICS minds in our country, remain mindful of Chief Anthony's creed: be On-Scene and In Charge and embrace your inescapable obligation while in the throes of chaos to provide clarity, organization, effectiveness, and safety to the people under your command and the citizens they're protecting.

As Chief Castro pointed out, early ICS training was relegated to memorizing organizational charts and understanding the difference between a division and a group, for example. Application was primarily to larger wildland events.

Beyond wildfires, several large-scale natural disasters, including the Loma Prieta (1989) and Northridge (1994) earthquakes, demonstrated the speed and efficiency of expanding and contracting the system to the incident size and on all levels of local, state, and federal agencies. This component is what made FIRESCOPE and ICS the precursor to the National Incident Management System (NIMS).

The Shift in Models

Organizations that fully adopt command and control as policy become both commanding and controlling.

—Gary A. Kreps, 1989[12]

In the 1970s, the field of psychology began to present more qualitative and quantitative data on decision-making models. Conflicts in Korea and Vietnam also presented more dynamic battle spaces that challenged traditional linear campaign war planning and experiences of senior staff. The military branches recognized they had to change. This led to an extensive review of U.S. Army approaches and comparisons of problem-solving decision models to traditional command and control. A 1970 study of professionalism conducted by the Army War College created 31 recommendations to shift the Army from the MDMP to a more modern and adaptable problem-solving model.[13] These recommendations were specifically connected to operational and organizational issues identified in post-Vietnam reviews. While the 1970s served as a period of reflection and study for the military on lessons learned and modern models, most changes were updates. It wasn't until after 9/11 that cultural change was forced.

Immediately following 9/11, the United States launched a war on terrorism and targeted an enemy unlike any previous foe. The new enemy was ideologies, cells, and tribes rather than clearly identified states or tangible actors. The new environment of an unclear battlefield and counter insurgent combat quickly became identified as fourth-generation warfare. The generational identification starts with first-generation, Revolutionary and Civil conventional and linear war; second-generation campaign wars of World War I and II; third-generation guerrilla warfare of Vietnam; and now the fourth-generation battlespace is still being defined.

Just prior to Operation Iraqi Freedom, the U.S. Army Chief of Staff charted a study of operations in Afghanistan. It was found that the most successful operations were smaller units, working quickly at the local level taking opportunities to build support, collaborate or coordinate information, and integrate into informal systems. The report discovered that traditional power hierarchy and the command and control model countered this need and was developing an oppressive culture.[14] A system driven by top-down explicit orders increases the communication demands between subordinates and superiors because any variation, or deviation, from an order requires that consideration communicated back up the chain and permission granted back down.

In his text *Developing Adaptive Leaders: The Crucible Experience of Operation Iraqi Freedom* Leonard Wong explains that it was because of conditions associated with fourth-generation warfare in Afghanistan and Iraq, where time and information are at a premium, that the Army could no longer rely on an industrial-era management model. In order operate successfully in the current pace and state of warfare, the Army had to transition from a command and control model to mission command. Mission command is a command structure where operators understand the commander's intent and the mission at hand. Through their training, education, and experience they are trusted to use initiative and creativity to solve problems and adapt to circumstances.[15]

The example of Napoleon shifting an approach but failing to continue to evolve was used to start this section and demonstrate how the failure to extend trust to developing subordinate leaders and adapt to the new pace of war led to his downfall. The shift in the U.S. military to a greater focus on mission command and extending trust to subordinate leaders in the conflicts that followed 9/11 has yielded tremendous success and marked rapid progress. In less than a twenty-year period (one generation) from the initial commitment to mission command to today, the principles of mission command have been institutionalized in our military. Attributions of success in operations to mission command models are so clear that countless texts are now in publication, in similar efforts to those of Chief Layman during the 1950s, to provide guidance to fire departments, the private sector, and even educational institutions on how to develop what could be called a highly mature combination of independence and obedience in subordinate leaders.

In short, we must adapt from the traditional command and control model of centralized command to the mission command model of decentralized command. ICS is the tool from which to implement a decentralized command model.

As the military recognizes four generations of warfare with increasing complexity and has adapted, so too must the fire service. Our first generation was the Revolutionary and Civil War eras when, as in the military, a centralized form of command existed and tactical tools like water buckets and steam pumpers came into existence. Next were the World War I and World War II eras that were characterized by continual centralization of command, while portable radios, internal combustion engines, and centrifugal pumps came of age. Then, the Korean War and Vietnam eras gave way to glimmers of a more formalized crossover from the military with the works of Lloyd Layman and others. *America Burning* in 1973 gave way to immense advances in fire prevention, Self-Contained Breathing Apparatus (SCBA), Personal Protective Equipment (PPE), and resources.[16] The 1970s also brought paramedics. While these tactical advances came about, the centralized command and control model remained. Where was the advancement in this arena? Despite the advent of ICS, we stagnated. ICS became perceived as a behemoth and myriad of complicated nomenclature, paralyzed by process, clip boards, vests, and processes that bog down the fire department's ability to get to work.

We now enter the fourth-generation. Like the advances during the past 20 years in warfare, so too must we advance at the highest levels on the fireground. We must approach ICS differently, train differently, empower differently, and transform officers from subordinates to leaders who are trusted and trained to make decisions in real time in the field.

Continuity, Coordination, and Cooperation

It would be possible to argue that with no planning, social units would likely evolve a more adequate effort in emergency response than they would burdened by the assumptions inherent in the command and control model.

—Russell Dynes, 1994

The scale of the initial response challenges at the World Trade Center incident in Manhattan on September 11th, 2001 was beyond the scope of control of any commander, command team, or command system. As discussed in the previous section, any system driven by top-down orders increases the communication demands between subordinates and superiors. The complexity and size of the World Trade Center attacks paralyzed all communication systems. Nearly all initial and early response actions were independently, locally, or organically initiated. While the number of people killed was 2,152, it is estimated that almost 16,000 people were evacuated or rescued from the buildings and exposures along with tens of thousands more from the Lower Manhattan area (fig. 1–8).

The extraordinary actions taken by the Fire Department of New York City (FDNY) officers, fire companies and individual firefighters, untrained civilians, business owners, and individual police officers on that morning were examples of what many in the field of emergency management like Dynes had been explaining and researching for decades prior. Traditional civil defense and command and control are rooted in thinking that people and panic will compound a problem. Research, history, and the experience of 9/11 shows that in modern society where education, training, and communication are at a high level, responders—civilian or sworn—are resources with the capacity to act as problem solvers, not problems themselves.

The events of September 11th, 2001 resulted in the broadest and most in-depth review of emergency response in American history. In 2002 the Department of Homeland Security was formed and Federal Emergency Management Agency (FEMA), along with 21 other federal agencies, was moved under the direction of the new cabinet. This restructure and reinvestment broadly expanded the field of emergency management, requiring a clear establishment of expectations, roles, and a common terminology across disciplines that had never interacted before. In March of 2004, the Department of Homeland Security released the NIMS with the goal of establishing a uniform guide for all levels of government.

The fire service took the introduction of NIMS in stride as most of the structure was based on the familiar Incident Command System (ICS) framework. Unfortunately, seeing the change from ICS to NIMS in name only as an exchange of the word command for management is short sighted. Command remained a title in NIMS, but it was abandoned as a belief. Dynamic incidents and the people involved in them as victims or responders are not commanded, they are managed. More specifically, they are led, and entrusted to lead others closer to the problem than the incident commander.

Fig. 1–8. 9/11 showed that large incident management was required well beyond wildland fires.
Source. Photo courtesy of FEMA Urban Search and Rescue (US&R), Sacramento, CA Task Force 7

The professionals in the field of emergency management, disaster study, and social sciences have concluded that in groups that were educated, trained, and prepared, the function of command and control was undercut. Like soldiers, responders take on their role willingly and participate in the development of their physical and mental capacities to respond, intervene, and mitigate.

Chaos, command, and control begins with a belief that there is an acute departure from normalcy that creates chaos, requiring a body to command attention, regain focus, and take back control. In actual practice, emergencies are not interruptions to life or acute departures from normalcy for the responders. Emergencies are their work and therefore a continuation of expectation, preparation, operation, and mitigation. Where the military advanced adaptability and efficiency in operations by separating strict vertical power structures of command and control to a more decentralized model committing to mission command, the field of emergency management is adopting a doctrine of continuity, coordination, and cooperation. In short, we are no longer looking to command and control, but to command and *coordinate*. As the saying goes, "You may be in command, but God is in control."

The military does not wait for the "big one" to implement decentralized mission command; it has become woven into daily work and organizational culture. This is type of thinking with regard to ICS that "we don't need big ICS on a small house fire" is a phrase relegated to the past, because how we practice today is how we perform tomorrow (fig. 1–9).

ICS today, as it will be presented in this text, is not your grandfather's ICS. Today, ICS is nimble, agile, mobile, adaptive, and quick. Today, incident command and coordination involves human decision making, tactical prowess, leadership development, empowerment, and a knowledge of construction and fire behavior like never before. This is the only way to outpace the enemy in the fourth generation.

Fig. 1–9. Modern house fires are more volatile than those of 20 years ago.
Source. Photo courtesy of Rick McClure

Summary

In her research for the Naval Postgraduate School, Chief Cynthia Renaud uses an excellent analogy for the differences of approach between command and control models and incident management systems by using a carousel.[17] Successful incident managers approach a moving carousel and observe what is happening until they find the right time to get involved (continuity). They know that once they are able to get on, they can work with those already on the carousel (coordination) on a plan to collectively slow it down (cooperation). A command-centric approach would see the carousel and everyone on it as a problem. They have been taught to believe that they are the only ones with the knowledge and power (command) to act (control). Unfortunately, if a carousel is stopped immediately by someone, everyone else is thrown off.

The purpose of this chapter was to provide a detailed lineage from the source of the command and control model through the channels that have influenced the fire service using examples of proper and improper application over history. Ultimately, it has been demonstrated through seminal events and evolution of theory in application that in the modern threat environment and society, strict adherence to the command and control model with centralized authority and concentrated power has negative functional, organizational, and cultural impacts.

Today's fire departments, both career and volunteer, are hiring and developing professionals. Failing to trust the quality of today's human resources, the level of our professional standards of performance, and the extent of our training to operate is a fatal flaw in meeting the expectations of our citizens and devalues our members. The research, case study, experience, and history show that emergencies do not reduce the capacities of properly trained individuals; instead, they only present them problems to solve.

> *Emergencies are characterized by decentralized and pluralistic decision-making; therefore autonomy of decision making should be encouraged not the centralization of authority.*
>
> —Russell Dynes, 1994

Chapter Review

Review Questions

1. Describe the origin story of command and control.
2. Explain the Military Decision-Making Process (MDMP).
3. Describe the influence of Chief Llyod Layman in regard to fire officer development.
4. What does FIRESCOPE stand for?
5. Explain the generations of warfare and their influence.

FESHE Strategy and Tactics (C0279) Related Content

Course objective #2 is to ensure the student possess a working knowledge of the Incident Management System and NIMS. This chapter provides an in-depth history of how the IMS and NIMS was developed. Section V, Basic Division of Tactics in the model curriculum course outline utilizes FPODP as the size-up method. This chapter includes initial presentation of FPODP with the origin story of how it came into fire officer training and development.

Endnotes

1 Wilson A. Shoffner, *The Military Decision Making Process: Time for a Change*, School of Advanced Military Studies Monograph, United States Army Command and General Staff College, 2001, https://apps.dtic.mil/sti/pdfs/ADA381816.pdf.

2 Alan V. Brunacini, *Fire Command*, 2nd ed. (Quincy, Massachusetts/Phoenix, Arizona: National Fire Protection Association, 1985).

3 Russell R. Dynes, "Community Emergency Planning: False Assumptions and Inappropriate Analogies," *International Journal of Mass Emergencies and Disasters* 12, no. 2 (August 1994): 141–58.

4 U.S. Marine Corps, *Marine Corps Doctrinal Publication 6: Command and Control* (1996) ch. 1. https://irp.fas.org/doddir/usmc/mcdp6/toc.htm

5 Peter J. Dean, "Napoleon as a Military Commander: The Limitations of Genius," The Napoleon Series, The Waterloo Association, 2000, accessed October 25, 2023, https://www.napoleon-series.org/research/napoleon/c_genius.html.

6 Roger S. Fitch, *Estimating Tactical Situations and Publishing Field Orders (*Fort Leavenworth, Kansas: U.S. Army Staff College Press, 1909).

7 "U.S. Army Field Manual 101-5, Staff Organization and Operations" (1932).

8 Lloyd Layman, *Attacking and Extinguishing Interior Fires* (National Fire Protection Association, 1955).

9 Lloyd Layman, *Firefighting Tactics* (National Fire Protection Association, 1953).

10 Robert Stafford Disaster Relief and Emergency Assistance Act of 1974, GovInfo, https://www.govinfo.gov/content/pkg/COMPS-2977/pdf/COMPS-2977.pdf

11 Department of the Interior and Related Agencies Appropriations for Fiscal Year 1973, Fire and Atmospheric Sciences Research, H.R., 92nd Cong. (1972).

12 Gary A. Kreps, ed., *Social Structure and Disaster: Symposium on Social Structure and Disaster, College of William and Mary, Williamsburg, VA, 15–16 May 1986* (University of Delaware Press, 1989).

13 U.S. Army War College, Study on Military Professionalism (Carlisle Barracks, PA, 1970), https://ssl.armywarcollege.edu/dclm/pubs/study1970.pdf.

14 Donald E. Vandergriff, "One Step Forward, Two Steps Back: Mission Command versus the Army Personnel System," *The Institute of Land Warfare*, no. 84 (August 2011).

15 Leonard Wong, *Developing Adaptive Leaders: The Crucible Experience of Operation Iraqi Freedom* (U.S. Army War College Press, 2004).

16 National Commission on Fire Prevention and Control, *America Burning: The Report of the National Commission on Fire Prevention and Control* (U.S. Fire Administration, 1973).

17 Cynthia Renaud, "The Missing Piece of NIMS: Teaching Incident Commanders How to Function in the Edge of Chaos," *Homeland Security Affairs* 8, Article 8 (June 2012). https://www.hsaj.org/articles/221

18 Louis Maurer, *Rushing to the Conflict*, 1858, hand colored lithograph, $22^{1}/_{16}$" × $17^{13}/_{16}$", Currier and Ives.

2

Command Philosophy and Mindset for Modern Incidents

Fig. 2–0. Military doctrine must be part of modern incident command.

Introduction

Every battle is won or lost before it is ever fought.

—Sun Tzu, *The Art of War*

This ancient quote from *The Art of Warfare* by Sun Tzu applies to the modern fireground and emergency environment.[1] We must be able to see the battle in our minds, train, forecast, extrapolate all permutations, and be ready for unknowns. The effort put into knowledge of building construction in your area, fire behavior, tactical decision-making, hands-on training, pre-planning, teambuilding, and drilling in realistic, stressful environments is what drives a winning battle. These are all elements that are factored in before the bell ever rings (before the battle is ever fought). This means everyone! Not just the incident commander (IC), but all officers and crew members.

Modern incident command and management is a team sport requiring exponential thinking, multi-dimensional preparation, and decentralized decision-making. Rather than linear thinking, where we expect an incident to follow our algorithms and play nice in the sandbox, modern incident command must prepare for simultaneous actions, exponential growth, violent events, ambushes, and unprecedented challenges.

Decentralized Command versus Centralized Command

The *Span of Control* is the number of resources a commander can effectively manage. Historically, the optimal span of control is 5:1. Modern incident commanders are not just engaging with companies working, they are also engaging with dispatch, bystanders, traffic, mapping, and pre-plans while also managing radios. In reality, 3:1 is more realistic.

Centralized Command means that all resources report directly to the IC and all tactical decisions are made by the IC. The result is often a central point of contact (the IC) for all resources, who is tethered to multiple companies *only* by an overloaded radio frequency, and who is likely operating outside a manageable span of control. This results in an IC who is redlining and unable to get or think ahead of the incident in the "what-if" mindset. If another alarm is called, an unknown victim is found, a Mayday is declared, or if any number of other unknown situations unfold, the IC who is taxed by normal operations becomes overwhelmed and is no longer in command or control.

A centralized command ethos slows down the response and allows the fire (or any incident) to *outpace* the fire department operations. We cannot stop the clock, put the fire in staging, or call another alarm of time. We must decentralize!

Effective incident command, coordination, and management cannot be accomplished by a centralized command philosophy with a single incident commander at the command post who has not trained and prepared the *entire alarm to work as a team*, nor given out tactical supervision (decision-making authority) to multiple officers who are in the tactical gap.

In short, no IC, no matter how experienced or well-trained, can keep up with, let alone stay ahead of, an incident like a team of officers working together and sharing the division of labor and decision-making authority can.

No IC (even with an incident technician or aide at the incident command post [ICP]) can be everywhere at once assessing the risk. Nor can they account for all the crews and resources on scene, rotating out and coming in. Other processes that tap the IC's mental bandwidth include forecasting changing building and fire conditions that could lead to tragedy; maintaining clear communication on multiple channels with dispatch and multiple resources; processing bystander inputs; and reconning valuable information regarding victim profile, additional hazards, and construction features. If this sounds like a lot, it is! And that's without a victim trapped, a Mayday, or exposures burning.

Often, the answer to these challenges is more technology or tools used by the IC at the ICP. The Fireground Accountability and Tracking System (FATS) tags and passports, global positioning system (GPS), drones, Google Maps, and countless other tools flood our industry in an attempt to help bridge the tactical gap and keep the IC from drowning. While there are some excellent tools, nothing replaces the use of ICS and a decentralized command philosophy whereby tactical supervisors are trained and empowered to help the IC drive the incident.

Some fire agencies have the luxury of a large overhead support system of chief officers, both operational and staff, who can respond. The challenge is often that these officers are out of position, on other alarms, or not available after business hours. Some systems have enough companies to assign an engine company to the ICP to assist the IC. Most agencies do not have either of these luxuries.

There is a new normal in emergency response that is redefined daily. Modern incidents have become immensely complex from factors including, but not limited to:

- Increased response levels due to rapid intervention team, EMS, and other requirements
- Decreased response levels due to closures and staffing
- Building construction
- Low-mass synthetics
- Wildland urban interface (WUI)
- Drought
- Weather events
- Terrorism
- Civil unrest, rioting, and mass arson
- Political and socioeconomic divide
- More complex hazardous materials
- Violence against first responders
- COVID-19
- Population increase
- More families co-living
- Social media
- Inexperienced firefighters
- Hoarding
- Commercial buildings over-filled with flammables
- Commercial buildings used as dwellings
- Excess fire loading, inventory, and high-piled stock
- Arson due to poor economic conditions

At Sacramento Metro Fire, the span of control was 15:1 for a battalion chief (BC) responding to a house fire from the time of dispatch: 4 engines, 2 trucks (that routinely split into 2 teams each), a medic unit, and another BC. In addition, the BC had no driver and was navigating code 3 driving in city traffic, talking on a command channel to dispatch, a tactical channel to responding companies, using the mobile data computer (MDC) to track resources and updates, and referencing the map to get to the scene.

Meanwhile, the tactical channel is crackling with companies arriving, going to work, and giving updates on conditions, while the command channel is buzzing with other units, incidents, and dispatch calling the BC with updates. This is all before arriving! The notion of 5:1 is *not* just the resources you are managing, it includes anything that is demanding your attention (even driving), requiring your brain to process information. 5:1 gets blown out very quickly.

Organizations should ensure reasonable span of control, which refers to the number of subordinates or activities under the control of a single commander. A commander's span of control should not exceed that commander's capability to command effectively. The optimal number of subordinates is situation-dependent. The more fluid and fast-changing the situation, the fewer subordinate elements a commander can supervise closely. Within this situation-dependent range, a greater number of subordinates allows greater flexibility, and increases options and combinations. However, as the number increases, commanders, at some point, lose the ability to consider each unit individually and begin to think of the units as a single, inflexible mass. At this point, the only way to reintroduce flexibility is to group elements into a smaller number of parts, creating another echelon of command. Narrowing the span of control—that is, lessening the number of immediate subordinates—deepens the organization by adding layers of command.[2]

The *Fog of War* is a phenomenon whereby the individual is confused or uncertain as to the situation, including status of the enemy (fire), troop strength, and positioning on the battlefield (building). Firefighters often find themselves in situations where they do not know where they are, do not know the location or extent of fire, and do not realize that their situation is perilous. This is an extremely common factor in Maydays and close calls.

War is the realm of uncertainty; three quarters of the factors on which action in war is based are wrapped in a fog of greater or lesser uncertainty. A sensitive and discriminating judgment is called for; a skilled intelligence to scent out the truth.

—Carl von Clausewitz

Decentralized command means the IC has divided the incident into manageable parts and given tactical divisions or group supervisors the authority and resources to complete objectives, account for crews, communicate the strategy, watch out for safety issues, manage the risk, and keep the span of control manageable. Rather than all companies tethered to the IC for direction, the IC gives the priority and strategy and tactical and control objectives, and lets the division or group supervisors carry out the objectives in the manner they see fit. The IC's job is to support the tactical supervisors, ensure the incident is in the correct priority and strategy, and maintain incident alignment.

A decentralized command allows the IC to step back, see the big picture, and be in the "then/what-if" level of thinking. This means considering unknowns like a Mayday, a victim found, rapid fire growth, exposures involved, and so on.

Mission Command—The Decentralized Military Model

Decentralized command structure allowed me, as the commander, to maintain focus on the bigger picture. . . Were I to get embroiled in the details of a tactical problem, there would be no one else to fill my roll and manage the strategic mission.

—Jocko Willink

According to Army Doctrine Publication 6–0, *Mission Command: Command and Control of Army Forces*, "*Mission command* is the Army's approach to command and control that empowers subordinate decision making and decentralized execution appropriate to the situation."[3]

The military lives by the principle of decentralized command. The key to success is trust. Trust can only be developed through training *and* time.

> *Mission command is based on the Army's view that war is a human endeavor and inherently dynamic and uncertain. No plan can account for every possibility, and most plans must change rapidly during execution if they are to succeed. The principles of mission command are:*
>
> - *Competence*
> - *Shared understanding*
> - *Mutual trust*
> - *Mission orders*
> - *Commander's intent*
> - *Disciplined initiative*
> - *Risk acceptance*
>
> *Mission command requires an environment of trust and shared understanding among commanders, staffs, and subordinates. It requires building effective teams and a command climate in which commanders encourage subordinates to accept risk and exercise initiative to seize opportunities and counter threats within the commander's intent. Through mission orders, commanders focus leaders on the purpose of the operation rather than on the details of how to perform assigned tasks. Doing this minimizes detailed control and allows subordinates the greatest possible freedom of action to accomplish tasks. Finally, when delegating authority to subordinates, commanders set the necessary conditions for success by allocating appropriate resources to subordinates based on assigned tasks.*[4]
>
> *The Commander's Intent is a clear and concise expression of the purpose of the operation and the desired military end state that supports mission command, provides focus to the staff, and helps subordinate and supporting commanders act to achieve the commander's desired results without further orders, even when the operation does not unfold as planned.*[5]

For fire department IC's, *Commander's Intent (Leader's Intent)* is the Incident Priority/Strategy declaration upon the assumption of command. When an IC states, "We are in life priority, offensive strategy. . . " that sets the left and right boundaries for the incident. Tactical division or group supervisors then execute the mission objectives of search, fire attack, ventilation, forcible entry, and so on.

When an incident stabilization/defensive strategy is declared, the same holds true. The supervisors keep everyone out of the building (parameters), then focus the battle on protecting exposures and preventing further growth (objectives).

Developing Your Officers for Mission Command

When placed in command—take charge!

—General H. Norman Schwarzkopf

The *mission command mindset* means command through controlling standards, expectations, and performance. This is a 24-hour-a-day enterprise consisting of training, drilling, pre-planning, coaching, empowering, communicating, analyzing, operating, debriefing, reviewing, and revising. Rather than waiting for the alarm to go off, we must develop our team and build trust through training and time. This empowers your company officers to step in as division/group supervisors or retain command if the chief is delayed. We must pre-plan our target hazards and train in realistic conditions. We must also remain vigilant throughout the shift to keep situation/resource status awareness around us.

> *Because mission command decentralizes decision-making authority and grants subordinate leaders s significant freedom of action, it demands more of subordinates at all levels. Commanders must train and educate subordinate leaders so they demonstrate good judgment when exercising initiative. Subordinates must be competent in their respective fields, and they must be confident they will have the commander's support to make and implement decisions. They must embrace opportunities to assume responsibility for achieving the commander's intent.*[6]

Sacramento Fire Battalion 7–A Platoon utilized mission command. Training included simulations, tabletop exercises, and realistic hands-on multi-company drills. In addition, we reviewed video and audio of our fires. Company officers were constantly trained, equipped, empowered, and encouraged to take initiative in all aspects of their job, *especially* on the fireground.

In the beginning, Battalion 7–A Platoon was comprised of many older, salty officers who were accustomed to a more traditional model of centralized fire command. We trained regularly on multiple aspects of leadership, tactical decision-making, construction, fire behavior, and everything we could think of to make us better as a team. I leveraged the senior captains' knowledge and experience and empowered them to lead training. We trained on the roles and responsibilities of division and group supervisors. Although they were experienced officers (some were company officers for 15 years or more), none had ever been trained in how to be a division or group supervisor. This new level of responsibility, supported by clear direction and training, was embraced by each of them.

I was blessed to be the BC in Battalion 7A for over 13 years. In the beginning, I was the youngest member of the entire battalion. Of the forty members, every captain, engineer, and firefighter was 10–20 years my senior. So, we started slow and easy. It was critical to assess what their operational norms and tempos were. We needed to learn from each other and trust each other to develop a baseline level of mutual respect as a foundation from which to build.

As the years went by, I became the most senior member of the battalion. New, young probationary firefighters came into the department, were assigned to Battalion 7A, gained experience over years in other battalions, and often returned as captains.

There was a mission command ethos that was reinforced at every turn. The philosophy was preached by me, and I always looked for opportunities for my officers to take ownership and initiative, from the most mundane daily functions to remaining in command on fires upon my arrival when the incident was going well. They knew that I trusted them, and they leaned into that trust by taking on more and more leadership decisions, 24-hours-a-shift. It was beautiful!

Every fire was painstakingly reviewed. No one was exempt from being held accountable, including me. This was *never* punitive. The intent was always to learn so that the team would be more effective next time. As a leader, it was imperative that I go first, sharing what I thought I could do better and the lessons I learned. I asked them what they thought I could do to help make them more successful. By holding myself accountable and setting the example of humility and vulnerability, it set the tone for everyone in the room to be honest and real with their own performance.

For officers who wanted it, or needed it, I would spend extra time doing simulations and reviews one-on-one. In some cases, I would see trends in behavior over several fires that needed to be addressed with a particular captain. This was done in a supportive tone, with the intent to learn and improve, never with the intent of punitive discipline. The key was to handle these issues privately.

> *Commanders develop a basic level of control within their organizations when they create a culture that embraces mission command at every level. The time spent inculcating mission command into training, education, and problem solving prior to operations saves time and simplifies command and control during operations. Commanders cannot expect subordinates to respond effectively to a mission command approach once operations commence if they have not developed subordinates comfortable in its use beforehand.*
>
> *Effective mission command requires well-developed subordinates capable of decentralized execution of missions and tasks. Training must create common, repetitive, shared experiences that build trust and allow commands to acquire competence in shared understanding. Trained teams are able to communicate explicitly and implicitly, conduct decentralized operations, and achieve unity of effort in uncertain situations.*[7]

The tenets of mission command are competence, shared understanding, mutual trust, mission orders, commander's intent, disciplined initiative, and risk acceptance. From a fire service standpoint, these may seem either overly simplistic or extremely esoteric. They can be both. On the one hand, these tenets are simple to understand and self-defining. On the other hand, they may be a collection of obscure military terms and the furthest thing from what is happening in a fire department.

Competence should be self-explanatory. It is the expectation that all members are skilled in their jobs and can act at the task level with the utmost skill and professionalism. If the company officer, or a chief officer, must intervene to assist with additional training or a performance improvement plan, then this should be done as soon as possible.

One captain was struggling with the notion of having more autonomy in decision-making at the tactical level. He had always worked in battalions with a more centralized approach, and he was not trained nor accustomed to what was being asked of him. We set up one-on-one meetings, simulations, and reviewed fires together. As the company drills ensued, he became more comfortable in the multi-company environment. When he was placed in a division role on a duplex fire, he did a fine job and often asked for more training and opportunities to grow. The key was that the training was not punitive, and he had a supportive learning environment.

Shared understanding means that the IC, tactical division or group supervisors, and task-level companies are all on the same page regarding the incident priority and strategy, tactical and control objectives, and the situation and resource status. In simple terms, they knew the incident action plan (IAP) for every incident. Shared understanding also meant that we did after-action reviews as an entire alarm. Whether a brief tailboard hotwash or a formal training session with video and audio, all members attended. This provided a shared understanding of the process and areas for improvement. This was also an excellent succession planning process for aspiring officers.

Mutual trust is the glue that holds the team and operation together. It does not happen overnight. Only through *training and time* does trust emerge. It is tested on the fireground. The officers must trust that the IC will keep the incident in the correct priority and strategy, call for resources when asked, and not steal back authority once it has been delegated to the tactical level. The IC must trust that their company officers can handle the burden of division or group supervision, will lean into the opportunity to lead, and will execute tactics in a similar fashion that the pre-planning and training has provided.

The El Camino Fire had minimal radio traffic, yet crews performed vent, enter, isolate, search (VEIS) and deployed multiple hoselines to varying locations without notifying me (the IC) of these tactics. They saved a 4½-year-old boy without having to ask for permission.

Tactical group supervisors led the operations and were entrusted to execute without asking permission or getting bogged down with constant conditions, actions, and needs (CAN) reports. Scan the code in figure 2–1 for the El Camino Fire.

Fig. 2–1. QR code for the El Camino Fire

Mission orders is a military term for tactical/control objectives. According to U.S. Army Doctrine Publication 6-0, Mission Orders are directives that emphasize to subordinates the results to be attained, *not* how they are to achieve them.[8] For fire IC's, these are tactical objectives like Rescue, Exposure Control, Confinement, Extinguishment, Overhaul, Ventilation and Salvage (RECEO-VS), and control objectives like confining the fire to the room or unit of origin.

Commander's intent (often called *leader's intent*) for fire officers is the clear priority/strategy declaration that tells units what the overarching plan is. The three components are *task, purpose,* and *end state.* By understanding these three factors, subordinates at all levels can act on behalf of the IC by whatever means available, within the parameters of the commander/leader's intent.

> *Never tell people how to do things. Tell them what to do and they will surprise you with their ingenuity.*
>
> —General George S. Patton, Jr.

Disciplined initiative is the characteristic to take action that is consistent with the commander's intent. *Initiative* is instigating action when none is being taken. This is synonymous with *bias for action*, in which an individual has a tendency or bent toward initiating action and engaging. This is crucial if you expect your officers to either assume command in the absence of a higher-ranking officer or aggressively assume a division or group tactical supervisor role when so ordered. As we have stated, the key is training and time to develop mutual trust if you expect to have initiative become a core value.

Finally, *risk acceptance* is the willful toleration of unknown circumstances in which the outcome cannot be assured. Strategic and tactical decision-making require officers to accept risk and make decisions when: (1) Life and death may depend upon the decision; (2) Incomplete information may be the only intelligence available to make the decision; and (3) The time in which to make the decision is compressed. Some officers have a hard time accepting the risk of making these high-stakes life-and-death decisions with limited information in a compressed timeframe; however, that is what leading in battle is all about. We will never have all the information and time to make the perfect decision. We must make the best decision we can given the limited time and information we have.

One barrier is fear of making the wrong decision and the ramifications of that decision. Most of the time, making no decision is worse than making the wrong decision. *Discretionary Time* is when you have enough to make a decision and can weigh various options. In general, the more discretionary time, the better. However, this is usually not the case on the fireground, so risk acceptance and disciplined initiative go hand-in-glove.

Given the same amount of intelligence, timidity will do a thousand times more damage in war than audacity.

—Carl von Clausewitz

Rather than think of an incident as beginning upon dispatch and ending when command is terminated, we must consider many pre-incident factors that will set our team up for success prior to the incident, and afterwards, many others that will set us up for success in the next incident. Building this mission command culture is a full-time job and success is truly developed between the dispatches.

Mission Command for the Incident Command System

Fortunately, the Incident Command System (ICS) is designed to manage any type of incident, from the simplest public assist single-resource response to a Type 1 national-level historic event. Unfortunately, ICS is often relegated to Type 1–3 incidents and not utilized and practiced on more frequent Type 4–5 incidents, where it can be easily applied to mitigate the common issues that plague these daily incidents. Often, ICS ends at the Type 3 Incident Management Team (IMT) or incident. This lack of usage of ICS at the more common Type 4 (multi-alarm) and Type 5 (single alarm) incidents has created an extremely disparate philosophy toward incident command and management, especially regarding structure fires.

Proper application of ICS can affect and improve common struggles associated with Type 4 and 5 incidents, including communications, accountability, risk management, span of control, safety, fog of war, and many other challenges.

A decentralized philosophy treats all the officers on a single or multi-alarm response as a Type 5 or Type 4 IMT, respectively. This requires a tremendous amount of training and time to build that team where the IC has enough trust in their officers to bridge the tactical gap and make decisions on behalf of the IC. This means trusting that a quiet radio frequency is a good thing because there are less task-level companies talking on the radio and instead talking face-to-face to their division or group supervisors.

One reason for the inconsistency in ICS application for structure fires is that ICS was developed in the wildland arena to manage massive fires. Since ICS was born out of necessity in the 1970's from large-scale wildland fires, one of the stigmas is that it is only good for large-scale incidents. Nothing could be further from the truth.

The Soberanes Fire in California burned 132,127 acres, 57 homes, killed a dozer operator named Robert Reagan III—a husband and father of two precious little girls—and had over 5,000 personnel assigned (fig. 2–2).

No one would argue whether ICS is needed to manage this immense incident. However, the use of ICS for structure fires is still a bit of a challenge for some. The common theme in those who struggle with ICS is simply a lack of proper training. These are smart, thoughtful, experienced, genuinely concerned firefighters who were never taught how to effectively apply ICS to structure fires.

Most ICS training is relegated to defining the difference between groups, divisions, and branches. Even then, many officers struggle. As we all know, applying a tool on an actual emergency scene requires vastly more training than simply reciting definitions. As we mentioned previously, no one argues the benefits of ICS on large-scale incidents. One of the problems with training is the failure to recognize the critical differences between wildland fires and structure fires and that applying ICS in a wildland method to a structure fire is doomed to fail.

Fig. 2–2. The Soberanes Fire on the second day. The fire started July 22, 2016 in Garrapata State Park.

Let's start by looking at some fundamental differences between wildland and structure fires that are of particular importance to an incident commander (see table 2–1).

The most critical factor overlooked is that wildland fires have little to no *interior* structural firefighting. For this reason, structure fires have much different complexities than wildland fires, in which structure protection is performed. Interior structural firefighting and exterior structure protection on wildland fires are two completely different animals.

Table 2–1. Differences Between Wildland Fires and Structure Fires

Factor	Wildland Fires	Structure Fires
Work Period	• 12–24 hours	• 12–24 minutes
Personal Protective Equipment	• Lightweight • No SCBA	• Cumbersome • SCBA
Communications	• Multiple frequencies • Not hindered by SCBA • Relatively open workspace allows clearer communications	• One frequency • Hindered by SCBA • Feedback due to confined workspace
Situational Awareness	• Supported by field observers • Visibility typically good	• Hindered by smoke and heat • Visibility typically bad
Conditions	• Relatively more predictable • Low thermal insult	• Relatively less predictable • High thermal insult
Incident Progression	• Relatively slower	• Relatively rapid
Life Loss Historically	• Less	• More
Duration of Incident	• Days to weeks	• Minutes to hours
Overhead Support	• Plentiful	• Minimal

No More Bread-and-Butter Fires

We call house fires "bread and butter." This is a lack of respect for our greatest enemy. *Anytime you do not respect your enemy, your enemy is much more likely to defeat you.*

Since house fires are our most common structure fire, we get arrogant and label these fires as bread-and-butter or routine when they are not. Modern house fires are more lethal than ever. We must recognize and respect house fires for what they are: the fire that claims the greatest loss of civilians and firefighter lives year in and year out. Therefore, we should be using ICS on house fires, but we often do not because of our view of them as easy, routine events.

Let's Look at Some Facts

In 2021, local fire departments responded to an estimated 1.35 million fires in the United States. These fires caused 3,800 civilian fire deaths and 14,700 reported civilian fire injuries. The property damage caused by these fires was estimated at $15.9 billion. On average, a fire department responded to a fire somewhere in the US every 23 seconds in 2021. A home structure fire was reported every 93 seconds, a home fire death occurred every three hours and eight minutes, and a home fire injury occurred every 47 minutes. More than one-third of the fires (486,500, or 36%) occurred in or on structures. Most of the fire losses were caused by these fires, including 3,010 civilian fire deaths (79%); 12,600 civilian fire injuries (86%); and $12.7 billion in direct property damage (80%). Major fires in the Colorado wildland urban interface (WUI) caused $648 billion in direct property damage (4%). Unfortunately, losses from WUI fires were not broken out by incident type. A substantial portion of the loss was undoubtedly due to structure fires. Only one-quarter of the fires (25%) occurred in home properties, including one- or two-family homes and apartments or other multifamily housing. However, these fires caused three-quarters of the civilian fire deaths (75%) and injuries (76%). One of every five fires (20%) occurred in one- or two-family homes, yet these fires caused nearly two-thirds of the civilian fire deaths (64%) and over half of the civilian fire injuries (54%). The 6% of fires that occurred in apartments caused 11% of the civilian fire deaths and 21% of the injuries. Vehicle fires accounted for 15% of the fires, 18% of the civilian fire deaths, and 10% of the civilian fire injuries. Neither structures nor vehicles were involved in half of the fires (49%) reported in 2021. These fires, not including structures nor vehicles, were brush, grass, or wildland fires—excluding crops, timber, and other properties of value (20%); outside rubbish fires (17%); outside fires involving property of value (6%); and other fires (6%). The 2021 overall estimate for total fire incidents was 55% lower than in 1980. Additionally, property loss, adjusted for inflation, was 20% lower in 2021 than in 1980. The 2021 estimate of total fire deaths was 42% lower than in 1980, home fire deaths were 50% lower, deaths in one- or two-family home fires were 47% lower, and apartment fire deaths were 66% lower. Because the U.S. population has grown since 1980, population-based rates have dropped even more than the estimates have. Less progress has been made in preventing the deaths and injuries associated with reported fires. ***For overall home fires, the 2021 rate of 7.9 deaths per 1,000 reported home fires was higher than the rate of 7.1 in 1980. The rate for one- or two-family home fires was 35% higher than in 1980****, while the rate for apartment fires was 31% lower. Most of the reduction in the number of reported fires and fire losses occurred more than a decade ago. There is still more work to do, particularly around home fires.*[9]

House fires are the greatest fire threat to civilian life and the most common structure fire for firefighters. These high risk and high frequency events are a perfect opportunity to utilize ICS on a regular basis. This develops officers at all levels with repetitions on the more commonly occurring fires (that claim the most lives). Then, when a larger fire comes along in an apartment or commercial building, your officers are experienced and ready to jump into tactical division/group supervisor positions.

The Operations Triangle and Core Values

The *Operational Triangle* is composed of the elements critical to successful operations: strategy & tactics, ICS, and standard operating guidelines (SOGs , also referred to as standard operating procedures [SOPs]) (see fig. 2–3). All three are required and work together in a symbiotic interwoven fashion. Rather than look at operations from a silo, with engine, truck, rescue, medical, and command all working independently, we must realize that all must work in concert for the best possible outcome. Strategy & Tactics are at the bottom of the triangle and represent the foundation and the most critical element. This is where incident priorities, strategy, and tactics like search, fire attack, vent, and forcible entry all happen. Depth of understanding in tactics, fire behavior, building construction, the latest Underwriters Laboratories/National Institute of Standards and Technology research, case studies, and experience come into play (fig. 2–3).

ICS is the national standard for commanding incidents, and it works very effectively in the hands of a well-trained IC and responding officers. ICS can fluctuate to fit your incident, whether food on the stove of a stand-alone 1200-square-foot home, an apartment building with multiple units, a commercial structure fire with Mayday, or a conflagration that burns down thousands of structures.

SOGs are your playbook. Many varying philosophies exist regarding SOGs. Some believe that they are a necessary evil and occupy computer space, that they have no real value, and check the box. Others believe that a fire department (FD) cannot have enough SOGs and want one for every possible permutation that could exist. This is usually followed by a stringent philosophy where the SOGs *must* be followed at all times. In addition, some believe that utilizing SOGs for resource pre-assignments for a structure fire is negligent and that the IC must orchestrate every move each company makes (centralized). Conversely, many FDs have SOGs that list pre-assignments and considerations for companies based upon the incident type and arrival order.

Mission command and decentralized \ both dictate that pre-assignments are a fundamental aspect of the success of outpacing the enemy (fire). Pre-assignments or SOGs are simply a playbook. Similar to how each team in the National Football League (NFL) has a playbook for different teams and circumstances, so too must an FD have a playbook (SOGs) that lays out the plan, including pre-assignments. Pre-assignments allow all responding officers to know the gameplan going into battle. Then, as inevitable variables unfold, adjustments can be made. Consider an NFL head coach who has studied the opposing team, trained, and prepared his team, and develops a game plan with scripted plays. Everyone knows the plan and has

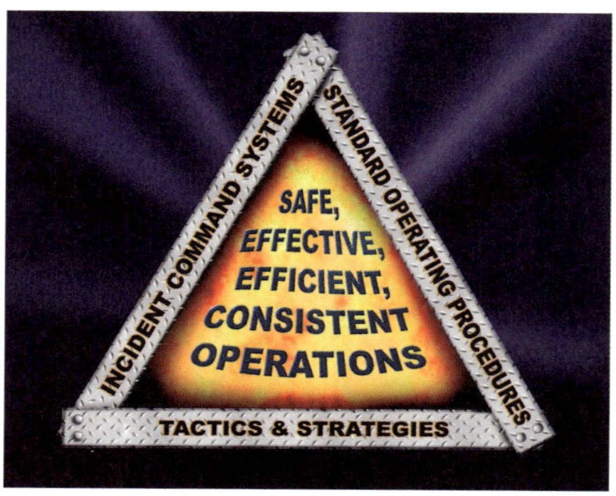

Fig. 2–3. The Operations Triangle

practiced and watched game film. Then, the quarterback steps up under center and sees a blitz coming, or another unexpected variable and calls an audible to the offense. They adjust. Then, split-second decisions are made by all the players to carry out the play.

This same process is applied to the modern fire department that recognizes how fast a fire grows and how quickly the window of time to save a civilian shuts. Therefore, pre-assignments play a *vital* role in staying ahead of the incident and in deploying mission command. Just as the NFL coach prepared his team for a game, a chief prepares their battalion or fire department for battle. The whistle blows on the field just as the alarm goes off in the firehouse. The play is called by the coach on the sidelines just as the chief declares a life priority/defensive strategy with the tactical objectives of fire attack, vent, search, and medical. Then, just as the quarterback calls "Omaha" to signal an audible for a blitz, the IC gets information back from their tactical supervisors that they must switch to a defensive strategy.

In this analogy, the IC is the head coach and the quarterback is a division supervisor. We need a plan that is flexible and can be adapted in the field by skilled officers who are empowered and trained to carry out the mission. Otherwise, we have everyone stage and ask for direction from the IC. This centralized process will not keep up with the fire, let alone outpace it.

The IC still has the latitude to override an operation for safety reasons. This is not the same as undermining the tactical supervisor, it's simply due diligence. As the IC on a large apartment fire, I was asked by Division 3 on the fire floor for blowers. From his vantage point, the Division 3 supervisor didn't see what I saw from the ICP: fire still blowing out of two apartment windows that were being fed by a gas leak. I simply replied, "Negative Division 3, you have heavy fire on the alpha/delta corner. Once you get that contained and we get the leak stopped, you can have all the blowers you want."

The outputs, or results of the operations triangle, are the operational core values: safe, effective, efficient and consistent operations.

The priority is in that order, top to bottom:

- *Safe*—Is the operation done without unnecessary risk? This job is dangerous. There is risk. Make sure there is no *unnecessary* risk. There is not a lot that is safe about going into a burning building. Perhaps the saying should be, "be smart," or "don't be dangerous," or "don't take unnecessary risk."
- *Effective*—Is the operation making a positive difference? If not, why? Either adjust or back out.
- *Efficient*—Is the operation done with as little wasted time and resources as possible? More resources, more radio traffic, and faster actions do not always mean more efficiency. High speed/low drag is the key to efficiency on the fireground.
- *Consistent*—Are our operations as an organization/region consistent across all shifts and companies? This is one of the most common challenges in the fire service; however, it is not insurmountable. High-impact, realistic training, as a battalion or full alarm assignment, over time, will result in more consistency.

The Operations Triangle was utilized in the consolidation of the two 20-station fire districts that created the new 40-station Sacramento Metropolitan Fire District. One agency was very loose and didn't have much in the way of SOGs. The other was the opposite and had more SOGs than anyone could count. In the wake of the merger, there was operational confusion and conflict as these two converging and opposing philosophies came to bear on fires. As we developed SOGs, one side wanted none, and the other wanted more. So, to meet in the middle, the operational core values were developed. They were the leader's intent for the new fire district. The operational core values were also the means from which to call an audible on the SOGs. Does deviating from the SOG (calling an audible) make the operation more safe, effective, or efficient? If so, do it.

Unacceptable Reasons to Call an Audible on an SOG

- I didn't know the SOG
- I don't like/agree with the SOG
- I don't like SOGs in general
- I don't like the people who wrote the SOG

SOGs were still developed with pre-assignments; however, the core values provided a succinct leader's intent.

The Ten Commandments of Incident Command

As we begin to delve into greater detail regarding incident command, an overview will set the tone for the rest of this book. "The 10 Commandments of Incident Command" is a list of fundamental rules of engagement that can be applied to any incident.

1. **Know the Enemy**
   The enemy is a structure fire, a vegetation fire, a hazardous material, an active shooter, a technical rescue, an earthquake, a flood, etc. We must know our enemy, respect our enemy, and outsmart our enemy. Each has unique challenges that we will explore in sections 2 and 3.

2. **Know the Battlefield**
   The battlefield can be a legacy home, a lightweight modern home, a multi-family fourplex, a center hall apartment, a multi-use taxpayer, a big box warehouse, a high-rise, a field full of light flashy fuel, a forest, a river, an industrial complex, a mountainside, a collapsed building, or any other possible location that humans occupy. We must know these different theaters and how they are built and designed, and how they evolved, function, and behave.

3. **Know the Weapons and Tactics of War**
   Each incident type and kind has specific tactics associated with it. We must know the tactics and resources available for each and every incident that we could encounter in our jurisdictions, and often beyond. Tactics evolve. VEIS was once VES. VEIS was once only for one room, now it has evolved into window initiated search. As an IC, you do not have to be a technician on each and every piece of equipment and apparatus in your battalion. You must know the capabilities and tactics, the resources, and the needs to support them.

4. **Perform a Thorough, Ongoing Size-up**
   As you will see in chapter 5, size-up is a critical, continuous, mental process to access risk versus gain and forecast how the incident will behave. Mission command is vital to a thorough, ongoing size-up.

5. **Stay ahead of the Incident Power Curve**
   The IC must remain ahead of the incident in the "then/what-if" mindset. Here forecasting, contingency planning, and preparing for an incident within the incident (IWI) occur. Do *not* get caught behind the curve! Again, mission command is vital to remaining ahead of the curve.

6. **Communicate in a Calm, Clear, Confident Fashion**
   The #1 problem in all incidents is communication. Learn and train others how to communicate effectively. "As the first line goes, so goes the fire" is a common saying. "If the IC panics, so will everyone else" is also true. Do not make it worse. Your incidents should all

sound the same. Whether a smoldering cigarette butt in the middle of the Nevada dessert or a 5-alarm apartment fire on a windy day with a busload of hemophiliacs in a glass truck, you should sound calm.

7. **Give Clear Objectives**
   Do not assume that all responders have clear leader's intent. Focus the troops on the most important tactical and control objectives and then support them to accomplish those objectives.

8. **Establish ICS Divisions and Groups Early**
   Proactive use of ICS will fulfill a mission command ethos and prevent you from getting behind the incident power curve. Do not wait until you are overwhelmed and beyond your span of control before assigning division and group supervisors. We will go into this in much more detail throughout the book.

9. **Have a Plan B and a Plan C**
   Contingency and emergency plans are crucial to have established in your back pocket before you need them. A contingency plan would be switching to a defensive strategy. An emergency plan would be a building collapse or a Mayday.

10. **Have the Guts to Say No**
    The IC is the least popular person to the troops on the fireground when they say no to continuing the firefight due to possible collapse, severe winds, lack of resources, deteriorating conditions, etc. As the Division C supervisor on a strip mall fire, I was told by command that we were switching to a combination strategy. I concurred, as the building was all clear and several units were untenable. We retreated to an offensive space and wrote off the areas involved. The crews were not happy about it, but it was the right thing to do. I just told them to yell at the IC.

Summary

The command philosophy of the 2020s and beyond represents a paradigm shift away from the centralized processes of the past. The military evolved decades ago into a decentralized ethos, and the fire service is well overdue to follow. Several key components will provide the foundation from which to move forward: mission command, ICS, and operational core values.

Mission command is the primary command philosophy of the modern military and is required of today's fire service. The key principles of mission command are competence, shared understanding, mutual trust, mission orders, commander's intent, disciplined initiative, and risk acceptance. Command and tactical decision-making authority are decentralized to officers in the field of battle who are charged with leading smaller groups of warriors. Those officers are trained, equipped, empowered, and entrusted to carry out the leader's intent. This gives left and right boundaries to these field officers and the latitude to fulfill the objectives in any matter they see fit, within the leader's intent. This reduces lag time, confusion, and communication overload, and encourages empowerment, ingenuity, and initiative to outpace the enemy, rather than be bogged down by a process-driven, centralized, and antiquated model.

ICS is built to apply mission command to the fire service. Decentralization, modular organization, and flexibility are all characteristics of ICS; however, some agencies either apply a wildland mindset to ICS or have not trained enough to apply it in the structure fire arena with success. Therefore, ICS has been relegated by some to larger, less common incidents due to the misconception that it is too large and cumbersome to be used expeditiously on structure fires, especially house fires, our most common and lethal foe. This book aims to solve these challenges.

Operational core values are also vital to mission command in that they give a condensed, standing order list of leader's intent: be safe, effective, and efficient. This empowers field officers to have disciplined initiative while accepting risk, both of which are vital to mission command.

Chapter Review

Review Questions

1. Define span of control.
2. Define and compare centralized and decentralized command.
3. Explain what the commander's intent would include on the fireground.
4. What are the tenets of mission command?
5. Describe the operational core values of the Operations Triangle.

FESHE Strategy and Tactics (C0279) Related Content

Course outcome #6 is the expectation that the student will be able to identify roles in IMS/NIMS as it relates to strategy and tactics, and #7 is that they can demonstrate knowledge of the various roles and responsibilities in ICS/NIMS. The content in this chapter introduces the purpose of utilizing the IMS and how the structure translates to delegation of tasks and decentralization of decision making and action. Understanding the mission (the why) behind the organizational charts is the keystone to comprehension and appropriate utilization.

Endnotes

1. Sun Tzu, *The Art of War.*
2. *ADP-06 Mission Command: Command and Control of Army Forces* (Washington, D.C.: U.S. Department of the Army, 2019), https://armypubs.army.mil/epubs/DR_pubs/DR_a/ARN34403-ADP_6-0-000-WEB-3.pdf
3. *ADP-06 Mission Command: Command and Control of Army Forces.*
4. *FM 5-0 Planning and Orders Production,* (Washington, D.C: U.S. Department of the Army, 2022), https://armypubs.army.mil/epubs/DR_pubs/DR_a/ARN36775-FM_5-0-001-WEB-3.pdf
5. *FM 5-0 Planning and Orders Production,* (Washington, D.C: U.S. Department of the Army, 2022), https://armypubs.army.mil/epubs/DR_pubs/DR_a/ARN36775-FM_5-0-001-WEB-3.pdf
6. *ADP-06 Mission Command: Command and Control of Army Forces.*
7. *ADP-06 Mission Command: Command and Control of Army Forces.*
8. *ADP-06 Mission Command: Command and Control of Army Forces.*
9. Shelby Hall and Ben Evarts, "US Fire Loss in 2021," *NFPA Journal*, December 7, 2022.

Section I

Fundamentals of Incident Command

3

Incident Command Professional Qualifications, Development, and Standards

Fig. 3–0. These are just a few of the standards that drive command.

The fire officer faces a broad range of challenges in the performance of their duties from the fire house to the incident scene. Many of the variables to these challenges are local, such as culture, resources, organizational makeup, and even building construction. While there are differences in performance and action at a local level, there are clear and consistent national standards and professional qualifications to hold these positions. It is critical that organizations and officers can demonstrate awareness and competence of national guidance not only to comply, but to have the confidence that your department or personal standards meets or exceeds them.

As a text of incident command for fire officers, the content of this book will match or support the various national standards, rules, and regulations regarding performance of duties for command officers or conduct of command on incident scenes. The term *professional* in the text is not intended to exclude volunteers or indicate a connection to career or compensation. The definition of the term professional within these pages is one who engages in a pursuit professionally and conforms to technical and ethical standards.

Professional Qualifications

National Fire Protection Association (NFPA) 1021: Standard for Fire Officer Professional Qualifications defines four levels of fire officer. It is generally accepted that Fire Officer I and Fire Officer II are company officers and Fire Officer III and Fire Officer IV are chief officers.[1] During the 1997 revision process, NFPA established clear job performance requirements (JPRs) for certification to each level of fire officer. In the 2009 revision, in an effort by NFPA to clarify that qualification must be continually demonstrated, the term certification was removed and was replaced by qualification. To support this, the 2009 standard was the first edition to add a skill maintenance requirement. Today (2020 edition of NFPA 1021) NFPA has eight areas of fire officer duties with associated JPRs:

1. General
2. Human resource management
3. Community and government relations
4. Administration
5. Inspection and investigation
6. Emergency services delivery
7. Health and safety
8. Emergency management

From NFPA 1021, these areas and JPRs are "intended to assist the users of the standard with the implementation of the requirements and development of training programs."[2] The concentration of this text is incident command, and the NFPA requirements in the areas of emergency services delivery and emergency management have been implemented into the text to provide specific job performance demonstration and ongoing skill maintenance training for organizations or individuals. Throughout the text you will find collar brass icons associated with each of the fire officer levels. These icons will identify sections of the text or simulations that connect to the qualifying JPRs.

NFPA 1021: Standard for Fire Officer Professional Qualifications for Emergency Services Delivery[3]

Fire Officer I

- 4.6 Supervising emergency operations and deploying assigned resources in accordance with local emergency plan.
  - 4.6.1 Develop initial action plan, given size-up information for an incident and assigned emergency response resources, so that resources are deployed to control the emergency.
  - 4.6.2 Implement an action plan at an emergency operation given assigned resources, type of incident, and preliminary plan so that resources are deployed to mitigate the situation.
  - 4.6.3 Develop and conduct a post-incident analysis given a single unit incident and post-incident analysis policies, procedures, and forms so that all required critical elements are identified and communicated, and the approved forms are completed and processed in accordance with policies and procedures.

Fire Officer II

- 5.6 This duty involves supervising multi-unit emergency operations, conducting pre-incident planning, and deploying assigned resources.
  - 5.6.1 Produce operational plans, given an emergency incident requiring multi-unit operations; the current editions of NFPA 1600, 1700, 1710, and 1720; and authority having jurisdiction (AHJ)-approved safety procedures so that resources and their assignments are obtained and plans carried out in compliance with NFPA 1600 and approved safety procedures, resulting in the mitigation of the incident.
  - 5.6.2 Develop and conduct a post-incident analysis given a multi-unit incident and post-incident analysis policies, procedures, and forms so that all required critical elements are identified and communicated, and the approved forms are completed and processed in accordance with policies and procedures.
  - 5.6.3 Prepare a written report, given incident reporting data from the jurisdiction, so that major causes for fire service demands are identified for various planning areas within the service area of the organization.

Fire Officer III

- 6.6 This duty involves managing multi-agency planning, deployment, and operations.
  - 6.6.1 Prepare an action plan given an emergency incident requiring multiple agency operations so that the required resources are determined and the resources are assigned and placed to mitigate the incident
  - 6.6.2 Develop and conduct a post-incident analysis given a multi-agency incident and post-incident analysis policies, procedures, and forms so that all required critical

elements are identified and communicated, and the approved forms are completed and processed in accordance with policies and procedures.
    - 6.6.3 Develop a plan for the organization given an unmet need for resources that exceed what is available in the organization, so that the mission of the organization is capable of being performed in times of extraordinary need.

Fire Officer IV

- 7.6 This duty involves developing plans for major disasters.
    - 7.6.1 Develop a comprehensive disaster plan that integrates other agencies' resources given risk, vulnerability, and capability data so that the organization can mitigate impact to the community.
    - 7.6.2 Develop a comprehensive plan given data, so that the agency operates at a hostile event, integrates with other agencies' actions, and provides safety and protection of members.

NFPA 1021: Standard for Fire Officer Professional Qualifications for Emergency Management[4]

Fire Officer III

- 6.8 This duty involves polices, procedures, and programs for the role of the fire service in the community's emergency management plan and the roles of local, state/provincial, and national emergency management agencies.
    - 6.8.1 Develop a plan for the integration of fire services resources in the community's emergency management plan, given the requirements of the community and the resources available in the fire department, so that the role of the fire service is in compliance with local, state/provincial, and national requirements.

Professional Development

Since the publication of *America Burning* and the National Fire Protection and Control Act of 1974, the professionalization of the American fire service regarding education and training has been a priority at the federal level. In the modern era, professional development has advanced to include higher education. When the National Fire Academy established the FESHE model, they intended to create a consistent path and curriculum for the fire service and higher education institutions to work together on professional development degree programs. Within the National Fire Academy Professional Development Model, the transition from purely certificate-based education to blending higher education begins at the Fire Officer I level.

The FESHE model provides curriculum guidance to include course outcomes, objectives, outlines, and references for the discipline-specific courses for Fire and Emergency Services associate's, bachelor's, and graduate degree programs. The content of this text supports the outcomes, objectives, and outlines for the FESHE associate's degree Strategy and Tactics (C0279) course.

FESHE Strategy and Tactics (C0279)[5]

Description

Provide principles of fireground control through utilization of personnel, equipment, and extinguishing agents.

Outcomes

1. Discuss fire behavior as it relates to strategies and tactics.
2. Explain the main components of pre-fire planning, and identify steps needed for a pre-fire plan review.
3. Identify the basics of building construction and how they interrelate to pre-fire planning, strategy and tactics.
4. Describe the steps taken during size-up using the FPODP model.
5. Examine the significance of fireground communication.
6. Identify the roles of the Incident Management System as it relates to strategy and tactics.
7. Demonstrate the various roles and responsibilities in ICS/IMS.

Objectives

1. Create a strategy and implement appropriate tactics.
2. Possess a working knowledge and execution of IMS at the incident.

In Appendix D, a detailed FESHE curriculum model for Strategy and Tactics (C0279) is provided to include page numbers, sections, and simulations that directly connect to the course outcomes, objectives, and outlines. This serves as an additional tool for both implementation and validation for the individual, organization, or institution utilizing this text as a component of professional development for the fire officer.

Professional Standards

The first two sections of this chapter concentrated on professional qualifications and development of the individual. For this section, the focus shifts to professional standards and expectations associated with the practice and operations of incident command and incident management.

National Incident Management System (NIMS)

As the first chapter demonstrated, the history and practice of command is centuries old, whereas the concept of incident management as it is known today has only matured in the past few decades. It was the 1970s when the first movement began that moved away from the civil defense centralized command models to the initial decentralized emergency management systems like FIRESCOPE. FIRESCOPE evolved into the Incident Command System (ICS) and the Multiagency Coordination System (MACS). In 1982 FIRESCOPE and the National Wildfire Coordinating Group (NWCG) worked together to create the National Interagency Incident Management System (NIIMS). As more disasters and events were managed and planned using these systems, they quickly became the dominate model for large scale incidents. While incident management practices were broadly adopted during the 1990s, it was not universal. The events of September 11th, 2001 galvanized the need for a unified approach, language, and practice of incident management, as well as federal accountability to it.

In 2004 the Federal Emergency Management Administration published the National Incident Management System and established the concept of command and coordination. The priorities of incident management are saving lives, stabilizing the incident, and protecting property and the environment. NIMS defines the guiding principles of successful command and coordination as flexibility, standardization, and unity of effort. Most of the fire service still views NIMS as an incident management model only for large scale, longer-term, multi-agency events. When NIMS is evaluated in greater depth with the historical context provided in chapter 1, it is clear it is a purposeful and powerful (federally directed) shift in models. The National Incident Management System is the catalyst for change from the traditional chaos, command, and control model to continuity, coordination, and cooperation that has been discussed for decades yet not collectively practiced. As proof that the guiding principles of command and coordination are flexibility, standardization, and unity of effort, the words of Chief Brunacini are echoed: "The modern fire service demands a highly mature combination of independence and obedience."[6]

Educational institutions, fire departments, or public safety organizations might question if the command practices contained in this text are compliant with the National Incident Management System. At the time of publication, there is no official certification or issuance of NIMS compliance. Programs and organizations are considered in compliance with NIMS by adopting the Incident Command System consistent with NIMS principles, terminology, and policies.

NIMS defines these as the foundational characteristics of incident command and coordination that contribute to the strength and efficiency of incident management:[7]

- **Common terminology:** Prior to the establishment of NIMS in 2004, regional influence or organizational slang terminology created confusion and conflicts in multi-agency or jurisdictional incidents. NIMS defines common terminology for organizational functions, resource descriptions, and incident facilities.
- **Modular organization:** NIMS is a management system designed to expand or contract in a planned and predictable manner to support the size, scope, and complexity of any incident.
- **Management by objectives:** Incident management is driven by the objectives established by the incident commander or incident management team informed by the most recent assessment, intelligence, or information.
- **Incident action planning:** incident action plans are the tools to clearly communicate the means and actions of achieving incident objectives. Every incident should have an action plan, but not all incidents need it to be formal or documented, especially in the initial phases. Some specific incident types or jurisdictions do have legal requirements surrounding IAP documentation; however this is not contained within NIMS.
- **Manageable span of control:** The ability for management to direct and coordinate subordinates, as well as the efficiency by which information can be communicated, is dependent on maintaining a manageable span of control. The type of incident, nature of the task, hazards and safety factors, experience of the supervisor and subordinates, and communication access between the subordinates and the supervisor are all factors that influence manageable span of control.
- **Comprehensive resource management:** Maintenance of an accurate, up-to-date, and real-time inventory of resources available for assignment or allocation is critical to mission success. From the first 5 minutes of the first alarm to the 25th day of a Type 1 incident, knowing what you have available to immediately address conditions, support ongoing operations, or create a tactical reserve is critical.

- **Integrated communications:** Integrated communications provide and maintain contact among and between incident resources, enable connectivity between different jurisdictions and various levels, achieve situational awareness, and facilitate information sharing.
- **Establishment and transfer of command:** The command function should be established at the onset of the incident. When command transfers, the transfer process includes a briefing that captures essential information for continuity of operations and communicates the transfer to all involved in the incident.
- **Unified command:** When no one jurisdiction, agency, or organization has primary authority or the resources to manage an incident on its own, Unified Command may be established. In Unified Command, there is no single commander. Instead, the Unified Command manages the incident by jointly approved objectives.
- **Chain of command and unity of command:** Chain of command refers to the orderly line of authority within the ranks of the incident management organization. Unity of command means that each individual only reports to one person. This clarifies reporting relationships and reduces confusion caused by multiple, conflicting directives, enabling leadership at all levels to effectively direct the personnel under their supervision.
- **Accountability:** Effective accountability for resources during an incident is essential.
- **Dispatch/deployment:** Spontaneous deployments overburden the recipient and compound accountability challenges. Resources should deploy only when appropriate authorities request and dispatch them
- **Information and intelligence:** The incident management organization establishes a process for gathering, analyzing, assessing, sharing, and managing incident-related information and intelligence
- **Incident facilities and locations:** Typical facilities include the Incident Command Post (ICP), incident base, staging areas, camps, mass-casualty triage areas, points of distribution, and emergency shelters.

The National Incident Management System is the compass by which the direction of all incident command and coordination models must be set, from the local to the federal level. Each of these core incident command and coordination characteristics will be spoken to directly, demonstrated in practice, and connected to key points in lessons-learned contributions throughout this text to ensure it is a relevant and reliable resource.

NFPA 1561: Standard on Emergency Services Incident Management System and Command Safety[8]

The first edition of NFPA 1561 was issued in 1990, born out of NFPA 1500. In 2000, the standard was expanded to better incorporate the shift in models from primarily command and control to incident management systems, and with it added Incident Management System to the standard title. 2005 brought a complete upgrade of the standard to align the content and direction of the standard with the National Incident Management System and the National Response Plan. In chapter 4 of the standard, System Implementation, the expectation is set that emergency services organizations shall adopt the National Incident Management System. In 2014 the standard title changed again to include the term Command Safety, as a new chapter under that title was added to the standard. This chapter included information and direction for the management of incidents with Mayday events and added clarification on the use of emergency medical services at all incidents, including when EMS shall be used and what minimum level of EMS shall be used. The 2020 edition of 1561 is the standard to which this text is written, since NFPA added language to address command practices to prevent the incident within an incident (IWI). "It is imperative that the incident commander stay in control of the entire incident and not become overly committed to the emergency." The tone of this language and direction mirrors one of the core principles of this text. The incident commander must stay ahead of the incident power curve to prevent the potential command lag created through the reflex time of addressing new conditions, challenges, or emergencies.

Fire or emergency services organizations composed of career, combination, or volunteer employees should familiarize themselves with NFPA 1561 as a bridge text that focuses in from the all-discipline, all-agency, all-hazards scope and language of NIMS and the National Response Framework towards more specific fire and emergency services applications. While National Fire Protection Association standards are not law, in most areas they serve in department polices as recommendations. NFPA standards can be trusted sources as they have stood both the test of time and legal challenge as national best practices.

NFPA 1561 clearly defines the national best practices for fire and emergency services regarding incident management and the role of command. Consider the 1561 standard to be the "what," or expectations, of incident management and command. This text will fill in the "how to" and "why" we do incident management and command.

The value of NFPA 1561 as an incident management and command guide is not limited to the constraints of the standard. The annexes of NFPA 1561 provide a greater depth of information to the chapter sections of the standard. It is encouraged that the reader also cross reference applicable annexes from NFPA 1561 to this text. It is the belief of the authors that the following annexes work purposely with the content in this text as supporting sources in both directions.

- Annex F—Division Designation: Tactical Assignments for a Multistory Incident
- Annex H—Functional Assignments for High Rise Building Incidents
- Annex I—Development of Subordinate Officers for Implementing a More Efficient Management System
- Annex J—Incident Management for the Fire Service on Type 5 or 4 Incidents
- Annex K—Structural Fire Fighting: Risk Assessment and Operational Expectations

NFPA 1710: Standard for the Organization and Deployment of Fire Suppression Operations, Emergency Medical Operations, and Special Operations to the Public by Career Fire Departments[9]

From the NIMS to NFPA 1561, and now on to *NFPA 1710: The Standard for the Organization and Deployment of Fire Suppression Operations, Emergency Medical Operations and Special Operations to the Public by Career Fire Departments*, expectations and definitions of the role and function of incident command are narrowed. From all the aforementioned qualifications, curricula, and standards, the common thread is that the modern approach to handling incidents is in a professional, informed, and managerial style. Out of this change, utilization of command teams has continued to expand. Early incident management systems were born out of large-scale events, and the publication of NIMS in response to the events of 9/11 has contributed to a perception by many in emergency services that the utilization of these systems is not necessary for routine incidents. This idea could not be further from the truth. In nearly every publication related to incident management referenced in this text, utilization of some form of incident management is recommended for every emergency event.

To inform emergency services on the application of IMS to the more routine events, the 2020 edition of NFPA 1561 includes Annex J—Incident Management for the Fire Service on Type 5 or 4 Incidents. To support the ability for emergency service organizations to stand up a command team at an incident that has expanded to the Type 5 or Type 4 level, recognition, and preparation for it, begins at the incident onset. Modern incident command is a team sport. It is accomplished through division of labor, manageable spans of control, and tactical coordination. This is where NFPA 1710, as the standard for organization and deployment, works to apply this to the fire service down to the initial incident organization at the first alarm level.

In chapter 5 of NFPA 1710, Fire Department Services, under the deployment section of 5.2.4, the tactics and tasks most likely associated with a full alarm working incident at four different occupancy classifications is defined.

- NFPA 1710—5.2.4.1.1 Single Family Dwelling (Typical 2,000 square foot, two-story single-family dwelling without a basement and with no exposures). Total effective response force with a minimum of 16 (17 if an aerial device is used).
- NFPA 1710—5.2.4.2.2 Open-Air Strip Shopping Center (Ranging from 13,000 to 196,000 square feet in size). Total effective response force a minimum of 27 (28 if an aerial device is used).
- NFPA 1710—4.2.4.3 Apartment (Typical 1200 square foot apartment within a three-story, garden style apartment building). Total effective response force a minimum of 27 (28 if an aerial device is used).
- NFPA 1710—4.2.4.4 High Rise (Building with the highest floor greater than 75 feet above the lowest level of fire department vehicle access). Total effective response force a minimum of 42 (43 if the building is equipped with a fire pump).

For the establishment of command at a single-family dwelling fire with no exposures, no basement, and an initial effective alarm response of 16 to 17 members, NFPA 1710 has a minimum of one member dedicated to this task. For responses to all other occupancies, to ensure overall coordination, direction, and safety of the initial full alarm assignment, a minimum of two members should be dedicated to managing the task of incident command. The change of occupancy size and exposure potential associated with the other classifications, combined with the increased number of members and units deployed to these occupancies, require an associated appropriate command level deployment.

This early command team must be competent and capable of initially establishing division of labor, manageable spans of control, and tactical coordination, not just the position of command. Too often, the fire service considers an initial command team as an IC and a scribe, or an IC and a safety officer, and while these two positions are extremely valuable to an incident commander and contribute to incident management, neither of those roles directly improve span of control or the efficiency of communication and information transfer in the initial stages of an incident. NFPA 1710 does not define the roles of these two members dedicated to managing the task of incident command. This allows the incident or authority having jurisdiction (AHJ) to determine if one serves as IC and the second as an early functional group or division supervisor which does improve span of control and reduces communication demands.

Summary

Proper use of incident management and the role of command can be viewed through two lenses: what is proper by the professional standards and mandates, and what is proper in practice. The fire service in the 21st century has reached a point where these two are in alignment. For too long the fire service has lacked strong, substantiated professional guidance for the practice of incident management. At the same time, in practice the fire service has had an almost egotistical low-trust view of incident management, with limited decentralization and delegation, retaining power at the upper limits of span of control. Command systems under stress and tension during a normal load will fail when the force of unexpected extension to an exposure, a victim being located, or a Mayday is added to them. Early establishment of division or group supervision and the support of tactics as they are initiated progressively and proactively reduces span of control and communication barriers. In the event of an incident within an incident, attempting to offload resources and reassign units to supervisors on the fly while simultaneously assessing and collecting information on the new emergency chokes communication paths and is a recipe for disaster.

When *1710, 1561* and NIMS are viewed collectively, the professional road map of continuity, coordination, and cooperation become clear. When the balance of this text is utilized to inform on how and why you as an individual or an organization must practice an incident management and command model to meet and forecast the needs of a more dynamic and time compressed emergency scene, your practices will also align.

Chapter Review

Review Questions

1. Explain the importance of the 2009 revision to *NFPA 1021: Standard for Fire Officer Professional Qualifications.*
2. What does FESHE stand for and what is the purpose?
3. Describe some of the foundational characteristics of the National Incident Management System (NIMS).
4. Explain NFPA 1561.
5. According to NFPA 1710, for what occupancies is it recommended that more than one member be assigned to the duty of managing command?

FESHE Strategy and Tactics (C0279) Related Content

Course outcome #3 is the expectation that students identify the basics of building construction and how they interrelate to pre-fire planning, strategy, and tactics. This chapter introduces the NFPA 1710 occupancy type recommendations to evaluate and establish appropriate deployment models and operations for departments and incident commanders for single-family dwellings, apartments, open-air strip shopping centers, and high-rises. Course outcome #6 is the expectation that the students identify the roles of NIMS and IMS as it relates to strategy and tactics, and course outcome #7 is the expectation that the student can demonstrate knowledge of the various roles and responsibilities in ICS/NIMS. This chapter connects NIMS to the fire service-specific expectations and recommendations within NFPA 1021, 1561 and 1710.

Endnotes

1 *National Fire Protection Association (NFPA) 1021: Standard for Fire Officer Professional Qualifications* (NFPA, 2020).

2 *NFPA 1021.*

3 *NFPA 1021.*

4 *NFPA 1021.*

5 Fire and Emergency Services Higher Education Initiative (FESHE) https://www.usfa.fema.gov/nfa/about/feshe/

6. Alan V. Brunacini, *Fire Command*, 2nd ed. (Quincy, Massachusetts/Phoenix, Arizona: National Fire Protection Association, 1985).

7 *National Incident Management System*, 3rd ed. (Federal Emergency Management Agency [FEMA], 2017), https://www.fema.gov/sites/default/files/2020-07/fema_nims_doctrine-2017.pdf.

8 *NFPA 1710: Standard for the Organization and Deployment of Fire Suppression Operations, Emergency Medical Operations, and Special Operations to the Public by Career Fire Departments* (NFPA, 2020).

9 *NFPA 1561: Standard on Emergency Services Incident Management System and Command Safety* (NFPA, 2020).

4

Anatomy and Physiology of an Incident

Fig. 4–0. Every incident has three primary levels of operation: strategic, tactical, and task.

Introduction

Regardless of the size, type, or nature of the incident, all have a common anatomy and physiology. Viewing it this way helps us approach any type of response with a consistent methodology and nomenclature to organize, manage, and mitigate the event. While no two incidents are ever the same, we can anticipate common characteristics in each, so we are equipped to apply a standard decision-making model, speak the same language, and operate consistently over time. As the variables specific to the incident at hand unfold, we plug them into the system to minimize chaos, keeping the troops calm while allowing the maximum ability to anticipate variables and adjust to unforeseen circumstances. This chapter gives an overview and context to the anatomy, terminology, and physiological characteristics of incidents. Subsequent chapters in the book are referenced and go into much greater detail. Consider this an introduction.

Incident Types

There are five incident types: Types 1–5. *Type 1 incidents* involve large, federal-level events like 9/11 (fig. 4–1), large wildfires, and massive hurricanes, for example. *Type 2 incidents* are typically state-level incidents like smaller fires, floods, and events that are located or impacting resources within that state. Examples would be smaller hurricanes or earthquakes, and flooding that impacts multiple counties. Type 1 and 2 incident management teams (IMTs) are composed of overhead personnel who will typically take over the management of larger incidents as they escalate.

Type 3 incidents are county or operational area incidents that last more than a day (operational period) and involve more than one agency (fig. 4–2). *Operational areas* are typically synonymous with county jurisdiction. Type 3 IMTs have become extremely common throughout the United States since 9/11. Incidents like local flooding, or multiple incidents operating concurrently like floods, fires, or human-made acts, would likely be managed by a Type 3 IMT until relieved by a higher-level team if escalating. More on IMTs in Chapter 16: Unified Command.

Type 4 incidents can be considered multi-alarm/multi-agency events (multi-alarm structure fire) that last up to 24 hours (fig. 4–3). A larger fire department will likely mitigate these incidents without any automatic/mutual aid, while smaller agencies may have automatic/mutual aid agreements in place that are required to mitigate the same-size incident.

Type 5 incidents are single alarm, initial response events that happen every day, from medical aids to house fires, and can be mitigated in a few hours or less (fig. 4–4). The scope of the majority of this text is on Type 4 and 5 incidents. Chapter 16 does discuss Type 1–3 incidents.

Pre-Incident Considerations

Modern incidents demand that we think more globally to ensure success. Rather than only considering the period between dispatch and close of the incident, pre- and post-incident considerations will set up the fire department for greater success. Therefore, they should be factored into the overall incident anatomy and physiology.

Pre-incident preparation includes pre-planning target hazards, high life-threat occupancies, and multi-family dwellings. Training and the capabilities of your team are critical. This includes considering who is staffing companies during the shift and considering their

Fig. 4–1. Example of a Type 1 incident after September 11, 2001
Source. Photo courtesy of FEMA US&R Sacramento, CA Task Force 7

Fig. 4–2. A Type 3 incident
Source. Photo courtesy of Greg Doyle, LAFD

Fig. 4–3. A multi-alarm structure fire
Source. Photo courtesy of Cody Weireter, LAFD

Fig. 4–4. A single alarm house fire
Source. Photo courtesy of South Metro Fire (CO)

training and experience. You may need to move personnel to set up the team for optimal performance.

The deployment model should be factored into your command philosophy and preparation. Countless communities and cities have endless permutations of deployment models, staffing, resource levels, automatic and mutual aid arrangements, volunteer/combination/career departments, and dispatch centers. On any given day, you may have staffing shortages, company closures, or units out of service for training, maintenance, special events, and so on. Consider traffic patterns, weather conditions, run volume, and special events. A good incident commander knows what companies are out of service or out of position throughout the shift. Situational awareness and preparation before the dispatch will keep you from being surprised when the bell rings.

Incident Anatomy and Physiology

Upon dispatch, the actual incident has 12 components, each with specific characteristics:

1. Dispatch/Response
2. Arrival/Size-Up
3. Priority/Strategy
4. Tactics
5. Tasks
6. Resources
7. ICS/Organization
8. Communications
9. Re-evaluation
10. Support
11. Transfer/Close
12. On-Scene After-Action Review (AAR)

Dispatch and Response

Upon dispatch, you will consider the type, size, and nature of the incident. Your first priority is to get to the incident safely. If you are a chief or a staff officer and do not have a driver, incident technician, or a company you can follow out of your station, you must take the time to map yourself into the incident before you respond code 3. Fortunately, most systems have an automatic mapping function.

As a company officer, you have the luxury of a driver, engineer, or apparatus operator. Many would-be first-due companies miss a turn and get delayed because of over-zealous officers who do not route their driver into the scene or look up the pre-plan. Consider who will be first-in and if the correct number and type of resources are responding. Is the normal first-due company delayed? Is the weather a factor and will there be additional delays due to traffic? What is the potential? Should any special resources be added to the incident response? Cues like multiple calls to dispatch, a column of smoke, a report of visible flames, and a report of a fire across from a particular address likely lead to a working fire. If you have a pre-plan of the location, only open it if you have a driver. Otherwise get to the scene safely and open the pre-plan later. This includes satellite mapping systems. If you know the occupancy or location and have important information, share it. Otherwise, keep the radio quiet. Standard operating guidelines (SOGs) for communications are extremely varied. Ensure all units are on the proper channel and that you have clear communication with dispatch.

Additional considerations for response are listed in sections 2 and 3, where specific incident types are discussed.

Arrival and Size-Up

This is often the most critical step in the incident because we start on the right foot, or not. A systematic, institutionalized method of size-up must be adopted and trained upon by your organization if you expect to have maximum effectiveness and consistency day-in and day-out.

Size-up is an ongoing mental process, from pre-planning, SOGs, and training to after-action review, more training, and SOG revision. As related to arrival of an incident, *size-up* is the mental process of rapidly gathering information, assessing risk, gain, available resources, strategic and tactical options, and developing an initial incident action plan. This process is continuous throughout the incident (fig. 4–5).

One misconception is that size-up and arrival reports are synonymous. This is not true. Size-up is a mental process that precedes the verbal process of giving an arrival report on the radio. An *arrival report* happens *after* a mental size-up and is the first-due company or chief officer's initial radio report to incoming units to communicate arrival, conditions, actions, and needs, including the assumption or passing of command for subsequent resources responding.

We delve much further into both size-ups and arrival reports in chapter 5: Size-Up, Risk Management, and Communications.

Priority/Strategy

The *incident commander* (IC) determines the incident priority and strategy, develops the incident action plan, and orders resources to mitigate the problem. The *strategic level* of the incident is where the IC resides who is typically furthest away (over 50 feet) from the incident and thinking furthest ahead of the incident (fig. 4–6). The mindset of the IC is in the "then/what-if" level of thinking. The IC must think downstream (then) regarding next stages and resource needs like additional alarms, special resource needs, transitioning into a salvage and overhaul, or investigation. The IC must think ahead of the current situation regarding potential risks and plan for and anticipate unknown and unplanned (what-if) events such as a Mayday, an unknown victim, exposures involved, or rapid fire progression. The strategic level is the furthest away from the incident and the IC must be ahead of the incident in the "then/what-if" mindset.

Fig. 4–5. You may have a limited view upon arrival to perform a size-up. What does this image tell you?
Source. Photo courtesy of Sacramento (CA) Metropolitan Fire District

Fig. 4–6. The IC literally looks at the big picture with a strategic mindset ahead of the incident.
Source. Photo courtesy of Rick McClure

The *incident action plan* (IAP) is the amalgamation of the priority, strategy, tactical objectives, communication, resource allocation, and organization of the incident. Type 1–3 incidents have the luxury of a planning section that produces a written plan for a given operational period. A Type 4 (multi-alarm) or Type 5 (single alarm) response such as a structure fire does not have the discretionary time or resources to develop such an IAP. The *starting point* for an IAP at these common, daily incidents is a consolidation of your agency SOGs/SOPs, policy, response levels, pre-plans, dispatch protocols, radio frequencies, training, and so on. These pre-established factors give us the opportunity to rapidly respond to, organize, and mitigate a Type 4 or 5 incident where discretionary time is minimal. Unlike Type 1–3 incidents, the true IAP on a structure fire or any other single or multi-alarm incident is verbalized and recorded as the incident unfolds.

The *incident command post* (ICP) is where the IC and other aids or overhead personnel reside. The location of the ICP has been open to debate for decades. Generally speaking, the front of a building is a good location; however, this is not always possible due to apparatus, conditions, hoselines, and so on. Also, a corner of a building gives the IC two sides of the building to see.

Two overall philosophies exist regarding the location and positioning of the ICP: inside the command vehicle or outside. There are benefits to each. Inside the command vehicle has the benefits of a quieter working environment. Fireground noise like apparatus, saws, radio traffic, sirens, horns, and inclement weather make it difficult to hear the radio, think, and not be distracted with bystanders and side conversations. Subtle nuances like the sound of a vibra-alert system, the sound of low-air self-contained breathing apparatus (SCBA), and the tone of voice of the crews give the IC insight. The benefits of being outside the vehicle are that the IC can see and hear conditions more readily and support staff can easily access the IC.

The *incident priority* is the most overarching urgency that determines the incident strategy, objectives, resources needs, and level of gain or risk to be considered. It is the first and biggest decision made by the incident commander. The three options for incident priorities, *regardless of the type of incident*, are life saving, incident stabilization, and property/environment. All incidents, regardless of type, nature, and size, will be in one of these three priorities.

A *life saving priority* indicates that there is civilian human life at risk and savable. Therefore, a higher amount of risk may be taken by responders. This would drive incident objectives like search and rescue operations, interior attack, or evacuation of civilians from hazardous areas where injury or death is imminent. The term Life "Safety" has been commonly used in the past. This creates confusion on what lives are the priority: the civilians or the firefighters? Safety is always part of our mindset as firefighters and we do not want to take unnecessary risk. That said, a life *saving* priority is for the civilians.

An *incident stabilization priority* indicates that all civilian life loss risk has been mitigated and the operations are focused on preventing further escalation, property loss, and bringing the situation under control. A moderate to low level of risk is allowed in this situation. This would drive incident objectives like exposure protection, confinement and preventing extension, extinguishment, and containing or stopping a leak.

Finally, a *property/environment priority* indicates that all civilian life is free of risk, the incident is controlled and stabilized, and now the focus of the response is on preventing further damage to property and the environment. Minimal risk is allowed in this situation. This would drive incident objectives like salvage and overhaul, mopping up a vegetation fire, preventing runoff from reaching a storm drain, or calling in special resources to clean up hazardous material.

The *incident strategy* determines the position from which resources will engage the problem. Every incident, *regardless of type*, will have one of three strategies: offensive, defensive, or combination.

An *offensive strategy* indicates that personnel are engaged in a forward moving posture to engage in more aggressive operations to accomplish incident objectives. Structural fireground examples would be interior fire attack, search, vertical ventilation, or salvage and overhaul. Vegetation fire examples would be direct attack operations (remove the fire from the fuel) that extinguish the fire with hoselines. An MCI example would include triage, treatment, and transportation. Hazardous material incident examples would include a team entering a hazardous area to contain a leak or rescue an exposed civilian or entering a building to evacuate.

A *defensive strategy* indicates that personnel are engaged in a backward-moving posture to engage in less aggressive and lower-risk operations to accomplish incident objectives and prevent further loss or damage. Structural fireground examples would be exterior fire attack for a prolonged period with large caliber streams or elevated master streams. Vegetation fire examples would be indirect attack operations (remove fuel from the fire) such as back-burning, firing out, cutting handlines, cutting in dozer lines, and performing structure defense. An MCI example would include remaining outside an area threatened by a hazardous material or active threats prior to performing triage, treatment, and transportation. Hazardous material incident examples would include restricting access to the area around the hazard, denying entry, and gross decontamination.

A *combination strategy* indicates that the incident has both an area where an offensive strategy is being employed *and* a defensive strategy is being employed simultaneously. An example would be a strip mall with collapse of the center unit (defensive space); however, units are able to enter the B and D exposures to contain the fire and evacuate (offensive space). A vegetation/wildland fire example of a combination strategy would be in the wildland urban interface (WUI), where direct attack is being utilized on the flanks of the fire while structure defense is utilized in front of the fire to protect residences. An MCI example would be an active shooter confined by law enforcement to a building in an office complex (defensive), while fire and EMS personnel triage, treat, and transport victims outside the building (offensive). A hazardous material incident example of a combination strategy would be an engine and truck company evacuating a commercial building (offensive) from the A side, while the leak on the C side of the structure is isolated and entry is denied until the arrival of a technically qualified hazardous material team (defensive).

Incident priorities and strategies can change throughout the event. For example, a structure fire may begin offensive, then deteriorating conditions warrant a retreat to defensive. Once an all clear is declared for the involved building, the priority will change to incident stabilization. It is certainly common for a fire to be in a stabilization priority with an offensive strategy. For example, the fire is knocked down, secondary search is all clear, yet crews remain inside to perform salvage and overhaul. The risk threshold is less, so if conditions deteriorate, the IC can easily switch to a combination or defensive strategy more quickly without regard for potential civilian life.

Tactical

The *tactical level* of the incident is between the strategic-level IC and the task-level crews working. *Tactical officers* make operational decisions, assign crews to specific tasks, account for their personnel, and assess safety issues to accomplish the incident tactical objectives. Decisions are made in the field by tactical officers closer to the work and hot zone (6–50 feet away) than the IC. The mindset of the tactical officer is in the next level of thinking. The tactical officer must think ahead of the current situation they are seeing and order resources in anticipation of needs to accomplish incident tactical objectives. Since they are closer to the hot zone than the IC, yet further away from the work than the task-level companies, they can recognize and anticipate changing conditions more quickly regarding collapse, vent-point ignition, flashover, and fire beneath or above companies working, for example. Their primary means to communicate with their crews is face-to-face, which reduces radio traffic but enhances communication quality and maintains accountability (fig. 4–7).

While a company officer will supervise a single crew, they are often too task-saturated to supervise multiple crews or companies while standing back from the anticipated work in a manner to truly fulfill the functions required of a tactical supervisor. Limited staffing and response levels force most company officers to perform tasks alongside their crews and prevent them from being very effective tactical officers.

Division and group supervisors will most effectively fill the role of tactical supervisor. *Division supervisors* focus on a geographic area, such as the front of a building, third floor, roof, or flank of a fire. *Group supervisors* focus on functional components of an incident, such as fire attack, search, ventilation, medical, or rapid intervention crew (RIC). *Consolidating multiple companies under one division or group supervisor is a critical factor in creating safe, effective, and efficient emergency scene operations.*

Fig. 4–7. The division supervisor communicates face-to-face with crews in his division.
Source. Photo courtesy of South Metro Fire (CO)

Examples of tactical officer actions for varying incidents include, but are not limited to:

- **Structure fire:** Ordering additional hoselines, ensuring primary and secondary searches are complete, coordinating fire attack and ventilation, checking for fire extension, recognizing collapse potential, anticipating/preventing hostile fire events, preventing access or evacuating crews, and prioritizing areas for salvage and overhaul.
- **Vegetation fire:** Ordering additional resources for their division, sending companies to get more water, coordinating with other divisions, prioritizing structures to be defended, and recognizing downed power lines, watch out situations, and changing wind direction.
- **Hazardous materials incidents:** Defining the exclusion zone, ordering additional resources, tracking entry and egress, supervising evacuations, and setting up gross decontamination.
- **MCI:** Setting up triage, treatment, and transport areas, establishing traffic patterns for ground ambulances, and identifying an area for a landing zone/helispot.

Tactics are employed under the priority and strategy established by the IC. *Tactical objectives* are the incident's operational and functional goals to be accomplished. Examples for structure fires are rescue, exposure control, confinement, extinguishment, ventilation, salvage, overhaul, RIC, rehab, medical treatment/transport, and support. Examples of vegetation fire tactical objectives are perimeter control, structure defense, establishing handlines or dozer lines, air attack, mop up, and so on. Examples for hazardous material incidents are identify hazard, deny entry, stop a leak, contain a spill, back up, decontamination, and so on. Examples for MCI are triage, treatment, transport, and so on.

Control objectives are based upon operational/geographic boundaries for the incident. A structure fire may have a control objective of containing the fire to the room or unit of origin. For larger fires, it may be to the wing, floor, or building of origin. For vegetation fires it is common to put a box around the fire so all units know the control objectives/boundaries for the fire. Boundaries are likely based upon geographical landmarks where the fire can be stopped (something like, "south of Elder Creek Rd, west of Excelsior Rd, north of Florin Road, east of Bradshaw Rd") (fig. 4–8).

The *tactical gap* is the space between the strategic-level IC and the task-level crews working where risk assessment, communications, accountability, and situational awareness become distorted. *This is the tenderloin or sweet spot of the incident, where the battle is truly won or lost.*

The view, conditions, and sense of situation/resource status is very different between the strategic and task levels. The IC is too far away and the crews are too close/inside the hazard area. Therefore, the need to assign tactical division/group supervisors is paramount. While the crews are up close, inside the environment working, they are subjected to smoke, noise, thermal insult, auditory exclusion, tunnel vision, increased cardiovascular stress, SCBA,

Fig. 4–8. Control objectives can be physical boundaries, like roads for a vegetation fire.

cumbersome PPE, confined spaces, and a host of other bad things. They are trying to survive a potentially lethal environment that wants to kill them.

Meanwhile, out at the command post, the IC is far away from the work and only seeing one side or corner of a building. The only tether to any situational awareness or accountability to those whose lives the IC is responsible for inside the building is through a radio. The IC is talking to a bunch of workers who really don't want to chat, are hard to understand if they do answer the radio, and don't hear most of the radio traffic in the first place. Their reality is very different than the IC's.

This is centralized command. Unfortunately, the fire service has been in the habit of this type of command structure for decades, where everyone reports to the IC. The IC's span of control is blown out, the IC gets overwhelmed and behind the incident power curve, and the communications get sloppier and harder to understand. And accountability is out the window. Sound familiar?

I was once in my battalion office, listening to a commercial structure fire across town in another jurisdiction. The engine officer transmitted some incomprehensible message through a SCBA to the IC from inside the fire. The IC responded with the usual, "Repeat?" The engine officer transmitted again, this time louder and slower. Still unable to understand the transmission, the IC said, "Engine X, I can't understand you, but you are talking which means you're alive, and that's good enough for now." At least the IC was honest instead of responding with the usual "copy" from the ICP when the IC has no clue what was just said.

Good luck continues to breed bad habits into the next generation. Many new chiefs are just trying to learn to drive code 3 while talking on multiple radios, navigating the route to the scene, getting updates from dispatch on the Mobile Data Computer/Terminal (MDT) and somehow trying to account for a bunch of adrenaline-filled firefighters who are running into a building that might kill them.

The tactical gap is usually overlooked. it's the space where you can really see what is going on in that area of the incident. It's the front or back of a building, 6–50 feet away. It's seeing fire in the overhead, or underneath your firefighters. It's the lap of the structure, when possible. It's the stairwell outside the 7th floor landing. It's seeing vent-point ignition, potential collapse, or a lack of coordination between the attack crews and vertical vent team. It's where you get a victim profile from occupants. *We must bridge the tactical gap ASAP with division/group supervisors. This is mission command and decentralization* (fig. 4–9).

It's also where the NIOSH 5 live (see chapter 5). It's where you can really see the risk and conduct active eyes-on and hands-on accountability. It's where you can talk to the crews face-to-face instead of on the radio and form a tactical plan while checking air status, company tags, and names on turnout coats. It's where the battle is won or lost.

I have been the division supervisor countless times and I *always* had a better level of situational awareness than the IC at the ICP. Not because I was smarter than anyone, but because I was standing in the tactical gap.

Task

The *task level* is where the work gets done. *Company officers* directly supervise the tasks of their crews and account for their personnel. Decisions are carried out by task-level officers and their crews directly within the hot zone (0–6 of situational awareness). Tunnel vision and auditory exclusion are maximized in the task level, causing common challenges with situational awareness, accountability, and communications (See chapter 5).

The mindset of the tactical officer is in the "now" level of thinking. Examples of tasks include but are not limited to stretching hoselines, forcing a door, searching, cutting a hole, pumping an engine, donning an SCBA, throwing a ladder, treating a patient, entering a hazardous material area, and cutting a handline.

As an IC, you must remember that time moves much slower at the command post than at the task level, where the work is getting done. Your tactical supervisors will see, in real time, progress being made, where crews are, and how conditions are changing for the crews working. With tactical supervisors in place, the need to request updates directly from crews is minimized. This will keep radio traffic minimized, and the crews won't feel harassed by the IC when they are just trying to do the work in a violent environment.

Resources

Resources must be called early. As a general rule, it is much better to have the resources and not need them than need them and not have them. The second priority upon response to a dispatch (after arriving safely) is ensuring the right type and number of resources are dispatched. This is accomplished by updates from dispatch, knowledge of the building and location, pre-plans, and other signs like a large column of smoke, heavy winds, or power outages. One critical question to ask dispatch when responding to a structure fire is if everyone is reported out of the structure or not. While we still perform a search and only *we* will declare an all clear for a building, an update that three confirmed victims are trapped is critical information that

Fig. 4–9. Look at the view this company officer has on the C side of this house fire. His head is up able to see the tactical view while all the firefighters are head-down working at the task-level.
Source. Photo courtesy of Rick McClure

must be shared immediately. Unfortunately, sometimes this information is not relayed to responding units.

There are three types of staging areas: level 1, level 2, and forward staging. *Level 1 staging* occurs when resources stand by in the direction of travel, within a block of the incident (fig. 4–10). This usually occurs with a first alarm unless assignments are given. *Level 2 staging* occurs when a separate and specific location has been established away from the main incident, usually when greater alarms are called (fig. 4–11).

Finally, *forward staging* occurs when personnel who are fully equipped with PPE and SCBA come with tools and equipment to a forward area to be imminently assigned (fig. 4–12). The location is usually the ICP or a division/group area. The benefit of forward staging is that it greatly reduces reflex time to get additional or replacement crews into service. Forward-staged crews can be used for anything the IC deems appropriate. For example, a crew can stand by to relieve another crew, deploy an additional hoseline, perform a secondary search, or set up RIC.

ICS/Organization

The incident must be organized as soon as possible to ensure accurate accountability of personnel and resources. This will prevent the IC from getting behind the curve and overwhelmed with an unmanageable span of control. When this happens, there is no longer an IC, just an OW (overwhelmed), chasing the incident on the radio and asking for a lot of updates.

Fig. 4–10. Example of typical level 1 staging

Fig. 4–11. Example of level 2 staging

ICS is the national standard and primary means of organizing incidents, whether small or large, simple or complex. Setting up divisions and groups proactively will bridge the tactical gap. All too often, incident commanders resist the use of ICS or setting up of divisions and groups. As discussed in chapter 2, this phenomenon is usually a result of a lack of training, or developing complacency on smaller, more manageable fires. *ICS is like a parachute: Don't wait until you hit the ground to pull the chord.* We will delve much further into ICS throughout the rest of this book.

For the Type 5 (single alarm) and Type 4 (multi-alarm) arena, ICS must be swift, simple, and quickly adaptable. Unfortunately, ICS is often applied in a wildland mindset to a structure fire, or the inappropriate level of ICS is applied. For example, the first level of ICS that is typically utilized would be a division or group. Many times, a fire department or IC will set up an operations chief (OSC). This is a much higher level than necessary and has virtually no benefits. Utilizing an OSC essentially passes the buck from the IC and dumps all the problems and challenges onto the OSC. The only one who benefits is the IC who now has very little to do. This is NOT the intent of ICS.

Notice the specific hierarchy in the operations section in figure 4–13. The bottom of the totem pole is the single resource (engine, truck, etc). Next is a *strike team* that is comprised of

Fig. 4–12. Example of typical forward staging at the ICP
Source. Photo courtesy of South Metro Fire (CO)

Fig. 4–13. The hierarchy of ICS within the operations section

5 of the same kind and type of resource, with a leader. A *task force* is a combination of kinds and types of resources with a specific task/objective. A Federal Emergency Management Agency (FEMA) Urban Search and Rescue Task Force would be a common example. In either case, the boss of a strike team or task force is called a leader, which delineates that both

strike team leaders and task force leaders are at the same level of authority. Or, that strike teams and task forces, while different entities, are still considered the same level and are each *one* resource.

Strike teams are primarily utilized in the vegetation/wildfire arena. Or, they may be utilized when requesting a lot of resources for incidents like an MCI, where a strike team of ambulances nearby may be called. It depends upon your SOGs. Task forces may be used there, or in the urban setting. For example, Los Angeles City Fire Department utilizes task forces comprised of two engines and a truck.

Above this single resource (unit, strike team, task force) level in ICS is the division and group. Again, these are both the same level of authority, each led by a supervisor, meaning that they do not work for each other. They work *with* each other. *Divisions* are geographic. For example, Division A, B, 1, 2, roof, and so on. *Groups* are functional, such as RIC, medical, and rescue.

Above the division and group level is a branch which can be either geographic or functional. For example, North Branch or Fire/Rescue Branch. The boss is called a director. At the top of the operations section is the operations section chief. As with the other sections (finance, logistics, and planning), the boss is called a chief.

So, in review, the lowest level of authority/boss is a single resource company officer, then a leader (strike team or task force), then a supervisor (di vision or group), then a director (branch), then a chief (section).

Do not let this confuse you. *The point is that most structure fires and other Type 5 and 4 incidents will likely be organized by divisions and groups.* Jumping up to a section literally passes the buck. It does not reduce the span of control, nor does it reduce radio traffic or enhance accountability. We will get much more into this throughout the book.

Communications

Your jurisdiction will have a communications plan or policy and relative SOGs. Commonly, dispatch is performed on a separate frequency from command and tactical channels. A *dispatch channel* is a frequency dedicated to outgoing incident dispatches. In some jurisdictions, the channel may also be utilized for command or tactical radio traffic. A *command channel* is a frequency dedicated to two-way communication between dispatch and responding units, or between units transmitting to each other. A *tactical channel* is a frequency dedicated to operations for a particular incident or region.

While policies and practices are extremely varied, a common methodology is for units to be dispatched on the dispatch channel, respond and communicate with dispatch on a command channel, and then switch to operate on scene on a tactical channel.

Direct frequencies (simplex) are line-of-sight channels that have limited distance capability; however, they may have other benefits such as greater privacy. They are subject to obstruction and terrain interference. Repeated frequencies have much greater distance capability.

As the IC, you must decide if your incident will require more than one or two channels to effectively communicate. Training is critical in the usage of your brand-specific mobile and portable radios, as well as radio procedures for communication. There will be more on communications in chapter 5.

Reevaluation

As the IAP unfolds and companies arrive and make progress, the IC must re-evaluate if the plan is working. Reports back from companies or division or group supervisors will determine if the plan is working or not. Perhaps an all clear was made by companies inside the structure and you switch to an incident stabilization priority while maintaining an offensive strategy. Tactical objectives would include salvage and overhaul.

Conversely, you may see or be told of deteriorating conditions and concern for structural stability and switch from an offensive to a defensive strategy. You may decide to call additional resources, add overhead personnel, or call special support units. Tactical objectives would include exposure protection with the tasks of establishing elevated master streams and water supplies. This in turn would drive the resource request for two engines and two trucks.

Support

Once the IAP is verified, the incident must be supported properly to execute the plan. Support functions may take on many forms, including but not limited to the following:

- Additional units/alarms
- Overhead personnel (safety and public information officers)
- Specialty units
- Lighting and air
- Food/water and rehab
- Utility companies
- Red Cross (for displaced occupants)
- Law enforcement
- EMS
- Building management
- Board up crews
- Investigator

Citizen support is vital to excellent service. You may have displaced renters with a landlord who is out of town. You may have an apartment complex with dozens of occupants out in the cold of night with an apathetic manager and no spare units. Perhaps the fire was confined to a specific area and the owner wants to continue to occupy their home and you have shut off the power. Heaven forbid, you have a child whose parents are not capable of caring for them. Whatever the case, you are only limited by your imagination and empathy. Be sure to go the extra mile and ask yourself, "What if this was my home and my family?"

Transfer/Close

As the IC, you will either command the incident until the end and turn the property back over to management/owner/occupant, or you may transfer command to another officer. Transfer of command requires specific information to be relayed to the new IC regarding the IAP. This includes current priority, strategy, tactical/control objectives, benchmarks accomplished, ICS (resource assignments), communications plan, and resource needs. Many tools are utilized to accomplish the command transfer effectively and efficiently. More on this will be discussed in chapter 5.

On-Scene After-Action Review

An excellent learning opportunity is conducting a brief, tailboard after-action review with companies prior to them clearing the scene (fig. 4–14). If the fire has not been completely overhauled, ensure that one company is keeping an eye out for flareups with a hoseline in place, especially in the attic or cockloft. Then, after the review, companies can be released back into service until overhaul and investigation is completed and the property is turned back over the owner/occupant/manager. We will study after-action reviews, reports and the full cycle to maximize learning for the future.

Fig. 4–14. A brief tailboard AAR is a great way to get any issue out early, before leaving the scene. *Source.* Photo courtesy of Michael Meadows

One AAR method is called the chainsaw AAR. Simply gather all members around. If you can have all members present, it will benefit the younger firefighters to hear the discussion first-hand. Consider letting the younger firefighters go first. This prevents them from refraining from telling the truth at the end if everyone was patting each other on the back.

The following should be addressed in order, one at a time as a group:

- What went well?
- What didn't go well, or what would you do differently next time?
- What is your biggest lesson or takeaway from this incident?

The IC should start by sharing an overall good thing about the incident. Then, the IC should also go first when answering the question: "What didn't go well, or what would you do differently next time?" This sets the example of humility and will encourage others to share and be honest as well.

See Appendix B for an example of a full Green Sheet.

Summary

This chapter reviews the fundamental anatomy and physiology of an incident, whether a structure fire, vehicle accident, vegetation fire, MCI, hazardous materials incident, or otherwise. The 12 components of any incident are dispatch/response, arrival/size-up, priority/strategy, tactics, tasks, resources, ics/organization, communications, re-evaluation, support, transfer/close, and on-scene after-action review. This framework will be built upon throughout the rest of the text, which will focus on the Type 5 and 4 incidents, and single and multi-alarm incidents, respectively.

Upon arrival, size-up must take place and continue throughout the incident. During a structure fire, conditions change constantly, buildings are under demolition, and gravity constantly pulls at the weakened structure. All the while, your troops are running headlong into harm's way. Understanding the big picture and how to set the incident up with the correct priority and strategy; identifying tactical objectives; decentralizing while organizing divisions and groups; maintaining clear communications; and constantly re-evaluating to properly support the incident or switch directions all come from knowing incident anatomy and physiology. Only then can you drive the incident, command, coordinate, and lead others to lead in the battle.

Chapter Review

Review Questions

1. Describe the five incident types.
2. Explain the difference between size-up and an arrival report.
3. Explain a combination strategy.
4. Describe tactical objectives and control objectives.
5. Explain the tactical gap.
6. What are the three levels of staging?

FESHE Strategy and Tactics (CO279) Related Content

The two course objectives for CO279 Strategy and Tactics are to create a strategy and implement appropriate tactics through a working knowledge and execution of IMS/NIMS at the incident. This chapter establishes the foundations of strategy selection and tactical support for an incident both operationally and through command functions. Course outcome #2 is an expectation that students understand the importance pre-fire planning and associated phases. Through this chapter, incident types and order are broken down by section providing a detailed review of opportunity for pre-incident preparation and planning. Course outcome #4 is the expectation that the student can describe the steps during size-up. This chapter details size-up as a prearrival, arrival and ongoing process providing the student with guidance at each step. Course outcome #5 is the expectation that the student understands the significance of fireground communications. This chapter contains a section on fireground communications through formal systems such as dispatch, command, and tactical channels as well as the differences between communication as interactions on the fireground with regard to face-to-face communication, environmental and mental stressors, span of control, and situational awareness. Course outcome #6 is the expectation that the students identify the roles of NIMS and IMS as it relates to strategy and tactics, and course outcome #7 is the expectation that the student can demonstrate knowledge of the various roles and responsibilities in ICS/NIMS. This chapter takes ICS/NIMS and begins to associate roles and responsibilities with their strategic or tactical location and role.

NFPA 1021 Job Performance Requirements

The content in this chapter covers incident typing and, incident structure and incident timelines. This information can be utilized to support training and educational programs associated with the Emergency Services Delivery Fire Officer I JPR 4.6, Fire Officer II JPR 5.6 and Fire Officer III JPR 6.6.

5

Size-Up, Risk/Gain Management, and Communications

Fig. 5–0. Sacramento Metropolitan Fire District works a house fire.
Source. Photo courtesy of Sacramento (CA) Metro Fire

Introduction

Predictable is preventable.

—Gordon Graham

Size-up, risk management, and communications are intertwined. To adequately assess risk, officers must be able to continuously communicate changing conditions to each other from different geographic areas of the same incident. The ongoing mental act of size-up is brought to life through the physical act of communicating. Both have specific key points that have proven effective. We will also present assessing gain, which is what we are saving. The most important thing we can save is human life. This is our job, and what we have sworn to do. At the same time, we must bring our troops home safe from every call while fulfilling our duty to the public. Risk and gain are at constant odds for the IC, and communication is vital to risk and gain management.

Size-up has several stages. First, training and pre-planning provide the IC and tactical officers a framework from which to maximize discretionary time benefits. Here, before the shift ever begins, endless wells of information can be processed prior to an incident to maximize strategic and tactical advantage. Second, at shift change, many other factors and variables are available and changing, increasing the IC's situational awareness and analysis prior to a dispatch. These factors change and continue to remain dynamic throughout the shift or tour. Third, the dispatch comes. Discretionary time is instantly minimized, if not eliminated, and the IC and tactical officers are plagued with physiological obstacles like auditory exclusion, tunnel vision, adrenaline dumps, heightened blood pressures and pulses, and limited capacity to process information while constantly being bombarded with changing conditions, communications, widening spans of control, and unknown factors that are constantly looming. Fourth, upon arrival, the scene presents itself. More injections come, including visual stimuli like the building, smoke, flames, companies, occupants, bystanders, law enforcement, and exposures. More communications flood multiple radio frequencies and the combination of useless and inconsequential information is interspersed with critical, life-saving information. The IC must remain ahead of an incident and somehow outpace this event that is growing, evolving, and becoming more complex as each second ticks by.

Meanwhile, the crews deploying into the unknown face the same physiological obstacles, as well as physical ones like thermal insult, trauma, asphyxiation, and cumbersome personal protective equipment (PPE) and self-contained breathing apparatus (SCBA) that make risk assessment, size-up, and communications even more daunting. As the incident continues, escalation causes physical and mental fatigue of the troops. Auditory exclusion and tunnel vision increase. The building has been under demolition, and gravity, hydraulic mining, and forcible entry continue to erode structural integrity. Everything is in a constant state of change: smoke, fire, building, troops, incoming resources, information, and the incident action plan. Meanwhile, time stops for no one. We cannot put time in staging to stop it or call another alarm of time to get more of it.

Hopefully, conditions present the illusion of becoming stable. Yet the combination of weakened and burned-out structural members, the weight of tons of water, and the constant tug of gravity makes for a dubious stabilized situation at best. Too many have succumbed in this phase of the fire, during overhaul, after exiting the structure for another bottle, or even to conduct an interview.

Finally, with all members safe and accounted for, we begin the review process. We identify what we did well, not so well, what we need to improve, and lessons learned. This happens informally within companies, semi-formally within the alarm, and perhaps more formally if warranted throughout the department or region—usually due to something going bad. The worse it goes, the wider the swath of things we want to learn. We learn, train, adjust, and repeat. This is the cycle, the big picture that requires constant size-up and continual communication. Each of the phases has specific tools and pitfalls that must be learned to not repeat the past, let alone prepare for the future.

WISDOM FROM THE MASTERS

Situational Awareness

CHIEF STEPHEN MARSAR, FDNY

Situational awareness (SA) is a concept that has been used in the military and aviation fields for over a century. SA is generally defined as knowing where you are and what is going on around you, allowing individuals and organizations to be more alert and informed and to make better decisions. A relatively new concept adopted by the fire service, SA includes the important element of reality. After conducting the initial scene size-up, SA should kick in and be applied to tell us if things are getting better or worse—not based on how we wish it to be, but, rather, based on reality. While size-up tells us what we have, what is going on, what resources are required, what delays or difficulties we have, and what may be likely to happen next, SA utilizes reality as a benchmark that allows firefighters, officers, and chiefs to recognize when things are not getting better or not going well. Some of those elements may include the fire and smoke conditions, radio reports, and even listening to our sixth sense.

For example, at a fire scene firefighters and commanders expect the incident to get better as the firefight progresses. As the initial attack is being performed and we locate the seat of the fire and apply water, we expect to see the smoke get lighter in volume, velocity, and density. We expect radio reports that the primary and secondary searches are in progress or completed (and either positive or negative for victims and fire extension). We expect the building to be opened up and salvage and overhaul to begin. However, if command is not receiving those messages, they cannot assume that those tactics are being employed and are getting done without complications.

Sometimes no radio reports can be just as dangerous as numerous radio reports being received at the same time. Is there nothing going on, or are the troops so heavily engaged that they simply failed to report the conditions they're encountering due to being laser-focused on the tasks at hand? Should the IC prompt them, or should the IC give the members a chance to do their jobs and get back to them in a reasonable period of time? What is that reasonable period of time? It may be different for each individual or department. However, the IC cannot assume that the lack of communication means that things are going well. That may simply not be the reality.

Conversely, if the IC is getting reports of great headway being made from the interior but the smoke and fire conditions that they see are not getting better, then that is not the reality. In that case, other steps must be taken immediately, rather than hoping that things will get better shortly. Tactical supervisors bridge this gap very effectively.

Size-Up versus Arrival Reports

Size-up is the mental process of rapidly gathering information, assessing risk, gain, available resources, strategic, and tactical options, and developing an initial incident action plan. Size-up is *the* most critical element to safe, effective, and efficient operations. Beginning operations on any incident without an accurate and proficient mental size-up is like building a house on sand without a plan. You will not have a strong foundation from which to build, and your operations will be extremely unstable.

Conversely, an *arrival report* happens *after* a mental size-up and is the first-due company or chief officer's initial radio report to incoming units to communicate arrival, conditions, actions, and needs, including the assumption or passing of command for subsequent resources responding.

The challenge with size-up is that you must make potentially life-and-death decisions with limited information in a compressed timeframe. In the emergency environment, you will rarely have all the information and time needed to make the *perfect* decision. Rather, you must make the *best* decision possible, with the limited information currently at hand, then adjust as conditions change or more information, resources, or options become available.

The term size-up is mistakenly considered synonymous with an arrival report. This common misconception leads to officers who fail to think before they talk on the radio, thus resulting in thinking out loud with the pressure to sound competent on the radio rather than take a few seconds to truly mentally process the situation before them. The radio vernacular "push to talk" regarding radio operations is not supposed to be "push to *think*"! Yet, officers do it every day. Arrival reports with a lot of "uuuh's," "ummm's," and "at this time's" peppered in the transmission are usually giveaways that the officer speaking has not performed a mental size-up before keying the mic. *Think* before you talk.

When incident operations are predicated upon inaccurate information or assumptions, bad things usually happen. Unlike a verbal arrival or on-scene report, a size-up is a mental process that begins *well before* the incident is ever dispatched and must continuously be emphasized until units have left the scene. And size-up is everyone's job, *especially* the IC and tactical supervisors. That said, most firefighters with their heads down, working at the task level, are not sizing up; they are working, doing, engaging, and focused on the task at hand.

Accurate and continual size-up saves civilian and firefighter lives, prevents injuries, and minimizes losses. A fire with an unsupported mansard on a strip mall may be overlooked by task-level crews, but a heads-up tactical supervisor or company officer may identify the collapse potential due to fire in the overhead of the scabbed-on facade. Individuals and officers, no matter how experienced or savvy, will not perform adequate ongoing size-up at the task level. When your head is down, stretching a line, breaching a door, doing a search, cutting a hole, throwing a ladder, donning an SCBA, pumping an engine, treating a patient, or entering a hazmat scene in a Level A suit, you are *not* in a good position to size up. You are working with little regard for anything beyond your immediate field of view, within arm's reach, and you are likely not hearing the radio, either (fig. 5–1).

Tunnel vision kicks in when you only see what is immediately in front of you, missing critical visual signs of changing conditions regarding smoke and fire, construction features, access and egress points, victims, bystanders, overhead lines, hazardous materials, and so on. This

Fig. 5–1. The smoke conditions clearly indicate that the fire is in the attic space and beginning to involve structural members. Do the firefighters on the balcony see this from their vantage point? *Source.* Photo courtesy of Sacramento (CA) Metro Fire.

becomes exacerbated by heavy smoke conditions inside or outside the building. *Auditory exclusion* is equally perilous. When performing tasks, especially with an SCBA facemask donned, radio traffic is commonly missed by task-level personnel working. External factors like background noise, air horns, sirens, pumping engines, revving saws, screaming bystanders, and roaring flames are just a few of the common obstacles to effective hearing and listening. To compound the issue, many officers are task-saturated due to the amount of work to be done, the lack of staffing to do the work, or the habit and nature to gravitate to the work.

CASE STUDIES IN COMMAND

Nothing Showing Means Nothing

CAPTAIN NICK PAPA, NEW BRITAIN (CT) FIRE DEPARTMENT

On a cold, winter morning, just prior to shift-change, the tones dropped for a structure fire, reported as smoke showing from the building. The address was only a few blocks from the firehouse, within a densely populated neighborhood we regularly frequent. While en route, no further details were provided verbally over the radio or written in the computer-aided dispatch (CAD) notes on the apparatus mobile data terminal (MDT). As we pulled past the 3-story balloon-frame 6-family multiple-dwelling, no indication of fire was noted from our 3-sided view through the cab of the engine. These occupancies have a center, front stairway, directly accessing two Railroad Flat apartments (configured with all rooms aligned like the cars of a train) on each floor, sharing a common center hallway in the rear that leads to a large wood porch/stairway.

Despite giving an arrival report, which ended with the classic phrase of ". . . nothing showing," we did not throttle back and alter our approach. The motor pump operator (MPO) positioned the apparatus and shifted into pump gear, the nozzle firefighter stepped up on the tailboard and prepared to stretch, and I darted down the D-side of the building to get a view of the rear. As I was making my way, I immediately saw a bystander pointing up at the last window on the second floor. Even at this point, there was no smoke discharging from the building or even the faintest odor.

On reaching the bystander, he matter-of-factly stated, "The back bedroom is on fire." Taking a closer look, I confirmed his claim, noting the window was blackened from soot-staining. Because the apartment was sealed up, the fuel-rich fire quickly consumed the available oxygen and became ventilation-limited, entering into a premature decay stage. The fire lacked the heat output to generate enough pressure within the apartment to discharge the smoke. Had the bystander not been there to point out that subtle clue, it could have very easily been missed.

I immediately got on my portable radio and informed the incident commander (the deputy chief who had arrived right behind us) that we had a fire in the back bedroom on the second floor of the D-side apartment and instructed my nozzle firefighter to initiate the stretch. The nozzle firefighter and I, with assistance from the MPO, dry-stretched up the front stairs and flaked out the working length on the second-floor landing, as the fire apartment door was controlled—isolating our position—and the enclosed stairway was smoke-free. After calling for water, completing a flow check of the handline, and masking up, we opened up the fire apartment door, "staying low and letting blow," prompting a dump of thick, dark, volume-driven smoke out into the stairway. After placing a spring-clamp to prop open the door and making entry, visibility was zero from our tripod position as we advanced the handline diagonally across the living room toward the kitchen. Because he had conducted his own size-up from the tailboard and possessed the skillset to read the layout/floorplan of the building (including the spans and angles), the nozzle firefighter instinctively knew where to go.

At this point, the ladder company outside vent (OV) firefighter—responsible for conducting horizontal ventilation—who was positioned on a ground ladder on the D-side, contacted his officer and asked for authorization to take the window in the back bedroom. Because the inside team of the ladder company made entry from the rear, they were uncertain as to our progress and instructed the OV to hold off. Since we were coming up to the threshold of the kitchen, which is in-line with the doorway to the back bedroom, putting us in the attack lane and on the final approach to the seat of the fire, I immediately got on the radio and instructed the OV to vent. With the window trimmed out opposite our advance, the nozzle firefighter could see the glow of the fire in the back bedroom and opened the bale, flowing-and-moving the 10 feet across the kitchen.

The nozzle firefighter initially directed the nozzle high and overhead, in a wall-ceiling-wall fashion, coating and cooling the surfaces, as well as contacting the gases. Using the reach of the stream and the fresh air which it entrained, he worked the stream downrange and opened his application pattern into a squashed O, sealing off the doorway to the back bedroom. Because the fire room was vented, this hydraulically induced a conversion of the ventilation profile—shifting from a bidirectional flow to the ideal unidirectional flow, the front apartment door becoming a full intake and the vented window in the fire room becoming a full exhaust. This prompted a more effective and efficient air exchange and enabled conditions to improve more rapidly. After moving into the back bedroom and completing extinguishment, the nozzle firefighter immediately began hydraulically venting, creating a negative pressure and accelerating the removal of smoke while increasing the intake of fresh air. These actions greatly enhanced the tenability of the space, which not only provided us and the ladder company inside team with relief to facilitate our operations, but, more importantly, increased the survivability of any occupants that would have been trapped.

While this operation was nothing more than a bread and butter room-and-contents fire and may appear insignificant from the surface, it highlights the importance of possessing positional discipline, operational doctrine, and brilliance in the basics. Because we had maintained our readiness and conducted ourselves as if there was a fire inside the building until proven otherwise, we were able to act decisively and swiftly. This level of diligence is critical, as the changes to the fire environment can create conditions where no glaring signs are detectable when we arrive. Due to our consistent and progressive training, in conjunction with clear standard operating procedures (outlining unit and riding assignments, as well as the execution of critical tasks), we were well prepared to address the situation implicitly and instinctively.

By controlling the openings within the building, as well as our performance, and communicating directly between the ladder company, the ventilation was coordinated in support of the fire attack and the search, which ultimately had the best possible impact on the fireground.

Training and Pre-Planning

It is by scoring many points that one wins the war beforehand in the temple of rehearsal of battle.

Sun Tzu, *The Art of War*

To enhance the ability of you and your team to perform size-up, much work can be done well before the incident is ever dispatched. The investment of time and effort *before* the incident will yield tremendous dividends by allowing you to have set established factors that are already baked into your decision-making and give you a tremendous tactical advantage. Size-up starts during training and pre-planning. During a drill, the astute officer will size up the capabilities of the crew (experience, training, confidence, skills, aptitude, fitness level, initiative, technical expertise, etc.). A battalion chief (BC) can learn the time it takes a particular engine to get out of the barn, make a water supply connection, stretch a pre-connect on a house fire, dead load on an apartment fire, or high-rise pack to an upper-floor fire condition. The truck companies can be assessed on the timing to get a vent hole on a residential, commercial, or multi-story roof.

Station 109 was an aggressive house and knew their jobs well. They were the ones you would want if, Heaven forbid, your home was ablaze and your family was trapped. They could be counted on to get to their first-due fires quickly and make extremely fast stretches, searches, and jump into the fray. Simultaneously, they were equally skilled at backing up other first-due companies in neighboring areas. They set the standard and other companies were expected to meet or exceed those expectations. Friendly rivalries arose within the battalion, and everyone was held accountable and made everyone better. The BC was no exception!

The truck companies were also competitive yet supportive of each other. When one arrived first to a 2-truck roof (a fire and building requiring more than one truck company to perform vertical ventilation), the first-truck officer was assigned roof division supervisor and the subsequent truck companies to that roof worked for the roof division supervisor from the first-arriving truck. There was no racing to the roof, which could lead to conflicting or redundant spotting of the apparatus or ladder placements. They supported each other and communicated constantly.

Pre-planning is critical to successful operations, especially in apartment complexes where dozens, if not hundreds, of lives are at risk. Meanwhile, access, egress, construction variances, narrow passageways, hydrants, false dormers, common attics, parked cars, carports, and a host of other challenges could require a change of plans. Figure 5–2 shows a dramatic zero clearance drop off on the C side of these apartment buildings. Better to know ahead of time! Pre-plans and walk-throughs of structures should include, but not be limited to:

- Occupancy load, type of occupants, and trends during certain hours
- Construction type, including various additions and modifications, common attics, balloon framing, unreinforced masonry, roof assembly, and presence of trusses
- Construction features such as basements, chases, false dormers, facades, or unsupported mansards (fig. 5–3)
- Access, including narrow passageways, fencing, and overhangs for apparatus placement. Also, alternative access points and gates and aerial apparatus and laddering considerations
- Water supply, including hydrants, fire pumps, pump stations, fire department connections (FDCs), standpipes, sprinkler systems, and control rooms
- Additional fire protection like fire walls, self-closing doors, and communications systems
- Annunciator panel location

Fig. 5–2. These buildings have significant hazards and access issues on the C side that should be discovered during pre-planning.

Fig. 5–3. Notice the false facade on the A side of this building. This could be obscured by smoke during a fire and missed.

- Special hazards, including batteries, electrical vaults, heavy machinery, high-piled storage, rack systems, or special processes
- Hazardous materials stored on site
- Forcible entry challenges, including bars, doors, security fencing, alarm systems, and dogs
- Security personnel, building engineers, private fire protection, and communications interface on site
- Length of stretch for different access points for fire attack
- Best means of performing fire attack, search, and ventilation for fires of varying locations and sizes
- Location of manager office, and/or onsite residence or manager/maintenance personnel
- Location of all utility shutoffs, including electrical, gas, water, and utility companies serving the building/facility
- HVAC and elevator systems and control rooms
- Confirmed victim profile scenarios
- Past history of incidents at the address

Other pre-incident size-up considerations for the start of shift or tour and throughout your shift may include:

- Weather
- Traffic
- Road closures and detours
- Construction
- Water mains or other utilities shut down
- Fire hydrants sparsely available
- Holidays
- Special events
- Company closures for training, special events, inspections, maintenance, and so on
- Short staffing
- Personnel not normally assigned to the company
- Normal occupancy loads
- Changes in occupancy load due to recent trends like COVID-19, the economy, people working from home, and so on

When combining pre-planning with realistic on-site training and a solid team, the wise and forward-thinking IC will have a tremendous strategic advantage. The officers and crews will also gain confidence and think of several permutations of situations and tactical considerations like attack locations, distances of stretch, obstacles, hydrant locations, apparatus placement challenges, and a myriad of other factors that will benefit them tremendously when faced with a fire and trapped victims.

Battalion 7 had a 3-story center hall apartment building directly behind the battalion headquarters station. We knew that building inside and out, having performed several hands-on drills with charged hoselines, smoke machines, mannequins as victims, VEIS tactics, and various line stretch scenarios. In addition, we built a scale model of the building and built simulations with Digital Combustion software. We knew that the building had no sprinklers, no standpipes, carpeted center hallways, and many occupants with physical challenges. We also knew the construction and where each access, hydrant, stairway, fire door, and utility shutoff was located.

One night at 2 a.m., there was an intentional gas explosion with structural collapse and heavy fire involving several units (fig. 5–4). Winds were approximately 20 miles per hour and victims were reported trapped upon dispatch. The first-due engine was closed that day. The fire went to 3 alarms with only one fatality, miraculously. Without the pre-planning and on-site realistic training, the fire would not have been confined to the wing and floor of origin and many more lives may have been lost.

The pre-planning, training, education, studying, classes, books, articles, incident reviews, knowledge of fire behavior, building construction, strategy and tactics, hands-on drills, simulations, meetings, and relationships you build are going to pay off before the alarm is ever dispatched. The fuel that drives this motivation must come from within yourself.

Modern incident command is a team sport led by ICs who lean into the wind and jump into the fray, undaunted by critics, unwavered by apathy, and uncompromising in command of their teams, battalions, and fire departments (no matter how big or small), *before* the incident comes.

There are countless, priceless lessons learned from our fellow firefighters. Some of them paid the ultimate price to give us these lessons. Honor them by learning from them.

Command of Personnel

In today's nomadic culture, firefighters move between assignments and stations regularly. Your fire department (FD) may or may not have a memorandum of understanding (MOU) or collective bargaining agreement (CBA) between labor and management that determines how

Fig. 5–4. This fire had explosion with gas leak, high winds, and heavy occupancy with no sprinklers. Pre-planning made all the difference.
Source. Photo courtesy of Gary Lascelles

companies are staffed and when, if, or how personnel may be moved. Policy or standard operating guidelines (SOGs) may also dictate the means in which companies are staffed. It may be seniority alone or a combination of seniority and other training and certification qualifications. The point is that when a member is assigned to a company and you are not comfortable with their skills or attitude, you should address it at the beginning of shift, within whatever authority you have.

It was 7 a.m. and my BC shift began. The crews' shift change was at 8 a.m. My first priority was to check the roster of my battalion and make any necessary adjustments prior to shift change at 8 a.m. for the captains and their crews. This meant checking for any personnel who were not certified and qualified to ride a particular apparatus. I did also have management discretion, within reason, per policy and MOU.

I noticed that both regularly assigned captains at my battalion house were off and their overtime replacements were filled in on the engine and truck. It was clear that the captain on the truck was not the best choice, given history I had with him on a previous incident. In addition, my concern was that he normally worked a slow engine house and that a busy truck like this may be more than he bargained for. Having had a previous issue with him (and to use risk manager guru Gordon Graham's philosophy that "predictable is preventable"), I chose to swap the captains. They were at the same house, just swapping engine and truck roles.

As expected, I clicked the mouse on the staffing program and I hadn't even had the chance to call the captain in my office to notify him that he was moved to the engine when he came storming into my office, red-faced and shaking with rage. "How dare you move me! You can't do that. Why did you move me? I'm going to call the union!" I remained seated and calmly said, "First, calm down. Second, close the door. And third, have a seat if you would like to discuss this." He begrudgingly obliged. I continued, "It is perfectly within my province to move you. The union will agree. So, the real question is why I moved you. You are certainly entitled to an answer. My question is, do you want the truth?"

"Yes," he responded.

I continued further. "Remember when you and I worked the airshow together 18 months ago and a vegetation fire started and was moving rapidly toward approximately 10,000 spectators and six Air Force Thunderbird F-16's?" He acknowledged it so I continued. "Well, when I directed you to assist with fire attack, your response was 'you don't need us, looks like you got it,' do you remember that?" He agreed. "Well captain, that's not going to happen today when one of our engines needs a hole cut to relieve smoke and heat and you respond by saying, 'no, ya don't need it,' because my job is to put the right people in the right seats to accomplish the mission and keep everyone safe. Any questions?" He had none and didn't mention it again for the next 24 hours.

The NIOSH 5

The National Institute of Occupational Safety and Health (NIOSH) tracks line-of-duty deaths (LODDs) of firefighters. The physiological causes like heart attack, asphyxiation, trauma, thermal insult, or vehicle accidents dominate the pie charts.

If we re-examine the NIOSH reports and focus on the *operational* causes of LODDs, we begin to understand that the common events that lead up to LODDs come down to 5 things. It's one thing to say a person died of trauma; it's another thing to know that he was pushed off a cliff.

To be more effective incident commanders, we must understand the factors that commonly hamper fireground and other emergency scene operations. If we better understand them, we can begin to prevent them from coming into alignment. This is not just the job of the IC. As we said, modern incident command is a team sport. Most of the time, the tactical supervisors are often best positioned to see these items come into alignment, since they reside in the tactical space.

These factors are also present in near-misses, significant injuries, and excessive loss of life (both firefighters and civilians) and property. Bottom line: if the NIOSH 5 are present, bad things are likely to happen. The more present, the worse it will likely be.

The NIOSH 5 are a lack of or improper application of the following:

Risk assessment. Poor size-up. Size-up is a mental process that is ongoing. From pre-planning to the night before, to morning of, to dispatch, to en route, to arrival, and through the whole incident, size-up is ongoing. This is different from arrival reports. Risk assessment is the job of every officer. Unfortunately, most NIOSH LODD reports have the same narrative: "Once upon a time an FD went to a fire. They didn't know what they really had. A firefighter died." This entire chapter is dedicated to preventing the NIOSH 5 from coming into alignment on your fires, or any other incident you command or respond to.

Communications. This is the greatest problem on most emergency scenes. Apparatus, equipment, SCBA, background noise, auditory exclusion, tunnel vision, and task saturation all contribute to this enormous problem. Improper or lack of communication paralyzes the team and a snowball effect ensues.

Incident command. Unfortunately, incident command is a common factor in the NIOSH 5. This can be an unmanageable span of control, diving into the task level, getting in the weeds and becoming distracted, not utilizing ICS, or getting behind the incident power curve. It is also commonly not understanding the size, scope, and severity of the incident until it is too late.

Accountability. Effective accountability requires discipline—from the individual, the company officer, the tactical supervisor, and the IC. Freelancing or knowing neither your job nor your boss is a common factor in disastrous events.

THE NIOSH 5

Lack of or improper application of:

- Risk assessment
- Communications
- Incident command
- Accountability
- SOGs or failure to follow SOGs

SOGs or not following the ones you have. We must have a playbook and follow it. Deviate when needed based upon the operational core values. All champion football teams have a playbook and a quarterback who knows when to call an audible. That's what an FD must do: have modern, relevant SOGs executed by officers who know when to call an audible. Not every possibility can be forecasted and planned upon in an SOG. That said, we must be on the same page for the common and predictable stuff. The key is smart officers who know when to deviate to make operations more safe, effective, or efficient.

In chapter 4, we introduced the tactical gap as the space between the strategic level IC and the task level crews working. This is where the battle is won or lost. The wider the tactical gap, the more of the NIOSH 5 will fall in. To bridge the tactical gap, we must assign tactical supervisors. They prevent the NIOSH 5 from coming into alignment by performing ongoing risk assessment, reducing radio traffic by increasing face-to-face communications, continually accounting for crews with active eyes-on, hands-on accountability, and reducing the IC's span of control.

In July of 2022, the Leadership Under Fire Team published "A Human Factors Review of NIOSH Fire Fighter Fatality Investigation and Prevention Program Reporting." This report does an excellent job of expanding upon the NIOSH LODD reports and 5 factors and shows how human elements come into play during these events. We encourage you to read this report.

The tactical gap is largely plagued with human factors that impact decision-making. This report is an excellent resource to better understand them.

Dispatch and Response

Once dispatched, your size-up continues, building upon the training, pre-planning, team building, and situational awareness throughout your shift. You should immediately be aware if normally responding companies to this alarm are out of position or causing delays, and the potential for unfamiliar crews arriving to someone else's first-due. That said, several triggers for additional resources can occur en route that we will address here.

The first concern once you have mapped your route is victim profile (VP). If dispatch does not report everyone out or any other victim profile, ask them on the radio while responding. Ideally, this should be part of your dispatch center's SOGs. If you do have a confirmed VP, then call an additional medic unit or transporting ambulance. A good rule of thumb is one ambulance per fire victim known, plus 1. This gives each fire victim a separate ambulance (critical for burn patients and multiple caregivers in the rear with the patient). The added ambulance is for unknown victims or firefighters. *Always have one transporting ambulance available on scene.*

Multiple callers and an obvious column of smoke can confirm a working fire (fig. 5–5); however, you and your team must not have to be convinced to gain the utmost respect and readiness for the call's potential. Whether multiple calls with a huge column, or one caller with no column, no one can rest until we have units on scene to confirm conditions at the *correct* location. Fires are commonly dispatched to locations where the reporting party (RP) is calling from what may be behind, across the street, adjacent to, or in the vicinity of the actual fire. Some are even second-hand callers, so take nothing for granted.

Fig. 5–5. Crews see large column on the way to a commercial fire.
Source. Photo courtesy of Sacramento (CA) Metro Fire

We were dispatched to a report of smoke coming from an apartment. E103, TR106 and Battalion 7 arrived with no smoke visible from the end unit of a triplex. E103 arrived first and reported "no smoke visible, cancel the assignment." As I arrived with the truck, I assumed command and replied, "Negative, we have a report of the odor of smoke. Force entry as needed." The first-due company officer missed this report. Once they forced entry, smoke billowed out of the front door. They found a mattress fire in the rear bedroom. Fortunately, no one was in the apartment. Had there been an occupant, they could have easily been unconscious.

The company officer was solid. The problem is that, often, first-due company officers miss subtle updates on the mobile data computer (MDC) because they are busy routing in the engineer, looking at pre-plans, looking for hydrants, and potentially strapping on an SCBA. This is where the teamwork comes into play. I was backing that officer up, as I had come to rely upon my company officers to back me up when I missed something. Again, this is a team sport.

If your system allows, communicating critical updates to each other on the tactical channel while responding is vital to getting everyone on the same page. Perhaps one officer knows the building is currently under renovation, is currently vacant, or some other issue that is not on the pre-plan.

Consider the lag time between when the fire started to when you were dispatched and the subsequent response time. A fire at night or in a vacant building can easily remain undetected for extended periods of time, allowing the fire to grow. When you add dispatch and response time, the fire will likely have an extreme advantage before you arrive. We tend to think in our own perspective of time. In our minds, the fire is not burning for 10–30 or more minutes before we arrive. The fire started when we were dispatched. Therefore, as we arrive to find a working fire with heavy involvement of the structure, we do not calibrate to adding 10–60 minutes to burn time. We think in terms of response time. This may prove disastrous.

Time of day is just as important. Consider potential occupancy loads based upon time of day. According to Firefighter Rescue Survey (FRS), most civilian rescues occur within a three-hour window between 3:00 a.m. and 6:00 a.m.

Traffic congestion during rush hour may warrant additional resources called sooner due to response times being lengthened from traffic. We were dispatched to multiple reports of fire in an elderly care facility at 5:30 p.m. on a weekday. Traffic was heavy. The first-arriving unit was an ambulance reporting light smoke from the street. I balanced it to a second alarm immediately due to multiple factors: the occupancy type and life potential, multiple calls, smoke from the street of a large care facility, and traffic congestion.

Weather is a major factor. High winds may warrant calling additional resources early due to fire growth and exposure potential. Conversely, rain and high humidity may conceal smoke and fire conditions.

From your pre-planning, consider fixed fire protection systems like standpipes, sprinklers, pumps, and so on. Your SOGs should dictate what company goes to the fire department connection (FDC) in different alarm and structure types. Depending upon the conditions reported upon arrival, an audible may need to be called.

Arrival on Scene

Lloyd Layman developed the FPODP size-up system in the 1950s. He also developed RECEO-VS for tactical priorities. Both are still relevant today. FPODP is excellent for initial arrival *and ongoing size-up* throughout the incident. Begin before you are in front of the building and take note of overhead obstacles, such as power lines which could be involved, and other factors that may impede access or use of aerial devices (fig. 5–6).

Addresses can be easily obscured or confusing from the street. Look for power and gas meters along with addresses to ascertain how many units are in the building. Figures 5–7 and 5–8 show multiple addresses and power meters that can help your size-up.

Fig. 5–6. Keep an eye in the sky at all times.

Fig. 5–7. Many addresses can be redundant and cause confusion.

Fig. 5–8. In the absence of addresses, power or gas meters may tell you how many individual units are in the building.

FPODP stands for:

- **Facts** (What do you know for sure?): Time of day, weather, type of structure, conditions visible, and so on.
- **Probabilities** (What is the potential?): Victim profile, exposures, attic involvement, growth potential in acreage for a vegetation fire, number of patients for an MCI, and so on.
- **Own situation** (What is your resource level?): I am the first engine on scene with the balance of a residential alarm responding. Or, I am the first chief on scene with two engines on scene and another engine and truck responding.
- **Decision** (What are your first overarching priority and strategic decisions?): These three options are utilized for all-risk operations and any type of incident. This is the primary and first decision made by the first officer/IC on go/no go and any boundaries. This should include the overall incident priority (lifesaving, incident stabilization, or property/environment).
- **Plan** (tactical level): Company officers will consider what to do with their crew, like establishing a water supply and whether to assume or pass command. Chief officers will be thinking about setting up the incident command post (ICP), confirming the priority, strategy, and tactical objectives, and calling more alarms/resources.

Looking at figure 5–9, the FPODP would be as follows:

- **Facts**: Two-story lightweight house with smoke from the rear.
- **Probabilities**: Moderate victim profile, room-and-contents fire, could extend into the attic.
- **Own situation**: First arriving engine with two engines, one truck and one BC responding.
- **Decision**: Lifesaving priority, offensive strategy.
- **Plan**: Officer performs a lap, engineer stretches a line to the front door and makes hydrant connection, firefighter forces front door. Add a precautionary ambulance.

FPODP is excellent for first-arriving and ongoing mental size-up. You may also arrive as the first chief with a lot of activity and companies already working upon your arrival. The FPODP for fig. 5–10 would be:

- **Facts**: One-story duplex with working attic fire through the roof.
- **Probabilities**: Moderate victim profile.
- **Own situation**: First-arriving chief with two engines, one truck on scene with vertical vent and fire attack underway.
- **Decision**: Lifesaving priority, offensive strategy.

Fig. 5–9. Simulation of a two-story house for FPODP practice.

Fig. 5–10. Chief arrives to find crews working on this duplex fire

- **Plan**: Get companies off the roof now that vent is completed and hole is blowing fire. Complete primary searches. Establish Division A, RIC, and Medical. Add a precautionary ambulance.

FIRST

Another system of mental size-up for the *first-due* officer is "FIRST," (developed by Battalion Chief Drew DeFazio of the Redmond, WA Fire Department) focusing on just the initial size-up upon arrival:

- **Facts** (What do you know for sure?): As with FPODP, these are the obvious factors that are known and seen.
- **Intuition** (What do you think could be the potential?): What is your training, gut, and recognized prime decision-making telling you regarding potential victim profile, extension, avenues, and direction of fire spread, exposures, and potential size?
- **Resources** (current resources): Resources currently en route, and those available with additional requests, including the response time.
- **Strategy** (life saving, incident stabilization, or property/environment priority and offensive, defensive, or combination strategy): This is your first and most important decision.

- **Tactics** (initial tactical and control objectives): For a company officer, this is just the initial task/tactics of the first-in company. Examples would be the firefighter stretches a line while the officer performs a lap and the engineer establishes a water supply. For a chief, this would be tactical objectives like rescue, exposures, confinement, extinguishment, ventilation, etc.

Getting the Lap/360°

Performing a lap (360°) of the structure is important, if possible. Many older cities and neighborhoods do not allow this option due to attached buildings, terrain, congestion, access, and so on. However, a floor above, roof report, or higher terrain location may accomplish this task.

You may not be able to complete a lap due to the construction type. Figures 5-11a and 5-11b show the A and C sides of the same structure, respectively. You are *not* performing a lap on this building. And the C side has serious access issues. An alternative is to get companies to the A and C sides and a company to the roof from the A side. The roof division can give you a size up of all four sides from the roof. As you analyze the structure, make note of construction features like basements, fire walls, partitions, breezeways, bed and bathroom windows, FDCs, and

Fig. 5–11a. Note the hotel is attached on both sides. No lap.

Fig. 5–11b. The rear of the same hotel has limited/delayed access.

utility access. In addition, take note of tactical points of opportunity where you may mount a more effective fire attack, VEIS, or cut off the fire from extending. Do not forget the roof. When a company goes to the roof for vertical ventilation on larger multi-family dwellings, commercial buildings, and assemblies, a roof report should be conducted as soon as possible. A *roof report* is the statement to the IC from the roof division regarding the construction, conditions, and load (CCL) on the roof. Construction involves possibilities like conventional, lightweight truss, bowstring truss, panelized, etc. Conditions involve the amount of smoke, heat, or flame present. The thermal imaging camera (TIC) is a great tool for this. Load involves the dead load on the roof such as HVAC units. An example of a roof report would be: "Main Street command, roof division with a roof report: panelized roof, heavy smoke in toward the Charlie/Delta corner of the roof, no heat signature, with four large HVAC units."

Additionally, according to the International Association of Fire Chiefs' "Rules of Engagement for Structural Firefighting," the incident commander must obtain a 360° size-up of the incident—it does not say they must do it personally. In other words, the lap can be delegated and a tactical supervisor most likely will complete the task, and notifying the IC with their observations actually meets the ICs obligation to obtain the 360° size-up.

If possible, perform a lap. You will see tactical points of opportunity such as windows and doors that may indicate the rooms and layout inside (fig 5–12).

When possible, a lap can be critical to gaining valuable knowledge of the following:

- Potential victims
- Building layout
- Construction
- Access/egress points
- Basements
- Fire location, avenues, and direction of fire spread potential
- Obstacles and hazards
- Exposures
- Utility shutoffs
- Tactical points of opportunity

Tactical points of opportunity (TPOs) are building construction, access, layout, and other physical features that can be leveraged to increase the impact of resources to accomplish tactical objectives. Examples include:

- Fire walls
- Breezeways

Fig. 5–12. Note the barred windows, bedroom, and bathroom windows on the D side of this center hall apartment building.

- Stairwells
- Windows
- Fire escapes
- Side/rear doors and roll up doors
- Alternative access points for apparatus or hose stretches

Alternatively, obstacles should be noted and communicated when appropriate. Examples include:

- Barred or boarded up windows and doors
- Hoarding conditions or junk that make access and hose stretches more challenging or impossible
- Vicious dogs
- Hazardous materials
- Swimming pools
- Out buildings
- Overhead powerlines
- Exposures

Victim Profiling and Survivability Profiling—Related Concepts to Help You Save Lives

As firefighters and officers, we are ingrained to think of risk management from the academy. We are constantly trained to wear our PPE properly, don our SCBA quickly, and limit our exposure personally. As incident commanders and tactical supervisors, we are now charged with the lives of many crews simultaneously. They and their families expect us to take care of them and get them home in as-good or better shape than when they showed up at shift change.

As a battalion chief, I knew I had 24 members going to a house fire. That was the risk. What I did *not* know was what we had to gain. How many, if any, civilian lives were present? What was the return on investment (ROI).

Think of investing your money. The first thing you ask yourself is, "What is my ROI?" That determines the level of risk you are willing to take. The higher the ROI, the higher the risk tolerance.

No IC or officer can allow any firefighter to be injured for mere property, especially property that is already lost due to contamination, let alone burned beyond recognition. The IC must be thinking and remaining logical in the face of impassioned firefighters who have unknowingly invested emotionally in the fight. Sometimes, this investment is appropriate to the ROI because one or more lives are at stake to be saved. Other times, their emotions and investment are not reasonable because there is no life to save.

Today's dwellings are far more occupied than in the past. In addition, dwellings typically have more people occupying them throughout the full 24 hours of the day, not just nights. Between COVID-19, the economy, and more families living together, your chances of having confirmed victims trapped is much higher at all hours.

Strategic pillars of risk management:

- Risk a lot to save a lot (save savable lives)
- Risk nothing to save nothing (life and property that have already been lost)

Victim Profiling transfers these strategic pillars into a useful tactical tool in the street.

Victim profiling (VP) is the practice of determining the presence, status, and location of civilian victims. VP brings strategic risk management to life through a practical, tactical tool for use in the field. This allows us to rapidly assess what we have to gain, what we are willing to risk, and what our priorities, strategy, and tactics should be.

All incidents start with a victim profile. Whether a structure fire, vegetation fire, hazmat, MCI, or something else, the victim profile drives everything: gain, risk, priority, strategy, tactics, tasks, resources, and ICS. On a fire, for example, if you arrive and a credible source (sane and sober) tells you everyone is out of the building, then you are likely going to risk less than if they said, "My family is still inside!"

Don't be mistaken. Low victim profile is **NOT** an all clear on the building. Only we, the fire department, determine that. This is simply a starting point from which to drive the priorities, strategy, tactics, and risk that we are willing to take with our troops.

Victim profile has four levels:

- **Low**: A *credible source* (bystander, occupant, dispatch, sane and sober) tells you everyone is out. Ask them, "how do you know?" If they clearly are just guessing or not credible due to mental, alcohol, or drug issues, skip to moderate. If they say, "because I live here," you are going to be much more confident. *Again, this is not an all clear. Only we determine that.*
- **Moderate**: There are no sources of additional information. The fire is during the day and no other signs of life nor credible sources are present. Again, in the wake of COVID-19, there is a more likely chance of occupants. That is why it is moderate. If conditions allow, we still search!
- **High**: The same structure is now burning at night with a car in the driveway. Or an apartment building is involved. These indicate a higher level of life potential. Today's apartment buildings are almost always occupied. Even if the involved unit is evacuated, the exposed units may be occupied.
- **Confirmed**: The opposite of low, here we have a *credible source* that tells us that there are human lives inside the structure. This could be "My baby is trapped" or "My kids are still inside!" Again, ask "How do you know?"

Get the SIGNAL

With a *confirmed* victim profile, you need more information and you need it *fast* to establish your tactical plan to get to the victims as soon as possible! Rather than just run into the front door, the SIGNAL method can give you the tactical advantage to be as efficient and swift as possible, when seconds count.

When confronted with a confirmed VP, ask the bystander the following (it should take no longer than 10 seconds):

- **Species**—Confirm the victim is human. How many of you have heard "My baby is trapped" and find out the "baby" is a cat?
- **Incapacity**—Obese, mentally challenged, deaf, blind, non-ambulatory, and so on. This will help determine the best tactic of choice and if you need more resources to make the rescue.
- **Gender**—If you are told it's a male and you find a female, you have more than one victim. According FRS, in house fires where a victim is reported (confirmed VP), on average there is a high probability that there will be *two* victims. In an apartment with a known victim reported, there is a high probability that there will be *three* victims. Generally, female victims tend to go towards life and natural exits (children, elderly parents, and doors) where male victims tend to try to handle the emergency and therefore may be found closer to the point of origin.

- **Number**—How many victims? This again will drive resource orders and help medical get ready for treatment and transport.
- **Age**—Behavioral patterns are predictable for different age groups during rescue:
  - **Adults**—Between the fire and means of egress (front/back door). At night, in bed.
  - **Teens**—Bathrooms because that is their safe space, and they think the water in the shower will keep them from burning to death.
  - **Children**—Under the beds, in the closet, under covers, under toys—hiding.
  - **Babies**—In the crib. Where is the crib? Parents' room, baby's room?
- **Location**—Where were they last seen? Educated guess where they could be? Don't settle for "upstairs," or "inside," or "in the bedroom!" Have them point and be specific.

When facing a low or moderate VP, support functions like 2-out and/or rapid intervention should be in place, based upon your SOGs. Conversely, a high or confirmed VP would indicate that 2-out and/or rapid intervention can be deferred until more resources are available.

VP will also help you be much more efficient in your tactics. For example, units arrived on scene of a working fire in an apartment that was well-involved and impeding upon other units. The father confirmed the location, age, and status of his son in the back bedroom. This resulted in coordinated fire attack, VEIS, and vertical ventilation. Had the crews not taken the time to get this critical information, they may have searched from the front door after knockdown, which would have undoubtedly taken much longer, and the child would not have survived.

Survivability profiling (SP) is the educated art of examining a situation and making an informed and intelligent decision based on known events or likely circumstances to determine if civilians can survive the existing smoke and fire conditions, and whether to commit firefighters to lifesaving and/or interior operations.[1] According to Chief Stephen Marsar from FDNY, "SP helps to quantify the term *Risk Versus Reward* in a realistic, boots on the ground and applicable way. SP does not say that primary searches are not necessary or, that they are not extremely effective. Especially when searching firefighters have good VP information and apply a targeted search tactic (going to where the victims are reported or, expected to be) rather than doing a blind orientated search and hoping to find them. In fact, when the SP concept is applied correctly and the answer is that trapped victims *can* survive the exiting fire and smoke environment, then go for it!"

Victim profiling and survivability profiling go hand-in-glove. They are *not* mutually exclusive concepts or efforts. While VP is for the civilians, SP is for the firefighters. Both are needed to form a scale to balance risk versus gain in an intelligent way so we may save those civilians we are capable of saving and not lose firefighters to unnecessary risk.

Remember, civilian victims can be in very tenable spaces behind hollow core doors, in bathrooms, closets, bedrooms, and so on. Just because it looks well-involved from the outside does *not* mean that the entire interior is untenable for life. What if it was *your* family? That said, we must ask questions of credible sources like other occupants, bystanders, neighbors, and family subsequently arriving (get the SIGNAL). So, we must be more thorough to quickly establish our gain versus risk.

Other factors like level of involvement, collapse potential, flashover, and a host of others will determine our ultimate incident action plan and SP may dictate that we cannot accept the risk; however, it all starts with victim profile—what we have to gain.

Reconnaissance

Reconnaissance, also known as "recon" is the act of further surveying the battlefield (commercial building, apartment complex, industrial facility, area, etc.) to better ascertain the enemy (fire, hazmat, threat, etc.) strength, location, routes of travel and obstacles to access. The objective is to gain more information than a size-up from the street alone would allow. During recon, the goal is *not* to extinguish a fire, for example. The goal is to find where the fire is, how big it is, how it is traveling through the building, what is exposed, and the best means and tactics to initiate attack. Obstacles like long passageways, locked doors, stairwells, and limited access for apparatus should be identified. The length of a possible stretch of attack should be estimated and alternative access points identified if the stretch distance is excessive. Recon allows you to identify TPOs before committing resources to a more complex operation so you can gain tactical advantage.

As a minimum, tools for recon should include:

- Portable radios
- Forcible entry tools
- Long-handled tools
- Thermal imaging camera (TIC)
- Water can
- Flashlights

Construction

Construction is extremely critical to size-up, so much so that we have dedicated several chapters to different construction and occupancy types. Each type has specific construction, command, and tactical considerations, and key points associated, and we delve into more detail in each, along with case studies.

The fundamentals of construction provide a foundation from which to delve into specific types. As an IC, your knowledge and that of your tactical officers specific to the buildings in *your* jurisdiction will prove vital in victim and survivability profiling; making tactical decisions regarding forcible entry, fire attack, search, and ventilation; assessing risk versus gain; predicting the behavior of the building (*fire spread and collapse tendencies*); and ensuring everyone goes home safe.

We *highly* recommend *The Art of Reading Buildings* by John Mittendorf and Dave Dodson (Fire Engineering Books & Videos, 2015), as required reading for any incident commander. In it, the authors detail how the size-up of a building is a function of that building's NFPA 220 type (I-V and much more), era (when it was built), use (occupancy), and size. The book also includes 52 building size-up guides that include tactical concerns regarding fire spread, collapse, ventilation, forcible entry, and search. In addition, specific hazards are listed.

Construction types and styles vary greatly throughout America. As mentioned above, era, use, type, and size should be considered. Historically, the fire service has looked at construction from Uniformed Building Code (UBC) definitions. NFPA 220 (Type I–V construction) categories were often used as part of initial arrival reports. This alone cannot provide much for size-up, as hybrids of construction are commonplace and many buildings that should not have burned, have. For example, the Roseville Galleria Mall in Central California had a massive fire even though the construction was composed mainly of concrete and steel.

From an incident command standpoint, it is most important to know how fire behaves in different buildings and construction types so you can forecast and add that knowledge to your incident action plan (IAP). Some construction is widespread and must be known by all fire service ICs, like balloon frame, in which the studs run continuously from bottom to top, often causing vertical fire extension in walls. Therefore, a basement or cellar fire can vertically extend into the attic of a multi-floor building.

Another common type is tilt-up construction, in which the concrete walls are likely to fail when the roof burns for extended periods. Rapid and catastrophic failure due to bowstring truss assembly and hazards associated with lightweight trusses with gusset plating should be common knowledge. That said, each region of the United States has trends, and, more specific to your jurisdiction, there are specific buildings and construction that you must be familiar with to adequately size up and forecast. Examples would be brownstone multi-family dwellings in Brooklyn, NY, or McKuen fourplexes in Central California. The point is to know the nomenclature, construction features, and fire behavior characteristics and tactics of choice in *your* area.

CASE STUDIES IN COMMAND

The Senior Nose

CHIEF BRETT LOOMIS, CORVALLIS (OR) FIRE DEPARTMENT

We had a good save tonight. No... we had a GREAT save tonight! I'm proud of my crews. Their tenacity. Their commitment. Their drive to get it right. It was a smells-and-bells call. One where we could have said, "We can't find anything, Mr. Smith. If anything changes, call us back." But we didn't. We kept on working. We used some tools of the trade, for sure. But most importantly, we used our brains, senses, and experience to make a difference and save a million-dollar home.

The tones dropped around 2230 hours. It started as an odor investigation: "E132 respond to the 3300-block of NW Van Buren Avenue." I listened to the dispatch from my desk. As I signed into the computer to look at the call details, dispatch ramped the incident to a residential structure fire. They filled the box with two more engines, a truck, and medic unit, and the BC (me). It was a report of smoke inside the structure.

That neighborhood is tight. Tree-lined streets with parking on one side designed in a time when most families only had one car. Legacy homes built in the early to mid-1900's. Lots of craftsman-style homes. White picket fences surrounding well-manicured yards. Cobblestone pathways and garages off alleyways that a brush engine could barely fit down. The other thing the streets were lined with were power and utility lines. There was no room for a ladder tower in there. I met the Station 1 lieutenant (LT) on the apparatus floor and told him to bring the engine, not the truck (we cross staff between the engine and the truck at headquarters).

The town was alive tonight. First good-weather weekend of the year. All the fraternities were hosting parties, we had a large rave party happening at the fairgrounds along with a paranormal circus. The town was hopping. We had extra staffing on, as did our colleagues at the police department and sheriff's office. We even had the state patrol in town.

E132 arrived and reported nothing showing from a two-story craftsman. They were investigating. I arrived about the same time as Station 2's medic, M125. I established Van Buren Command and assigned the medic crew to assist with the investigation inside. By this time, we had several law enforcement officers on scene and the neighbors were out in force. I directed the next engine to stage on the plug and had the other two engines stage at the next major intersection.

E132 reported no visible smoke or fire inside the home. They were chasing an odor. They also reported the home had been evacuated and a primary search completed. I know what you're thinking. Why are you reporting a primary search complete if you have a confirmed evacuated structure? We do that to drive home the point of searching everything, *all the time*. If we practice it on the small incidents, on the large ones it will be second nature. E132 recommended an under control but wanted to keep their medic on scene. The LT also asked me to face-to-face with him. I reported the incident under control, released staging companies, threw my coat and helmet on, and headed into the house.

I hadn't taken three steps inside the house, and I said to the LT, "There's something burning in this house." I immediately regretted releasing the staged companies. I should have held one additional company until we knew for sure. The LT said, "I know, Chief. It doesn't smell electrical, but it doesn't smell like burn either. We don't know what it is." I said, "It's wood smoldering. There's some slow burn going on somewhere."

The interior of the house was stunning. The house had just gone through a nearly half-a-million-dollar renovation. High vaulted ceilings with beams. A commercial kitchen with a range that was probably near the value of a half-year's salary. Immaculate hard wood floors. An outdoor space with a kitchen and a brand-new brick gas fireplace. Certainly not a place that we want to randomly open walls and ceilings to try and find an odor.

I asked the crew about attic spaces. The reply was, "No attic spaces. All vaulted ceilings." "Any issues with power?" I asked. The LT said the occupant reported no power issues. I was thinking about a basement. It's an older neighborhood and most houses were at least built on pier blocks with a crawl space. For whatever reason I didn't ask if they'd found a basement. So, I went for a walk to see the whole house for myself.

The smell was strongest in the B/C corner of the house, in the main bedroom on the first floor. I walked through the first floor opening every door on the way towards the D side of the house. By the time I made it to the door to the garage, I lost the smell. The LT joined me in the garage with the owner of the home to check the circuit panel. I saw a door and asked the owner about it. He said, "That's the basement." I felt it first with the back of my hand (old training dies hard), cracked it, and saw nothing and smelled nothing. There were stairs down to a finished, unfurnished basement. . . and a crawlspace.

Then I walked upstairs to the second floor. Three bedrooms, one with its own bathroom and two with a Jack and Jill. Just a faint odor upstairs, but still present. I looked in every closet and what did I find? An attic access scuttle.

Returning downstairs, the odor was stronger. After pointing out that there was an attic space upstairs and a crawlspace in the basement, the owner shared that there was also an attic access in the main bathroom off the bedroom where the odor was the strongest. I asked the LT to assign two people to check the crawlspace and two more to check both attic spaces, focusing on the one in the bathroom first.

I asked the owner to tell me the story that he had told the crew. He said that around 6:00 p.m. he and his wife had had some wine on the new patio. He said that they used the new gas fireplace for about an hour. They turned off the fireplace and then went next door to their daughter's house for a couple of hours. When they came home, he smelled the odor. He tried to find it for about an hour before calling us. He said over the hour that he searched, the smell got stronger. He never saw any smoke or fire.

We returned to the bedroom. The odor was still there and seemed to be getting stronger. The cover plates for the outlets in the bedroom were off. The owner told me he had done that trying to see if there was an electrical problem. The LT reported no tripped breakers, and they felt no heat around the outlets on either side of the wall. We all know that a lot of times when we're checking on odors, we can become accustomed to them. Almost numb to them. Often, we'll step outside, take in some fresh air and then return to see if we can detect it again. No need here. It was definitely present.

The crews reported no odor in the crawlspaces or attic spaces. No signs of smoke or fire. The next plan was to set up a fan at the front door, blow the odor out, and see if it would re-create itself. It was my plan, and I didn't love it because somewhere in the back of my mind I knew we were dealing with something burning. So, I asked the crew to put the fan in place but not start it.

Back in the bedroom, I got down on my hands and knees and smelled each of the outlets that were missing their cover plates. There was a slight odor from the outlet right inside the door. The outlet towards the back wall was a different story. No doubt there was some slow combustion close to that outlet. It was strong and undeniable. I had the lieutenant do the same thing and he agreed. One more TIC of the wall and there was a change. We could clearly see the studs, but this time we could also see the drywall screws. The nozzle operator on E132 said, "Chief, that is not what we saw when we came in. We couldn't see the drywall screws." We TIC'd around the outlet again and didn't see a significant heat signature (about 70° Fahrenheit), but it was diffuse. A large spot to the left of the outlet, also not there upon the initial investigation. Time to open the wall.

We spread a tarp to protect the carpet. We got a utility knife from the engine and borrowed a straight edge from the homeowner. One more use of the TIC to find the studs and mark our area. With precision the crew opened the wall. They left clean edges so the homeowner would have a simple and straight forward patch job on the wall (fig. 5–13). The probie on the medic unit stood by with the can in the event of a flare

Fig. 5–13. Crews take their time to ensure the patch job would be clean and easy.

Fig. 5–14. This situation could have been a lot worse. *Source.* Photo courtesy of Brett Loomis

up. Within seconds of making the first set of cuts the nozzle operator said, "There's a fire in here." A couple of minutes after that the wall was opened. And there it was.

There was burn hole completely through the plywood from the exterior wall. About a stud bay wide and a foot high. Some discoloration on the studs on both sides. Full depth chart on the R15 insulation (fig. 5–14). No real smoke and certainly no active flame. Slow combustion was still in progress, however, on the perimeter of the hole. All on the shared wall with the exterior gas fireplace. The homeowner said, "You guys just saved my house!" The best feeling of all.

We re-examined the exterior fireplace and determined that the contractor had installed it without fire brick. It was a thin layer of decorative brick only. There was a backer board that we could see through the plywood that looked like it was cracked. We could touch the brick with our bare hand and not detect anything unusual. The teaching moment here is that the enemy knows how to infiltrate weaknesses and establish its battlefield. It's our job to outthink the enemy and we did. Had the homeowner ignored the odor or had we swept this incident under the rug, this would have been fire through the roof at 4:00 a.m.

Every incident is a learning opportunity and this one was no different. A few of the takeaways here are:

Teamwork—We all bring different experiences to the job. Don't let your pride get in the way of the mission. Bring in others to help solve the problem. This job was not about me finding the problem. It was me helping to find the problem and, in the process, imparting some of my experiences on my crew. They are now more prepared for an incident like this next time. In this case I was able to contribute to the team to achieve the mission. I can tell you unequivocally that I rely on my company officers and crews to contribute as much, if not more, than they rely on me in these situations.

The Art of Warfare—Sun Tzu's philosophies have been applied throughout the world in various conflicts, and the fire service is no different. Know your enemy. Know your battlefield. Know the strategies and tactics of our war against fire. Prepare your firefighters, teach your firefighters, lead your firefighters. Know the tools and weapons of war. Support your warriors in their fight. "The line between disorder and order lies in logistics"—Sun Tzu.

Tools of the trade—The fire service has some great tools. The pickaxe, forged Halligan tool, New York hook, the list goes on. We also continue to develop amazing technology, such as thermal imaging cameras, infrared thermometers, and atmospheric monitors, to just name a few. All of these tools, however, are nothing without our most valuable tools: our firefighters with their brains and five senses. The ability to size-up a situation, develop a plan, and act on that plan does not start with an axe or a TIC. It comes with a properly trained and supported firefighter. Show them the unharnessed power they possess as human beings absent of any tools of the trade and you will be amazed at the result. Lt. Colonel Oliver North in an address in 2009 described the current day American soldier as "... trained to use his body like a weapon and his weapon as an extension of his body ..." Imagine if all firefighters trained like that!

The crew cleaned up the debris and vacuumed the carpet in the bedroom. I checked out with the LT and shook the hand of each member of the crew and told them what a great job they had done. I thanked the homeowner for his support, and he said to me, "You knew the moment you walked in the door there was a fire. How did you know?" I said, "The firefighters knew it too. They were the first ones to know it. I just helped them find it. It's a team effort." This was a great night.

Smoke and Fire Conditions

The Art of Reading Smoke Fire Engineering DVD training series by Dave Dodson should be fundamental study material to any IC and tactical supervisor and more thoroughly augment what we present here. Chief Dodson describes the smoke as, "the fire talking to you and it's telling you the future."[2] He identifies four critical aspects of smoke: volume, velocity, density, and color.

Volume is the amount of smoke, relative to the building. *Velocity* is the pressure (speed and flow) in which the smoke is escaping the compartment or structure. This is driven by heat. The more heat, the greater velocity the smoke will emit from the structure. Rapid or boiling smoke is hotter and closer to the seat of the fire. Slow or seeping smoke that is not under pressure is either farther away from the seat of the fire or it has traveled a distance from the fire and is already starting to cool.

Turbulent smoke is a telltale sign of high heat and impending flashover, as the inside of the box is heat saturated and unable to absorb any more energy, thus reflecting back the heat and causing turbulent smoke.

Density is the thickness of the smoke. Smoke is a fuel and can be ignited if conditions are right. Modern smoke is laden with hydrocarbons, and the denser the smoke, the greater potential for ignition. Dense smoke may be too rich to burn; however, vent point ignition may occur when the smoke reaches the outside environment and comes within its flammable range.

Color determines the stage of heating. Lighter colored smoke may indicate early-stage heating as moisture is greater and evaporating. Also, lighter smoke may be due to a longer distance the smoke travels. Smoke succumbs to friction loss like water in that the pressure (velocity) is reduced and color lightened the longer it travels and the tighter passageways it flows through. That said, muddy brown smoke can indicate wooden structural members involved, especially in the overhead. Yellowish gray puffing and sucking can indicate backdraft while thick black can indicate more hydrocarbons, which is highly common in modern fires due to low-mass synthetics. Thin, flame-pushed smoke means the fire is nearby and burning more cleanly. This can be from a window, door, or hole in the roof.

Out of the four signs of reading smoke, color is the least reliable predictor of fire progression and heat conditions. In tightly sealed occupancies, the darker, solid parts of the smoke may be filtered out as the smoke pressure forces it out of small openings. For example, if there is light gray or even white smoke coming from a structure but it has volume, velocity, and density, you may still be dealing with an advanced fire—especially in colder weather, where the smoke is cooled almost immediately upon leaving the structure.

Bi-directional flow from a door can indicate the fire is starved for oxygen and any horizontal ventilating may cause rapid fire progression. Vent-point ignition occurs when thick, fuel-laden, hot smoke that is too rich to burn reaches the outside atmosphere and ignites. This indicates high heat, thick laddering of fuel in the smoke, and the potential for rapid fire progression (fig. 5–15).

Tunnel vision is easy during size-up. Figure 5–16 draws your eyes to the flames; however, learn to look at other signs, like the window to the left. Is that searchable?

Fig. 5–15. This smoke is telling you that high heat is present and structural elements are burning. Do the firefighters see it?

Source. Photo courtesy of Sacramento (CA) Metro Fire

Fig. 5–16. Look at what is not burning for searchable spaces, like the window to the left.

Interior Firefighting "Watch Out" Situations

The Los Angeles City Fire Department (LAFD) has identified the following "Watch Out" situations for interior firefighting following the Western Fire in 1998 that claimed the life of Captain Joseph Dupee. We thank the LAFD for their contributions to this text. The following is from *LAFD Book 101*:[3]

> *There are eighteen "watch out" situations that have been identified when working in the wildland. The failure to recognize "watch outs" by firefighters has on more than one occasion caused injury or death to firefighters. With this in mind the following list of twelve "watch out" situations have been developed for interior structure firefighting. These were situations that were identified as contributing factors in the LAFD Western Fire Significant Incident Report.*
>
> *Although not all-inclusive, the following twelve structure "watch outs" should certainly provoke communication among members in the hope that it could prevent a similar tragedy. Experienced firefighters should recognize having been in these situations and share the experience so that others might live.*
>
> *Interior firefighting "watch outs":*
>
> - *You have a working fire and your entry will be delayed. This may occur for any number of reasons, not the least of which are the forcible entry challenges facing the Department.*
> - *Multiple companies have been assigned to enter through one entry. Single door entry/exits are designed for a single person to pass through in a non-emergency environment. Consider the impact of trying to evacuate multiple companies through a 36" opening in a flashover condition.*
> - *Roof division is being driven off as you prepare to go inside. Coordination between roof and interior are critical for making decisions on offensive versus defensive operations and the extent of interior commitment by engine companies. Radio communications are a necessity.*
> - *Air is being drawn in rapidly in zero visibility and the heat is banking down. Oxygen is being fed into the fire. The temperature is rising rapidly and radiating downward. Your location will soon become untenable and require a rapid retreat.*
> - *You can hear the fire burning above you but can't see it. The sound of a burning building is very distinctive as it burns and comes apart. Lack of ventilation will allow a smoke layer to generate below the flame level reducing visibility to zero. This is an indication that members should retreat to a safer working location until conditions improve.*
> - *You realize you are working underneath a mezzanine. Mezzanines (and more recently facades) have been recognized as significant threats to the safety of firefighters. This is discovered in interior firefighting operations when a member attempts to open a ceiling with a pike pole only to discover that there is additional decking overhead. Often this decking is used for storage of materials above the office area. In addition, mezzanines are often constructed illegally and often use inferior construction techniques. Keep in mind that it would be unusual to find a mezzanine inside a rolling steel doorway. Consequently, rolling steel doorways should be utilized for interior operations, if possible. They are easier to see, offer a much larger exit, and will normally not have a mezzanine or façade above them. If*

confronted with an interior rolling steel fire door (fusible link), make sure to block the door open so as to prevent accidental closure due to heat buildup.

- *You feel uncomfortable. Listen to your instincts, because they are generally correct.*
- *Your SCBA alarm bell sounds and you still haven't found the fire. Time stamping an incident that involves interior operations at a working fire is very important. One of the most effective ways to time stamp is to recognize the amount of time it typically takes before your low air warning sounds. If in the 10 to 20 minutes it takes for this to occur you have been unable to find a working fire in zero visibility, serious consideration needs to be given to a change in tactics. In some cases, allowing additional ventilation to occur prior to entry will assist in locating the seat of the fire and identifying improved access.*
- *You flow water for several minutes and make no progress. Additional and/or larger lines are needed if insufficient gallons per minute (GPMs) of water is not reaching the fire.*
- *You hear the sound of roof ventilation being conducted behind you. Conditions on the roof, including lightweight construction, may require the ventilation teams to initiate operations further from the seat of the fire. Coordination is critical to prevent ventilation holes from pulling the fire over the interior companies.*
- *You are unable to communicate with the incident commander, or your division or group supervisor. The incident commander has overall responsibility for both resource and situation status. If you are unable to communicate interior conditions and the status of your crew, or if you are unable to receive incident status information, you need to retreat to a safety zone until communication is established.*
- *You are working with unfamiliar members. Responsible officers must ensure that any members not normally assigned to the company are familiar with all standard operating guidelines. This information needs to be passed on immediately upon reporting for assignment.*

Communications

Communication is the conduit from which size-up travels to keep personnel safe and the IAP relevant and functional. Communication problems is one of the NIOSH 5. Almost *every* incident will have communication challenges. Communication problems may often take place without us knowing it. For example, companies were dispatched to a commercial fire with all civilians reported out of the structure. This was never relayed to incoming companies. A subsequent collapse occurred with firefighters inside and on top of the building. By the grace of God, none died for what was a vacant structure that was a total loss.

In addition, communications will often begin to decay with the first arrival report. The officer has not practiced mental size-ups and pushes the radio button to think, instead of thinking first and pushing to *talk*. Clear, effective communications start with clear expectations. SOGs, policy, training, and officers all sharing expectations with their crews and each other are critical.

Leader's intent is a clear, concise statement about what people must do to succeed in their assignments.[4] It delineates three essential components:

1. **Task**—The objective or goal of the assignment.
2. **Purpose**—Why the assignment needs to be done.
3. **End state**—How the situation should look when the assignment is successfully completed.

The majority of leader's intent can and should be done before an incident ever occurs. Through training and time, trust can be established, and the team can get on the same page. This is a tremendous tactical advantage that can be in place prior to response.

Modern incident command is a team sport, requiring all officers and crew members to understand the potential roles they may fill on any given incident. As a BC, my company officers and their crews always knew my standing leader's intent and intent related to a specific incident.

For example, standing leader's intent regarding incident command for Battalion 7A was:

1. All company officers will be proficient in fulfilling the role of IC, or tactical division/group supervisors (task).
2. This will bridge the tactical gap and prevent the NIOSH 5 from coming into alignment on our incidents, thus making operations more safe, effective, efficient, and consistent, whatever type they may be (purpose).
3. All incidents will have a well communicated IAP, ongoing size-up, clear communications, minimal radio traffic, active accountability, and manageable span of control, while saving the most civilian lives possible and ensuring all members go home upright and in one piece (end state).

We have a lot going against us regarding communications, which is why it is such a major factor in operational problems. External factors include:

- Unclear expectations or not adequately communicating the IAP (from the IC down)
- Inaccurate information or lack of information
- Apparatus, siren, and horn noise en route
- Saws and other on-scene tools revving
- Engines pumping,
- Warning devices on aerial apparatus
- Firefighters yelling and bystanders screaming
- Radio feedback
- Too many people trying to talk on one channel
- Not having or forgetting the portable radio
- Being on the wrong channel
- Dead or low batteries
- SCBA
- Smoke and confined spaces in the structure
- Weather

Communication is one of the biggest problems on any emergency scene (fig. 5–17). Internal factors include:

- Unclear expectations or not adequately communicating the IAP. Notice this is both external *and* internal. Internally, this is not understanding or valuing what is coming in or from whom.
- Adrenaline
- Tunnel vision
- Auditory exclusion
- Misinterpretation
- False assumptions
- Personal feelings, disagreements, or lack of trust
- Pride
- Fear and inexperience

Fig. 5–17. Radios are important for communication in operations.

Ideally, functional and effective communications should have six characteristics:

1. **Clear**: Speak using clear wording and common terminology for your agency or operational area. Do not use technical jargon, ten codes, or words that are not part of your normal vocabulary.
2. **Complete**: Give complete information. For example, many CAN reports are missing the needs: "E5, Division 3, give me a CAN report."

   "Division 3, E5, it's hot and we are attacking the fire." These are called "California" reports because CAN without needs is CA. Don't give CA reports, give CAN reports. Another example would be, "There's a wall collapse on the inside courtyard." This has no sender or receiver identified. A more complete transmission would be, "All units from Division 3, safety message: We have a wall collapse in the interior courtyard."
3. **Calm**: Do *not* let your emotions get the best of you. Stay calm and professional. Do not escalate and make the situation worse. As the leader goes, so goes the team. You should be known for remaining calm in the chaos and cool under pressure. Take a deep breath before you key the mic to transmit. Practice during realistic training.
4. **Concise**: Say what you have to say in as short a transmission as possible. Some people love to hear themselves talk on the radio. First, ask yourself, "Is this important enough to transmit on the radio?" Then, ensure you remain short and sweet. One way to ensure this is to think before you speak (sound familiar?).
5. **Confident**: Speak with confidence in your voice. Just as the IC can either calm the scene or escalate the stress and chaos, so too can professional and confident demeanor instill confidence in the team that it's going to be okay.
6. **Closed**: Close the loop. Ensure that your message was understood and vice versa. Confirm transmissions and follow up on crew assignments, safety messages, accountability reports, CAN reports, and so on. It takes too much time to repeat transmissions verbatim; however, the key factors should be repeated for the benefit of the entire team.

   For example, "Division A from E2."

   "Go ahead E2."

   "E2 has advanced 100 feet in with a 1¾" line. We have heavy fire and need vertical ventilation. Also need a second line to assist with fire attack and another crew for search."

   Rather than repeat the entire transmission verbatim and tie up radio traffic, an efficient response would be, "Division A copies E2 at 100 feet, heavy heat. Requesting additional line, search, and vertical vent."

Tactical versus Task Assignments

Words matter, just like details matter. When assigning a company or crew to a task, the term should be "initiate" to clarify that they are working at the task level. For example, "E5, command, initiate fire attack on the second floor." This means that the IC wants E5 to pull hose and begin attacking the fire on the second floor, not assume Division 2 or Fire Attack Group Supervisor.

Conversely, when assigning an officer to take over a tactical assignment and supervise a division or group, the term "assume" is used to clarify the tactical level. For example, "Engine 2, command, upon arrival, assume Division 2, your objectives are fire attack and search. You have Engine 5 working for you."

Radio Channels and Frequencies

A *dispatch channel* is the frequency from which incident alarms are initially transmitted to alert units to respond. These are typically not utilized for resources to speak to dispatch, nor each other. A *command channel* is the frequency utilized for resources to speak with dispatch and each other for routine, daily radio traffic. Command channels can also be utilized for the IC to communicate with dispatch and/or tactical supervisors during an incident.

A *tactical channel* is a frequency utilized for units assigned to an incident to speak to each other and/or the IC or their tactical supervisor.

As communications become more complex due to the nature of an incident, command and tactical channels become overloaded with radio traffic. More channels may be required to divide the radio traffic so frequencies can be opened up and transmissions can be heard. Otherwise, critical transmissions involving a firefighter Mayday, civilian victims found, or other urgent safety messages will not be heard. Such was the case on a Mayday at a warehouse fire. The captain of the downed firefighter could not call a Mayday on the tactical channel due to too much radio traffic. He switched to the command channel to call the Mayday.

A *communications plan* is utilized to add additional command and/or tactical channels to an incident to divide the radio traffic among tactical supervisors and functions.

Arrival Reports and Assuming and Passing Command

Once on scene, the arrival report from the first company officer and/or chief is essentially a form of leader's intent. Again, it must *follow* the mental size-up. *After* you perform a mental size-up, you are much better prepared to give a clear and concise arrival report. This sets the incident off on the right foot.

An excellent system for arrival reports is identify, object, conditions, actions, and needs (IOCAN), which can be used for any type of incident. We will show several examples below for varying incident types (fig. 5–18):

- **Identify**
  - Your unit identifier
  - Location of the incident (if different than the dispatched address).
    Example: "E3 is on scene at the corrected address of 125 Main Street."
- **Object**: What is it? Use clear wording where possible (McDonald's). Include size, construction, and occupancy type. For vegetation fires, include size (in acres), and fuel type (light, moderate, heavy).
  - "One-story, medium, lightweight house."
  - "Large, commercial, tilt-up warehouse."
  - "One-acre grass fire in light, flashy fuels."
  - "Large industrial facility."
  - "MLK High School."

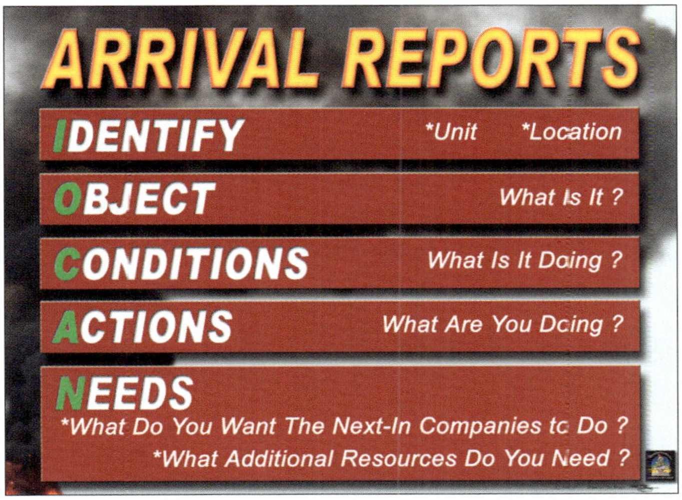

Fig. 5–18. IOCAN is an effective and concise arrival report system.

- **Conditions**: What's it doing? Include smoke/fire, victim profile (if known), and exposures. For vegetation fires, include rate of spread (ROS), values at risk, and potential size. For hazmat, give direction of hazard, potential, and exposures. For MCI, give type (trauma, hazmat, medical) and estimated number of patients.
  - "Heavy fire from the A side windows with moderate victim profile."
  - "Heavy smoke from the C side, evacuation in progress."
  - "Moving at a rapid rate of spread toward homes to the north. Potential for 5 acres."
  - "Unknown white gas from the C side traveling to the south toward Northridge Mall."
  - "Trauma MCI. Law enforcement on scene reporting 5 victims shot."
- **Actions**: What are you doing? Assuming or passing command. Tactics. ICP location if assuming command.
  - "E3 is passing command, initiating fire attack, and has a water supply."
  - "E3 is passing command, initiating attack with 2½" line from the C side."
  - "E3 is passing command, initiating fire attack with a progressive lay on the right flank."
  - "E3 is assuming Main Street Command. ICP is on the A side. Continuing evacuation, isolating, and denying entry to the C side."
  - "E3 is assuming MLK command, ICP is mobile, initiating triage."
- **Needs**: What do you need? Resources/alarms, assignments to incoming companies.
  - "Next in, assume command, stretch a backup line, calling one ambulance."
  - "Next in, assume command, bring a water supply to E3. First truck to the C side for forcible entry and search."
  - "Next in, assume command, and initiate structure defense ahead of the fire. Calling two additional Type 3 engines."
  - "Balancing to a Level 3 hazmat. Calling two Type 1 hazmat teams, two ambulances, and law enforcement for evacuation."
  - "Balancing to a trauma MCI. Next in, assist with triage and treatment, calling five additional ambulances, an EMS supervisor, and two additional engines for treatment."

An update should be given ASAP. This may be after a lap is complete (if possible, you cannot always get a lap in certain cities or construction types); after you get a victim profile from bystanders/occupants; after you get a face-to-face with other responders (law, EMS, etc.); or you see something new that you didn't before (fire in attic, exposures, basement, power lines down, etc.).

EXAMPLE (house fire): "E3 update to incoming units. Moderate victim profile. Fire is extending into the attic. Truck 1, upon arrival, initiate a search."

EXAMPLE (commercial fire): "E3 update to incoming units. Low victim profile. Fire is extending into C side from rear loading dock. Truck 2, upon arrival, go to the roof for a roof report."

EXAMPLE (vegetation fire): "E3 update to incoming units. Fire is anchored on the right flank at Main and Rural Route 2. First mobile attack unit, bump past E3 on the right flank to catch the head of the fire."

EXAMPLE (hazardous materials incident): "Main Street Command update to incoming units. Approach upwind, from the north."

EXAMPLE (MCI): "MLK Command, update to incoming units. Law states that threat has been neutralized in the gym."

CAN (conditions, actions, needs) reports are a derivative of the IOCAN arrival and are a quick effective way to transmit updates throughout the incident. Remember to give the *needs*. This is often forgotten.

First Company Officer

The first company officer to arrive on scene must decide whether to retain or pass command. If you can prevent injury or save life with your crew, pass command. *You are still in command until passed to the next officer on scene.* While you still have command until relieved, you have notified incoming units that you are task-saturated and need the next officer to take command off your shoulders ASAP. This can be identified as working command.

In contrast, if the incident is so massive and complex that you will *not* make a difference with your crew, then retain command. You will be better utilized calling additional resources and developing an IAP to start getting ahead of the incident power curve. Often a second company officer will arrive before the first chief and assume command from the first company officer. This is appropriate when the first officer is, again, task-saturated and working (rescue, fire attack, etc.), having passed command. Your SOGs will drive that operation.

First Chief

If another company officer is on scene and has given an accurate arrival report of the situation, the first chief does not have to regurgitate everything that was already said, thus tying up more radio traffic. If the previous arrival was accurate, then the chief may transmit, "Battalion 7 on scene, confirming E110's arrival. Battalion 7 will assume Arden Command with the ICP across the street from the A side. Continuing an offensive strategy and moderate victim profile. E110, you are now Division A with the objectives of fire attack and search of the house. You have TR106 and E109 working for you."

Perhaps a correction was needed due to an inaccurate or inadequate arrival by the previous company officer. For example, E110 arrived at a two-story fire in a balloon-framed house, but transmitted, "E110 on scene of a house fire. We are going in." The chief may ask for more accurate information en route. Upon arrival, the chief may transmit, "Battalion 7 on scene, this is a two-story house with possible balloon frame construction with fire in the basement and smoke from the attic. Battalion 7 is assuming Arden Command. E110, confirming you are performing fire attack or search, and do you need a water supply?" Once E110 confirms with a CAN report, Battalion 7 may continue, "Copy, E110 is initiating search and has a patent water supply. Confirming an offensive strategy, moderate victim profile. ICP is located behind E110. Next engine and truck. . . "

The first chief officer to arrive and assume command from a company officer has several choices. One is to get a transfer of command from the company officer via radio. This often takes a lot of time and bogs down the incident, tying up radio traffic. The second option is to have the company officer come out and meet the chief at the ICP for a transfer. This may pull the officer away from critical work and the tactical gap will widen. Often, both will miss radio traffic during the face-to-face dialogue. Third is to relegate the company officer back to their crew at the task level. By doing so, the chief has given away the tactical advantage that they had in the brain of the initial IC who had accountability, sized up the risks, and was developing a tactical plan. This results in more radio traffic as the IC starts to ask for CAN reports.

An extremely effective option is to make the initial IC a tactical division or group supervisor to rapidly bridge the tactical gap (See chapter 7 for more). This is incident- and SOG-based and may depend upon arrival sequences and the level of task-saturation when the chief arrives. **EXAMPLE**: Upon arrival, the chief would state, "Battalion 7 is on scene, assuming command from Engine 23. The ICP is on the A/B corner across the street. Confirming an offensive strategy and low victim profile. E23 you will assume Division A, with Engine 24 and Truck 5 working for you. Objectives are fire attack and search."

E23 already has active accountability of those resources; therefore, there is no need to regurgitate that radio traffic. The chief would have heard the assignments made by the previous IC. By decentralizing and allowing that previous IC to now become tactical, the need for the chief to ask a lot of questions is minimized. Active accountability now rests with the Division A Supervisor. Very little has changed.

A heads-up company officer in command can drive the incident and help bridge the Tactical Gap for the chief at transfer of command. **EXAMPLE:** Upon arrival, the chief would state, "Battalion 7 is on scene, ready to assume command from you, Engine 23." This would be a good choice if the company officer in command is too task-saturated and the chief realizes that the timing is not ideal at the moment. Engine 23 could respond, "Battalion 7, recommend I assume Division A with Engine 24 and Truck 25 working for me. We have fire attack and search of the house. Need an additional hoseline."

The point is to bridge the tactical gap ASAP, based upon the conditions and resources at hand. By decentralizing command, the need for radio traffic is minimized, yet accountability is enhanced. Note that in the second example, the chief did not immediately assume command, rather, the chief simply notified the company officer/IC that he had arrived and was *ready* to assume command. This has several benefits in certain circumstances. First, if the scene is chaotic and a lot of radio traffic is clogging the radio, the chief may not want to interrupt so the IC can finish an order or operation. Examples would be an apartment fire with multiple victims being rescued, or multiple buildings involved, and the current IC is deploying hoselines on the radio upon the chief's arrival.

Second, the incident may be so large and escalating that a face-to-face transfer of command is more appropriate. An example would be a mass shooting or large commercial building fire with a second alarm already called. Rather than stay tied to a rigid form or procedure of communicating, the chief arriving to assume command should size up the situation, how best to utilize the current IC (tactical or task), what resources are arriving, and how best to communicate with the company officer who is undoubtedly wanting to get command off his shoulders.

Still, other situations may dictate that the chief allow the company officer to remain in command so they can get additional experience and is on the promotional list. This opportunity for training and mentoring will yield big dividends downstream. As a BC, I routinely had several options available upon assuming command and those options predicated my communications. For example, if the fire was contained in a room or garage of a house upon my arrival, I would often ask the captains in command if they would like to retain command for the experience. They often obliged. The communication was, "Command, Battalion 7 is on scene. ICP will be located across the parking lot on the A side. Come to the ICP at your convenience." Once the IC arrived at the ICP I would just ask if they would like to retain command. When they did, I would transmit, "All units, E105 is retaining Northrop Command. We are continuing an offensive strategy with low victim profile."

When a fire was still escalating upon my arrival, as was often the case, I would typically assign the current company officer/IC to a tactical division position. The hot division or area would be the first one established and the IC would bump up to that forward position (Division A, for example). This would allow a continuum of supervision and accountability by the officer who was already supervising the area as the IC anyway. The communication was, "Battalion 7 is on scene, assuming Watt Avenue Command from Engine 103. The ICP is on the A/B corner across the street. Confirming an offensive strategy and high victim profile. E103 you will assume Division A, with Engine 101 and Truck 106 working for you. Objectives are fire attack and search of the apartments facing Watt Avenue."

If the incident was much larger, escalating with additional alarms, chances were that the IC was very task-saturated and wanted/needed to get back to their crew. Here, the need to set up a tactical division/group was great, however, with immeasurable tasks needing to be completed, like multiple hoselines and rescues, the second chief may be the better choice to assume a division role. The communication would be, "Battalion 7 is on scene, assuming Marconi Ave Command, ICP is located across the street on the A/D corner. This is an offensive strategy with confirmed victims trapped on upper floors. E109 captain, re-join your crew. Calling a second alarm." Then crews would be accounted for via radio with CAN reports as conditions allowed, followed by division/group supervisors once established.

WISDOM FROM THE MASTERS

You're Not Seeing What I'm Seeing

CAPTAIN BILL GUSTIN, MIAMI-DADE (FL) FIRE DEPARTMENT

Miami-Dade Battalion 5 companies were operating at a fire in a large, old, wood-frame rooming house. Companies were having difficulty locating the fire due to limited visibility. Eventually, at the 10-minute incident time mark, one company encountered some flame on the floor that appeared to be a pool of accelerant or melted plastic. The company officer advised the IC, Battalion 5, that they had located the fire and were knocking it down.

What the company officer didn't tell command was that every time they stopped flowing water, the small fire on the floor would reappear. Something else was odd: How could such a seemingly small fire fill the entire house, from floor to ceiling, with heavy smoke? Shortly thereafter, command ordered all companies out of the house. One of the company officers acknowledged the chief's order with a report that they were making "good headway" and should have the fire under control in a few minutes. Command replied by saying "that's negative; you're not seeing what I'm seeing. The order has been given; back your company out of the building."

What command could see that the company officer could not see is that the smoke was getting progressively darker, more turbulent, and pressurized. Minutes after crews evacuated and were accounted for, they heard a "whoosh" and the rooming house became fully involved. After a defensive attack, crews discovered that the entire floor of the house collapsed into a fully involved basement filled with mattresses and upholstered furniture. Crews had no idea that they were operating directly above a raging fire. The elusive fire that kept reappearing was actually from a hole burned through the floor. Had crews remained in the structure they most likely would have ended up in the burn center or a cemetery. The preceding scenario is not fictitious, it was real. I was there. I was the company officer who requested to remain operating in the building.

Two Size-Ups on Every Fire

There are two size-ups that must be performed continuously on every fire: the inside size-up of conditions observed by companies operating in the fire building and the outside size-up by the IC. If a department's command procedures require the IC to remain in his vehicle or mobile command post, the IC must designate someone to be their eyes on the fire building (tactical supervisor). If the inside size-up and outside size-up are not in agreement, a red flag should go up because somebody has it wrong and it is *almost always the companies operating inside, because of their limited perspective.*

For example, a company officer reporting on conditions on the floor where they are operating may not be aware that they have fire beneath them on a lower floor or overhead in an attic. Fire in the concealed space between a rain roof and the original roof will most likely remain undetected by interior companies and first observed by the IC. Similarly, according to Chief Vincent Dunn, FDNY (ret.), "There are six sides to a fire: The four sides plus the floor above and floor below. But a wood-frame building has a seventh side, the combustible exterior walls. Beware of fire spread via the exterior."

When an IC orders companies out of a building and a shift to defensive operations, he doesn't want to hear radio transmissions such as, "Just a few more minutes and we'll have it, chief," or, "We're making good headway." Companies who advise the IC on their progress and conditions do so from a limited perspective. They are not seeing what the IC is seeing.

Ego Skews Judgment

I wonder how many firefighters have lost their lives because they chose or were ordered to do the wrong thing for too long. When firefighters are performing a task, such as forcing entry or advancing a hose line, they have skin in the game. As a result, they are not just physically involved, but it is natural for them to be *emotionally* involved. Ego is a powerful motivator and can skew judgement. Ego can cause firefighters to the do the wrong thing for too long, even though it is apparent to the IC that their tactics are no longer effective and the intended benefit is not worth the risk. As a result, they can't bear the thought of failing or being ordered to change tactics. Hence, when they are told by the IC to back out, it is common to reply to the order with a dissertation, justifying why they should remain, such as "We're making good headway" or "We almost got it." This begs the question, in terms of time, exactly how much time is "we almost have it, Chief"—minutes, hours?

"Something Isn't Right!"

Miami-Dade companies were attempting to locate the source of smoke in second-floor bedrooms above an attached garage. Before ascending, the first-arriving company officer did the prudent thing; he checked the garage, finding it was perfectly clear with no odor of smoke. Firefighters operating on the second floor were not aware of two critical factors that could have taken their lives: First, the second floor was supported by light weight, engineered parallel chord wood trusses; Second, the fire was burning below them in the truss void (the space above the one-hour fire-rated garage ceiling and the second floor).

Firefighters found smoke that was increasing in density but no heat. Accordingly, their TIC showed no indication of heat. To improve visibility, the companies operating on the second floor requested the venting of second floor windows. The IC, Battalion 13, sensing something wasn't right, ordered the companies on the second floor to back down the stairs to the first floor before venting the windows. When the first window was vented, it was followed by a blast of flame and the sudden collapse of the bedroom floors into the garage. This scenario compels the question: How many firefighters' lives have been saved because the boss acted on gut intuition and pulled his troops out of a dangerous situation?

At the horrific Boyd Street fire in May of 2020, the LA City Captain directing roof operations stated that something wasn't right upon reaching the roof. Following are conditions that could lead a fire officer to determine that something isn't right:

- Inability to locate the fire.
- Unexpected or unusual fire behavior, such as smoke being sucked back into a building or, at the Boyd Street fire, smoke and flame were blowing out of a building like a rocket's exhaust.
- Extended incident time with companies making little or no progress.
- Construction type is prone to collapse or conceals fire.
- A chimney emitting dark smoke; indicative of smoke from a basement fire entering a furnace flue through a draft diverter.
- Capable, reliably strong companies not making progress.
- Size-up conditions observed by the incident commander do not agree with reports from companies operating inside.

Reports from companies operating inside are inconsistent. For example, companies reporting fires on different floors or portions of a building or companies operating on a roof reporting smoke or fire, indicating heavy fire in an attic, whereas inside companies report clear conditions or just a light haze of smoke.

Chief officers, listen carefully to radio reports from company officers; not just *what* they say but *how* they say it. If their transmissions sound distressed or give you an uneasy feeling, trust your gut! It may be time to regroup, reevaluate tactics and strategy, and reassess the level of risk.

Command Post Location

The ICP must be well communicated to all personnel on scene and en route. This includes subsequent alarms. Staging areas must be well communicated with the staging area manager and a tactical channel for staging must be identified. The location of the ICP is based on the incident type, conditions, access, weather, time of day, traffic, and the manner in which other apparatus are staged on scene or going to be arriving and staging.

For house fires, parking directly across the street in a driveway of another home may work well in parts of the U.S. that have driveways. However, in older cities, it is less likely that a driveway will exist. Also, long driveways and those serving multiple set-back residences may present unique challenges.

A long-standing debate continues between those who remain in the command vehicle, or in a remote ICP behind the command vehicle, versus those who physically move forward to a position directly in front of the building. Again, many of the trends are geographical in nature. The East Coast, or older cities like San Francisco on the West Coast, may not be conducive to a remote ICP due to congestion, sidewalks, and narrow streets and alleys, whereas much of the Midwest and West Coast may have more direct access, wider roads and streets, and more options available.

One benefit to remaining more remote is that it prevents the IC from tunnel vision and getting sucked into the task level. In a controlled environment, like the inside of a command vehicle, the noise and distractions of the scene (personnel, bystanders, weather, engines pumping, sirens, horns, saws, screams, etc.) can be minimized. The IC should be able to clearly hear the radio and detect subtleties like personal alert safety system (PASS) devices or vibra-alert noises in the background.

Regardless of whether in or out of the command vehicle, or up close or further away, some considerations will help you. First, you do not want to move the ICP, so think ahead. Do not set up too close, where radiant heat, smoke, potential collapse, or other hazardous conditions may change and compromise your position. Second, get the best view you can, within reason. Give the prime real estate to the companies working. Try to see two sides of a building fire, at the corners. The benefit to a decentralized command philosophy is that the ICP and IC view does not have to be as ideal since the tactical supervisors are filling the tactical gap as eyes and ears for the IC. Third, locate the ICP in a place that is logical, close enough for access, yet far enough for escalation. Think of it like setting up an aerial. You only get one shot at it and once it's up, it stays. Nothing is worse than trying to move aerial apparatus due to changing conditions like collapse, heat, or weather. Think of your ICP the same way (fig. 5–19).

For commercial fires, expect a lot of hose to change the landscape. If able, drive around the structure to get a decent size-up prior to setting up the ICP. Many downtown areas may not give you this luxury. Consider a quick drive down an alley to get a C side view prior to setting up the ICP.

For apartments, the challenge is if the fire is deep into the complex, as is often the case. Again, the ICP should be close enough, but not prevent subsequent ladder and engine companies from getting access or prime real estate.

For hazardous material incidents, think upwind and uphill. Also consider weather changes and potential unified command with law.

For MCIs, the chief will most likely be in unified command and should plan ahead by contacting the law enforcement counterpart on scene and suggesting they set up a unified ICP together.

For smaller vegetation fires, the ICP is likely to remain stationary. As the fire progresses, however, there may be a need to move. This is not ideal, but, if necessary, should be communicated. Typically, a vegetation fire ICP will be located near the anchor point, upwind and near the fire's origin. With topography, a higher location may be ideal; however, the base of a populated hillside where the fire originated will likely give the IC a greater view of the values and structures at risk. In addition, it will be a safer location in the long term.

Fig. 5–19. The ICP for this commercial fire is at a good distance to see the big picture and allows for escalation.

Source. Photo courtesy of James Hedlund-Kaiser

Personnel Accountability Reports

Personnel accountability reports (PARs) have historically been conducted at time intervals, like 10 or 15 minutes, per the SOG of the FD, or when a strategic shift or sentinel event occurs, like switching to a defensive strategy or a roof collapse, respectively. In any case, the conventional centralized PAR from each company and resource on scene to the IC on one frequency ties up the radio frequency far too long. A decentralized methodology would have a PAR go through the division and group tactical supervisors. This causes far less radio traffic, and the tactical supervisors will likely already have active *hands-on, eyes-on ongoing active accountability*, thereby reducing reflex time. For example, a roof collapse occurs on a commercial fire. The IC would transmit: "Command to all divisions, partial roof collapse on the A/B corner. We are switching to a combination strategy; Division A will be defensive. Evacuate Division A. Division C will remain offensive. Stand by for a PAR by division... Division A, Command, PAR." Each division/group supervisor would contact the companies assigned, many by face-to-face contact, others by their respective tactical channels if multiple were in use.

Summary

Size-up and communications are interlinked components of risk management. Modern incidents move more quickly than ever before. The more time spent sizing up before dispatch through training and pre-planning, the more likelihood of success during the incident.

Weather, construction, smoke, fire, resources, time, and many other factors must be constantly sized up mentally, and forecasting communicated continually. These issues present themselves before shift, at shift change, upon dispatch, during response, upon arrival, and throughout the incident.

The NIOSH 5 reside in the tactical gap and include lack of or improper: risk assessment, communications, incident command, accountability, and lack of or failure to follow SOGs. These predictable phenomena can be identified by the use of tactical supervisors who see risk, reduce radio traffic by increasing face-to-face communications, conduct ongoing active accountability, and help the IC maintain a manageable span of control.

FPODP and FIRST are two tools to help you with effective and efficient mental size-up, while IOCAN helps with arrival reports. CAN reports provide an efficient means of updating tactical and strategic bosses.

Reconnaissance is vital to deeper size-up and situational awareness for larger incidents like commercial structure fires.

Victim profile and survivability profile go together to measure gain versus risk. The overall goal is that all firefighters return safely, *and* all savable civilians are rescued. These are not competing, mutually exclusive goals. The IC's job is to balance these two functions during changing conditions throughout the entire incident. The only way to ensure these two goals are accomplished is through constant size-up.

The conduit of size-up is communications. Leader's intent is foundational to excellent communication and mutual understanding before a fire is ever dispatched. Communications must be practiced to be effective. Whether face-to-face or over a radio, communications must remain calm, clear, confident, complete, concise, and the loop must be closed. This includes the means by which the first-arriving company, and subsequently arriving chief officers, communicate arrival, transfer of command, CAN reports, and PARs.

Chapter Review

Review Questions

1. Explain tunnel vison and auditory exclusion and how they can impact situational awareness.
2. Describe the importance of pre-planning and walk-throughs, including some of the components.
3. What are the NIOSH 5?
4. Explain the difference between victim profiling and survivability profiling.
5. What are the four critical aspects of reading smoke?
6. Explain leaders intent.
7. Describe the command transfer process.
8. What are some differences, benefits, and challenges associated with different command post locations?

FESHE Strategy and Tactics (CO279) Related Content

The content contained in this chapter provides detailed information to assist with instruction and education for the first five expected course outcomes of students in the CO279 Strategy and Tactics course.

1. Discussion of fire behavior as it relates to strategy and tactics.
2. Provision of the main components of pre-fire planning and the steps needed to perform a pre-fire plan review.
3. Identification of the building construction basics and how they interrelate to pre-fire planning, strategy, and tactics.
4. Description of the steps taken during size-up.
5. An examination of fireground communications.

Chapter 5 also aligns with the CO279 Strategy and Tactics course outline as follows:

III. Visual perception

- a. Pre-planning
- b. Size-up

IV. Pre-fire planning

- a. Concept
- b. Phases
- c. Methods
- d. Format
- e. Occupancy classifications
- f. Building types

V. Basic division of tactics

- a. Size-up
    - i. Facts (F)
    - ii. Probabilities (P)

iii. Own situation (O)
iv. Decision (D)
v. Plan of operation (P)

VI. Rescue

a. Life safety problems of fire
b. Determination of life hazard
c. Rescue resources and operations

NFPA 1021 Job Performance Requirements

The information in this chapter can be utilized to support training and educational programs associated with the Emergency Services Delivery Fire Officer I JPR 4.6, 4.6.1, 4.6.2, Fire Officer II JPR 5.6, 5.6.1, and Fire Officer III 6.6, 6.6.1. The content in the chapter regarding the NIOSH 5 and incident reviews also aligns with post-incident analysis related job performance requirements of Fire Officer I 4.6.3, Fire Officer II 5.6.2, and Fire Officer III 6.6.2.

Endnotes

1. Steve Marsar, "Survivability Profiling: Are the Victims Savable?" *Fire Engineering*, December 1, 2009, https://www.fireengineering.com/firefighting/survivability-profiling-2/#gref.
2 Dave Dodson, The Art of Reading Smoke DVD (Fire Engineering Books 2008).
3 Los Angeles Fire Department, *Book 101: Rapid Intervention Company*, Self-Study Material (2018).
4. National Wildlife Coordinating Group, *NWCG Incident Response Pocket Guide (IRPG)*, PMS 461, January 2022, https://www.nwcg.gov/sites/default/files/publications/pms461.pdf. *Wildland Fire Leadership Development Program*. https://www.nwcg.gov/wfldp. *Leading in the Wildland Fire Service*, PMS 494-2, January 2007, https://www.nwcg.gov/publications/494-2.

6

Aggressive Command and Tactics for Life

Fig. 6–0. FDNY firefighter transfers a young girl to Emergency Medical Services (EMS) for assessment.
Source. Photo courtesy of by Lloyd Mitchell

Introduction

It is possible that as recently as 15 years ago, a textbook with a chapter titled "Aggressive Command and Tactics for Life" would not even make it to publisher review. Fortunately, the environment and culture of the fire service has changed significantly in a very short period thanks to an era where access to information, research and evidence based practices is compounding.

Where the term *aggressive* was once deeply connected to recklessness, today it is more closely associated with anticipation in planning, assertive decisions, and proactive actions. This is how we stay ahead of the incident power curve. Education helps with anticipation and proper positioning at the tactical level through that deeper understanding of the dynamics of our environment. Collectively, we now have a much greater awareness of the degree to which time effects the fireground. Just a decade ago, discussion on tactics for life would surround firefighter safety, risk avoidance, and protective measures in operations. Thanks to advances in technology, funding for research, and the efforts of Underwriter Laboratories, Fire Safety Research Institute (UL FSRI), new tactical considerations for the fire service are being shared with detailed reporting on the impacts of tactics on victims.

The objectives of this chapter aim to inform the decisions of today's fire officers using the information of today. Relying purely on experience is not sufficient for the higher-accountability world in which we live, public trust in our profession, and expectations of the positions we hold as officers. This chapter and this text will provide the appropriate blend of experience and evidence that is necessary to support command competence and confidence in the explanation and execution of duties and decisions.

One thing is for sure: training saves seconds, seconds save minutes, and minutes save lives. This means everything we do costs time. When we save time, we can save lives. The window of opportunity to save lives shuts faster than ever before. Every firefighter, engineer, apparatus operator, chauffeur, company and chief officer must shave time off their individual evolutions and operations. Whether turn-out time, response routes, staging on scene, deployment of tactics, personal protective equipment (PPE), self-contained breathing apparatus (SCBA), and so on, it all matters. If we shave seconds off individual and company tasks, that will lead to time off tactics, and could add up to enough time to save a life. The only way to shave time off hose stretches, SCBA donning, laddering, vent, enter, isolate, and searching (VEIS), and countless other tasks and tactics is to train!

Training Saves Seconds, Seconds Save Minutes, Minutes Save Lives!

Fire Dynamics

While we clearly recognize a difference, it is challenging to explain the specifics between a legacy or traditional fire and one of today. When terminology like "modern fire behavior" or "today's fires" is used, the focus is put on the fire or era, when in fact the combustion process of fire has remained entirely unchanged.

What has changed is the chemical inputs of the fuels, the understanding of ventilation impacts, and the influence of the containers in which fires occur. More than any other influence, the changes in fuel packages at the chemical level has changed the pace and byproducts of fire (fig. 6–1).

The traditional terminology of fire behavior must be eliminated from our vocabulary as it implies that fire behaves. It is unfortunate that this graph was the instruction tool used in

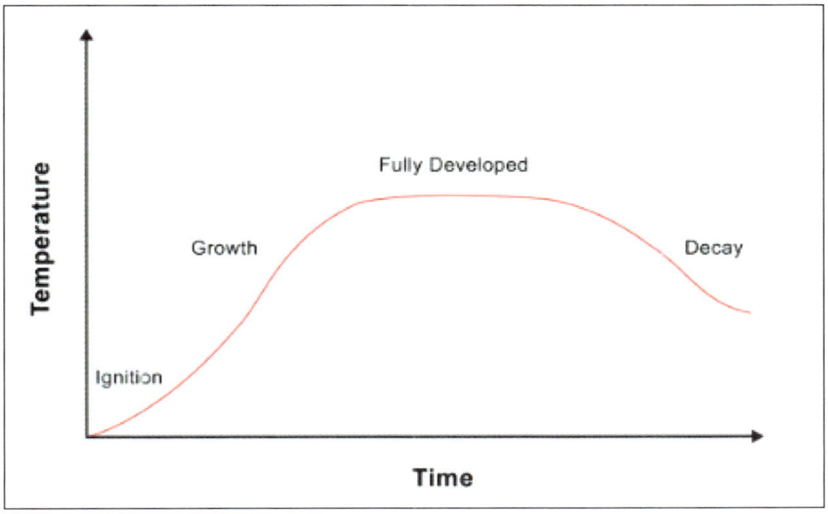

Fig. 6–1. Traditional fire behavior curve

textbooks published as late as 2015, more than 5 years after the release of the UL and NIST fire dynamics videos presented in real time the differences in modern and legacy furnishings as fire fuel packages.[1]

The historical fire behavior curve is constructed on the premise of a fuel-controlled fire. This would be an accurate presentation of a fire outside in the open air with a fixed fuel load—a campfire, for example. Fire behavior inside a structure is not a predictable bell curve but a dynamic, supply and demand wave. Only our most recent generations of firefighters have fire dynamics as their initial imprint for fire education. Most of our operating firefighters were indoctrinated under traditional terminology and are having to transition and understand the gravity of the differences. Accurate understanding of the fireground is built on a foundation of an accurate understanding of fire dynamics (fig. 6–2).

The greatest change in the modern age is not fire; it is the proliferation of petroleum-based synthetic materials in nearly all products available to the fire. When comparisons of legacy to modern fires or furnishings are made, the difference is the type of materials burning. When this information was first presented to the fire service in 2009 by Underwriters Laboratory, the specificity of that point was not made clear enough. Even though this display made a great impact on the fire service, it did not fully inform it.

In 2020, 10 years after the first delivery of real-time, high-definition fire dynamics education to the fire service from UL, they did it all over again. You will notice that the terms *legacy* and *modern* are removed, and the current presentation shifts the attention from the era to the

Fig. 6–2. UL FSRI legacy and modern burn comparison

actual materials. This ensures that the message is clearer that the difference in fire dynamics is related to materials, not decades (fig. 6–3).

To expand on this concept, it helps to know the law of conservation of energy. The law of conservation of energy states that energy can neither be created or destroyed, it can only be transferred from one form to another. When this is the base for comparison between natural and synthetic materials, the differences in burn characteristics become very clear. Natural products like, wood, cotton, and leather are grown over a longer period using natural forms of energy like sunlight, oxygen, soil, and grass. Synthetic materials are primarily petroleum-based, which must be extracted from deposits just to be made available for use. Regardless of the item, a significant amount of supplemental energy and process is put into converting oils to plastics and various forms and fibers in a short production time (fig. 6–4).

The difference in energy for development between natural and synthetic materials is mirrored in the fire development and energy release rate during the burning process.[2] The chart in figure 6–4 shows the time, and the videos shown in figure 6–4 (available by scanning the QR code) display the intensity. Using the numbers for this chart drawn from UL single-room burn tests, we can deduce that synthetic materials result in a room reaching flashover 8–10 times faster than those involving natural products. By watching the videos, you will also observe that this is a more explosive and violent process as well.

It must be recognized that these changes in the products found in our fire environment have been gradually occurring over the last 50 years. Unfortunately, the necessary modifications of

Fig. 6–3. UL FSRI home furnishing burn comparison[23]

Natural vs. Synthetic Times to Flashover (min:sec)

Experiment	Room with Natural Furnishings	Room with Synthetic Furnishings
1	29:30	3:40
2	> 30	4:45
3	> 30	3:20
4	> 30	4:50

Fig. 6–4. Natural versus synthetic table

professional development, education, and operations in our profession have only just begun. Every fire we respond to has a head start on our companies—we are, by default, reactive when the call comes in. The changes in our fire environment have a significant head start on our profession. Now that we know this clearly and are informed, anything less than an aggressive approach to improving, preparing, and learning about it would be negligent.

The world knows fire by its flame, but the most destructive element of fire is not the flame, it is the heat. Discussion regarding cause and effect of moving a fire from an open-air environment into a compartment or container usually begins with fuels and ventilation, as we did in this chapter. However, the extent of influence on conditions goes much further. For a stand-alone fire outside, fuel is controlled, ventilation is unlimited, and the intensity of the fire does not affect either, which results in that predictable curve. When a fire is moved into a compartment, the three forms of heat transfer become major influencers on the environment and the availability of fuels. Additionally, ventilation profiles are dictated by the construction of the container and the rate of heat transfer by the fire's development.

Heat transfer is the exchange of thermal energy to the fuel by the mechanisms of conduction, convection, and radiation. In an exterior, fuel-controlled fire, that exchange of energy goes into consuming the single fuel and the open-air environment, thus displaying growth, development, and decay on a consistent path. A fire in an enclosed structure does not allow for any transfer of energy to an open environment; it is all transferred into the structure and its contents. Heat energy impacts the ceiling, wall surfaces, and other items inside the structure, breaking down more fuels in more areas. This is why fire development inside a structure, regardless of fuel type, is more rapid than a fire on the exterior: the enclosure improves the efficiency of energy transfer to available fuels.

Heat release rate (HRR) is the rate at which heat energy is generated by burning. It wasn't until the early 2000s that the operational fire service began to really take note of heat release rates. This is not the fault of the fire service; it was a technology barrier. Until recently, actual measurements of heat release rates were conducted on very small samples due to the instrumentation available, and it was only through calculations that determining heat release rates for larger items of combined materials was possible. It was the 2009 UL videos that provided a real-time visual display of heat release rate differences and enhanced the impact of the numbers. One can explain on paper that a small trash can fire produces 300 kW of heat energy and a sofa produces 3 MW, or 10 trash cans of heat energy at once, but seeing is believing.

The law of conservation of energy explains that energy can neither be created or destroyed, it can only be transferred from one form to another. Dr. Vytenis Babraukas states that HRR is not just a variable of fire, it is "the single most important variable in describing fire hazard."[3] He describes HRR as the driving force of fire activity due to the positive feedback of heat transfer breaking down other materials, feeding further heat development. Because of this mechanism, all other fire variables, such as smoke production, presence of toxic gases, and oxygen depletion directly correlate to HRR.

To watch the UL videos is to observe in real time how the high-energy investment and rapid production associated with making synthetic products becomes a high-energy release and rapid breakdown when burned. When comparing an exterior fire to an interior fire, the heat release for an exterior fire does not impact environmental temperatures. Due to the containment of an enclosure, the heat released increases environmental temperature. Increased temperatures in a compartment lead to higher gas pressures, increase the speed of convective currents, and accelerate the pyrolysis process, breaking down more solid fuels to gasses.

While there is a visual, nearly explosive representation of these actions in the UL videos, what is not seen is how the fire activity also increases the demand for oxygen to sustain combustion. It is shortsighted to think of ventilation for fire as just air. A fire in a house does not become vent limited and create a vacuum. Oxygen is consumed by fire in the container while heating and expanding other gases; there will be plenty of air in the container, so it is the reduced oxygen percentage in the air that is the limitation. To further complicate this vent limited concept, UL-conducted burns have determined that while the atmospheric 21% oxygen provides for complete combustion of natural products like paper, it is insufficient for synthetic materials. To clarify this, a couch burning in a wide-open parking lot would be vent limited because the demand for oxygen to complete combustion cannot be satisfied by the air around it.

To better demonstrate this, consider an acetylene torch. When you first light an acetylene cutting torch, you ignite the acetylene only (fig. 6–5). Even though the flame is in the open air, it is very orange in color, and it gives off soot and smoke, all of which are indicators of incomplete combustion.

It takes the addition of supplemental oxygen to raise the percentage well beyond that available in the natural atmosphere to achieve a clean burn and complete combustion (fig. 6–6). Acetylene is a petroleum-based fuel, as are the plastics in our homes.

Synthetic products are made from toxic materials that require supplemental energy in production, which means that they will release more intense energy levels and toxic byproducts when burned. When Dr. Babraukas explains that the HRR is the most important variable in describing the fire hazard, he is not just speaking to fire damage—he is also referring to fire lethality. Complete combustion is clean, whereas incomplete combustion is a dirty burn, resulting in increased particulate smoke, more byproduct gases, and increased carbon monoxide levels, to cite a few toxins. To know that at any time, under any conditions, burning synthetic materials in a structure are in a state of incomplete combustion, is to know they are making the environment in that structure more toxic, fuel-rich, and oxygen-depleted with every passing second (fig. 6–7).

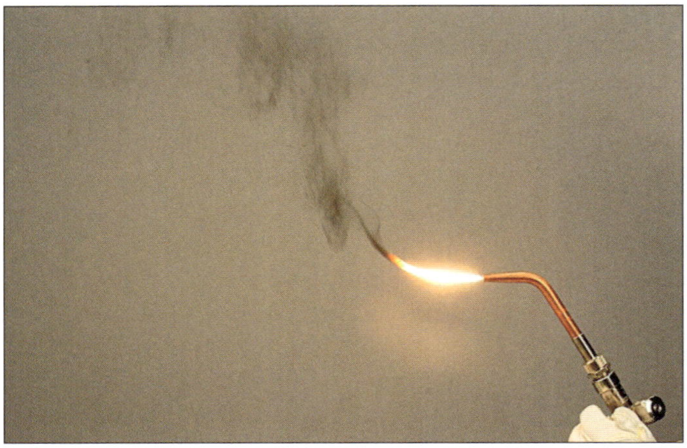

Fig. 6–5. Acetylene-only flame

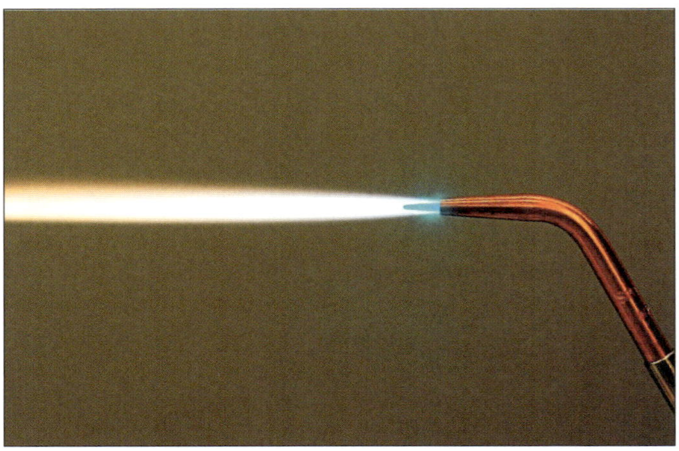

Fig. 6–6. Acetylene-and-oxygen flame

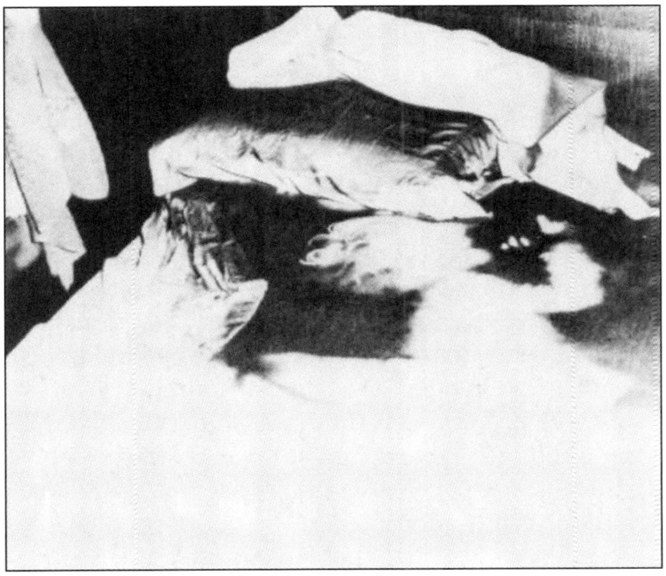

Fig. 6–7. Threat of smoke toxicity identified in the 1973 publication *America Burning*

The Exposure Problem

This picture is from the 1973 white paper *America Burning.* The picture itself is of a tragedy; a soot silhouette of a child who died in a residential fire. There is no thermal damage anywhere in the picture, only the black byproducts of incomplete combustion. In the soot you can see the details of curls of the child's hair and hand position, as if they are still there sleeping. The caption confirms what we know just by looking at this image. It was smoke, toxic gasses and lack of oxygen that claimed this child's life. The greater tragedy is that it wasn't until 45 years later that a public safety message to sleep with closed doors was developed.

Today we not only know better, but we also know more. Advances in technology and fire sciences have connected the anecdotal phrase "fires burn and smoke kills" to empirical evidence of the thermal and toxic threat inside a working structure and the relationship between the two. In 2015, when UL FSRI started the *Study of the Impact of Fire Attack Utilizing Interior and Exterior Streams on Firefighter Safety and Occupant Survival,* they initiated research into the effects of tactics on the survivability of unprotected occupants. The lack of previous work in this area once again was not due to a lack of interest, but a lack of technology. It was not until this time that it was feasible to incorporate the instrumentation necessary to measure temperature, pressure, heat flux, gas concentrations like carbon monoxide and oxygen, and moisture content in a live fire environment at multiple locations. UL FSRI researchers even instrumented sections of pig skin to evaluate tissue damage differences. This work continues to develop. In the 2020 Search and Rescue Studies, UL FSRI became the first in the world to utilize new technology to measure hydrogen cyanide concentrations in the live fire environment.

The difference between a fire outside and inside a compartment is everything. For centuries, fire service experience has told us that we have a fire problem based on the understanding of observations from outside of fires. The last few decades of evidence from research inside fires has helped us understand that the truth is that we have an exposure problem. While the fire is the source, it is the cumulative, compounding, and dynamic thermal and toxic exposure throughout the structure that is so damaging and deadly. This knowledge is what is now driving tactical decisions towards mitigating the thermal and toxic threats as soon as possible to improve conditions, and away from the myopic focus on extinguishment alone.

Simultaneous versus Sequential

The chapter led with the statement that accurate understanding of the fireground is built on a foundation of an accurate understanding of fire dynamics. For more than a century, the foundation of understanding the fireground has been built on fire behavior that follows an order of incipient growth, development, and decay, and a belief that heat is the greatest threat and extinguishment is the priority. Most fire departmental procedures which establish fireground priorities and order of operations will demonstrate these influences; first engine is fire attack, second engine secures water supply and backs up fire attack. Between these first two resources, extinguishment is being initiated, supported, and set up for continuous operation: initial attack, first engine tank supply, back-up hoseline, and establishment of a hydrant source. The reflection of this procedure is that fire attack follows an order, heat is the greatest threat, and extinguishment is the priority. This commitment and order to extinguishment is an example of sequential tactical operations. To clarify the objective of this section of the text, sequential tactical operations are encouraged, and a tactic is often initiated at the task level then supported with more resources and, when possible, supervised and coordinated by a tactical supervisor. The problem is when the sequential order of a tactical operation unintentionally drives the incident priorities and assignment of incident resources. When the aforementioned order for the tactic of fire attack becomes the driver of the incident, it isn't until the third unit that a second assignment can be initiated.

Today's knowledge and evidence shows that, in a structure fire, the thermal and toxic threats are cumulative, compounding, and dynamic. A greater priority should be placed on the direction of simultaneous threat mitigation through cooling and compartmentalization. The use of the word simultaneous here does not explicitly mean *occurring at the same time*; it means to consider and address them together. The last four major studies from UL FSRI have one common tactical consideration. A tactical consideration is an evidence-based concept for the fire service to consider implementing into their department to enhance efficiency, effectiveness, and increase knowledge to accomplish their mission. From the *Study of the Impact of Fire Attack Utilizing Interior and Exterior Streams on Firefighter Safety and Occupant Survival* (2018), *Analysis of the Coordination of Suppression and Ventilation in Single-Family Homes* (2020),[4] *Analysis of the Coordination of Suppression and Ventilation in Multi-Family Dwellings* (2020)[5] and the "Study of Fire Service Residential Home Size-Up and Search & Rescue Operations" (2022), the recommendation is "when resources permit, interior search and rescue operations can and should proceed simultaneously regardless of the fire attack selected."

As stated, this is not the literal meaning of occurring at the exact same time, as it is understood that arrival order, available staffing, and even state or local regulations are variables. The intent of the language in the tactical consideration and the reason for emphasis is that in many jurisdictions, resources are limited and delayed. The unintended consequence of prioritizing extinguishment alone with procedural sequential assignments is that other tactics are not initiated until that checklist is complete. Given the "first engine is fire attack, second engine secure water supply and establish back-up" scenario, it is possible that search would not be initiated until the arrival of the third apparatus. In some organizations, the first two engines are dedicated to attack, supply, and back-up with the first truck company prioritizing ventilation. In this example, the assignment of search may fall to the fourth-arriving unit. By shifting from the sequential procedure language to "when resources permit, interior search and rescue operations can and should proceed simultaneously, regardless of the fire attack selected," there is more flexibility in the use of early resources and a greater emphasis on getting firefighters into the structure for search operations.

The Fire Department of New York City (FDNY) is the largest fire department in the nation. Due to the size and density of the population served, they successfully execute more rescues and save more civilians from fire every year than any other department in the United States. It is believed by many that the success in the operations of the FDNY is in their size and resource availability. However, most FDNY members will tell you that the majority of fires are extinguished with the first line from the first engine and most victims are rescued by the first truck company initiating primary search. In the situations where the truck company locates the fire before fire attack, prioritizing the act of controlling the door to the fire compartment has the greatest impact in mitigating thermal and toxic exposure to all other areas and occupants, sometimes minutes ahead of extinguishment. The FDNY has very detailed sequential procedures for second-, third- and fourth-arriving units with assignments to reinforce and support tactics, but they all fall in line behind the initiation of fire attack and primary search, never at the expense of either. Current fire service research is challenging the current and past practices of many organizations. With that said, it is also validating some. Many urban departments with a history of responding to high-density residential occupancies have the experience and exposure to situations that demand the prioritization of simultaneously addressing extinguishment and rescue, and their practices reflect this. This has embedded "aggressive command and tactics for life" not only in practice but also department culture.

The Firefighter Rescue Survey

Launched in 2016, the FRS is a web-based survey tool composed of about 40 questions used to collect data points on civilian rescues on the fireground by firefighters. Through the National Fire Incident Reporting System (NFIRS), the National Fire Protection Association (NFPA), and the United States Fire Administration (USFA), the fire service has decades upon decades of fire loss information. The Firefighter Rescue Survey is the only source collecting and reporting fire saves. As of the publication of this text, the Firefighter Rescue Survey has collected the operational details of how, where, and when from over 3,000 fireground civilian rescues, and the information included in it is incredible. For the 3,005 total recorded rescues of civilians on the fireground by firefighters, 1,916 (64%) were still alive at the time of reporting.[6] With so much information flooding in on how fire dynamics and building construction are working against us, the knowledge that firefighters are still consistently saving civilian lives everyday must serve as a beacon of hope for our citizens and a spark of motivation for our firefighters.

Much of this chapter has been dedicated to the demonstration of data from research and science-based evidence, but the Firefighter Rescue Survey demonstrates data from real-world experiences. Anecdotally, we can claim that the FDNY is operationally successful with civilian rescues through the simultaneous actions of dedicating a company to fire attack and dedicating a company to search before any operations are reinforced, but the Firefighter Rescue Survey data set provides an opportunity for tactical actions and victim survivability to be evaluated.

The fact that fuel and compartment conditions are progressively accelerating and intensifying fire conditions cannot be debated. What has been questioned is what this means for fire victims. Around 1960, Dr. R. Adams Cowley, known as the father of trauma medicine, coined the concept of the *golden hour* for victims of significant trauma.[7] This golden hour was rooted in his experience as a physician and the data collected through both civilian and military patient populations, showing that getting a patient to definitive care within 60 minutes of the injury was critical to survival. The supportive documentation of improved outcomes when time to care was reduced through interventions such as improved prehospital care, rapid transport systems like ambulances and helicopters, and even the use of hydraulic rescue tools, were the catalysts for modern shock trauma programs, EMS services, and rescue squads.

Prior to the Firefighter Rescue Survey, the fire service had only collected data on civilian fire fatalities and injuries. There has never been a method to report or evaluate condition, intervention, and patient outcome as a process like the work of Dr. Cowley did for trauma patients and care timelinnes. For this reason, our profession has been unable to show how the changes in fire conditions over time have affected fire victims. With that said, the FRS can show current survival rates and, if the work is continued, it can demonstrate trends in the future (fig. 6–8).

From the FRS First 3,000 data set, we as a fire service have found that the golden hour for civilian fire victims is about 10 minutes. Fire victims located and removed within 10 minutes of arrival have a survival rate of greater than 50%. After 10 minutes, there is a significant decline. Similar to how EMS utilized the work of Dr. Cowley to support implementation and expansion of interventions that expedite access to definitive care for trauma victims, the fire service must use the FRS information to support operations and training which expedite the location, removal, and EMS access for fire victims.

FRS First 3,000 data set

With this 10-minute window of survivability displayed in the FRS data, the UL FSRI tactical consideration that, when resources permit, interior search and rescue operations can and should proceed simultaneously regardless of the fire attack selected, becomes much clearer. Addressing the thermal threat through fire suppression is extremely critical to improving interior conditions, but an active nozzle is a tactical intervention that supports interior access.

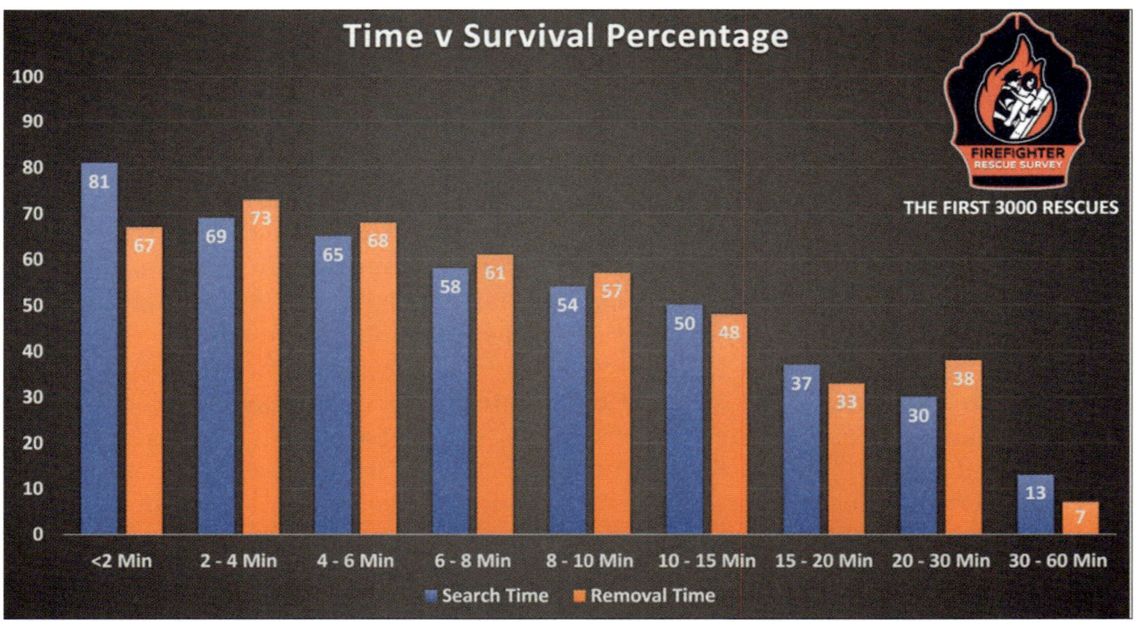

Fig. 6–8. Firefighter Rescue Survey data slide for time versus survival percentage

Search and rescue is the tactical intervention with the primary function of locating, removing, and getting victims to EMS care. This is not exclusionary. Fire attack may, and often does, locate victims before search, and search may, and often does, locate the fire before fire attack. But in both cases, management of a secondary duty will never be as efficient as execution of a primary assignment (fig. 6–9).

When the FRS data is evaluated on the tactical level, it shows that companies assigned to fire attack make 24% of the reported rescues and crews dedicated to the assignment of primary search make 57% of the rescues. Combined, these two initial assignments account for 81% of the civilian rescues on the fireground. If they are occurring together, within the first 10 minutes of arrival, fire victims have the greatest odds of survival.

Knowing these details, reconsider the first-engine fire attack, second-engine water supply and back-up model provided earlier in the chapter. While this supports and reinforces suppression, it does nothing to improve the likelihood of rescue, because the only crew operating on the interior of those two resources is the single fire attack crew (24%). When sequential order of the tactic of suppression becomes the driver of the incident priorities and resources, primary search (57%) is relegated behind fire attack support and redundancy and by default will be delayed. The increasing time to assignment increases our time to intervention and cuts into our golden 10 minutes.

UL FSRI has a very powerful statement in the online training program for "Search and Rescue Tactics in Single-Family Single-Story Residential Structures" which absolutely clarifies that the greatest threat to the unprotected civilian, operating firefighters, and damage of property is the passage of time. They conclude that, given all the details of the data they collect, speed in the execution of operations and improving time to task is the most powerful intervention. "Speed is not recklessly moving forward, speed is seeking the shortest amount of time you can make an impact. Speed is taking ownership of the time you can control".[8] Every second by every firefighter and officer counts. If everyone is operating with minimal wasted time and effort, then we outpace the fireground and can save a life.

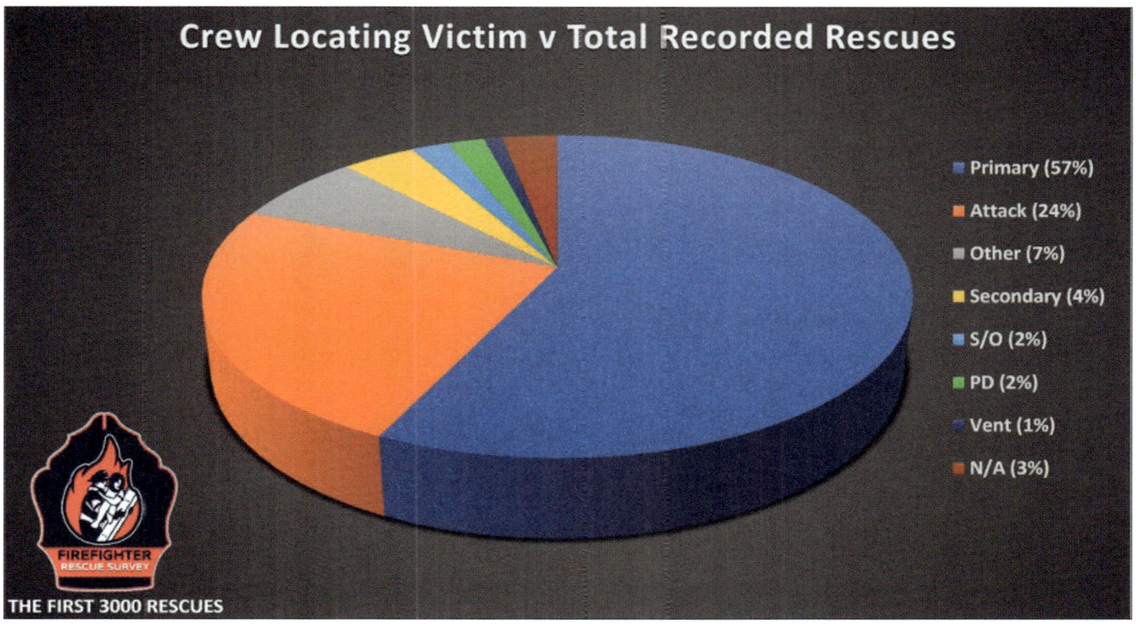

Fig. 6–9. Firefighter Rescue Survey data slide crew locating victim

It's About Life—The Rescue Mindset

As an officer, you may be required to make life-and-death-decisions, with limited information, in a compressed timeframe. You will rarely have all the information, or all the time required to fully process a decision or handle a situation when lives hang in the balance, but you must be capable of rapidly assessing the situation, prioritizing needs, and executing initial actions of available resources. The only way to perform this successfully is through a very clear organizational expectation and communication of what the priority is. It is unfortunate that, through extensive interaction with departments around the nation, when asked "what is the first priority of the fire service," we find the first point of confusion.

On June 25th 2022, Sacramento Metro Fire E42 responded to a law enforcement assist. An estranged dad held his son at knifepoint and was barricaded in a bathroom of an apartment. While sheriff deputies attempted to negotiate, the dad lit the apartment on fire. The sheriffs evacuated and called in E42 into the scene, who balanced the alarm. The sheriffs and E42 worked together to breach the common wall from the next-door apartment. Knowing that the dad was armed with a knife and possibly a gun, E42 went through the wall, undaunted, to find the dad and child in the bath tub, unconscious from the smoke. They successfully pulled them through the wall and they both lived. This is the epitome of the rescue mindset. Maximum risk was taken in an intelligent way for maximum gain. See chapter 16 for more details and a first-hand account of this incident.

The National Incident Management System (NIMS) is the federally mandated foundation of command expectations both in performance and terminology. NIMS lists emergency incident priorities as "save lives, stabilize the incident and protect property and the environment." [9]The International Fire Service Training Association (IFSTA) Chief Officer and Company Officer textbooks, which are the foundation for many certification classes, lists emergency incident priorities as "life safety, incident stabilization and property conservation."[10] We see in the two most influential sources in the development of organizational and professional command priorities two different priorities: one is life saving and the other is life safety. If the life in both definitions is the unprotected civilian, there is no difference in these priorities. It is when some see the life as that of the unprotected civilians and others define the life as fully protected and trained firefighters where the difference arises.

The USFA and NFPA have been collecting data on fires, fire losses, and civilian and firefighter injuries and fatalities since the late 1970s. When the two most recent and comprehensive reports on civilian fire fatalities and firefighter fire fatalities are compared, there is a divergence which cannot be ignored.

NFPA Research U.S. Fire Service Fatalities at Structure Fires: 1977–2018[11]

- Since 1977 the number of U.S. firefighter deaths at structure fires has dropped by more than 80%. Over the same period, the annual number of fires has declined by 53%.
- In the late 1970s, traumatic deaths for firefighters inside structures occurred at a rate of 2.5 deaths per 100,000 structure fires.
- Since 2015 the overall death rate for firefighters operating inside or on structures has been less than 1 per 100,000 fires.

NFPA Research Firefighter Fatalities in the U.S.—2019[12]

- Fires and explosions claimed the lives of 13 firefighters in 2019, with 10 occurring at structure fires and 3 on wildland fires.
- This is the lowest number of deaths ever reported to this data set and the third time the past four years there have been fewer than 20 fireground deaths.

- This continues a clear downward trend since data has been collected in 1977, when the number of firefighter fireground deaths annually averaged more than 80 per year.

NFPA Research Home Structure Fires in the U.S.—2021[13]

- It appears that the only reduction in civilian fire deaths over the past decades has been due to a reduction of fires rather than the prevention of harm after a fire is reported.
- In 2021 the civilian single-family home fire death rate per 1,000 reported home fires was 8.4. This is a 19% increase since 1980.
- The rate of 9.5 deaths per 1000 reported one or two-family homes was 34% higher in 2021 than in 1980.

NFPA Research Home Structure Fires in the U.S., 2016–2020[14]

- While the number of reported home fires and the number of home fire deaths have been cut in half since 1980, both the death and injury rates for civilians are higher now than they were in 1980.
- The fire service is doing a better job of preventing fires than they are of coping with fires after they start, which is why the fire service needs to understand how these fires happen and how they can be prevented and mitigated.

To summarize this information is to conclude that more than 4 decades of data shows us that, when compared to 40 years ago, there has been a significant reduction in the number of fires. When fires do occur, statistically, firefighter lives are safer than they ever have been, and civilian lives are at a greater risk than they ever have been. While we have made tremendous and critical advances in firefighter safety, we must pay equal attention to the mission of saving civilian lives, and this is where the rescue mindset is developed.

This does *not* mean that we take unnecessary risk, or that we dive into a fire with little to no chance of finding a civilian while having a high chance of getting hurt. It *does* mean looking at risk and gain in a new way that maximizes both firefighter safety and civilian survival. These are not mutually exclusive objectives. With that said, it is very clear that making things safer on the fireground for firefighters to operate does not equate to making things safer for the unprotected civilian. It has only been since the publication of the 2018 UL FSRI *Study of the Impact of Fire Attack Utilizing Interior and Exterior Streams on Firefighter Safety and Occupant Survival* and the 2022 release of the First 3000 data set from the Firefighter Rescue Survey that our profession has had access to data on the relationship between fire conditions and occupant and firefighter survivability. We now know that when the mission is to save the lives of unprotected civilians and execute operations with speed and prioritize those tactics that can improve survivability for occupants, the higher standard will inherently mean safer conditions for firefighters.

Aggressive Command and Tactics for Life

To this point, a solid foundation of fire dynamics has been established to clarify that our fire problem is the compounding effects of the toxic and thermal exposure threat. It has been demonstrated that this environment demands simultaneously addressing extinguishment and search with the initial responding resources. The transition from the initiation of the tasks of fire attack and search by first arriving officers to supporting, supervising, and coordinating tactics with the mission of saving lives by the IC is where aggressive command and tactics for life is defined. Nothing is free on the fireground; there will always be opportunity costs. For every task or tactic you employ, time and resources are paid and not devoted to another objective. Every decision, action, task, and tactic must be as orchestrated and coordinated as possible to maximize the impact of those exposure threats.

This goes beyond the fireground: checking the rigs at shift change, not after breakfast, ensuring your portable radio batteries are fresh, confirming the radios are on the right channels, having forcible entry tools ready, ensuring your PPE is set up, and countless other details are the sum of success or failure. As you approach the scene, is careful consideration being given to apparatus placement of trucks, engines, medic units, and rescue companies? Are we blocking medic units from rapid egress? Is a charged 5-inch supply line going to block the transport of an unconscious fire victim? Once on scene, is every firefighter operating at maximum performance regarding SCBA, stretching line, forcing doors, and cutting holes? Are they moving at a fireground pace (moving at a light jog, with purpose), or are they lagging because of lack of fitness, confidence, or leadership? Incident commanders must be active in considering and evaluating all the ways to impact these prior to and on the scene. As UL FSRI so perfectly stated it, "speed is taking ownership of the time you can control."

Known Rescues and The Rescue Cycle

As presented in the section on simultaneous versus sequential, the fire service is very clear and at times overly committed to the sequence of suppression given a known fire. The order is typically initiate attack with initial supply, back up attack with additional line, and secure a continuous supply. If we agree that saving lives is the priority of fire service operations, we should work towards a similar and clear industry standard for the rescue. When training scenarios set a rescue as a single firefighter locating a rigid plastic dummy, dragging them out by a handle and dropping them in the front with no follow up, they set a commitment to failure. This is the reasoning behind the terminology of cycle and the definition of the rescue mindset. Search is a single assignment, rescue is what we are anticipating as a result, and with that comes the demands of removal, triage, transport, and resumption of search. We would fail everyone if we as incident commanders only made an assignment with no plan to execute the result. This would be the equivalent of stretching a dry hoseline with smoke showing because we weren't sure if the fire would be located.

The other reason to plan for a rescue cycle and prepare for repetition is from the data presented in the 2021 Fireground Civilian Rescue Research Project.[15] During the first quarter of 2021, the Fireground Civilian Rescue Research Project collected and confirmed the rescue of 881 civilians by firefighters from 454 fire incidents. This shows that for fire incidents with a rescue operation, there was an average of almost 2 victims. When the data is further broken down by occupancy type, there is even more valuable pre-incident planning information. There were 293 single family dwelling fires with rescues yielding 461 victims, an average of 1.6 victims per fireground rescue operations. For multi-family dwellings and apartments, there were 161 incidents with rescues yielding 420 victims, an average of 2.6 victims per fireground rescue operation. Whether we are dispatched to a fire with a report of a trapped victim, or a victim is located during a primary search, we must be planning for multiple victims. In the event you are responding to an apartment fire with a report of victims trapped, consideration should be given to preparing for a potential mass causality incident as presented by FDNY Chief Leeb in chapter 15 (fig. 6–10).

Regardless of the number of victims, the rescue operation is not complete until all areas of the structure are cleared and all victims are triaged, treated, and transported or released. The rescue cycle reminds us of this continuous process. The information of a known rescue could be received at any time. Getting this information from dispatch during the response provides

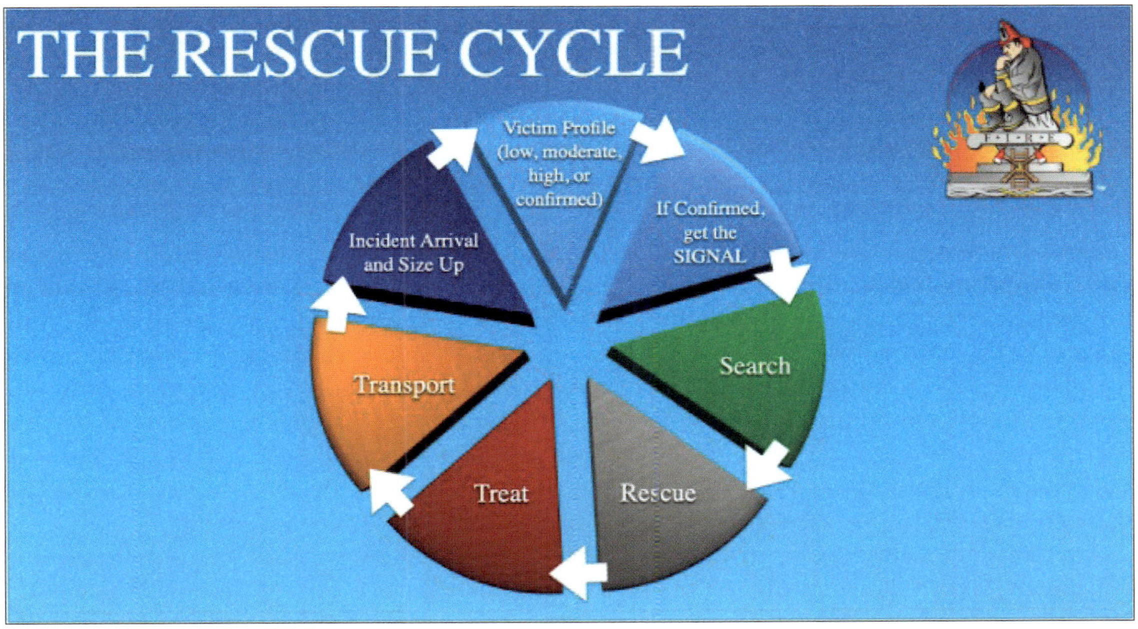

Fig. 6–10. The rescue cycle

an incredible opportunity in reducing reflex time of additional resources by requesting additional fire companies and ambulances to support the rescue cycle early. As the responding IC, the dispatch report can prompt you to anticipate quickly establishing the two functional groups of fire attack and rescue and leveraging tactical supervision to reduce radio traffic and accelerate the pace of problem solving. Information of a known rescue can also come in the middle of a very active scene when a victim is located on primary search or by an advancing hoseline. While the time to react and availability of resources is much different, the process remains the same. The victim will need to be removed, triaged, treated, and possibly transported, and at the same time search will need to be resumed for the potential of additional victims, while crews now occupied with rescue and medical will need to be backfilled.

Tactics For Life

Four critical objectives must be successfully employed to give a victim the greatest chance of survival: fire attack, rescue, ventilation, and medical. While all are important, they will be initiated based upon conditions present (fire, building, SIGNAL, etc.) and resources available. This means not checking boxes for tertiary objectives like utilities, exposures, rapid intervention crew (RIC), or even a hydrant connection. If the first two companies have enough water to rapidly knock down the fire and search, then perhaps a hydrant connection can be completed by the third engine. If you have a 3-person truck, securing utilities during a known rescue may take that firefighter away from the higher priority. We cannot say it enough. It all starts with FPODP and VP!

Notice that the term *objectives* is used, not *tactics* or *assignments*. In an ideal setting, every fire department would have the resources available to formally assign companies to each objective. The fact is that many departments do not have this luxury and even the largest departments can't ensure all their resources are arriving fast enough to be deployed simultaneously. Knowing there are four critical objectives on the fireground to support the action of saving lives and knowing resources will at some point be limited, we must start with prioritizing objectives and initiating tasks. Tactics are the coordination of tasks, and strategies are the coordination of tactics. Incident command as an organizational chart is built from the top down, but the fireground scene is most often built up from those initial task-level actions taken.

For the last several decades of the modern incident command model, a great deal of attention has been paid to ensuring tactical coordination, and this is very important at the strategic level. As it has been presented in this text, incident command is not suffering at the strategic level, it is suffering at the tactical level, which is where we have identified the gap. Our profession is more informed than ever on how tasks and tactics impact the survivability of occupants and firefighters. It is incumbent on us as officers that we apply this knowledge. When resources are limited and demands are high, efficiency is the key to success. We have identified the four objectives that are critical to improving survivability: fire attack, rescue, ventilation and medical. The UL FSRI recommends that, "when resources permit, interior search and rescue operations can and should proceed simultaneously, regardless of the fire attack selected." When we do not have enough companies available to give tactical assignments, we still initiate tasks with individuals because speed is critical and "speed is seeking the shortest amount of time you can make an impact. Speed is taking ownership of the time you can control."[16]

Tactical coordination is very important as the scene develops, but task interaction is a critical concept when determining initial actions and ensuring efficiency of resources. The 2018 UL FSRI *Study of the Impact of Fire Attack Utilizing Interior and Exterior Streams on Firefighter Safety and Occupant Survival*[17] was the most in-depth study and scientific analysis of fire attack ever conducted. The study highlighted that even though the fire service has almost 200 years of experienced with pressurized fire attack for extinguishment, our understanding of water application as an intervention was limited. In 2021 the NFPA published the first edition of *NFPA 1700: Guide for Structural Firefighting*, which was prepared by the Technical Committee on Fundamentals of Fire Control Within a Structure Utilizing Fire Dynamics. Within the guide is an entire chapter titled Tactical Considerations for Fire Control and Extinguishment. This language is very important, indicating that extinguishment is the ultimate goal but fire control is the initial priority.

The following picture shows water application at the start of a hallway that leads to a well-involved bedroom (fig. 6–11). Ceiling temperatures in yellow with patches of red are in the 600- to 900-degree range. The light grey areas are around 200 to 300 degrees and black is less than 100. NFPA 1700 refers to this step of fire control as interior advancement. For years it was taught that water should not be applied inside the structure unless flames were visible, because there was concern that this would delay extinguishment and potentially cause water damage

Today, we know that the faster we address the thermal and toxic threat environment inside a structure, the faster survivability potential is improved for unprotected civilians and firefighters. This thermal image is a powerful way to show how just one fraction of a second of a

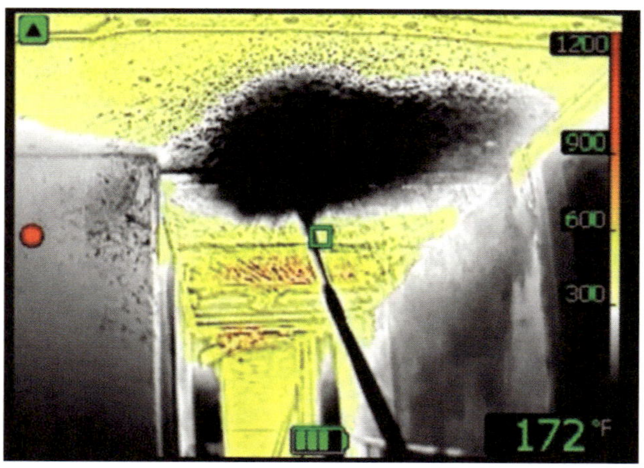

Fig. 6–11. Thermal imager view of solid stream water mapping a hallway

fire stream of adequate volume and proper placement can immediately reduce temperatures. What are not seen in this picture but which are well articulated in NFPA 1700 are the accessory benefits to this.

- Firefighters should avoid advancing under a superheated thermal layer without cooling as they advance.
- Utilizing the reach of the stream to wet ceiling surfaces forward of the crew position means a safer interior progression to the seat of the fire.
- The flow rate should be appropriate with heat release rates and areas of involvement.
- The primary goal is to cool hot compartment surfaces, which allows those surfaces to absorb more thermal energy.
- Water cooling surfaces interferes with the pyrolysis process, which halts flaming combustion, prevents fuel conversion and contribution, and reduces the heat release rate.
- As surfaces are cooled, deflected water droplets cool hot gases, resulting in gas constriction, dilution, and lift within the compartment, reducing flammability and radiation.

NFPA 1700 Chapter 10: Tactical Considerations for Fire Control and Extinguishment

A fire stream does more than put out fire, it improves survivability and prevents fire extension as soon as it begins to address the interior environment. A crew advancing a hoseline can also close doors that it passes on the way through the occupancy to limit extension and exposure. At the simplest, most efficient level, in the absence of any other resources, the task of opening a nozzle, if operated by an educated and trained firefighter, can address fire attack (cooling of atmosphere, coating interior exposure surfaces, and extinguishing fire) and to some extent ventilation (gas constriction and lift, introduction of fresh air) at the ground level. While this is a powerful concept, the adage that "if you put the fire out everything else gets better" is only a half truth. While the open nozzle can address fire attack and ventilation, it cannot remove occupants (fig. 6–12).

This illustration provides a context for what could be expected of the conditions at the end of a hallway like the one in the thermal imager picture. These numbers were documented prior to suppression operations. Like the nozzle discussion, the action of rescuing a victim, when performed by a by an educated firefighter, can have a significant effect on the survivability for the occupant. The data from the "Study of Fire Service Residential Home Size-Up and Search & Rescue Operations" (2022) found that, in some of the experiments, when a victim was located on the floor and their airway was raised to the 3-foot level for the duration of removal, they

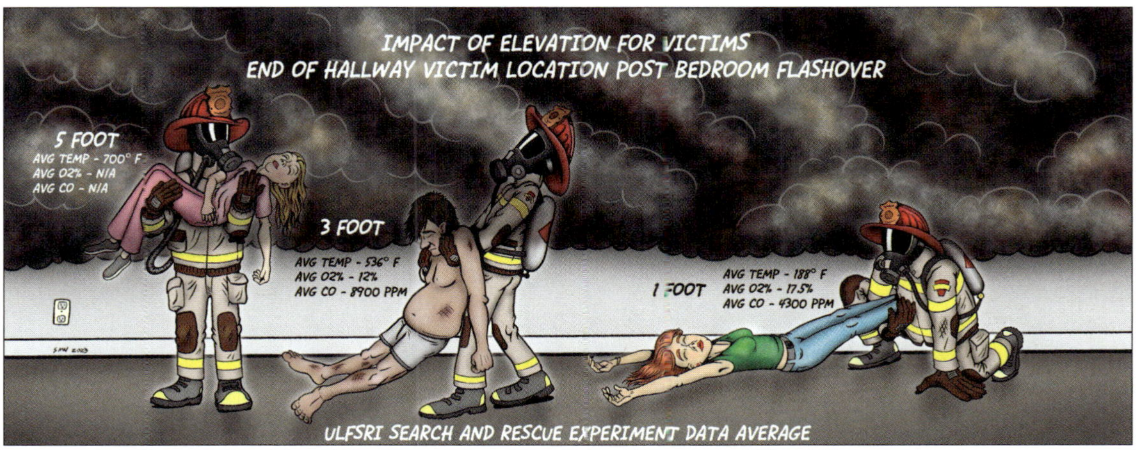

Fig. 6–12. Illustration represents data from an end-of-hallway measurement average from the Residential Search and Rescue Study.

would have been in a lethal atmosphere and worse-off than had they been left in place during suppression, whereas the data for the victim located on the floor and removed at the 1-foot level improved on the removal.

> *The 1 ft elevation resulted in negative relative FED, or lower toxic gas exposure compared to remaining in place. Conversely, the relative toxic FED increased for the occupant in the hallway at the 3 ft elevation as the occupant was removed through the smoke layer.*[18]

FED refers to the fractional effective dose. FED can be used to describe the percentage of the population for which conditions become untenable.

In the simplest form, the message is that if you can't see, the victim can't breathe. It is a terrible thing to consider how many fire victims were potentially compromised not because of the fire, but during the process of their removal from it. It has been stated repeatedly in this text and will continue that we now know better and, with that, we must operate better. Evidence-based practices drive the medical field, and when our mission in the fire service is to save lives, they must also drive ours. The task of victim removal, if performed by an educated and trained firefighter, can address rescue (removal) and medical (reduced exposure). Just as an open nozzle cannot remove a victim, a firefighter dragging a victim cannot cool the environment or get smoke and heat to lift. It takes the work of both to provide the greatest opportunity for survival for victims and firefighters.

It is ideal to dedicate full crews to performing tactical assignments of fire attack, rescue, medical, and ventilation simultaneously. To reduce the UL FSRI recommendation that "when resources permit, interior search and rescue operations can and should proceed simultaneously, regardless of the fire attack selected" to the task level, an initial attack crew of just two members should also plan for and be trained in the need to simultaneously perform fire control and rescue. If a victim is anticipated or encountered by the initial attack crew, to abandon fire attack would abandon protection, and to delay removal would extend exposure (fig. 6–13).

If one firefighter can keep an open and continuous stream high to cool the environment, gases will constrict and lift, and the water mapping and surface coating will prevent fire extension into the immediate area. The lifting and cooling will improve conditions for the victim at

Fig. 6–13. Firefighter initiating fire control high in the atmosphere to cool the environment and coat surfaces with water

the ground level and provide protection for the other firefighter working to remove the victim at the lowest possible level. When resources permit, consideration of tactical coordination is crucial. When resources are limited, consideration of task interaction is critical.

Fire Attack, Rescue, Medical, and Ventilation

Fire attack is the most common first action taken at structure fires. It has been demonstrated that, given a scenario of locating a victim in an entry or hallway, there is an opportunity for fire control and rescue to occur simultaneously with a two-person crew—but this is a contingency plan, not a primary mode of operation. The data from the Firefighter Rescue Survey shows that only 24% of rescues were made by fire attack. This is not due to lack of effort; the primary duty of fire attack is to quickly locate confine and extinguish the fire. Advancing a hoseline is a requirement of this operation. Fire attack members are typically task-saturated, moving hose and addressing the thermal threat. The footprint they cover is primarily limited to the "ways," or the entryway, hallways and stairways to the fire compartment. To search anything else would delay the primary objective of fire control and extinguishment. To understand that fire attack is likely working less than 25% of the occupancy is to understand why they are only making 24% of the rescues.

A dedicated primary search crew is responsible for making 57% of the rescues, documented through the FRS. Unencumbered by a hoseline, with the primary duty of locating and removing victims, primary search should be covering more areas and have a much higher success rate for rescues. Combined, these two initial assignments account for 81% of the civilian rescues on the fireground, once again supporting the UL FSRI recommendation of ensuring they are prioritized as soon as resources permit. Establishing medical early in an incident anticipates the potential for a rescue of a civilian and it also is a provision of safety for all fire department members operating on the scene. Ventilation as an objective may occur at any time and be provided or restricted by any crew, as it is a very important support function of all fireground operations. Once a specific method is determined, decisions as to the extent of support and supervision can be decided. As the initial company officer or IC, your priority is initiating these actions. As the formal incident commander, it is your responsibility to support, supervise, and coordinate them. It is not enough to make assignments; the incident commander must anticipate outcomes by operating in the "then, and what if."

Fire Attack

For fire control and extinguishment methodology and tactical considerations, the best resources are the UL FSRI fire stream, water mapping, and coordinated fire attack studies and online UL FSRI Fire Safety Academy Platform, as well as *NFPA 1700: Guide for Structural Fire Fighting.*[19] For the formal incident commander, the objective of fire attack has the most detailed professional expectation and operational definition set by *NFPA 1710: Standard for the Organization and Deployment of Fire Suppression Operations, Emergency Medical Operations and Special Operations to the Public by Career Departments.* NFPA 1710 sets out the baseline performance and support expectations of four occupancy types.

NFPA 1710 Chapter 5—Fire Department Services

Typical Single Family Dwelling Fire: 2,000 square feet without a basement and no exposures

- Effective water application rate of 300 GPM from two handlines (attack and back-up)
- Establishment of an uninterrupted water supply of a minimum of 400 GPM

Typical Apartment Fire: 1200 square foot apartment within a three-story garden-style building

- Effective water application rate of 300 GPM from three handlines (attack,back-up, and exposure)
- Establishment of two uninterrupted water supplies of a minimum of 400 GPM each

Typical Open-Air Strip Shopping Center: 13,000 to 196,000 square feet

- Effective water application rate of 500 GPM from three handlines (attack, back-up, and exposure)
- Establishment of two uninterrupted water supplies of a minimum of 500 GPM.

High Rise: Any building with the highest floor greater than 75 feet above the lowest level of fire department vehicle access.

- Effective water application rate of 500 GPM on the fire floor from two handlines (attack and back-up)
- Effective water application rate of 250 GPM from one handline on the floor above (exposure)
- Establishment of an uninterrupted water supply to the building standpipe/sprinkler connection sufficient to support fire attack operations.

The single-family dwelling fire is our most common incident, and it supplies our greatest index of experience. For the single-family dwelling fire, the NFPA 1710 performance expectation is the only one with a single supply and no exposure line. This is least challenging to meet with a standard first alarm assignment. All other occupancies recommend an additional handline be deployed for the exposure potential, and the establishment of two water supplies to the scene. The ultimate responsibility of these expectations being met on the fireground is that of the IC. The professional incident commander is always in the game and anticipating needs in all areas, trying to stay ahead of the incident power curve. The intent of NFPA 1710 is to provide you a best practices model. Here is an NFPA 1710 "then, what if" example for the IC and the tactic of fire attack for a simulated apartment fire (fig. 6–14):

I am going to ensure fire attack, search, an exposure/ back up line are initiated for the fire unit 101.

Then I am going to have forth arriving Engine 4 lay in a second supply line and stretch an exposure line to Unit 201.

Next, if Engine 4 reports fire in 201 I will make them Division 2 Supervisor and start feeding them resources from the second alarm.

Fig. 6–14. First-floor apartment fire with second-floor exposures *Source.* Photo courtesy of Alameda County Fire

This is a practiced Plan A exercise which develops experience in communication, meeting NFPA 1710 best practices, and anticipation of proactively managing span of control; this is not the only plan.

Regarding span of control and fire attack, additional consideration must be given to apparatus operators. The base fire attack expectation for an apartment fire from NFPA 1710 is attack, back-up, and exposure handlines and two supply lines. Ensuring proper flow and accountability of multiple handlines is a hands-on, eyes-on task to be supervised by the apparatus operator. This scenario puts their span of control at 5:1 for hose lines. If you add to this the attention demand of crew communications on the radio and monitoring the apparatus overall, that operator is well beyond a manageable span of control given the pressure of the situation. As the incident commander, this must also be considered when establishing divisions with active fire attack. When possible, dedicate a separate source, apparatus, and operator to that division. If that is not possible, provide assistance to the single attack apparatus operator.

An effective tactic with a confirmed victim profile (VP) and no hydrant close by is to have the second-due engine perform a tank transfer to the first engine. This allows the second engine crew to get to the scene faster to assist with rescue and medical as needed, rather than laying in or out and potentially taking more of the second engine's crew away from the labor-intensive scene. This gets much-needed firefighters and water to the scene faster. The caveat is that this is a one- to two-tank knockdown. Obviously, with heavy fire, a patent water supply is important. Again, as we have said throughout the text, it all starts with size-up and the VP.

Search and Rescue

At the time of publication, the body of research from UL FSRI into search-and-rescue techniques and tactics is limited to a single study. However, the data in it is significant, as past collection on victim and firefighter survivability in structures is essentially nonexistent. In an open floor plan, proximity to the fire and elevation within the structure are the most critical factors in the severity of thermal and toxic threat exposure. A victim on the floor in the hallway just outside of the fire room may have a lesser thermal and toxic exposure than a victim on the top bunk of a bed in a more remote room with an open door to that same hallway, as was shown in the 1-foot, 3-foot, and 5-foot points in figure 6–12. The power of compartmentalization has been proven in study after study from UL FSRI, and is reflected in the Firefighter Rescue Survey data, which shows that victims located behind a closed door had an 82% survival rate. This information is challenging some traditional approaches, and when the tactical considerations from UL FSRI and the Firefighter Rescue Survey are combined, they can be applied to improving and supporting fireground search operations.

"Study of Fire Service Residential Home Size-Up and Search & Rescue Operations" (2022) Tactical Consideration 5.1 has been discussed at length in this chapter already. When presented side by side with the Firefighter Rescue Survey data, the need to quickly initiate both fire attack and search is evident (fig. 6–15).

Tactical Consideration 5.2 uses the terminology *window-initiated search* in place of what most in the fire service call *vent-enter-search*. Throughout many of the UL FSRI search-and-rescue experiments, the speed in which areas of the structure were accessed and victims removed or moved to an area of refuge by firefighters initiating searches from windows, as compared to the front door, was significant. The data from the FRS is demonstrating that the UL FSRI finding is consistent with fireground outcomes. Whether the victim is located in the first room entered from the window, the hallway, or the bedroom, or remote from the initial entry, both had a similar survival rate. The key takeaway here is that when conditions allow for a search to be initiated from multiple points or be continued beyond the initial room of entry, it reduces time to task, instead of entering, isolating, and searching single rooms one by one from the exterior at the cost of relocation time (fig. 6–16).

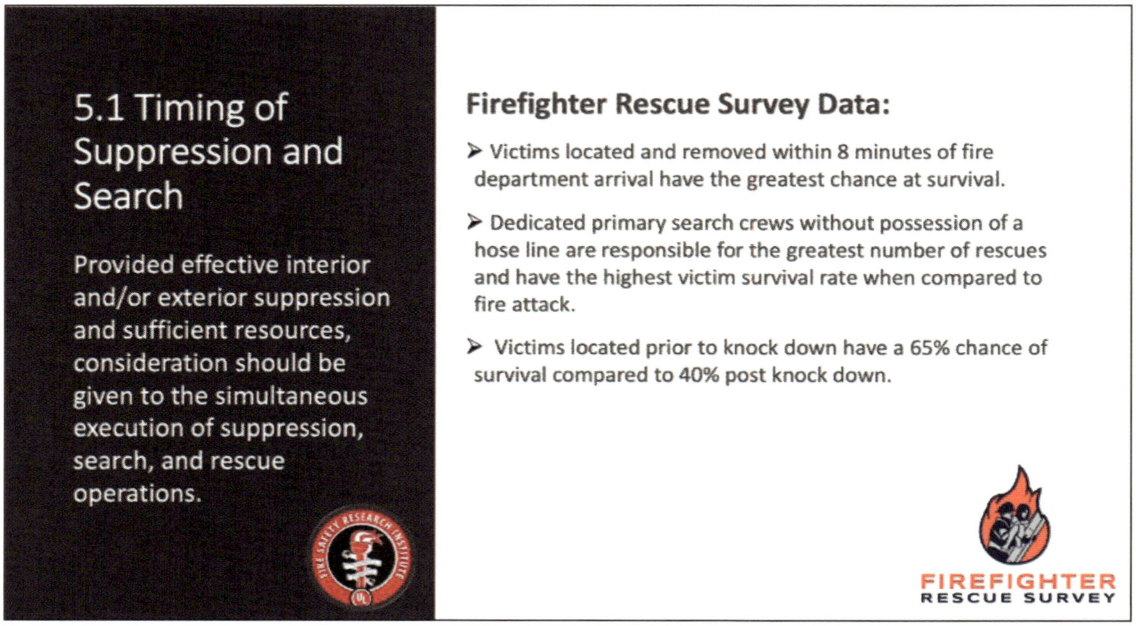

Fig. 6–15. UL FSRI Search Tactical Consideration 5.1

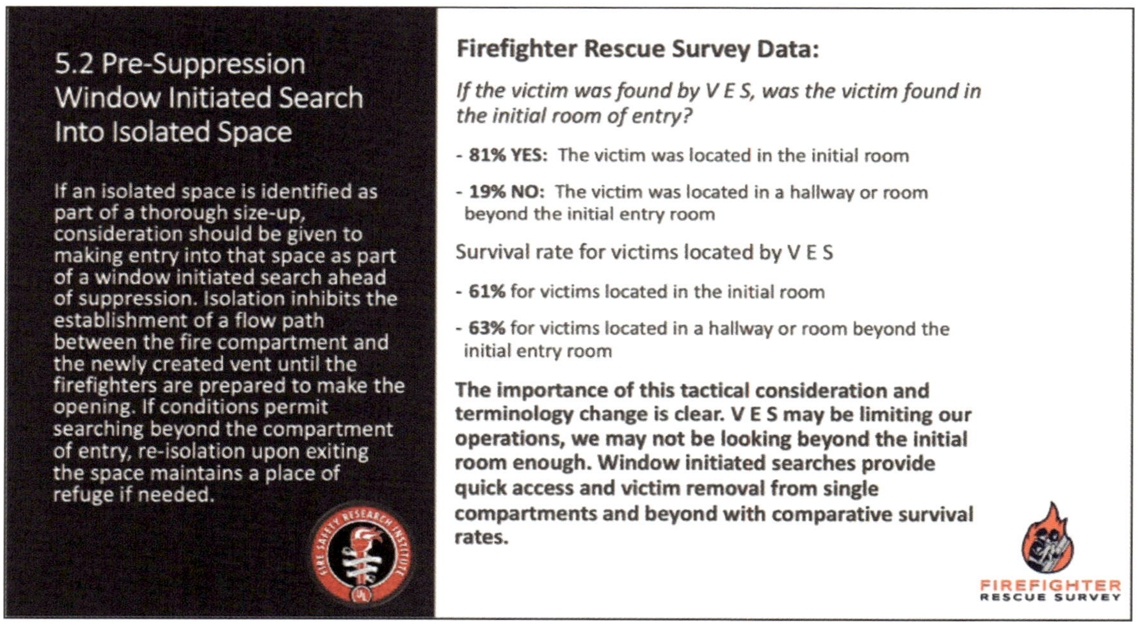

Fig. 6–16. UL FSRI Search Tactical Consideration 5.2

Tactical Considerations 5.7 and 5.8 speak specifically to the rescue/removal of the victim. The elevation of a victim in the atmosphere is very important to their exposure. In the illustration, which uses the pre-suppression hallway data points from the search and rescue study, we see that the victim was safer remaining in place when compared to being removed with their airway at the 3-foot level. This will translate to training firefighters that, when conditions are significant, dragging patients to an area of refuge is safer, and compartmentalization may be better than picking them up and running them out the front door. When searches of upper floors are being conducted and the Firefighter Rescue Survey data is considered, even if a search crew entered through the front door and ascended the stairs to conduct the search, laddering the second floor to provide a faster and removal to fresh air for any victims improves survivability (fig. 6–17).

Tactical consideration 5.9 in figure 6–18 will be one of the most difficult for firefighters to consider, as our profession is one that defaults to action. Given the data presented on fire dynamics and the compounding thermal and toxic threat, isolating and protecting victims in place until the fire is controlled and the interior is ventilated may be the best tactic to ensure survivability. While the sample size is small for victims protected in place from the Firefighter Rescue Survey (6%), the survivability rate is higher. This tactical consideration is a reminder

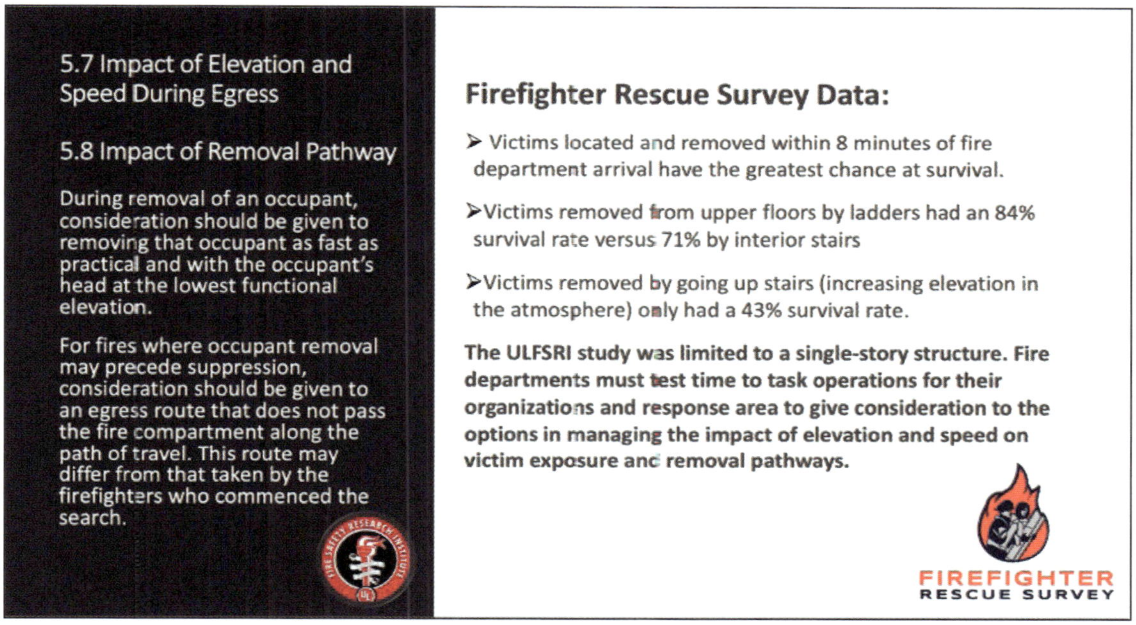

Fig. 6–17. UL FSRI Search Tactical Consideration 5.7 and 5.8

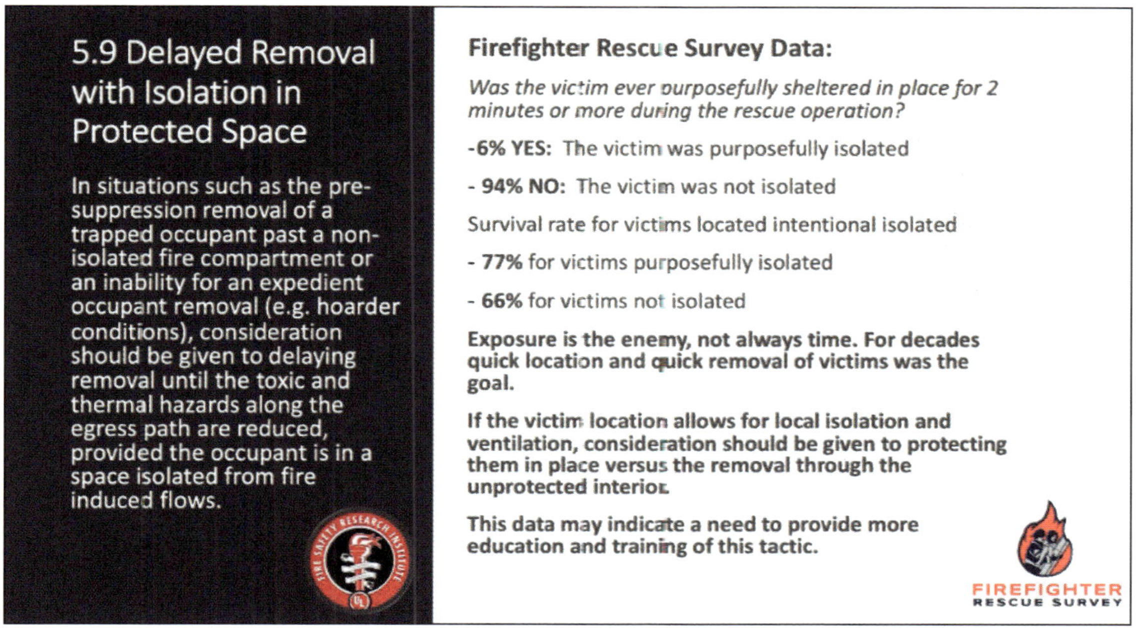

Fig. 6–18. UL FSRI Search Tactical Consideration 5.9

that today the greatest threat on the fireground is exposure, and in many cases, isolation is the most effective intervention against it.

From tactical considerations to command expectations, as with fire attack, NFPA 1710 also has deployment recommendations for search and rescue for each of the four occupancy types.

NFPA 1710 Chapter 5—Fire Department Services

Typical Single Family Dwelling Fire: 2,000 square feet without a basement and no exposures

- Provision of at least one victim search-and-rescue team with each such team consisting of a minimum of two members.

Typical Apartment Fire: 1200-square-foot apartment within a three-story garden-style building

- Provision of at least two victim search-and-rescue teams with each such team consisting of a minimum of two members.

Typical Open-Air Strip Shopping Center: 13,000 to 196,000 square feet

- Provision of at least two victim search-and-rescue teams with each such team consisting of a minimum of two members.

High Rise: Any building with the highest floor greater than 75 feet above the lowest level of fire department vehicle access.

- Provision of at least two victim search-and-rescue teams with each such team consisting of a minimum of two members.
- Provision of at least two evacuation management teams to assist and direct building occupants with evacuation or sheltering actions, with each team consisting of a minimum of two members.

Like the deployment to support fire attack, NFPA 1710 is recommending the deployment of multiple search teams for all occupancy types beyond the single-family dwelling (fig. 6–19).

The critical component to the success of multiple search teams is tactical supervision for the crew(s) performing the searches in the form of a rescue group supervisor (RGS). Tactical supervision for multiple search operations or multiple search points like a *pincer search*, where crews are entering multiple locations for known victims in unknown locations, is critical. In this case, the RGS must coordinate laterally both to ensure fire attack is underway to protect

Fig. 6–19. Search group supervisor coordinates window-initiated search

crews so overventilation does not increase fire activity, and that medical is standing by to receive victims, once found. Either through a division supervisor of a geographic area or a group supervisor with a focus on managing all searches, a tactical supervisor for search operations must be anticipating the outcome of search to be a victim located and a rescue. For this reason, regardless of known or unknown victim status, the group supervisor of search operations should be called rescue group supervisor. In the event of a rescue, the rescue cycle begins, and the RGS works directly/laterally with medical for a rapid patient transfer, as well as command, to get additional companies to assist with the removal or resume search operations where they were interrupted.

An RGS may even be coordinating the support resources of heavy rescue or truck companies to assist with removal access, such as ground ladders or *window-to-door conversions*, where a window is made into a door for victim removal. Another tactic would be *breach-enter-search* (BES), as described previously when Sacramento Metro Fire E42 went through a common wall to rescue two victims. This is essentially vent-enter-search through a wall. VEIS is not possible on upper floors in many cases, so BES is an alternative option (fig. 6–20 and fig. 6–21).

Medical

Fires burn but smoke kills. The U.S. Fire Administration estimates that approximately 80% of fire-related fatalities are the result of smoke inhalation.[20] The combination of synthetic materials and incomplete combustion is a one-two punch for dirty and toxic smoke. Carbon monoxide (CO) and hydrogen cyanide (HCN) are the two most common and deadly fire gases. The hemoglobin in our blood cells has an affinity for these two gasses that is over 200 times greater than that of oxygen, so when these gases are available in the air, they will quickly become a systemic problem for the victim. While the need to be prepared to treat burn patients at a fire is still a priority, the root cause of the unconscious or cardiac arrest fire victim is most likely to be the result of the chemical asphyxiation caused by CO or HCN. Limiting exposure through rapid removal to fresh air or isolation behind a closed door to limit the dose is in the scope of the firefighter; triage, treatment, and transport of these victims is medical. The compounding effects of CO and HCN poisoning should prompt an over triage by EMS personnel anticipating

Figs. 6–20 and 6–21. Window area is cut down to the floor and flapped out to create a door-size opening for improved victim removal

the potential for initial presentation to decline. Definitive care for these patients may include the need for hyperbaric treatment, and a transportation destination with this capability should be considered.

An emerging treatment for hydrogen cyanide poisoning is an antidote called hydroxocobalamin, or more commonly the trade name CyanoKit. Hydroxocobalamin detoxifies the cyanide and converts it into a form of vitamin B12, removing it from the hemoglobin and allowing for oxygen to attach again. The data from the Firefighter Rescue Survey shows that the difference in survivability for a standard smoke inhalation patient is only 3%, but in the unconscious, unresponsive, pre-arrest patient, the survival rate is 14% higher for those who received the CyanoKit and 9% higher for those patients with associated severe burns (fig. 6–22).

Aggressive command and tactics for life must be a continuum. From the fire floor to transfer of care at the emergency room door, proper care and treatment is the space between. Just as fire attack is supported by a back-up line and a continuous water supply, search is supported with on-scene medical and an available transport unit. When NFPA 1710 is referred to for direction on the provision of medical on the fireground by occupancy type, for the single-family dwelling there is no recommendation for initial medical. It does recommend that when the single-family dwelling fire incident escalates beyond the initial alarm assignment, or when there is significant risk due to the magnitude of the incident, the IC shall request an EMS crew. All other occupancies have a clearly defined medical recommendation.

NFPA 1710 Chapter 5—Fire Department Services

Typical Apartment Fire: 1200-square-foot apartment within a three-story garden-style building

- The establishment of an initial medical care component consisting of at least two members capable of providing immediate on-scene emergency medical support, and transport that provides rapid access to civilians or members potentially needing medical treatment.

Typical Open-Air Strip Shopping Center: 13,000 to 196,000 square feet

- The establishment of an initial medical care component consisting of at least two members capable of providing immediate on-scene emergency medical support, and

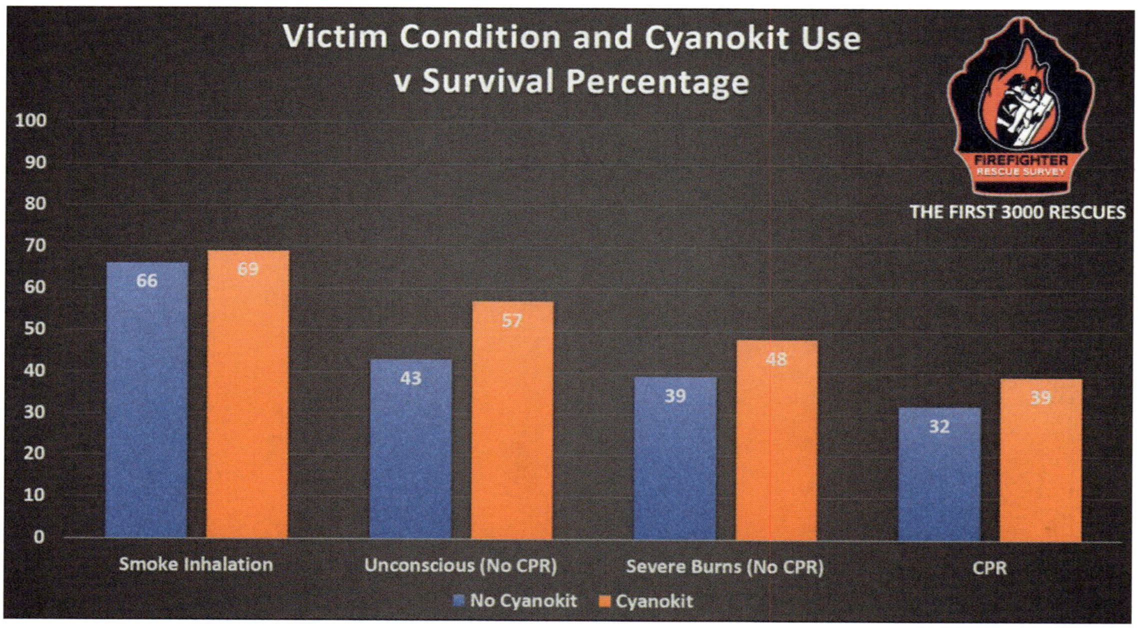

Fig. 6–22. Firefighter Rescue Survey data slide Cyanokit administration and survival percentage

transport that provides rapid access to civilians or members potentially needing medical treatment.

High-Rise: Any building with the highest floor greater than 75 feet above the lowest level of fire department vehicle access

- The establishment of an initial medical care component consisting of at least two crews (four people total), each with one member trained to the advanced life support (ALS) level capable of providing immediate on-scene emergency medical support, and transport that provides rapid access to civilians or members potentially needing medical treatment.
- Provision of a minimum of two members to manage member rehabilitation and at least one of the members to be trained to the ALS level.

In the absence of a recommendation for medical on scene for a single-family dwelling, it would be consistent with the message of this text that a proactive IC remain ahead of the incident power curve and request EMS at the time of dispatch. In the situation of a known victim, it is the recommendation of this text that an ambulance is requested for each known fire victim plus one unit to ensure the scene is not without medical support. From the NFPA 1710 recommendations, the apartment and the open-air shopping center recommend an EMS crew for treatment and transport, but it is non-specific for the level of care. The high-rise recommendation is two medical crews for patient care to the ALS level and a crew to the ALS level for member rehabilitation. At the incident command level, to understand this as professional best practices would prompt early establishment of a medical group supervisor, possibly with the first-arriving medic unit to coordinate directly with division supervisors, base, and staging.

It must be restated that just as fire attack is supported by a back-up line and a continuous water supply, search for occupants is supported with on-scene medical and an available transport unit. Organizations with fire-based EMS services often have these expectations built-in. For organizations reliant on mutual aid or a third-party EMS provider, the fireground in your jurisdiction is your responsibility, and with that the expectations of performance on that scene are your authority. Ensure that operational plans and tactical roles and responsibilities are shared with any entity involved. This includes EMS.

Ventilation

Before the tactic of ventilation is discussed, the objective needs to be clarified. The most common definition of ventilation in fire textbooks is the removal of heated air, smoke, and fire gases from a structure. When the definition of ventilation is reviewed in EMS textbooks, it is described as the exchange of oxygen and carbon dioxide through the movement of air between the lungs and the atmosphere. The fire definition of ventilation supports fire; the EMS definition supports life. Setting a higher standard of operations that support life will by default improve fire operations. The problem that has been identified with the definition that the purpose of venting is to remove smoke is that it lends itself to a more is better approach. Oxygen is fuel for fires, and uncontrolled or over-venting of a fire without coordinated attack will intensify it, increasing the amount of smoke, carbon dioxide, and hydrogen cyanide it produces. The awareness of the effects of ventilation on enclosed fires began to improve for our profession with the 2010 publication of the UL *Impact of Ventilation on Fire behavior in Legacy and Contemporary Residential Construction*.[21] Since this initial study on the impact on fire behavior, UL FSRI has placed a greater emphasis on the coordination of ventilation and the impact of ventilation on occupant survivability. The most detailed information available at the time of this publication in the "Study of Fire Service Residential Home Size-Up and Search & Rescue Operations" (2023) can help inform everyone operating on the fireground on their abilities to vent for life. If a search crew can isolate a room from the fire and open a window to exchange the smoke in the room for cool, fresh air, or a fire attack crew can flow water on the ceiling ahead of their advance, thus cooling the environment, stopping off gassing, reducing temperatures, and getting smoke conditions to lift by increasing oxygen levels at the floor, then they are both providing lifesaving local ventilation.

To look for examples of how focused the fire service is on the action of incident ventilation being centered on the removal of smoke, we need not look any further than NFPA 1710.

NFPA 1710 Chapter 5—Fire Department Services

Typical Single-Family Dwelling Fire: 2,000 square feet without a basement and no exposures

- Provision of at least one team consisting of a minimum of two members, to raise ground ladders and perform ventilation.

Typical Apartment Fire: 1200-square-foot apartment within a three-story garden-style building

- Provision of at least two teams (four people total), to raise ground ladders and perform ventilation.

Typical Open-Air Strip Shopping Center: 13,000 to 196,000 square feet

- Provision of at least two teams (four people total), to raise ground ladders and perform ventilation.

High Rise: Any building with the highest floor greater than 75 feet above the lowest level of fire department vehicle access.

- Provision of an officer and a minimum of three members (four people total) to conduct vertical ventilation operations.

In NFPA 1710 the provision of personnel for ventilation is centered around people and equipment to support vertical ventilation operations.[22] Vertical ventilation is a highly effective ventilation tactic that, when coordinated properly, reduces interior temperatures, provides lift of smoke and gases, and increases oxygen levels at the floor. However, it is not the only tactic. The risk of direction like this in a standard or policy is that when a specific tactic isn't right for the occupancy, fire, or resource availability, the objective is disregarded as well. It once took a mantra that safety is everyone's responsibility to begin to shift the idea of safety on the fireground from being the presence of a safety officer to the execution of safer operations. The fire service may be in a period where a similar message is stressed for ventilation. The objective of ventilation to suppress fire and support life is everyone's responsibility. This may be through the provision or restriction of air movement, and sometimes these opposite acts are occurring at the same time in different locations independently yet meeting the single common objective. Ventilation of a fire room may be coordinated between a company officer on an attack line calling a single outside vent firefighter to take a window. Ventilation may also be intentionally limited by the first firefighter off the truck closing the front door of the house after a quick check for victims to restrict fire growth while the engine is stretching a line. The summary is that making an assignment of ventilation to support the strategy of the incident is secondary to seeking opportunities to improve ventilation in a tactical space. When awareness of the provision or restriction of ventilation and its impact is improved at the task level, it will lead to greater coordination and communication of ventilation operations at the tactical level. Clearer communication and coordination of ventilation at the tactical level improves anticipation of outcomes and needs at the strategic level.

Prioritizing and Initiating Tactics for Life

The following series of pictures is from Colorado Springs Fire Department (CSFD). This is a metro-sized department with a population of about 600,000 people. They serve everything from urban areas to suburban neighborhoods to urban interfaces. Just as discussed with the FDNY before, do not get lost in CSFD resources. Focus on this residence, these incident priorities, and the fact that three of the essential fireground objectives are being executed simultaneously and in coordination by the first two companies. Medical is to be handled in this agency by third party EMS.

This first picture (fig. 6–23) is just after arrival of the first engine and truck. This incident was a fire that started on the exterior in the back of the residence, extended into the house, and was exposing the attic through the eaves. While the engine crew is stretching an attack line across the front yard (fire attack), the truck splits (2 members) will assess forcible entry and prepare to search (rescue). The second split (2 members) is going to the roof to vertically ventilate (vent). You can see how charged the occupancy is with smoke pushing from eaves around the entire structure (high heat highly toxic interior conditions).

Once teams make entry, the interior confirms the fire is inside the occupancy with difficult conditions. There is no time to delay actions. Until ventilation is performed on the exterior, flowing the nozzle into the upper atmosphere to address the ceiling will initiate interior compartment cooling, leading to gas constriction and smoke lift, and supporting attack line advancement and search operations.

On the outside, ladders are thrown and access is made to the area suspected to be closest to the fire. This was determined by smoke conditions of the eaves and of the roof vents near the ridgeline. Access to the cutting area is done utilizing the hip of the roof for stability and the hook to provide solid sounding of the roof decking. Once at the ridge, a roof survey can be quickly visualized to confirm fire location and roof stability.

The attic space has been heated by the fire in the occupied space and early extension in the eaves. Once the position is chosen, the roof decking is cut and louvered. That high heat wants to go up and the hot, expanded, high-pressure gasses seek the lower atmospheric pressure exterior. This powerful pair of fire dynamics is reflected in the intensity of the pressurized smoke out of the opening. If the operation is stopped here, only the attic space is relieved and there is no improvement or support provided to interior companies in through lift or temperature reduction (fig. 6–24).

Fig. 6–23. Arrival conditions with engine company stretching and truck company splitting

Fig. 6–24. Decking is louvered and ceiling is dropped.

The sounding firefighter drops to a knee, knocking in the lid of the occupancy, now providing a clear and direct vertical path for the interior of the occupied space to the lower-pressure exterior. When the higher heat and dirtier incomplete combustion smoke from the fire compartment makes the exterior, the smoke quickly ignites. This *vent-point ignition* can occur in any place where the combination of heat and fuel in the smoke meets an available oxygen concentration to support flaming combustion. Note that a dominant thermal column develops as a result of the intensifying fire over the pressure-driven smoke prior, but increased fire activity and energy is largely channeled up and out (fig. 6–25).

You can see in this picture that the overall presentation of the structure is cleaned up and you have free burning out the vent hole, lifting smoke conditions (fig. 6–26). There is no longer pressurized black smoke pushing from the compartment (decreased interior pressure and interrupted convection currents), and the front door is now a complete intake of fresh air (increasing interior oxygen percentage and decreasing temperatures). As seen in figure 6–26, less than a minute after the hole is cut, the truck company vent split is on the ground.

On the interior, the engine crew now has very clear access to the fire compartment for extinguishment, and from below they can hit the fire in the attic above them. All the area behind the nozzle is protected for the truck company search split, and the inflow of fresh air from the door is improving survivability for any occupants and visibility for operating firefighters.

Post-knockdown, the steam created through the heating of water into gas is now observed following the same channel up and out of the vent opening (fig. 6–27). This is a real-world example that textbook execution can occur outside of the textbooks. The take-home here is that while CSFD potentially has a bigger first-alarm assignment, the prioritization of the four fireground objectives and initiation of specific actions by eight people made an incredible impact, through leveraging the time we can control in training, preparation and education.

Conclusion

While our nation is suffering fewer fires, and our operations, education, training, and equipment are making our firefighters safer, when unprotected civilians suffer a fire, they are not suffering any less. We have made great strides in preventing fires and addressing firefighter safety. It is time that we seek to apply those same methods and mission to saving more savable civilian lives, as that is truly the reason we exist. Aggressive command and tactics for life is a thirst for knowledge, passion for training and preparation, anticipation in planning, assertive

Fig. 6–25. Thermal column develops

Fig. 6–26. Overall fire conditions are cleaned up.

Fig. 6–27. Post-control

decisions in command, and proactive actions. To approach the response to fires with a rescue mindset is to set the highest standard of performance. This is achievable only by chance and luck if you do not explicitly make it the goal, define the direction, and put in the work.

The premise of this chapter was to deepen the understanding of the four main objectives and how actions are initiated. It is from this foundation that the direction and organization to support, supervise, and coordinate these actions through an aggressive and decentralized command system is built.

Chapter Review

Review Questions

1. Explain heat transfer and heat release rates.
2. What are some of the complications of vent-limited fires?
3. Describe sequential and simultaneous operations.
4. Explain the differences in statistical trends associated with firefighter and civilian fire fatalities.
5. Explain how the term *fire control* differs from extinguishment.
6. What percentage of fire fatalities are associated with smoke inhalation?
7. What span of control considerations should be applied to apparatus pump operators?
8. Explain some of the NFPA 1710 recommendation differences between the four occupancy types with regard to tactical assignments.

FESHE Strategy and Tactics (CO279) Related Content

The content contained in this chapter provides detailed information to assist with instruction and education for three expected course outcomes of students in the CO279 Strategy and Tactics course.

1. Discussion of fire behavior as it relates to strategy and tactics.
3. Identification of the building construction basics and how they interrelate to pre-fire planning, strategy and tactics.
4. Description of the steps taken during size-up.

The intent of developing chapter 6 was to ensure not only that it aligns with section 1 of the CO279 Strategy and Tactics course outline, Fire Chemistry Terms and Concepts, but that this text provides the most current and detailed resource to support it. At time of publication, all available UL FSRI studies and reports were reviewed for content inclusion. Accurate understanding of the fireground is built on a foundation of an accurate understanding of fire dynamics.

Beyond section 1 of the CO279 Strategy and Tactics course outline, chapter 6 also provides foundational education with regard to sections V Basic Division of Tactics, VI Rescues, VII Exposures, VIII Confinement, and IX Ventilation.

NFPA 1021 Job Performance Requirements

The information in this chapter can be utilized to support training and educational programs associated with the Emergency Services Delivery Fire Officer I JPR 4.6, 4.6.1, 4.6.2, Fire Officer II JPR 5.6, 5.6.1, and Fire Officer III 6.6 and 6.6.1.

Endnotes

1. "Fire Dynamics," National Institute of Standards and Technology (NIST), last modified January 25, 2024, https://www.nist.gov/el/fire-research-division-73300/firegov-fire-service/fire-dynamics.

2. Bill, "Underwriters Laboratories Fire Safety Research Institute (UL FSRI): New Comparison of Natural and Synthetic Home Furnishings," *Firefighter Nation,* October 1st, 2020, https://www.firefighternation.com/prevention/ul-fsri-new-comparison-of-natural-and-synthetic-home-furnishings/#gref.
3. Vytenis Babrauskas, "Heat Release Rate: A Brief Primer," Fire Science and Technology Inc., https://doctorfire.com/heat-release-rate-a-brief-primer/.
4. John Regan, Julie Bryant, and Craig Weinschenk, *Analysis of the Coordination of Suppression and Ventilation in Single-Family Homes* (Underwriters Laboratories, 2020), https://d1gi3fvbl0xj2a.cloudfront.net/public/2021-07/Coord_Tactics_Single_Family_Homes.pdf.
5. Keith Stakes et al., *Analysis of the Coordination of Suppression and Ventilation in Multi-Family Dwellings* (Underwriters Laboratories, 2020), https://d1gi3fvbl0xj2a.cloudfront.net/public/2021-07/Coord_Tactics_Multi_Family.pdf
6. Firefighter Rescue Survey. https://www.firefighterrescuesurvey.com
7. R. A. Cowley, "Resuscitation and Stabilization of Major Multiple Trauma Patients in a Trauma Center Environment," *Clinical Medicine* 83, no. 1 (1976): 16–22.
8. "Search and Rescue Tactics in Single-Family Single-Story Residential Structures," Underwriters Laboratories Fire Safety Research Institute (UL FSRI): Fire Safety Academy Training Course, May 16, 2023, https://training.fsri.org/course/107/search-and-rescue-tactics-in-single-family-single-story-residential-structures.
9. *National Incident Management System*, 3rd ed. (Federal Emergency Management Agency [FEMA], 2017), https://www.fema.gov/sites/default/files/2020-07/fema_nims_doctrine-2017.pdf
10. International Fire Service Training Association (IFSTA), *Fire and Emergency Services Company Officer,* 6th ed. (Stillwater, Oklahoma: Fire Protection Publications, 2019).
11. Shelby Hall, "Fire Loss in the United States," National Fire Protection Association (NFPA), November 1, 2023, https://www.nfpa.org/education-and-research/research/nfpa-research/fire-statistical-reports/fire-loss-in-the-united-states.
12. Hall, "Fire Loss in the United States."
13. Hall, "Fire Loss in the United States."
14. Hall, "Fire Loss in the United States."
15. Brian E. Brush, "2021 Fireground Civilian Rescue Research Project," Firefighter Rescue Survey, 2021, https://www.firefighterrescuesurvey.com/uploads/2/5/3/6/25366566/firegroundcivilianrescueresearchprojpost.pdf.
16. "Search and Rescue Tactics in Single-Family Single-Story Residential Structures," UL FSRI: Fire Safety Academy Training Course.
17. Robin Zevotek, Keith Stakes, and Joseph Willi, *Study of the Impact of Fire Attack Utilizing Interior and Exterior Streams on Firefighter Safety and Occupant Survival* (UL FSRI, 2018), https://d1gi3fvbl0xj2a.cloudfront.net/public/2021-07/DHS2013_Part_III_Full_Scale.pdf.
18. "Study of Fire Service Residential Home Size-Up and Search & Rescue Operations," UL FSRI, May 16, 2022, https://fsri.org/research/study-fire-service-residential-home-size-and-search-rescue-operations.
19. *National Fire Protection Association (NFPA) 1700: Guide for Structural Firefighting* (NFPA, 2021), https://www.nfpa.org/codes-and-standards/1/7/0/1700?l=131.
20. U.S. Fire Administration, *Fire in the United States 2008–2017* (Federal Amergency Management Agency [FEMA], November, 2019), https://www.usfa.fema.gov/downloads/pdf/publications/fius20th.pdf.
21. Steve Kerber, *Impact of Ventilation on Fire behavior in Legacy and Contemporary Residential Construction* (UL FSRI, December 14, 2010), https://d1gi3fvbl0xj2a.cloudfront.net/public/2021-07/DHS_2008_Grant_Report_Final.pdf.

22. *NFPA 1710: Standard for the Organization and Deployment of Fire Suppression Operations, Emergency Medical Operations and Special Operations to the Public by Career Departments* (NFPA, 2020), https://www.nfpa.org/codes-and-standards/1/7/1/1710.
23. Los Angeles City Fire Department, "UL FSRI Home Furnishings Comparison Natural vs. Synthetic" YouTube, October 2, 2020, video, 4:13, https://www.youtube.com/watch?v=87hAnxuh1g8.

7

Incident Planning and Organizing

Fig. 7–0. The incident command post at a multi-alarm incident in Los Angeles (CA)
Source. Photo courtesy of Mike Meadows

Introduction

The incident commander (IC) is responsible for establishing the incident action plan (IAP) (fig. 7–0). This begins with a size-up and then establishing the incident priority, strategy, and tactical objectives to get the troops off on the right foot. As the incident moves forward, the IC must bridge the tactical gap by assigning tactical supervisors in key positions and roles around the incident. Fortunately, much of this can be pre-planned and trained upon with proven tools and techniques to rapidly remain ahead of the incident while ensuring safe, effective, and efficient operations that remain in alignment.

Incident Priorities, Strategy, and Alignment

Command decision-making goes in a top-down order. The IC must *first* determine the incident priority. Regardless of the incident (fire, hazmat, MCI, etc.), there is only one of three priorities: lifesaving, incident stabilization, or protection of property/environment. *Second,* this drives the strategy of offensive, combination, or defensive. Then, the tactical objectives are established which drive the tasks. This top-down order of decision-making is crucial to success. Unfortunately, it is counterintuitive to most firefighters, like swimming upstream. One of our greatest strengths as firefighters is that we are action-oriented, ready to jump into the fray. This is also our Achilles' heel and downfall. We sometimes act first, then think second. This can cause a push-pull between the firefighters and the IC.

The primary role of the IC is developing and executing the IAP to ensure the appropriate priority and strategy are communicated and that the tactics are in alignment with that strategy. When completed (early in the incident), your tactical supervisors will be better positioned to execute and lead the battle, allowing the IC to step back and think ahead. As discussed in chapter 4, while the tactical supervisors are thinking in the "next" mindset, you are thinking further ahead in the "then/what if" mindset. This *is* aggressive command, which allows you to stay ahead of the incident power curve and support your tactical supervisors and task-level crews.

Examples of being in the "then/what if" mindset include, but are not limited to: "Then. . ."

- Checking exposures and floors above for extension and evacuation of occupants.
- Calling additional companies, alarms, or specialty resources like light/air units, law, hazmat, Red Cross, investigator, utility companies, board up crews, and so on with staging areas.
- Requesting additional tactical channels/frequencies from dispatch.
- Considering salvage and overhaul needs after the fire is contained and secondary searches are complete and all clear is declared.
- Adding companies for rotation of crews due to excessive heat or cold.

"What if. . ."

- There is vertical or horizontal extension that is unseen, or into exposures?
- A Mayday is declared?
- An unknown victim(s) is found?
- Weather changes affect fire conditions?
- This goes beyond our control objectives?
- The troops need critical incident stress debriefing?

While the IC arrives to find firefighters jumping into action, they must look at the big picture and determine if what they are doing is right. As a young firefighter, I have often jumped into action, eyes rolled back, knowing subconsciously that I will just keep going until someone orders me to stop. I didn't really think of the tactical implications of what I was doing, I just wanted to get some. One commercial fire, I stepped off the truck and started breaking windows. I thought I was helping, venting. Why wouldn't I want to get smoke out of the building? Turns out it was cold smoke and all my window-breaking did was more damage.

Once on a house fire, I was assigned division A, and my boundaries were the whole house and my resources were all the crews working inside. As I performed my lap, I found a young firefighter on the C side pounding away at a window and about to jump in. No captain in sight. As he masked up, the window was getting ready to flash. I stopped him. There was no indication of a victim in the house, crews were already performing a primary search inside, and he was only doing it for the fun of it. He had just learned VEIS and could not wait to try it out. The problem was this was not the right place or time. Crews were already searching inside, of which he was not aware, and the window was clearly about to flash, which he didn't see. I love aggressive firefighters—not foolish firefighters.

Crews on the roof of a commercial fire were all heads down, cutting for vertical ventilation. This included the captain, who was *not* in the tactical "in the next" mindset. Conditions deteriorated and the officer was not aware as he was head-down, with a trash hook, attempting to create a louver out of the hole. The roof then collapsed, and the three members narrowly escaped tragedy.

Lack of tactical awareness and task saturation that causes tunnel vision is all too common. You must expect this as the IC. A crew can cut a great vent hole, right before the roof comes down. An engine can stretch a line very effectively under a mansard that is about to crush them. The IC must insert tactical supervisors to see what the task-level companies do not see.

As the IC, I've had to switch from offensive to defensive strategies. Crews didn't see the bigger picture. That's normal. I've also remained in an offensive strategy when one of my division supervisors wanted to get everyone off the roof. He didn't realize the construction was older, dimensional lumber, and a rain roof built over an older roof. The roof division saw things he didn't see. From his perspective, it may have looked sketchier. But the roof division saw it from a different perspective. I was the one to tip the balance and decided to remain offensive. This was an outside-in fire. The fire had started outside, went up the C side wall, and was traveling the space between old and new roofs. The inside was clear.

IC stands for incident commander. *IC also stands for incident chiropractor.* The IC must maintain alignment between the strategic, tactical, and task levels. This is critical. Division and group supervisors help maintain this alignment by carrying out the IC's strategy. Their job is to get crews back into alignment, so everyone is working on the same page. This is a combination of keeping the NIOSH 5 from coming into alignment while maintaining alignment with the incident priority, strategy, tactical objectives, and tasks of the operation, both laterally and vertically.

Most firefighters know that an offensive strategy is an aggressive or internal posture. Defensive is a regressive or external posture. This applies to all incidents, regardless of the nature of the operation. Incidents that get out of alignment can lead to potential problems. For example, a strip mall fire may be a combination strategy, with the center unit defensive and exposures offensive. Yet, Engine 10 is working in the defensive space under a roof that may come down quickly. The division or group supervisor should see this and fix it right away. I've been on several fires where crews had to be redirected out of the building or to the offensive area of operations.

When the incident is out of alignment, bad things usually happen. Similar to how the human body suffers numbness and tingling, an incident will start to have issues with risk assessment, accountability, and communications (fig. 7–1). As the alignment gets worse, the human body experiences pain. An incident will then have a sentinel event like a collapse or a Mayday. Finally, the human body could experience paralysis. So too an incident could have firefighter injuries or death.

Offensive versus Defensive Strategies

The *offensive strategy* is the primary forward-moving posture firefighters want to take when responding to an incident, especially a structure fire. As the IC, you must be disciplined to think before you act. Do not allow yourself (and the rest of the alarm) to react and simply jump into the fray before you perform an adequate size-up. Remember, you are leading a team of very aggressive human beings whose natural inclination is to run into battle unless someone restrains them. And restraint is much easier said than done.

As the IC and division supervisor, I have had to withstand the onslaught of criticism, frustration, and outright anger at the declaration of a defensive strategy. You must remove your emotions from the situation or you will become part of the problem. I have even had two captains almost go to blows at the command post and had to remind them that we had bigger fish to fry, like a multi-alarm commercial fire, and they had to put their egos in a box until we could get the fire under control. In the midst of chaos, the IC must have command maturity.

Command maturity is the ability to detach from the emotions of the situations and be calm, rational, focused, thoughtful, systematic, and professional. *Command presence* is the ability to

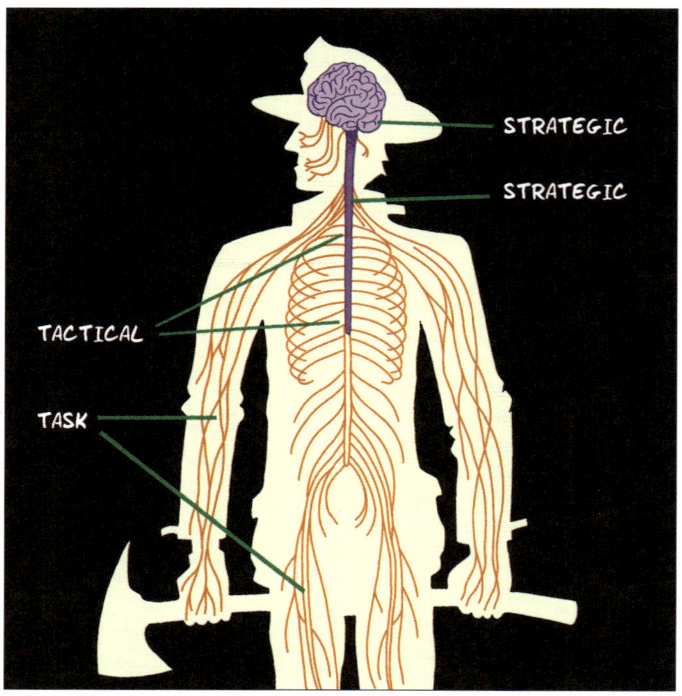

Fig. 7–1. Like the human body, the incident must be in alignment.
Source. Illustration courtesy of Sam Wallace

exude confidence that inspires your troops to follow you. You must have command maturity to have command presence.

An offensive strategy is indicated in the following situations:

- Victim profile can be low, moderate, high, or confirmed. The higher the profile, the more risk can be taken (we know what we have to gain and how much risk we are willing to take).
- Conditions allow interior/forward moving operations that will be *sustained long enough* to accomplish the tactical objectives, without injury or death to the troops (we can stay in the fight long enough to close the deal).
- Human and physical resources match the current and anticipated needs/conditions (we will or already do have enough troops, trucks, and tools to get the job done).

Conversely, a *defensive strategy* is one of retreat to a safer area whereby crews are in an incident stabilization or property/environment priority (fig. 7–2).

A defensive strategy is indicated in the following situations:

- Low victim profile/vacant
- No fire breaks, walls, partitions
- Fire involvement outstrips the available resources
- Heavy, violent smoke conditions
- There is not enough time to complete the objectives of an offensive strategy before conditions may deteriorate. Remember, it takes time to order resources, have them gain access, and then support them long enough to make a difference or complete objectives. This is not a simulation where things happen instantaneously.
- Lightweight construction
- Out of business
- One business without partitioning
- Excess fire load
- Heavy, hazardous materials
- Flammable liquids

Combination Strategy

Many do not know what a *combination* strategy entails. A *combination strategy* has both offensive *and* defensive operations/spaces in the same incident. The incident priority and strategy

Fig. 7–2. A defensive fire with elevated master streams
Source. Photo courtesy of Rick McClure

must be clearly known by all, and they may change during an incident. You can have a lifesaving priority with a combination strategy. For example, you have a commercial building with one section defensive due to collapse, with another area offensive with interior fire attack and search based on a fire wall separating the two.

The combination strategy is the most challenging for commander's intent because there are both offensive and defensive spaces on the same building(s). Here, more strategic direction is required to outline the specific parameters and objectives that are to be employed at the tactical level. As the IC of a large car wash fire at night, I was ready to go defensive as a portion of the roof was about to collapse (fig. 7–3).

This was 2:30 in the morning and an all clear had been declared. We were clearly in an incident stabilization priority. As I notified the division supervisors of my intent to switch to a defensive strategy, one called on the radio, "Chief, we have a fire wall in here. I am about 50 feet in from the D side with a 2½" line, inside at the mouth of the hallway. I think we can save half the building. We are having a positive effect and we will not go past this point." This was critical information from a heads-up experienced captain who was in a tactical mindset. His actions and tactical leadership made it possible to save the mechanical end of the carwash, the adjacent smog shop, and the upstairs office with all the business records. He went in and out of the car wash to ensure that his plan was working and conditions on the outside matched the inside (figs. 7–4a, 7–4b, and 7–4c).

The delineation between the offensive and defensive spaces was the fire wall he discovered. We continued an offensive strategy on one side with him supervising that operation. Toward the B side, we utilized elevated master streams and 2½" defensive lines in the defensive space. The job of the tactical supervisors was to keep the task-level companies from migrating and informed of the combination strategy. We struck a line at the fire wall (fig. 7–5).

The same could apply to a vegetation fire with direct attack and structure defense happening simultaneously. This was the case on a fast-moving vegetation fire in light, flashy fuels on a hot, windy day. The fire involved a large older two-and-a-half-story Victorian home with balloon frame construction. Initially, we attempted to contain the fire in the home while crews also were fighting the grass fire and performing structure defense to protect a neighborhood nearby. As the winds grew and water supply became more challenging, I ordered that the house be defensive. Meanwhile, direct attack and structure defense continued in other parts of the fire.

A combination strategy can also occur during an active shooter MCI where the threat is confined by law enforcement in the gym (defensive), and several patients are in the quad area (offensive).

Fig. 7–3. The car wash fire was about to go defensive.

Fig. 7–4a, 7–4b, and 7–4c. The tactical supervisor going out and back in to compare conditions

Fig. 7–5. Note the elevated master streams to the left (defensive) and interior stream coming out at the fire wall (offensive)

A hazmat combination strategy would be the hot/exclusion zone where isolation and denial of entry is the tactic, and the cold/support zone where occupants are evacuating the exposure. Conversely, you can have an incident stabilization priority after an all clear is declared, yet still be in an offensive strategy during salvage and overhaul inside.

Combination Considerations

- Moderate to high victim profile
- Fire breaks, walls, and partitions can be leveraged for tactical advantage

- Fire involvement versus resources
- There is time to complete objectives in the offensive space before conditions deteriorate
- Heavy timber construction or greater
- Business/building can be easily isolated/segmented

The Incident Power Curve

The incident power curve is the common sequence of events that occurs in most emergency incidents and fires in which the demand upon the IC and tactical supervisors is increased and the potential to become overwhelmed is greater (fig. 7–6). Initially, the activities on a structure fire begin chaotically as companies go to work initiating fire attack, search, forcible entry, and many other tactics. This causes the curve to go up steeply at first. This leads to the first plateau and is often a false plateau (first peak). As the impact of these efforts continues, conditions often improve; the IC will get reports of a primary search completed or fire knocked down, for example. This leads to a lull and the IC may believe that the incident is going according to plan and that it will not escalate (first valley). Then, an unforeseen event happens that drives activity up again to a second plateau (second peak). This can be something relatively benign like a burst hoseline or a hydrant out of service. It can also be more significant like fire in the attic space that was undetected, an unknown victim found, or a Mayday.

An *incident within an incident* (IWI) is a secondary event that happens during emergency operations that escalates demand significant enough to warrant additional resources and incident action planning. Examples would be a firefighter Mayday, roof collapse, or vehicle accident involving a fire department (FD) apparatus or vehicle. An IWI often requires adding tactical division/group supervisors, ordering additional resources, or rapidly switching priorities and strategy. One goal of the IC is to stay ahead of the incident power curve (fig. 7–6).

By decentralizing command to division and group supervisors, the IC can remain in the "then" and "what if?" mindset. The IWI is always looming. If the IC gets sucked into the tasks, the IC is not doing their job and will fall behind.

Command lag is the difference between an IWI occurring and the IC responding to it.

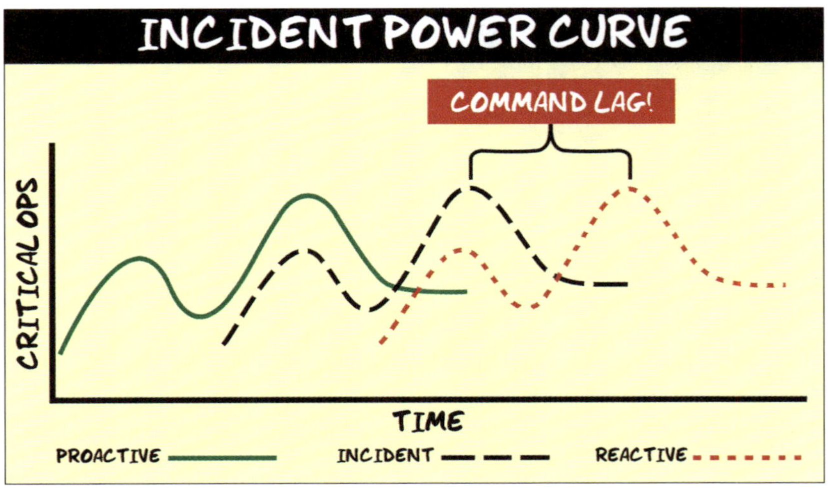

Fig. 7–6. A predictable sequence to assist you in staying ahead of the incident is the incident power curve.

Note. X axis is time; Y axis is bad stuff; Red line is reactive, behind the incident—a bad place to be; Black line is real time—the present; Green line is ahead of the incident—where you want to be

For example: You forecast (through your division/group supervisors) that the roof will collapse, so you go defensive (on the green line). Then the roof comes in on the black line. This kept you off the red line.

The Tactical

As mentioned previously, the tactical space and tactical gap is where the battle is won or lost. Regardless of the IC's plan or leader's intent, no IC can ever truly see what is really going on at all levels and in all places of an incident (fig. 7–7).

The right tactic, at the right time, in the right place is vital. Conversely, the wrong tactic at the wrong time and place can be catastrophic. A task-level crew can cut a great hole in a roof, but if that was not the right tactic, not coordinated, or the roof comes in, things can get worse. During a rescue, you may have a victim who needs to be removed using a window. If the crews initiate search from the front door, the victim may not survive if the search takes too long or the victim is brought out through the immediately dangerous to life or health (IDLH) environment. An effectively stretched 1¾" line may be too small a caliber where a 2½" was needed. A transitional attack may take too long and unknown victims may suffer.

No IC can see the entire incident and ensure that all these things are happening effectively, let alone account for all the crews. Task-level crews do not see what other companies are doing due to tunnel vision, and this can result in redundant or conflicting efforts. At the very least, time and resources will be wasted. At the worst, firefighter or civilian lives will be lost.

Tactics for specific incidents and structure types vary greatly. The vast body of knowledge regarding tactical decision-making fills many volumes of outstanding works by experienced and wise authors like John Norman and Anthony Avillo, to name just two. East coast and west coast tactics are a derivative of common construction features, eras of construction, layout, use, access through city streets, topography, tradition, and a host of countless other factors.

San Francisco and New York City have incredible differences in topography alone that require alternative methods of access, laddering, and fire attack. Topography and vegetation cause significant variances in tactics within the same state San Francisco and Los Angeles will have

Fig. 7–7. The IC will never see the entire incident.

vastly different operations. While San Francisco has very little vegetation, brush, or wildland, Los Angeles County has thousands of areas and communities that are threatened in the wildland urban interface (WUI), where heavier fuels along with the steep hills and mountainsides are commonplace.

The older parts of the United States will have legacy and balloon construction and likely will have dimensional lumber, while the newer cities and communities will likely have more lightweight, modern construction. Communities at elevation will have construction features to accommodate snow and mountain terrain, while parts of the country along the south and eastern seaboard that are susceptible to hurricanes will have unique fortification features of their own.

Each community will have unique construction specific to that area. Row houses like brownstones are common in larger, older cities, while tilt-up construction is more likely found in post-1960s areas. There may also be extremely specific construction based upon the builder, like a McKuen Fourplex, or an Eliot Tri-Level.

The point is to know the specific construction features common to *your* area, and the specific target hazards in your city, battalion, or first-due area. This knowledge and your resource levels drive associated tactics, and pre-planning is priceless.

Every fire department has different staffing levels, standard operating guidelines, response matrices, construction types from different eras, occupancy loads, terrain, weather, congestions, access/egress issues, and a host of other factors. We ask that you consider what is presented with an open mind. You may find that something new can be employed, what you have been doing is verified, or that you are presented with something you have not considered before. The goal is learning, not preaching.

One thing is certain: change is a constant, and we must always challenge our assumptions, try new things, and realize that the enemy and the battlefield are always in flux. The key is to balance experience with academia. Both are important.

UL FSRI studies continually give us new data to make more educated tactical decisions. In addition, Project Mayday and FRS have given us priceless data to ensure that our crews go home in one piece, while saving as many civilians as possible. That said, there are some universal tactical considerations that should be discussed regarding each type of structure fire or other incident. For the IC, the tactics employed by your FD must be well understood so that you can anticipate, choose, support, command, and coordinate them. This includes knowing the kind, type, size, and number of resources available to you. All fire departments, districts, and authorities have specific resources.

At the same time, the IC does not need to keep up on the latest chains used on the chainsaws, or the latest pump pressure cards on the engines, nor does the IC need technical certification in every possible area to be effective. The IC should know the amount of water in your engine water tanks, pump capacity, amount of hose, and reach of the ladders on your truck companies. If your agency does have dedicated truck companies, being assigned to a truck as a company officer prior to promotion to chief officer is certainly advantageous, if not crucial.

The IC straddles the strategic/tactical spaces. *A chief officer IC should not dip into the task level.* This can border on micromanaging and take the IC off the ball for the bigger picture. This will create more radio traffic and slow down the time-to-task interval of getting the work done. The result will be an IC who is behind the curve and the incident will outpace them. If an IWI occurs, the IC is instantly overwhelmed. Once that happens, there is no turning back. You will not catch an incident that has outpaced you.

Time-to-task (TTT) (also known as time-to-completion) is the time it takes to accomplish a task once it is identified as necessary to the incident tactical objectives. This includes communication from the IC to the tactical supervisors and vice versa. For example, crews arrived to a confirmed victim profile on the El Camino Fire. The victim was on the gurney at 4:32 from arrival. This TTT of 4:32 was within the 6–8-minute window to get a surviving fire victim out and to EMS, based upon FRS statistics of over 3000 rescues.

As stated previously, *modern incident command starts well before an incident and is a team sport*! The time and training required to build trust is vital if you are to outpace an incident and save civilian lives. Trust allows the IC to decentralize, thereby minimizing radio traffic, while the tactical supervisors and company officers act with the IC's intent by choosing the appropriate tactics at the right time in the right place. This minimizes radio traffic and TTT, allowing the officer closest to the problem to act quickly.

One way to get all officers on the same page is to train on shared recognition of construction features and tactics, so the intent of the IC is carried out in the field. *Tactical points of opportunity* (TPOs) are building construction, access, layout, and other physical features that can be leveraged to increase the impact of resources to accomplish tactical objectives. Figure 7–8 shows a common lightweight two-story home. Note that over the garage are two-bedroom windows and on the D side is a garage access door, a bedroom window, and a bathroom window. The door and bedroom window are TPOs. Examples include fire walls, breezeways, partitions, stairwells, windows, side doors, roll-up doors, and so on.

Reinforcing a fire wall can keep a fire contained with much less water and resources, for example. Attacking the fire that is close to a commercial roll-up door with large caliber hoselines, instead of attacking from the front and pulling two or three hundred feet, can quickly knock down a commercial fire while minimizing exposure and TTT of companies.

Some firefighters hate acronyms. The fact is that they are a great training tool to build muscle memory. Tactical worksheets are also a useful resource. Checklist firefighting is a derogatory term implying that there is no thinking when using an acronym or checklist. This can happen. Some aspiring (and incumbent) officers will regurgitate the same things no matter what is in front of them. For example, they will set up two-out on a known rescue or have the second engine secure a water supply with a small travel trailer on fire. I have heard IC's direct the second truck to secure utilities on a house fire with a *confirmed* victim profile because that is what the SOG says. Not good. Standard operating guidelines are just that: *guidelines*. They are not standard operating *laws*. That would make them SOLs and then we *would* be SOL! Having said all this, consider the military and aviation models of checklists and redundancies. The *Checklist Manifesto* by Atul Gawande is based upon the principle that "... highly trained, experienced, and capable people regularly make avoidable mistakes."

The acronym RECEO-VS has been around since the 1950s, when Lloyd Layman created it. He also created FPODP. Both are still very useful today and are timeless.

Fig. 7–8. Windows and doors on basic homes provide TPOs.

SAW-CSS-RECEO-VSS

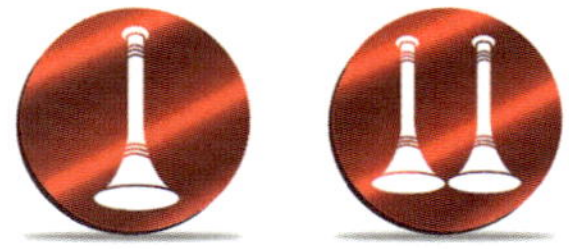

The IC and tactical supervisors must ensure the proper tactics are being employed. We have added front and back-end components to the RECEO-VS algorithm to include additional command considerations to give you a system that encompasses more of all the tactical considerations for structure fires in the context of the IC. Each has universal key points that must be considered. These are benchmarks that the IC and tactical supervisors should share and communicate. The SAW-CSS-RECEO-VSS system has been used nationwide for years. You will notice that italicized words correspond with key points on the tactical worksheet that follows.

Size-Up

All incidents start with size-up. Size-up is so crucial that we have dedicated an entire chapter to it (Chapter 5). *Remember, size-up is a mental, ongoing process.*

- Start with your *FPODP or FIRST* algorithm.
- Consider *7 parts of the structure* (four sides, top, bottom, inside).
- The *construction features* are extremely critical to assess early. Aside from the construction type (commercial, residential, legacy, lightweight, balloon frame, tilt-up, unreinforced masonry [URM], platform, Type 1–5, etc.), consider features such as an unsupported mansard, common attic, cockloft, or facade that could conceal fire or collapse.
- When you first look at the building, try to ignore the fire and smoke and truly focus on the construction type and features. Once you focus on the flames and smoke, you will not see the big picture and likely not notice these construction features. Things like overhead power lines, exposures, fire walls, and many other aspects of the building can be missed once you experience tunnel vision on the fire.
- If possible, it is critical to get a *360°, or lap* around the structure. This may be easier said than done. While standalone residences are relatively easy to lap, row houses, large commercial buildings, and limited access will prevent this luxury. Options include having companies approach from another direction to get a different view and identify key hazards or construction features. Another option is to get a report from the roof division, which may include exposures and hazards in the rear. A *roof report* is a short transmission to the IC from the roof division that includes the *construction, conditions, and load* (CCL) on the roof. More on this in chapter 10 on commercial fires.
- The *reporting party* (RP) is the individual who has reported the incident, typically by calling 9-1-1. A *responsible party* is the individual who either owns or manages the occupancy or property involved. This can be a homeowner, apartment manager, or retail manager at a store, for example. Be sure to contact either the RP or responsible party when possible, as they may have additional information like victim profile, fire location, location of access points, enunciator panels, and utilities, or may have critical information on the fire load and contents of the structure.
- *Assess the victim profile (VP) as soon as possible and report it to all units.* The VP may change from en route to arrival, or after completing a lap or talking to an RP. Keep all units updated. See chapter 5 for more on victim profiling.

- Your *initial size-up* will conclude with an arrival report to incoming units; however, remember that size-up is continuous. As you talk to the RP, bystanders, occupants, employees, and attempt a lap (chiefs should delegate the physical lap to a company officer/tactical supervisors), you will garner more critical information. The *IOCAN* (identify, objects, contents, actions, needs; see chapter 5 for more information) system is designed to give a quick arrival report. Your arrival report is likely dictated by your SOGs. Your agency may require additional details like building height, size, construction/occupancy type, VP, exposures, and many other factors.

Apparatus Placement

Apparatus placement should be considered early, as you typically do not get a second chance. The size and lack of maneuverability, combined with other units arriving, makes staging your apparatus in the right place the first time a daunting task. As supply lines are laid and attack lines stretched, the available real estate around a building can rapidly deteriorate. Plan ahead based upon your type of apparatus, mission, SOGs, order of arrival, and conditions, including how conditions may change.

- A first-arriving engine will likely attempt to see *three sides* of a smaller stand-alone structure, allowing the officer to have a tactical advantage before stepping off the rig. Exposures, access, fire location, and tactical points of opportunity can be identified early.
- A larger building, such as a commercial structure, row houses, strip malls, big boxes, high-rises, or industrial facilities will require a more methodical approach. This can include driving around the structure if subsequent units are delayed, having other units approach from other directions, or directing a unit to stage in the rear and give report on conditions from the C side of the building.
- The old adage that the truck always gets the address is based upon several factors. First, ladders have limited reach and do not have as much flexibility as hoselines that can be extended. An aerial master stream is only as good as the reach, and that cannot happen from two blocks away. Second, truck companies carry specialized equipment that may need to be readily available. And many agencies only have one truck company and want it close, to have a prime location before multiple engines and hoselines eliminate options.
- The *hot zone* is any area that will cause damage to apparatus, including, but not limited to, collapse, thermal insult, excessive smoke, hazardous materials, power lines, and so on. Ensure that apparatus are staged out of the hot zone, including potential changes in conditions. Many aerial master streams have been shut down and moved after conditions changed to a defensive strategy and collapse potential increased.
- As discussed in chapter 4, there are three *staging* levels: Level 1, Level 2 and forward staging. Identify a location for Level 2 staging if you call additional alarms/resources.

Water Supply

Water supply must be considered as soon as possible to take advantage of any hydrants while responding into the scene. A secondary supply must be considered in case the initial hydrant is out of service. Before you race to a fire without a water supply, ensure that the second-due company is close by and multiple hydrants are within reach of one engine. Many over-zealous company officers have bypassed hydrants to beat other companies into the fire, only to find themselves out of water quickly.

- *What* is the required fire flow? You can refer to chapter 6 for the NFPA 1710 recommendations by occupancy type, or some organizations still utilize the National Fire Academy formula: (building length × width)/3 = GPM.

- *Who* is securing the water supply? This will be dictated by your SOGs. Is this being accomplished with one or more companies? Will this require a relay pumping operation and the establishment of a water supply group supervisor (often in rural communities)?
- *Where* is the water supply? Is it reachable by one engine, or is this an extensive lay? Will water tenders be required?
- *What* is the supply? A dry or wet forward lay by the first-due engine from the hydrant to the fire? A wet forward lay by the second-due engine? A driveway/split lay? A reverse lay by the second engine from the first, out to the hydrant? A tank transfer? A water shuttle from a distant hydrant, or drafting operation?
- Once a patent water supply is established, it should be announced on the tactical channel.
- If an *FDC* is present, consider having a separate engine pump the FDC with a separate supply.

Command

Command must be established and announced. *There can only be one IC at a time, and the IC must be on scene to assume command.* While your SOGs will determine specific parameters, some fundamental best practices exist.

- When the first-arriving officer is a chief or staff officer with no tactical capability, assuming command is the most likely option. Developing an IAP and calling additional resources will typically be the best decision. Get the VP and announce it.
- *Command option/mode* determines one of three postures that the first arriving *company* officer is taking regarding incident command: assuming command, passing command (often called fast attack or rescue mode), and investigating. Each has specific considerations and potential triggers. The determination is made by the first-arriving officer based upon whether or not their company can make a significant difference in the incident that will save/protect civilian lives or gain a tactical advantage to prevent significant injury or loss.
    - **Assuming Command** is indicated for the company officer if the incident is so large and complex that developing an IAP, calling additional resources, and directing subsequent units will be more impactful and likely to remain ahead of the incident power curve than performing initial tasks with a single unit.
    - **Passing Command** is the act of notifying subsequent responding units that the initial company officer plans to relinquish command as soon as the next officer arrives and assumes command. This is indicated when the incident demands immediate action, and an appreciable difference can be made to save lives and prevent injury or significant loss. A key point is that the officer passing command is *still in command* until it has been passed to the next officer who has arrived on scene.
    - **Working or Mobile Command** is a term utilized by the initial IC to indicate that they are performing tasks while in command and not in a fixed location. This indicates that incoming radio transmissions should be limited to key information (arrivals, water supplies, etc.) so the IC can give updates and direction to arriving units and perform the transfer of command via radio.
    - **Investigation Mode** is the term utilized to indicate that the first company has no indication of fire or smoke (other than a possible audible alarm) and has chosen to engage in further investigation. Reconnaissance may be indicated with more than one company to further ascertain the situation (see chapter 5). Investigative mode may be accompanied by either retaining command or passing command to the next officer, based upon the situation.

- If you are a company officer arriving first on the scene, *you are in command*. Whether you like it or not, that's a fact. You may choose to *pass command* upon arrival; however, *you are still in command*, and command has not been passed to another officer until *they are on scene and has assumed command from you.*
- Once a chief officer or permanent IC has arrived, the incident *name* and the *ICP location* should be announced. This allows all members to find the IC and sets the template for the incident. For example, "Chief 1 on scene, assuming Main Street Command from Engine 10. The command post is located across the street from the A/B corner. . . "
- The IC should then announce the incident priority, strategy, and tactical objectives. ". . . This is a lifesaving priority, offensive strategy with the objectives of fire attack, search, and vertical ventilation."
- If an adequate description of the scene was given by the first-arriving company officer, then the chief does not need to repeat those conditions, thus saving valuable radio traffic. If, however, the first-arriving company officer did *not* give an adequate arrival report, the chief can clean up the description as part of the subsequent arrival report. For example, companies are dispatched to a house fire. Engine 1 arrives, stating, "E1 on scene. We have heavy fire. Initiating fire attack. Need a water supply. Passing command." Upon arrival of the chief, a subsequent arrival would be, "Chief 1 on scene, assuming Main Street Command from Engine 1. Command post is located across the street from the A side. This is a large 3-story, balloon frame house with heavy fire on the third floor. High victim profile. This is a lifesaving priority, offensive strategy. Objectives are fire attack and search."
- Call *additional resources alarms* early and confirm the staging location.
- Set up your *ICS* early.
- Do you have adequate channels for *communications?*

Strategy

Confirm your priority, strategy and *VP*, apparatus placement, water supply, and your command mode. Priority: *lifesaving, incident stabilization, or property/environment*; and strategy: *offensive, combination, or defensive*. Keep it simple. Remember, they may change as conditions change.

Safety

Safety is listed sixth in this list because you will not know what your safety issues are until you have performed/decided upon the previous five factors. You cannot possibly know what the safety challenges are until you have sized up the incident and determined your priority/strategy. Offensive and defensive strategies have very different considerations, as noted previously.

- *RIC and two-out:* Offensive and combination strategies will likely necessitate *rapid intervention and/or two-out* protection for your members. Known rescue situations eliminate the need for such measures. Fire departments across America fulfill the two-out mandate in countless ways. Your FD SOGs will determine the appropriate level of rapid intervention readiness and/or two-out implementation. That said, consider backup teams at all entry points where fire attack is initiated on a structure fire. These teams are separate from the rapid intervention crew (RIC).
- Rapid intervention is *not* the same as the minimum two-out. RIC was developed out of the Code of Federal Regulations 1910.134 Two-In, Two-Out standard. Stay ahead of the curve and ensure the proper protection for your crews. Your FD has SOGs on rapid intervention, Maydays, two-out, and so on. More on RIC in chapter 12.

- The old adage that safety is everyone's responsibility may be true, yet it creates potential for no one to truly be watching for safety if all members are task-oriented/saturated. One tragic case in point was a double drowning of two two-year-old children. They were at a backyard BBQ and everyone thought someone else was watching them. They managed to ascend a locked ladder to an above ground pool. By the time anyone realized the children were missing, both toddlers were floating face down in the pool.
- A *personnel accountability report* (PAR) is conducted on the radio periodically to ensure crews are safe and in their designated assignments. This can easily be decentralized to tactical supervisors for ongoing accountability. *Active accountability* is the act of tactical division and group supervisors being constantly aware of their crews' locations and assignments. This is done through visual, face-to-face, and radio communication.
- *Utilities* should be secured, and often this can be assigned to a RIC.
- Considerations for structure fire safety include, but are not limited to, the following:
  - Excessive fire load from contents inside the structure
  - Presence of hazardous materials and accelerants
  - High rack storage systems
  - Industrial processing systems
  - Void spaces, holes, open hatches, and compromised flooring
  - Construction type, including multiple re-models (especially against fire codes)
  - Conversion of commercial occupancies into living spaces against code
  - Heavy occupancy due to COVID-19 and economic pressures
  - Collapse potential of facades, unsupported mansards, parapets, and walls
  - Roof collapse of lightweight gusset truss systems, bowstring trusses, open web bar trusses, oriented strand board (OSB) I-beams, and other lightweight construction
  - Overhead power lines, including energized metal fencing
  - Fire above and beneath firefighters working inside the structure
  - Vent point ignition, flashover, backdraft, and smoke explosion
  - Rapid change in fire conditions
  - Increased collapse potential from deteriorated/unstable building construction during salvage and overhaul
  - Propane tanks, stored gasoline, or natural gas supplies exposed
  - Excess load on roofs including HVAC systems, concrete tile, and large signs or antennas
  - Apparatus in the hot zone (see apparatus placement mentioned previously)
  - Extreme weather conditions and dehydrated or overheated firefighters
  - Extremely inexperienced firefighters or officers
  - Highly emotional scene due to firefighter Mayday or excess victims

Rescue

Remember, the victim profile drives *everything*. Lifesaving is the first priority and rescue the first tactical consideration. That said, you may have to get attack lines in place to save the most people, based upon the building construction, layout, and fire location. For example, controlling the fire in an interior stair or hall tends to do the most good for the most people inside. All situations, incidents, and fires are different. Some key points are:

- A low victim profile does *not* mean an all clear. Only *we* determine all clear after searches.
- Use separate companies for *primary* and *secondary* searches. A fresh set of eyes will always see things differently and tend to notice different nuances of a search.
- Primary complete is not the same as a secondary complete all clear declaration. Ensure *all* areas are searched *twice* by different crews before declaring an all clear on the fire.

- *Medical* is part of rescue. Ensure you have advanced life support and transport capability on scene at all times. Always have one transporting ambulance available on scene at working fires for firefighters or unknown victims. Typically, fire victims require an ambulance for each patient due to the critical nature of burn victims. *A good rule of thumb is one ambulance per fire victim, plus one.* Most local EMS delivery models do not have the capacity that may be needed to sustain this level of coverage. Many are not part of the fire department. The key is to train together and ensure you have a communications plan *before* a fire. A working structure fire with civilian victims or firefighters down is *not* the time to find out that you cannot communicate with your EMS provider. Forward stage the ambulance crew with gear and gurney at the ICP so you can have face-to-face communications at all times. Their apparatus should be accessible quickly, and not blocked in by fire apparatus. Again, training is crucial *before* a fire.

Exposures

Typically, *interior* exposures are a higher priority than *exterior* exposures. While you may consider this as *confinement*, it is common to see fire departments attack exterior exposures when there is still a significant amount to be protected in the building of origin. Areas closest to the fire, adjacent rooms, halls, stairs, and floors above are typically of greater priority and victims have less time than those in separate buildings.

Exterior exposures downwind or with the greatest loss potential are usually your priority. For example, if you have a fire in a single unit of a strip mall with a common attic, the prevention of lateral spread to the side with more units to lose is generally your priority. Row houses are a particular challenge.

Confinement

Confinement is often much easier and faster than extinguishment, and can be accomplished with less water and fewer firefighters. We have seen fires allowed to spread to uninvolved areas because firefighters attempted to extinguish head on, and like taking a knife to a gun fight, the fire beat them. Confine first in the area to keep the fire off victims and away from uninvolved areas, interior exposures, paths of egress, hallways, and stairwells, *then* extinguish. Utilizing TPOs is a huge tactical advantage for confining a fire. Consider the *location of hoselines and flow path*. Both are critical to confining the fire quickly.

Extinguishment

What is your required fire *flow*? Unlike supply, extinguishment is about the *size and number* of attack lines. One common pitfall is to utilize commonly pulled 1¾" lines in fires that require 2½" lines to open. A general rule of thumb is it takes two to three firefighters to stretch a 1¾" attack line, and three to four firefighters to stretch a 2½" attack line. These are minimums. Distance matters. Generally speaking, it is more critical to get one well-stretched line in place before trying to get two lines simultaneously. This may not be an issue for resource-rich systems. However, most regions would be served well by this rule.

Overhaul

Do *not* rely on the thermal imaging camera (TIC) alone. Otherwise, you are asking for a rekindle. The one time in 13 years as a battalion chief that I had a rekindle was when I allowed crews to rely on the TIC and didn't look in the attic myself. The TIC is infrared spectrum, not x-ray. Open areas and look for smoke. Cellulose insulation, in particular, can conceal smoldering hot spots. Confirm *who* is doing the overhaul and *where* it is needed.

Ventilation

What is the *purpose* of the ventilation? Support fire attack, confine the fire, or rescue? This *must be coordinated* with fire attack, regardless of the type of *vent tactic*, otherwise the fire can worsen. Water on the fire is a prerequisite to ventilation. That said, old school vertical ventilation often was ahead of the fire and allowed lift to create visibility. Whatever the tactic, coordinate through your tactical supervisors. Your options are vertical, horizontal, positive pressure (PPV), HVAC, and hydraulic.

Salvage

Who is performing salvage and *what* is the priority? Talk to the owner/occupant/*RP* about pictures, pills, and pets. Opportunities to salvage may take place at any time, based upon conditions and priorities. Treat the occupant's belongings like your own and teach your troops to do the same.

Support

Here, we consider what is needed to support the IAP. Several agency-specific resources are certainly available. As a minimum, consider the following:

- Law enforcement (LE)
- Investigator
- Public information officer (PIO)
- Safety officer (SO)
- Red Cross
- Utility companies (gas, electricity, and water)
- Board up/fencing company
- Rehab
- Air/light unit
- Hazmat team

Tactical Worksheet—Structure Fire

S	Size-Up	☐ FPODP ☐ Seven Sides ☐ Construction	☐ 360° ☐ RP	☐ VP ☐ IOCAN
A	Apparatus	☐ Three Sides	☐ Hot Zone	☐ Staging
W	Water Supply	☐ Who ☐ Where	☐ What	☐ FDC
C	Command	☐ Name ☐ ICP	☐ Additional Alarms ☐ ICS	☐ Comms
S	Strategy	☐ Confirm VP ☐ Offensive	☐ Combination	☐ Defensive
S	Safety	☐ RIT	☐ Two-Out	☐ PAR
R	Rescue	☐ 1st ☐ 2nd	☐ All clear ☐ Medical T/T	☐ Utilities

E	Exposures	☐ Interior	☐ Exterior	
C	Confinement	☐ Location of lines	☐ Flow Path	
E	Extinguishment	☐ Flow Required	☐ Size	☐ Number of Lines
O	Overhaul	☐ Who	☐ Where	
V	Ventilation	☐ Purpose	☐ Type	☐ Coordination
S	Salvage	☐ Who	☐ Where	☐ RP
S	Support	☐ Law ☐ Invest ☐ PIO ☐ Safety	☐ Red Cross ☐ Utility ☐ Board Up	☐ Rehab ☐ Air ☐ Hazmat

The Tactical Supervisor

The tactical division/group supervisor is critical to the success of an incident. *The primary purpose of a tactical supervisor is to bridge the tactical gap between the IC and the crews working.* This is accomplished by the IC trusting the supervisor to be their eyes and ears in the tactical space. The decentralized authority given to the supervisor by the IC requires solemn trust that can only be accomplished through training and time (fig. 7–9).

To be an effective officer in the division/group supervisor role, tactical expertise is required. As with any officer, knowledge, skills, and abilities (KSAs) are paramount to success. KSAs to be an effective division/group supervisor include knowledge of FD SOGs, tactics, fire behavior, building construction, and signs of deteriorating conditions; ability to size up a changing situation, multi-task, exude command presence, and be decisive; and communication, accountability, and risk management skills. With training, any good company officer can become an effective tactical supervisor. In addition, clear leader's intent (without micromanaging) from the IC will ensure that the IC and tactical supervisors are on the same page, and can maintain alignment.

As discussed in chapter 4, the tactical supervisor must remain in the "then" mindset, ahead of the current situation. Task-level crews are working and their officers are in the "now" while executing the work. Simultaneously, the tactical supervisor is ahead, thinking in the future. What will be the CAN in the next 3–10 minutes and beyond?

Examples of being in the "next" mindset include, but are not limited to:

- Anticipating and seeing changing/deteriorating fire conditions increase in pressurized smoke during attack, vent-point ignition, flashover, backdraft, and so on
- Looking for collapse potential and fire in the overhead or under firefighters

- Considering retreating crews
- Getting additional attack lines in place
- Ensuring primary and secondary searches are being completed by different crews
- Coordinating ventilation and fire attack
- Seeing if conditions inside match what is seen outside. Are conditions improving?
- Rotation of crews
- Setting up salvage during the fire attack on an attic fire, for example

Vital information must be shared between the IC and tactical supervisors to ensure alignment, reduce radio traffic, minimize TTT, maximize effectiveness, and ensure tactical benchmarks are completed. The IC must become accustomed to less radio traffic by relying upon the tactical supervisor to keep the IC abreast of incident tactical benchmarks, rather than unnecessary chatter about tasks. In turn, the tactical supervisor must not overwhelm the IC with unnecessary task-level transmissions yet give the IC critical tactical benchmarks and CAN reports as needed.

Examples of *appropriate* radio transmissions/*benchmarks* from a tactical supervisor to an IC include, but are not limited to:

- Primary clear of a floor or building—"Command, Division A, primary search clear of the house."
- Secondary *all clear* of a floor or building—"Command, Division 3, secondary search complete, all clear on floor 3."
- Water on the fire/knockdown—"Command, Fire Attack Group, we have knockdown of the fire, checking for extension."
- Fire contained—"Command, Division C, fire is contained to two rooms on the Charlie side, first floor."
- Switching to salvage and overhaul—"Command, Division Sub 1, secondary search complete, all clear in the basement. Switching to salvage and overhaul."
- Need additional companies by type (engine, truck, squad, rescue)—"Command, Division B, negative fire extension on the B-1 exposure. Need one truck company for positive pressure ventilation."
- Objective completed—"Command, Roof Division, ventilation complete. Exiting the roof. Will be available for an assignment in 5 minutes."

Fig. 7–9. A division supervisor communicates face-to-face with one of his company officers.

Examples of *inappropriate* radio transmissions from a tactical supervisor to an IC include, but are not limited to:

- "Command, Division A, Truck 1 is searching the kitchen."
- "Command, Fire Attack Group, we are deploying another 1¾" line."
- "Command, Rescue Group, Truck 3 is performing VES."

While traditional command without tactical supervisors may require these transmissions, the benefit of utilizing division/group supervisors is that many previous transmissions become unnecessary, thus reducing radio traffic. *Benchmarks and needs are the vital transmissions.*

BRO-HAM and TACOS

When assigning a group or division supervisor role to an officer, give the following:

- **Boundaries**—What is the division/group area? (Whole house, A side of commercial, all fire attack functions, search of all units in the complex?)
- **Resources**—Who is working for them? (E101, E103, TR106)
- **Objectives**—What is their job? (Fire attack and search, just fire attack, prevent lateral spread and evacuate)

Division and group supervisors must be *highly*:

- **Hostile**—Strong command presence. Don't let companies go around you to the IC via radio or otherwise. Make sure they know that you are *their supervisor* (fig. 7–10).
- **Agile**—Seeing the bigger picture while being eyes and ears for IC. *Simultaneously*, being the guardian angel for the companies while seeing hazards and looming problems. Thinking in the "next" (secondary search, roof/attic conditions, coordinate vent, crew air and accountability), with a 6–50-foot sphere of awareness. Seeing what the crews and IC do not see in the tactical space.
- **Mobile**—In, out, and around the tactical area. Do not remain stationary. It requires a lot of movement to have the situational awareness and accountability needed to be safe and effective. *That said, do not put yourself in jeopardy too far into the hazard area.* Remain close or in physical contact with firefighters in your tactical area for short periods to communicate, size-up, and account for members. This requires the division/group supervisor to be in full PPE, including SCBA (fig. 7–11).

Fig. 7–10. Waco (TX) Chief Robby Bergerson jumps into the fray at the tactical level with full personal protective gear (PPE).

Fig. 7–11. Captain Evans conducts face-to-face communications with his crew at the threshold of the hot zone.

CASE STUDIES IN COMMAND

The Tactical Supervisor

CHIEF DAN COTRELL, BILLINGS (MT) FIRE DEPARTMENT

In late October 2020 I was an incumbent captain and aspiring to be promoted to battalion chief. I was assigned to the rescue company with the Billings Fire Department. As part of my BC development training, I was required to attend the "Calm the Chaos—Mastering Fire Ground Command" course that my department had brought in. As a twenty-plus-year veteran of the career fire service, I had seen several new trends come and go. I had little hope that this course was any different.

My skepticism was fueled by a long career with a very traditional fire department, where the incident commander had total control of the incident. In my old system, all of the personnel working for the IC were working at a task level. I refer to this as a "mother, may I?" system of command. This system of command and control had always worked well for us, and we had seen and expected success on incidents. However, in the words of Dr. David Griffin "you don't know what you don't know."

On day four of the class, we reported to the training grounds for a hands-on application of the course. I was still apprehensive of an approach where the IC had to relinquish a portion of his authority, but I decided that I was going to give the system a chance to prove itself. On scenario number one, I was the first-in captain on an engine company. My company arrived at a simulated single-family residential home with a working fire. Upon arrival, I gave my normal report, put my crew to work at a task level, made a lap, and then started to apply the ideology I had learned the past three days in the classroom. Instead of passing command and initiating fire attack, I retained command and started thinking at a tactical level. What is the incident priority? (In this case, lifesaving.) In order to meet the priority, what do I need next? I assigned my crew to pull two lines to the front door and re-set the fire from the outside. I assigned the second engine the objective of fire attack and gave my firefighter to the captain of the second engine so we would have both a fire attack crew and a two-out line.

The battalion chief arrived and I transferred command to the BC with the recommendation that I assume a division supervisor roll. I was assigned Division A with the objectives of fire attack and search. Engines 2, 1, and the truck were all working for me upon arrival. I was able to make quick assignments to the companies assigned to me, coordinate their efforts, and meet my objectives with little to no radio traffic and zero duplication of efforts or freelancing.

The way I felt after that first scenario concluded is hard to describe. For the first time in my career as a company officer, I felt that I was permitted to fully do my job. The previous three days of training clicked, and I was able to feel how empowering it is to allow the people on the ground, the people who first feel the heat, who first detect the change in conditions, to address those issues at a tactical level, without tying up the radio asking command permission to make adjustments.

Fast forward to July 26, 2022. I had been promoted to BC exactly seven months prior. My department had embraced assigning division and group supervisors to multi-company incidents, and we experienced a great deal of success with the system. In the early morning hours of the shift, we were dispatched to a possible structure fire in an area that is mixed-use, residential, commercial, and agricultural. The fire was located just outside the city limits but within the department's response area. A normal residential structure response was assigned, which consists of four engines, one truck, and one BC. Because the address is outside the city, a recommendation for a water tender is made as well.

The first engine (Engine 2) arrives and reports a large auto shop with light smoke showing from the alpha side. Because of the residential nature of the area, there is a reasonable likelihood of living quarters in the structure, and a moderate victim profile was assigned. After completing his 360°, E2 Captain suspected a vent-limited fire located in what he thought was a room located on the Delta side with an exterior-walk door for access. He had his crews deploy handlines to that door and initiated fire attack from the exterior at a task level.

I arrived and parked on the Alpha side and made contact with E2, who reported that they believed they had knocked the fire down through the Delta side door. I assumed command, confirmed an offensive strategy, and ordered additional resources. Conditions from the command post consisted of light smoke from the gable and eaves. I had a false sense that we could make an aggressive attack and extinguish the fire. Upon arrival of the next engine, I assigned the captain, a 33-year veteran of the fire service, as Division Alpha, with objectives of fire attack and search. He was assigned E1 (his unit), E2, and was told that he would also need the next engine on scene to complete his objectives.

When working as a division or group supervisor, you are expected to use the acronym BRO-HAM. You must know your *boundaries*, in this case the entire structure (for now); *resources*, in this case Captain Christensen had three engines (is this enough?); and *objectives*, or what tactics are in play and how do I as the supervisor assign the task-level crews to achieve the objectives? These are all action items.

The HAM is the attitude you must have. You must be *hostile*. Seek out the crews that are assigned to you and physically grab them, tell them who you are, that they are now working for you, and what their task is. You must be *agile*, not only physically, but more importantly *mentally agile*. You must know fireground tactics and the tasks needed to make them successful. You must be thinking about not only what do I need to do now, but what needs to be done next to meet the assigned objectives?

Last but most certainly not least is being *mobile*. In order to be an effective supervisor, you must stay mobile. This means making contact with your task-level crews (seek them out for updates), commit to frequent laps looking for expected changes, and be prepared to report to command early if the tactics you have assigned aren't producing those expected results.

Keeping the BRO-HAM attitude in mind, remember that from the command post I was seeing light smoke, and had received a report from the first-due crew that they believed they had reset the fire.

Division Alpha supervisor made a lap and observed that there was heavy smoke and heat in the Charlie-Delta corner of the building. Furthermore, when he came back to the Alpha side, he observed the fire attack crew opening the overhead door and attempting to make entry. What happened here? What's wrong with this picture? How was it fixed? It turns out that the incident was out of alignment without me, the IC, knowing it. The task-level fire attack crew thought that they had reset the fire but realized the room they initiated fire attack on was isolated from the rest of the building, so they were attempting to re-engage through the Alpha side overhead door. They were unaware that they had never really applied water to the seat of the fire, and they had now introduced a flow path when they opened the overhead door. The vent-limited fire was now in a free burning state at the opposite end of the structure.

Thankfully, the division supervisor followed BRO-HAM to a tee. He made his lap, observed conditions within his boundaries, considered how he could meet his assigned objectives with the units assigned to him, made a point to find those resources, and let them know that they were now working for him and gave them their assignment. In this instance, when the division supervisor checked in with the fire attack crew, they were so task saturated they had not heard any of the radio communications and did not know that they had been assigned to Division Alpha. He was able to inform them of the rapid fire growth and re-directed them to engage from the exterior until more recon could be conducted.

As the incident progressed and intel was gathered, we found out from the owners that there were no people in the structure and that there was a second-story mezzanine on the entire Delta side of the building, the exact area that the fire attack crew was attempting to access. Due to the limited water supply, we were unable to make progress extinguishing the fire, and a defensive strategy was called.

I am so grateful that I had a competent and experienced company officer on the ground assigned as a division supervisor that morning. He was able to fill the tactical gap that existed between me and the fire attack crew. With his supervision, we were able to re-align the incident, and most importantly his application of BRO-HAM kept that task-level fire attack crew from entering a structure that was in no way worth the risk.

The job of the tactical supervisor is:

- **Tactical decision-making:** Crew assignments to complete the objectives, including (but not limited to) size, number, and placement of attack lines; search tactic of choice and area of priority; coordination of fire attack and ventilation; priority for salvage and overhaul, and so on.

- **Active accountability:** The process of accounting for crews by being in close physical proximity, including, hands-on, eyes-on accounting of companies, members, location of entry, air levels, rotation of crews. FATS tags on turnouts help (fig. 7–12).
- **Communications between companies and command:** CAN reports; face-to-face with crews assigned preferred to reduce radio traffic yet increase efficiency of communications; radio transmissions when needed to companies further into the hot zone.
- **Ongoing size-up and risk/gain assessment:** Victim profiling and SIGNAL: reading smoke and fire conditions to anticipate fire behavior, location, and spread (roll-over, flashover, vent-point ignition, backdraft, smoke explosion); reading building conditions for potential failure and collapse.
- **Safety officer:** Ensure adherence to all safety policies and SOGs and watch out for deteriorating conditions.

Being an effective tactical division/group supervisor takes training, including realistic hands-on training (fig. 7–13).

Fig. 7–12. FATS identifiers are only good if they are worn and accurate.

Fig. 7–13. Realistic hands-on training is crucial to building effective tactical supervisors. *Source.* Photo courtesy of El Dorado Hills (CA) Fire Department.

Bridging the Tactical Gap

Chief and/or company officers can be assigned as division/group tactical supervisors. If your system has a lot of chiefs, then you may have the luxury of only assigning chief officers into tactical positions, thus allowing the company officers to be solely focused on task-level supervision of their companies. Most jurisdictions, however, do not have enough 24-hour coverage of chief officers to have them on scene in a timely or abundant fashion. Therefore, company officers can make great tactical supervisors as they are often on-scene prior to any chief officers, and likely multiple company officers may be on-scene before the second chief officer.

When a company officer is assigned as a tactical supervisor, they have several options from which to assign the crew. First, the crew may be assigned to another company officer to augment that crew and ensure proper supervision. This is a good option when the crew is inexperienced. Second, the company officer may assign the crew to a non-IDLH assignment like a water supply, exposure line, or EMS. Third, the crew may be assigned an interior IDLH position if the crew is senior and/or a senior member of the crew is capable of proper supervision. This happens regularly on senior truck and rescue companies. Finally, the crew may assist the company officer in the tasks of tactical supervision, such as pulling a second backup line, performing door control, accounting for crews, or changing bottles, for example.

The first-arriving company officer will be in command until the arrival of a second company officer or chief. Whether command is passed to the second company officer or retained by the first, the first-arriving chief may choose to assign the company officer in command as a tactical supervisor upon arrival. This quickly bridges the tactical gap by leveraging the tactical advantage of the prior IC's knowledge of the building layout, fire location, tactical plan, and accounting of crews. This helps keep the alarm ahead of the incident power curve. Rather than assign the company officer/IC back to their crew at the task level, the previous IC can bump up to a tactical position rapidly without a lot of delay in command transfer. The new IC (chief) can rapidly get into the "then/what if" strategic mindset while the newly appointed tactical supervisor remains in the "next" mindset. This bridging of the tactical gap from the *bottom-up* invests one company officer to maintain manageable span of control, maximize accountability, minimize TTT, continue risk assessment, and minimize radio traffic and delays in command transfer.

Task-saturation is the phenomenon in which an individual performing tasks is overwhelmed, unable to think tactically, hear radio traffic, multi-task, or see the bigger picture. *Tunnel vision* is the condition in which an individual is extremely focused visually and misses their surroundings. *Auditory exclusion* is the condition in which an individual is focused on a particular sound or voice and does not hear other background sound. Examples would be missing an exposure building because you are focused on the flames of a house fire, or talking face-to-face with someone and not hearing your portable radio, respectively. In general, the second company officer to an area is a better choice to become the tactical supervisor because the first company officer is task-saturated, focused on stretching an attack line, forcing entry, donning an SCBA, initiating a search, and so on (fig. 7–14).

Ideally, the tactical supervisor should be mobile, yet posted up at the threshold of the entry point to the hot zone in their division. This would be the front door of a house on division A, or the back door of a commercial building on division C. The tactical supervisor must be able to see the bigger picture and not get too far into the weeds. If they are in the hot zone with heavy smoke, they can no longer account for crews or step back and see conditions. They can go in briefly for face-to-face communication and accountability, but they must not remain in the hot zone or they will lose situational awareness. For this reason, even on a two-story house with fire on the second floor, the tactical supervisor will likely be division A if the attack is from the front door (fig. 7–15).

Fig. 7–14. The tactical supervisor steps back and looks at smoke conditions overhead. *Source.* Photo courtesy of Rick McClure

Fig. 7–15. This tactical supervisor is literally standing in the threshold to the hot zone.

In some cases, most or all of the company officers may be required to initiate tasks on a larger incident. Multiple lines may need to be stretched while supply lines are established and rescues initiated. Here, the chief arriving and assuming command may choose a *top-down* method of bridging the tactical gap by taking on more centralized tactical decision-making until the arrival of the second chief. This was the case on a large multi-story center hall apartment fire. We needed multiple attack lines from different locations with separate water supplies, search from two different stairwells, vertical ventilation, and medical all simultaneously. This required all the company officers working on the fire floor to remain task-oriented, as it took at least two companies on each of the two different stairwells to get attack lines in place on the same floor. This necessitated that I, as the IC, direct crews with more tactical details on stairwell and hydrant locations and on what side to position apparatus. By the time the second chief arrived, the majority of the first alarm was assigned and he went in behind them as the division supervisor of the fire floor.

CASE STUDIES IN COMMAND

Mayday and Mugging

BENTON STREET BASEMENT FIRE DIVISION SUPERVISOR
CHIEF FRANK RICCI, NEW HAVEN (CT) FIRE DEPARTMENT

Division supervisors help decentralize command, breaking up the "opinion brigade," as Anthony Avillo would say, putting your chiefs and officers in forward command positions to increase your influence and span of control. When assigning a division, the individual at a minimum should be assigned a radio designation, area of responsibility, and crews reporting to them. And in complicated incidents, their tactical objectives should be included. At a structure fire, when assigning a division to a competent chief, the objective can be as simple as put out the fire. How this sounds on the radio: "Car 81 is Basement Division, Truck 4, Rescue 1, and Engine 6 are reporting to car 81." If command or operations fails to put this over the radio after a face-to-face assignment, the division officer should make the announcement and/or follow up face-to-face with their crews..

On a stubborn fire that went to a second alarm, calling me in from home as a member of the command staff, I was assigned to the training division. Command assigned me the basement division. Prior to my arrival, crews were pulled out and defensive operations were attempting to reset and make a final push on the fire.

Division Supervisors

It is critical to understand that whether you are supervising one or multiple companies, *you are not a member of the company.* You are, in fact, operating alone while coordinating companies in a forward position. These officers are not responsible for your accountability. You are responsible for theirs. When resources come available in a division, supervisors should be assigned an aid or interior safety officer.

Initially, I was assigned one company with five members and the simple objective of putting out the fire in the basement. I had ordered three members (including their boss) to advance the line and told two others to push the line in. I ordered that as soon as the first firefighter comes out to change his bottle, the backup firefighters should follow the line in. I waited outside, checking the conditions of the advance. I could see that they were hitting fire and making good progress. This staggered my time of entry a few minutes behind the hose crew. I knew this would help me maintain continuity of command. While that first crew had to exit for air, the next crew would meet up with me to extinguish the fire.

I made entry crawling in with zero visibility. There was low heat, with about six inches of water on the floor and a loud constant noise of water flowing. I met up with the officer who thought they were at the far side of the basement. I directed him that they had about 10 feet more to go to reach the Alpha side.

Then, one member's air pack alarmed for low air. I figured two would go out and the other two would come in, allowing us to extinguish the fire. I failed to communicate my expectations of having one of the firefighters with air remain to staff the nozzle.

The firefighter handed me the nozzle, and I was going to pass it to the other firefighter.

"MAYDAY, MAYDAY, MAYDAY."

Instead, the entire crew of three went out as a team, and before they got to the hatch way door, things went bad and conditions deteriorated rapidly. They made it up the stairs, but the two other firefighters were ordered by the chief not to descend the stairs. I found myself operating alone. At first, I had the pipe. As a chief, this never happens and I figured this was kind of cool. Afterall, the other members will be here in a minute.

I could tell that I needed to advance the line another ten feet towards the Bravo side, down a little hall, to hit the base of the fire. The line was significantly caught up and I knew it would take another crew to make the final push, but no one came. The condition in my location was still favorable. Now, I knew my radio transmissions were not getting out and I could hear the chief call a Mayday for me. I tried to radio back to no avail.

Still with low heat, I started to back out with the pipe in my hand. Pulling the line back over itself, due to debris and water, I could not find the hose but had the pipe in my hand. I radioed that I needed another line and a crew, but there was still no answer. As I crawled about half way back, intermittently stopping to hit the fire in front and behind me, I fell forward from a crawling position into the water.

I lost the nozzle and looked up only to see an orange glow all around me. I calmly, but deliberately scurried to find the nozzle. Finding it allowed me to chase back a significant fire situation now clearly defined in front of me and behind me.

I later learned that my transmissions were not getting out due to a microphone short (this was confirmed after manufacturing testing). I knew I was close to the Charlie side, but not sure of the location of the stairs. I figured that once I hit the wall, I will go ten feet in one direction and if I didn't find the stairs, I would go ten feet in the other. I remember laughing to myself, because if I run out of air I might drown with all the water on the floor. At this point, Firefighter Matt Thomas located me in zero visibility as I was in a kneeling position, flowing water while keeping the fire in check. He was pulling me towards him and I kept asking if he knew where the door was.

I was adamant not to go with him until I confirmed he knew where the door was. At the Fire Department Instructors Conference, FDNY Lt. Eddie Cancro told a RIT story in which a downed firefighter was located, but the RIT lost their orientation and couldn't find the stairs to escape. Firefighter Thomas confirmed that the stairs were straight ahead. I started to follow him when the fire came back. I opened the line to chase it back. A few second later, he grabbed me and pulled me to the stairs. I kept saying that I was fine. As I hit the stairs, others started to grab and pull me up the stairs. They were pulling my shoulder straps. As they pulled, my bottle kept catching the cement stairs.

While this was done with the best of intentions, this was not a fun event for me. Due to having two previous surgeries on my shoulder, it felt more like a mugging than a rescue. Those firefighters all played a significant role to ensure I made it out, and I am extremely grateful.

The moral of the story is that you cannot be left alone in a structure fire when you are a division or group supervisor. It's one thing to step in and check conditions or quickly account for your crews. You should *not*, however, be operating alone. Be sure to remain with other crews or notify a backup team, two-out team, or RIT that you are going in and to expect you to be out within two minutes or less. This will ensure your safety, accountability, and prevent an unnecessary RIT deployment.

Remember to put this great book into practice. Checkbox command, where we check off that we have a company or two dedicated for RIT, doesn't mean anything if you don't ensure you get out of the firehouse and train with crews. This way you will be confident that they can competently fulfill the task at hand.

The ICS Formula and Templating

ICS never put a fire out, but it organizes the people and resources that do.

There seems to be a lot of confusion around America about when and why to use a division versus a group. First, they *are* the same level of authority, as both bosses are called supervisors, so they work with each other laterally. For example, a ventilation group may assist with PPV on divisions 4, 5, and 6. Or a rescue group may assist division 3 with multiple VEIS rescues. Another example would be a rescue group handing victims over to medical group for treatment and transport. So, a group does not work for a division, nor a division for a group. They work *with* each other.

So, when might we choose a group as a better means to tactically organize an incident? There are few easy situations. Functions like RIC and medical are almost always groups, as they are functionally organized. Several crews focused solely on search and rescue could work under a rescue group. Conversely, divisions A, B, C, or 1, 2, 3 are geographically organized. A roof division is usually a more appropriate choice for vertical ventilation than a ventilation group.

One of the misconceptions of ICS is that it is not suited for fast-moving Type 4/5 incidents like structure fires. This is not a flaw in the system, but a flaw in traditional training. Most ICS training is relegated to defining the difference between a division, group, or branch, and filling out an organizational chart. Classes like I-100, I-200, I-300 and beyond are excellent as a foundation; however, the nature of daily single and multi-alarm fires and other incidents requires a different type of training and application of ICS.

It is a fallacy to consider *any* modern-day incident routine or bread-and-butter, as the nature and complexity of daily Type 4 and Type 5 incidents has changed dramatically since 2001. Another common misconception is that the use of divisions A, B, C, and/or D is only for defensive operations. In fact, the use of geographic division is highly effective for interior, offensive operations. The portal and side of entry determines the division designation when using A-D. However, many other factors come into play for determining an effective and efficient ICS plan. Figure 7–16 shows a house fire to illustrate the strategic, tactical, and task levels working together. The span of control for the IC is currently 3:1, with division A, roof division, and

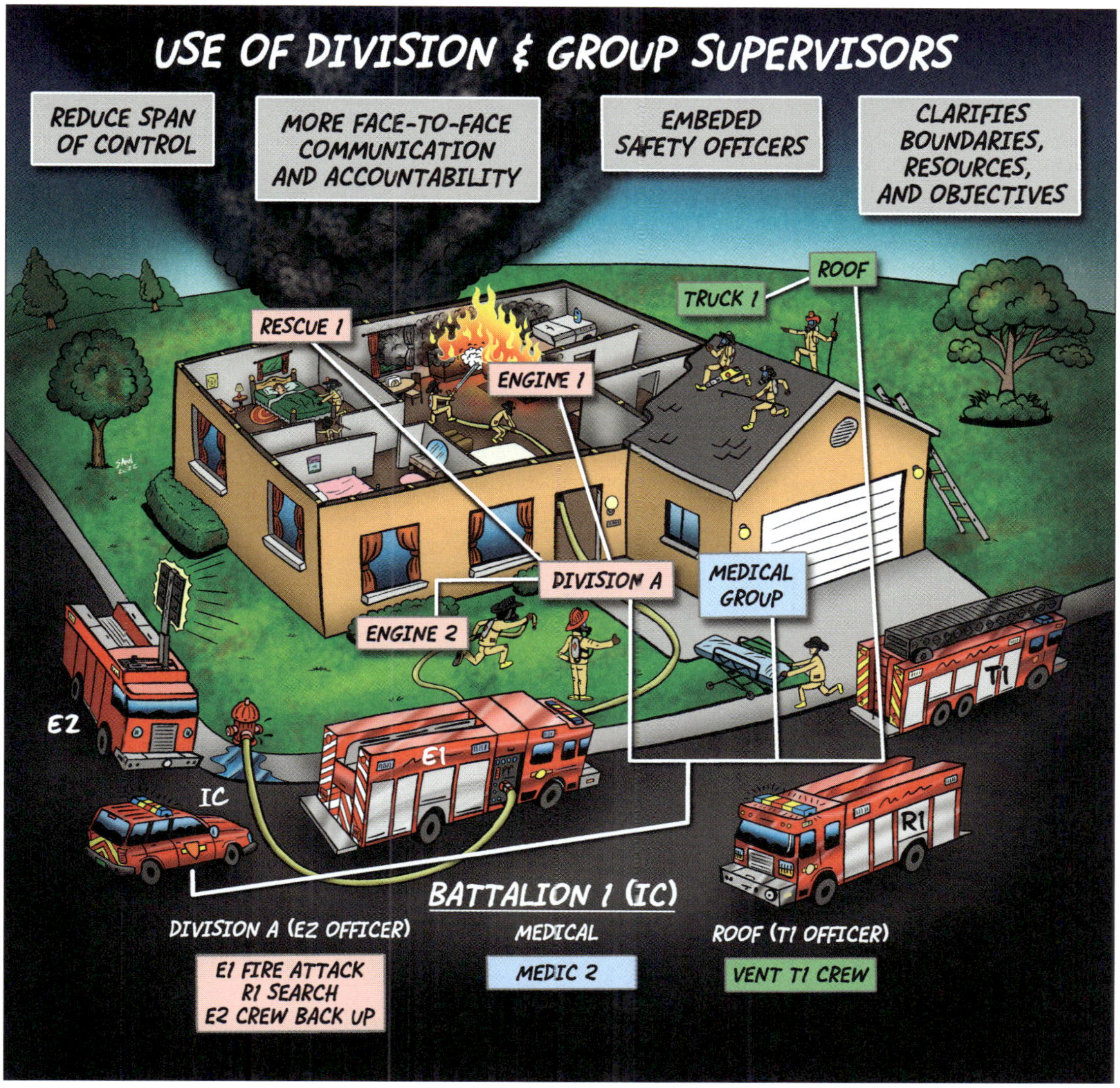

Fig. 7–16. The IC has a 3:1 span-of-control and additional tactical benefits.

medical all working directly for the IC. In addition to reducing the span of control, additional benefits include more face-to-face communication and active eyes-on accountability, embedded safety officers, and division of labor for objectives and resource allocation. This house fire diagram is a foundation for the following section on how to apply ICS rapidly.

ICS templating is the process used to rapidly apply ICS to an incident and remain ahead of the incident power curve. *The ICS formula* is composed of the building/situation + conditions + resources which determine your ICS template. All three factors should be considered in determining if, when, where, and how a group or division should be established (see fig. 7–17).

The ICS formula (B+C+R=ICS) is composed of the following three factors/inputs:

- **Building (B):** Large/small, residential/commercial, high/wide, big box, strip mall, center hall, garden-style, rowhouse, taxpayer, etc.
- **Conditions (C):** Food on stove, heavy fire, victim trapped, exposures, high winds, lateral extension, vertical extension

Standard Conditions		Standard Actions		
Building (Fixed Factor)	Problem (variable)	Response (Department Policy)	Priority Fire Occupancy Assignments IC-1	Additional Assignments IC - 2
Single Family Dwelling		4 Engines 1 Ladder Command – IC2 Safety - ISO	Fire Attack Primary Search Back Up Water Supply	Vent On Deck RIT
Apartments		5 Engines 1 Ladder Command – IC2 Safety - ISO	Fire Attack Primary Search CHECK EXPOSURE Back Up Water Supply	Exposure Resources Vent On Deck RIT 2ndry Water
Stand Alone Commercial		5 Engines 1 Ladder Command – IC2 Safety - ISO	Fire Attack Roof Report Back Up Water Supply	Exposure Resources Vent On Deck RIT Building System
Stripmall		5 Engines 1 Ladder Command – IC2 Safety - ISO	Fire Attack Roof Report CHECK EXPOSURE Back Up Water Supply	Exposure Resources Vent On Deck RIT Building System

Standard Conditions		Standard Actions		
Building (Fixed Factor)	Problem (variable)	Response (Department Policy)	Priority Fire Occupancy Assignments IC-1	Additional Assignments IC - 2
Single Family Dwelling	*Working Fire*	4 Engines 1 Ladder Command – IC2 Safety - ISO	E1 - Fire Attack L1 - Primary Search E2 - Back Up E 3 - Water Supply	L1 - Vent E3 - On Deck E4 – Lvl 1
Apartments	*Light Smoke*	5 Engines 1 Ladder Command – IC2 Safety - ISO	E1 - Fire Unit L1 – Check Exposures E2 – Spot hydrant	All other Lvl - 1
Stand Alone Commercial	*Nothing Showing*	5 Engines 1 Ladder Command – IC2 Safety - ISO	E1 - Investigating L1 – Complete 360 and roof Report	All other Lvl - 1
Stripmall	*Working Fire*	5 Engines 1 Ladder Command – IC2 Safety - ISO	E1 - Fire Attack L1 - Roof E2 - EXPOSURE E3 – Water Supply Second Line	Safety Off - Exp Supervisor E4 Crew– Back Up E4 Off – Fire Atk Supervisor E5 - RIT Building System

Fig. 7–17. Examples of ICS formula and templating from the Midwest City (OK) Fire Department

- **Resources (R):** How many engines, trucks, medics, and chiefs are responding? What is the staffing? What is the response time? Arrival order? You cannot have a very extravagant ICS with two-person engines arriving 20 minutes apart from each other.

Template examples:

- **Commercial/big box:** Divisions A, B, C, and D based upon entry point. Can be used for offensive, combination, or defensive strategies.

- **Strip malls:** Division A and C initially. May include divisions B and D if necessary to contain lateral extension.
- **Regional mall:** May use divisions A, B, C, and D based upon the need to contain.
- **Garden-style apartment:** If smaller, divisions A and C can get the front and back quickly organized.
- **Center hall multi-floor:** Divisions 1, 2, 3, 4, and so on.
- **Townhome:** Divisions A, B, C, and D.
- **Smaller house:** Division A. This includes two-story houses.
- **Larger house:** Division A and C, or others as needed. Portal of entry is the division. If crews enter the C side and go to the second floor, they still work for C.
- **Known Rescue:** In house, duplex, fourplex, condo, garden apartment—fire attack group, rescue group, medical group, roof division, or vent group. If operations are on the roof, a roof division clarifies to everyone on scene that roof operations are in progress.

Incident templating also applies to other incidents:

- **Vegetation fires:** Typically, division B (left flank), division O (head), and division X (right flank). This literally puts a BOX around the fire quickly. If needed, fill in the gaps with additional divisions as more resources arrive.
- **MCI:** A medical group can have the objectives of triage, treatment, and transport. If a larger MCI, then medical branch could have triage, treatment, and transport units. Other groups and/or branches like fire/rescue, law, evacuation, and others may be required.
- **Hazmat:** A hazmat group can have the objectives of entry, identification, decontamination, and so on. Others may be medical group, evacuation group, and so on. Again, size and complexity may drive the need for branches.

Fire Attack and Rescue Groups

When faced with a fire wih confirmed victims trapped, either notified en route by dispatch, or upon arrival from bystanders, things change. Firefighters will experience an adrenaline dump and have more tunnel vision and auditory exclusion. In addition, every firefighter leans toward rescue *or* fire attack as the *first* priority in a known victim situation. You are thinking of one or the other as you read this. That's normal.

Relegating multiple known victims *and* fire attack to a lone division supervisor can be overwhelming to that division supervisor. To overcome these predictable obstacles, splitting the load between a fire attack group and rescue group can easily organize an otherwise chaotic scene. I learned this the hard way on Morse Avenue with the three fatalities and had a chance to do it differently with El Camino. Again, the other factors matter. The building and resources are still part of the ICS formula. So, they must be taken into consideration. If your FD has few resources, you may stick with a single division. If the building is huge, like a Home Depot, or a high-rise, a single fire attack group supervisor cannot possibly manage all fire attack operations, so divisions are the way to go.

However, you may augment with a rescue group. For example, I had a three-alarm fire in a center hall apartment building with multiple victims confirmed upon arrival. The main fire was on the top/third floor, which became division 3. Because of the number of potential victims, I set up a rescue group to work *laterally* with division 3 to augment the search and rescue operations by utilizing VEIS from the exterior, while crews searched from the interior hallway.

A fire attack group and rescue group template is good for known victims in houses, duplexes, triplexes, fourplexes, condominiums, and garden apartments. Beyond those, known victims in larger buildings or with few resources responding will need a division template. *Again, this is not one-size-fits-all. Train, train, and then train some more.*

Branching Out

Typically, divisions, groups, or a combination of the two will work very effectively for the vast majority of structure fires and other incidents in the Type 5 and 4 arena. Occasionally, however, the size or type of incident may dictate the use of branches. *Branches* can be either geographic or functional and are comprised of multiple groups/divisions. They are very effective for larger, more complex incidents like a massive MCI, hazmat incident, or extremely large building fires like regional malls, massive apartment complexes, high-rises, or fires that involve multiple buildings or large areas.

When a division or group supervisor becomes too saturated and the span of control too great, they and the IC can either create a new tactical level (split one division into two) or branch out (create a new layer). The new branch can then oversee the previous units in the division/group by splitting them into new tactical units under the new branch. This keeps the IC's span of control the same, yet reduces the previous division/group supervisor's span of control back into a manageable size.

For example, a division supervisor has too many units on the tenth floor of a high-rise fire. They may notify command of the need to branch out by turning division 10 into *branch* 10 with division A composed of companies accessing floor ten from the north stairwell, and division C composed of companies accessing floor ten from the south stairwell. If five companies were coming from each stairwell, division 10's span of control was 10:1. By branching out, an officer from each stairwell could be identified as the new division supervisor, bringing branch 10's span of control from 10:1 down to 2:1.

A second way to reduce the span of control is to create a new division. For example, I assumed command at a structure fire in a large home that was over 10,000 square feet. This was clearly beyond the scope of the first division A supervisor working at the front of the structure. We created a second division D for crews entering from the right side of the home. While my span-of-control went up by one, it was still manageable.

Early one foggy morning, I responded to a 30-vehicle pileup on Highway 5 in central California. There was fire, hazardous materials leaking, occupants trapped, and countless people injured. We set up a fire/rescue branch, a hazmat branch, and a medical branch. Over 30 resources responded, yet the IC's span of control was 6:1 (PIO, safety, liaison officers included). Unified command was established with the California Highway Patrol. More about this in chapter 16.

ICS = B + C + R

The Safety Officer

Many agencies assign the second-arriving chief officer as safety officer. This can be very effective and get a fresh set of eyes on the incident to identify the safety issues listed earlier. Hazardous materials incidents and other technical incidents may require that the safety officer have technical qualifications and the position may even be mandated by law, depending on the jurisdiction or nature of the incident. For example, an assistant safety officer is often required on hazardous materials incidents and the officer must have the same qualifications as the hazardous material team making entry.

For structure fires, an alternative methodology is to have the second chief officer arriving assume the division supervisor role of the hot division. If a company officer is already filling the role of the primary division supervisor, the second chief can either take over the primary division or assume a new division supervisor position if the incident is expanding.

Remember, one of the primary responsibilities of any division/group supervisor is to be the safety officer for their tactical space. Therefore, if a house fire has division A, roof division, RIC group and medical group, there are actually *four* safety officers embedded in the incident. Rather than one safety officer who cannot be in more than one place at a time, multiple safety officers can constantly monitor their respective areas and crews working. This includes accountability as well.

As discussed in chapter 2, decentralized incident command or *mission command* allows for tactical supervision to be placed in a forward position in the tactical space. Rather than a single safety officer and/or accountability officer, each division/group supervisor fulfills these roles around the incident. In addition, a safety officer increases, not decreases, the span of control for the IC. Typically, the safety officer has no tactical authority and can only stop or identify safety issues.

WISDOM FROM THE MASTERS

Do Incident Safety Officers Really Help?

CHIEF DAVE DODSON (RET), EAGLE RIVER (CO)

Way back in 1998 I wrote the first book dedicated to the specific duties and responsibilities of an incident safety officer (o), and in 1999 it was published by Delmar Publications. My motivation to write the text came from an increasing trend of firefighters being caught in flashovers and collapses, but also from a misuse of safety officers—a shallow expectation that they were merely bunker-cops. In the book, I professed that the ISO must bridge any disconnect between the incident commander, changing conditions, and crew exposure, and better predict potentials such as flashover and building collapse. I also stated that the ultimate goal of any safety officer is to *eliminate the need for safety officers*. Since then, the book has gone through three revisions, a new publisher, and become the study reference for over ten thousand fire officers seeking Pro-Board certification as an incident safety officer. I share this history not as a celebration, but a preface to some discoveries I've made and some realities that all fire officers must embrace.

Twelve years as a duty safety officer and over twenty years networking with members of the Fire Department Safety Officer Association has led me to the belief that one of the first—and most important—contributions an assigned safety officer (now labeled SOF) at a building fire incident can make is to evaluate (read) the building and communicate fire spread and collapse/failure concerns to the incident commander. In most cases, firefighters are quite capable of taking action to quickly find and control fires inside a building. Those same quick actions may come at the expense of fully understanding the host of their efforts: the building itself. Add to this the realities that firefighters and fire officers may not possess the knowledge/experience/wisdom that years of building construction/collapse study can bring, and you have a formula for a negative outcome.

The SOF that has made a thorough and unending investment in studying building construction, fire effects on buildings, and building walk-throughs of those within their jurisdictions can—and will—help. Sure, there are other functions that the SOF must address, but reading the building and communicating issues and concerns is paramount!

There are other realities that I've discovered over the years. First, the use of a well-trained, articulate, and dedicated SOF can make a difference in preventing firefighter injuries and deaths. While no official SOF save data is captured, I receive hundreds of feedback stories of instances where the SOF's actions prevented a sure firefighter injury or death. Second, a dedicated SOF must evaluate and judge myriad environmental conditions, tactical operations, and potential outcomes to ultimately be effective. Third, a SOF must be acutely dialed into the IC's action plan and simultaneously aware of tactical crews' progress and risk exposure. Lastly, no single SOF can see everything that needs to be seen and communicate all the necessary suggestions and interventions to make sure everybody goes home. More work needs to be done.

That work starts with an institutional shift in the expectations of tactical officers and the way a dedicated safety officer is utilized. The functional expectations of the SOF are known:

- Monitor the environment/operating crews
- Evaluate risk-taking and the IAP
- Develop countermeasures
- Intervene for imminent hazards
- Communicate hazards and established zones

The SOF performs these functions to help bridge any disconnect or dysfunction between the IC, operating crews, and changing conditions. This may sound great—and some SOFs are good at it—*but NO SOF can cover all these bases and eliminate all the disconnect issues.* Chief Kastros has blamed similar disconnects on the reality of the tactical gap. I'm now on board with him and agree that the tactical gap can be addressed only if tactical officers, group/division supervisors, and branch directors prepare and perform more like an incident safety officer (truly embedded safety officers). This institutional shift of tactical officer expectations requires more SOF-like situational awareness training of the five reads:

- Reading the building (fire spread and collapse potentials)
- Reading smoke (fire spread potential and predicting/preventing hostile fire events)
- Reading risk (risk versus benefit)
- Reading hazardous energy (electrical, mechanical, chemical, weather, etc.)
- Reading firefighters (appropriate aggressiveness, task performance, and rehab)

Once this training and these expectations are culturally instilled, a dedicated safety officer can be utilized for overhead safety/health functions such as:

- Confirming building type/era/use/size and hazard potentials with the IC
- IAP evaluation (cause and effect–unintended outcomes)

- Rehab-effort evaluation
- Evaluation of firefighter exposures to contaminates and cancer threats
- Identifying and establishing hazard zones

It is noteworthy that some departments are already on this path, and the successes are encouraging. So, do incident safety officers really help? I think they do, and they will continue to help prevent needless injuries and death. Perhaps, though, I should make an adjustment to my stated goal: The ultimate goal is to embed safety officer thinking/actions for all fire officers, especially those in division/group tactical supervisor roles, so we can bridge any tactical gap issues and defeat the NIOSH 5.

Resource Tracking

The IC must be astute at resource tracking. Who are your division/group supervisors, who is working for them, and where are they located? This is vital to accountability. In addition to tracking tactics and the completion of tactical objectives through the SAW-CSS-RECEO-VSS system, we must also track who is doing what, where, and with whom. Fortunately, new tools are available to replace the pencil-and-paper methods of the past. Figures 7–18 through 7–21 show screens that are utilized for incident command, personnel, and resource tracking (photos courtesy of Tablet Command).

Summary

Incident planning and organizing requires pre-planning and training before an incident is dispatched. This allows the IC to quickly develop an IAP with rapid execution, organization, and assignment of resources on scene. An incident must remain in alignment between the strategic, tactical, and task levels. Otherwise, conflicting efforts and confusion will create potential for the NIOSH 5 to come into alignment. The IC must determine the incident priority, strategy,

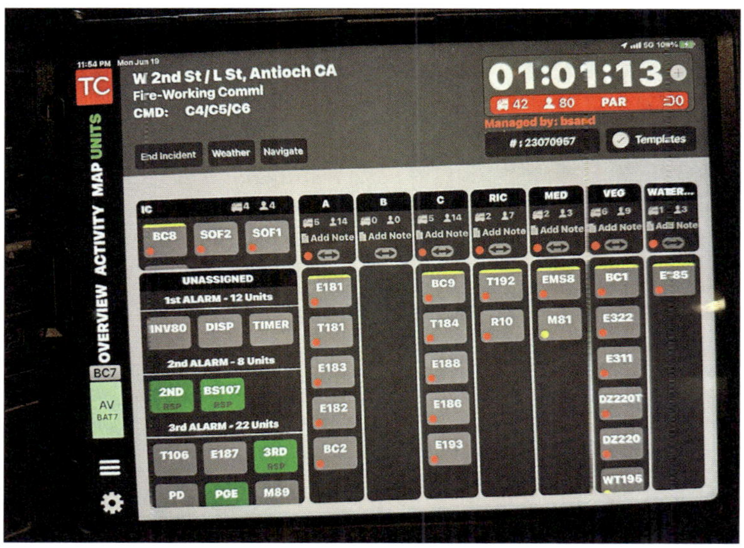

Fig. 7–18. A working commercial fire with divisions and groups established and alarms listed on the left. The unit listed at the top is the respective division/group supervisor.

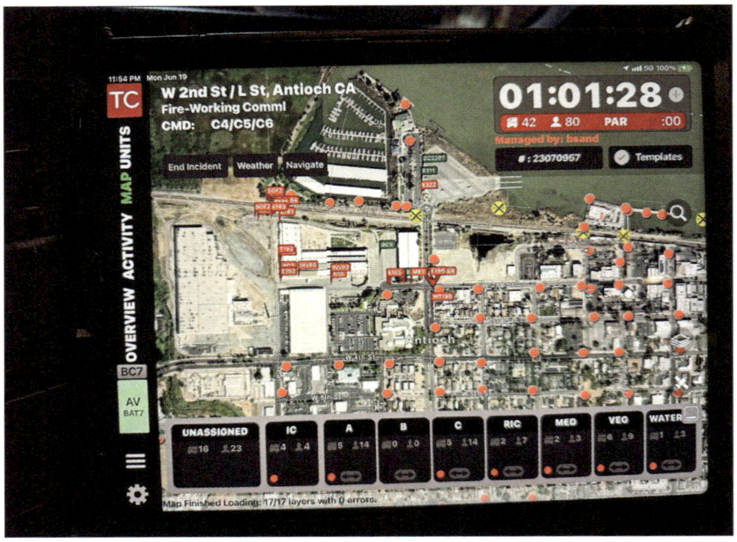

Fig. 7–19. Physical locations are identified with hydrants and links to pre-plans.

Fig. 7–20. Utilizing groups at a residential structure fire

Fig. 7–21. Personnel accountability for each unit, computer aided dispatch and automated vehicle location can be linked for real-time accountability of resources and personnel.

and give clear tactical objectives to the division/group supervisors to execute with their companies. Then, the IC and tactical supervisors must maintain that alignment throughout the incident. This makes the IC an incident chiropractor.

The incident power curve is the predictable and common sequence of events that ebb and flow through an incident. The incident will first be chaotic during initial size-up and arrival, and as crews get to work. Then, an initial improvement in conditions will cause a lull in the operations. This is often followed by a surge in demand and activity, such as fire in the overhead or below, an exposure involved, roof collapse, victim found, or Mayday, for example. The IC must remain vigilant for these shifts and remain ahead of the curve.

SAW-CSS-RECEO-VSS is a system that has been utilized for several years and that the IC can use to remain organized and fulfill tactical benchmarks throughout the incident. Rather than check box command, the intent is to help the IC train beforehand and remain organized during the incident. This does not alleviate the need for the IC to have requisite knowledge of fire behavior, building construction, tactics, SOGs, and so on.

Known as BRO-HAM, the tactical (division/group) supervisor must have the boundaries, resources, and objectives of their area and remain hostile, agile, and mobile to fulfill the job to bridge the tactical gap. The IC and tactical supervisors must be on the same page regarding BRO.

ICS can be quickly and effectively applied to any structure fire (or other incident) by beginning with ICS Templating based upon the ICS formula of Building + Conditions + Resources = ICS. This allows the IC to rapidly develop an IAP and utilize ICS to quickly organize the incident to remain ahead of the incident power curve.

One critical role of the tactical supervisor is safety officer. By embedding the role of safety officer into the expectations of each division and group supervisor, through training and SOGs, each tactical supervisor will become a safety officer and the IC will have the benefit of multiple sets of eyes around the incident constantly assessing safety issues.

Chapter Review

Review Questions

1. Describe the "then/what if" mindset.
2. Explain the difference between command maturity and command presence.
3. Provide two examples of a combination strategy.
4. Describe the challenges and preventative means associated with managing an incident within an incident (IWI).
5. Explain the command options/modes and situations for each.
6. What is the primary purpose of the tactical supervisor?
7. Define the acronym BRO-HAM.
8. Explain incident templating and the ICS formula.

FESHE Strategy and Tactics (C0279) Related Content

Chapter 7 provides detailed information to assist with instruction and education for expected course outcomes of students in C0279 Strategy and Tactics with regard to pre-incident planning (2 & 3), size-up (4) and the utilization of ICS/NIMS as it relates to strategies and tactics. This Incident Planning and Organizing chapter content directly supports the two course objectives: (1) Create a strategy and implement appropriate tactics), and (2) Possess a working knowledge and execute implementation of ICS/NIMS for an incident.

NFPA 1021 Job Performance Requirements

The information in this chapter can be utilized to support training and educational programs associated with the Emergency Services Delivery Fire Officer I JPR 4.6, 4.6.1, 4.6.2, Fire Officer II JPR 5.6, 5.6.1, and Fire Officer III 6.6 and 6.6.1.

Section II

Commanding Structure Fires

8

Single-Family Dwellings

Fig. 8–0. Firefighters work an attached garage fire in a house.
Source. Photo courtesy of Los Angeles City (CA) Fire Department

Introduction

A *single-family dwelling* is defined as a stand-alone residential structure built for occupancy of one family (fig. 8–0). Houses/homes can be a few hundred square feet up to tens of thousands of square feet in size. Houses/homes come in all shapes and sizes. Countless home types exist and can be described by era built, construction type, design, size, and builder. Craftsman, Victorian, Queen Anne, split-level, modern lightweight and mobile homes are just some of the categories of homes across America.

Large estate dwellings (aka McMansions) are houses that exceed 3,000 square feet (per NFPA 1700), often found in suburban areas (fig. 8–1). These homes are built in tracts for owners who want greater square footage but cannot afford a fully custom large home. They are characterized by open floorplans, tall ceilings, false dormers, and various décor that can be mass produced to cost-effectively enhance the look and style of the home. Other common characteristics include truss roof construction, tile roofs, stucco siding, brick facades and veneer, energy efficient windows, and grand archways, to name a few.

Conversely, truly custom large homes that are over 3,000 square feet may have characteristics not typically associated with a house. These may include, but not be limited to, wine cellars, theaters, elevators, walk-in refrigerators/freezers, multiple kitchens, and service entries. Some jurisdictions have residential fire sprinkler requirements for homes over a certain square footage, and they may have fire department connections (FDCs).

Fig. 8–1. McMansions are more common each year.

CASE STUDIES IN COMMAND

The Stilt Court Fire

CHIEF JEFF HELVIN, SACRAMENTO CITY (CA) FIRE DEPARTMENT

We sometimes think the unthinkable happens to the other guy. On October 7, 2008, I was that other guy. My engine company, Engine 15, was part of a full first alarm assignment dispatched to a reported structure fire. We arrived as the first due engine to find a newer two-story wood-frame single-family dwelling with heavy dark smoke showing from the second floor on the A, B, and D sides of the structure. An offensive attack was initiated with a 1¾" attack line while having to force entry through the front door (fig. 8–2).

I saw the first floor clear of any smoke and was able to see clearly from the front door to the rear sliding glass door that there was no obvious smoke downstairs. We committed the attack line up the stairwell to the second floor. As we advanced upstairs the smoke started to get thicker, and by the time we reached the second floor we were in zero visibility with no heat. I scanned down the long hallway with the thermal imaging camera (TIC) but was unable to see the screen due to the thickness of the smoke. We continued crawling down the hallway, feeling for some type of heat to indicate where the fire may be. At the end of the hall was a large master bedroom.

This is where we started our search. I was off the hoseline searching for the seat of the fire and any victims while remaining in voice contact with the firefighter on the nozzle. As we were conducting our search in the master bedroom, the second-due engine was arriving, securing a water supply, and pulling a secondary hoseline. The captain assumed command, updating the incoming battalion chief (BC) with situational awareness and completed a 360° view of the fire. . . all good things! Meanwhile, I continued

Fig. 8–2. Stilt court home

the search for life and fire in the walk-in closet when the nozzle firefighter stated that the hoseline "just went flat." I wasn't too concerned, as I figured it was the engineer switching from tank water to hydrant water.

Shortly after, I made my way back to the hoseline to check pressure myself, and the hose was completely flat. At that point, I knew something was not right. I started to crawl back to the hallway to check conditions when I was hit with a tidal wave of extreme heat. My two firefighters on the stairs, assisting with the stretch, were getting pushed up the stairs and down the hall from the heat and fire below them. It was confusing, chaotic, and terrifying. I did not know where the fire was or why it was blowing up, but I knew I needed to get everybody in front of me to exit the building and regroup. We all started to crawl back down the hall with me in the back of the line when we got bottlenecked at the top of the stairs.

This is where we all got separated. I was taking a beating from the heat by the open stairwell and had to turn back to escape what felt like the skin melting off my body. The first two firefighters tumbled down the stairs and made it out the front door. The third firefighter went head first out a second-floor window. My plan was to retreat to the master bedroom and exit out of a second-floor window.

I made several passes down a wall where I thought I had previously found a window, but could not find it again. I started to second-guess my location and thought maybe I got turned around; however, I remembered that I had passed over the abandoned hoseline when reentering the master bedroom.

It was at this point that I felt like I was not going to make it out alive. My air supply was diminishing and I was unable to find a secondary exit, so I huddled up in a corner to await my demise. I thought of three things: the firefighters that were with me and if they made it out and were okay, never seeing my wife and young son and daughter again, and the death process I would experience. I knew this was not acceptable, and had to make one more attempt at making it down the stairs. I reentered the hallway, got to the top of the stairs, and as I made it half way down I was enveloped in fire. I jumped over the banister handrail, landing in a small sitting room. I was completely disoriented and terrified, not knowing which way to go. I saw some outside light, which was an open sliding glass door on the B side of the structure. Once outside, I was assisted by other firefighters, assured my crew was out and accounted for, and transported to the local burn unit.

This fire started in the kitchen on the first floor. The kitchen is not visible from the bottom of the stairs due to a couple of interior walls. The smoke traveled up the open stairwell and out the open second floor windows, giving the illusion that it was an upstairs fire. Routine. . . right? As one can read, this was anything but routine. The fire service continues to, and always will, face many evolving factors that are unforeseeable, ever-changing, and out of our control. Below are the lessons learned. Some lessons are operational and department-specific, and some are lessons that are personal and unrelated to the incident but have changed the way I look at things moving forward.

Lesson Learned

Forcible entry and horizontal ventilation need to be synonymous. As we force entry this changes the flow path of the interior. We need to consider how this may impact our operation and fire spread.

We hear about performing a 360° on the outside of a structure, but need to consider a 360° interior prior to committing a hoseline upstairs. Had I performed a quick survey of the downstairs I would have noticed the fire in the kitchen.

While using the TIC upstairs to scan down the hallway, I couldn't see the screen and didn't understand why. In training, I always held the TIC about 6" in front of my facepiece, but this does not provide a view in heavy smoke conditions. I never thought to press the TIC to my facepiece to provide some type of view. This is not ideal, but it is better than nothing. Also, consider using the TIC from the outside when assigned to perform the exterior 360°.

When moving down a hallway searching for the seat of a fire, move together and do not leave unsearched rooms behind you. This could have been catastrophic.

Review Mayday and self-survival procedures at the beginning of each shift, and incorporate it into your daily radio and SCBA check. I was in a Mayday situation but never thought to call a Mayday or activate my personal alert safety system (PASS).

Now, from the standpoint of a BC, we should be reviewing how we will manage the Mayday and what we are going to say if a firefighter down or Mayday call comes in. On your worst day, do not assume that a drill from six months ago is going to have you respond at your very best. How we respond to a downed firefighter needs to be at the forefront of our memory, and we can only do this if it's part of our *daily* routine—part of our morning check.

Of the three firefighters that were with me that day, only one was on the right tactical channel. One firefighter had a dead radio battery and the other was on a different tactical channel. Give your radios the same respect as a paramedic does their heart monitor!

Have awareness of modern versus legacy construction and fuel types. Most of our fires and experience take place in the older sections of our district. Fires burn more rapidly and differently in newer construction. Are your operations and tactics adjusted when responding into newer built construction? Can we have modern fuels in legacy construction?

Lack of Respect

The greatest number of both civilian and firefighter injuries and fatalities on the fireground occur in house fires. The most common pitfall in house fires is thinking they are bread and butter or routine. This lack of respect for the enemy will lead to the enemy winning the battle and potentially causing injury or death to civilians and firefighters alike. *Remember, most Maydays, firefighter deaths or injuries, and civilian victims are in house fires.*

Many houses are often remodeled without proper permitting and adherence to code, and can have hoarding conditions or excess occupancy. Garages are remodeled into living quarters, and houses can be converted into makeshift commercial occupancies or healthcare facilities, often with no sign of such modifications from the outside.

Kitchen fires are the most common fire type. Other common occurrences involve improper heating, smoking, or use and storage of flammable materials. Low-mass synthetics make the smoke in today's house fires more flammable. The smoke itself is fuel (fig. 8–3). Garage fires and attic fires are a common characteristic of house fires and both can spur intense debate among firefighters across North America.

Fig. 8–3. House fires today pose a greater threat than ever, especially due to low-mass synthetics. *Source.* Photo courtesy of Rick McClure

The following statistics are from the NFPA:

- More than one-quarter (26%) of reported fires in 2015–2019 occurred in homes. Even worse, three-quarters (75%) of civilian fire deaths and almost three-quarters (72%) of all reported injuries were caused by home fires.
- During this five-year period, U.S. fire departments responded to an estimated average of 346,800 home structure fires per year. These fires caused an annual average of 2,620 civilian deaths; 11,070 civilian fire injuries; and $7.3 billion in direct property damage.
- Most home fires and fire casualties result from five causes: cooking, heating, electrical distribution and lighting equipment, intentional fire setting, and smoking materials.
- Over the five-year period of 2015–2019 as a whole, cooking was the leading cause of home fires and home fire injuries, while smoking was the leading cause of home fire deaths.[1]

From a command standpoint, it would be easy to dismiss house fires as routine or simple and therefore not in need of ICS application. Many firefighters have died and been injured in house fires, including in homes less than 1,000 square feet. *Size does not matter.*

Units in Sacramento, CA responded to a house fire in a structure that was no more than 1,000 square feet. The house was surrounded by debris and piled storage like a junkyard. Crews were unable to perform a lap due to the amount of hoarding in and around the home, requiring them to respond from two different streets to access the house. Within a few minutes, they had a Mayday from a firefighter and multiple civilian victims. Prior to the Mayday, the IC wisely set up Divisions A and C in the front and back, respectively. Not because of the size of the house, nor the number of resources. He did it to gain situational awareness from two tactical supervisors on either side of the house. In addition, this reduced the radio traffic on the tactical channel significantly.

The Mayday was not your textbook transmission of "Mayday, Mayday, Mayday. . . !" It was a quick, barely audible transmission, "I'm lost and out of air!" Due to the reduction in radio traffic by the reduced span-of-control, one of the division supervisors heard the Mayday transmission and crews were able to respond appropriately, and a disaster was averted—all before the IC could even ascertain what was happening. More on this fire in chapter 12.

Additional common pitfalls exist across North America regarding house fires. House fires are the most common type of structure fire you will encounter. It is easy to develop bad habits because of good luck. Some of these pitfalls may seem simplistic, but their results can be dysfunctional at least, and catastrophic at worst.

- Improper size-up
- No victim profile and delayed/missed search
- Lack of tactical supervision (divisions/groups)
- Lack of fire attack/vent coordination
- Garage fire tactical confusion
- Attic fire tactical confusion
- Misuse of transitional attack

The following sections look at examples and, more importantly, solutions to these common pitfalls of house fires.

Size-Up

Remember, lack of or improper risk assessment is number one on the NIOSH 5. On house fires, this can be catastrophic and can happen very quickly, as crews are eager to put out the bread and butter fire, often without sizing up what they have. According to Project Mayday, a 360° lap was not completed or only partially completed in 70% of the Maydays in career fire departments, and 62.7% in volunteer fire departments. Figure 8–4 shows a working attic fire that was confined to the attic due to electrical. This required a combination of vertical ventilation and fire attack from underneath.

Fig. 8–4. Modern house fires are no longer bread and butter.

I have had to remind my company officers many times that they must get a lap, when possible, before committing to the inside of a house fire. We often get into a hurry and focus on the task of stretching the first line and beating other companies to first water. Sometimes, a lap is not possible on a house; however, most of the time, it is. It becomes a matter of priority and training.

One of my captains kept bypassing a lap on his house fires. Finally, after several reminders, I told him that if he didn't do a lap on his next fire, I would send his engine back to the station. He had lots of excuses, like getting a pack on, helping with the stretch, locked gates, and so on. I get it, I've been there, but it's about making it a priority. We trained and he eventually got the point and on the next fire he performed a lap. Then, it became habit and he had a routine with his crew. The firefighter stretched the line, the engineer secured a water supply and irons to the front door, and the officer did the lap and update on the radio to incoming units.

Companies were dispatched to a house fire early in the morning. While en route, dispatch updated the BC that flames were visible from below the house, and this was repeated to companies on the tactical channel by the BC.

Two engines arrived simultaneously to find a two-story home with smoke visible and no fire showing from the A side. Neither engine officer on scene performed a lap and did not realize fire was below them as they entered the front door. Both companies raced to get water on the fire, and for a supply from the same hydrant. Approximately one minute later, fire was seen from the side of the house at the ground level, and this was not transmitted on the tactical channel.

Less than two minutes later, two firefighters fell through the floor into an involved basement. Both sustained significant injuries, as did other members who performed a gallant rescue of their comrades. If one of the first two officers would have slowed down the operation and performed a lap, they *may* have seen fire in the basement and performed an exterior attack. At the very least, they would have been able to size up the situation more thoroughly before committing to an interior attack.

The above is just one example. This same story has been repeated countless times across the North American fire service. Crews arrive, do not know what they really have, and bad things happen. Take the time to size up what you have before committing crews inside a structure.

We were dispatched to a house fire one night at approximately 0230. Unknown to any of us responding, the fire was reported by a fire medic ambulance that was transporting a patient to the hospital. The medic crew stopped at the fire and the driver got out and kicked in the front door without PPE or a SCBA. This created a flow path as the firefighter entered the structure to search, unbeknownst to anyone, without PPE, SCBA, nor a water can, let alone a hoseline. The house was abandoned, and windows were boarded up from the *inside*. The fire quickly grew due to the oxygen supply, and the firefighter narrowly escaped. The firefighter did not see this and never transmitted any updates to the responding alarm.

As the first BC, I noticed the fire medic unit leaving the scene when we arrived. Only later was I told of the actions of this crew. Needless to say, we could have had an unknown firefighter trapped in an abandoned boarded up home before any suppression units arrived on scene. I wish I was making this up; however, it happened and, thankfully, disaster was averted. Needless to say, the firefighter and his partner got a stern talk (putting it mildly) after the facts came to light.

First-due company officers must take the time to perform a size-up (FPODP or FIRST) prior to giving a verbal arrival report. Pulling past the house will give you three sides to see before getting off the rig. This should include a VP as early as possible. The VP drives everything else (priority, strategy, tactics, resource requests/assignments, tasks, etc.). An initial VP can be given upon arrival based upon the time of day, presence of vehicles, signs of life, notification from occupants, and so on.

After initial direction is given to the crew, the company officer should complete a 360° lap, if possible. Confirm the VP, look for TPOs (side doors, bedroom windows, breezeways), check for basements, see if the fire is in the attic, and look for forcible entry/egress challenges. Also, confirm the layout of the building and location of the fire. Transmit a 360° update to incoming companies.

Every medical aid call in a house to which you respond is an opportunity to pre-plan. You should gain awareness of the following, as a minimum:

- Layout of different housing tracts
- Occupancy habits and patterns
- Tactical points of opportunity (TPOs)
- Hording conditions of "frequent flyers"
- Location of utilities
- Neighborhoods with sparsely located hydrants
- Alleyways and alternative access routes
- Gates to private communities

Victim Profile and Search

Firefighters often experience tunnel vision on the flames and want to initiate fire attack without considering a VP early in the incident. Often, if a VP is not announced on the radio, search can be an afterthought. For FDs with SOGs, a search should be part of the initial alarm assignment. A VP gets everyone on the scene on the same page regarding the amount of risk versus gain and sets the tone for the rest of the incident.

Victims have been found in house fires after the fire is overhauled and even discovered by investigators. We had a fire in a mobile home park that was dispatched at 0630. as a "mobile home fire with elderly female trapped." Upon arrival, the first-due engine reported fire in a *motor* home and directed incoming units to cancel the water supply. Two engines were on scene upon my arrival as the first BC. I assumed command and was told by the first engine on scene that the primary search was complete, no victim found. I directed the second engine to perform a secondary search and check with bystanders to confirm an all clear for the motor home and trailer park. They responded a few minutes later, "Command, Engine 2, secondary search negative. No victim found. We have an all clear on the motorhome. We have also checked with neighbors, and no one has seen her. We are sure she is not here."

I then cancelled the ambulance and balance of the alarm. By this time, the local news had arrived to set up their van and boom for a morning news update. They interviewed me and I told them that we were dispatched to a fire in a mobile home. Upon arrival, it was a motor home, and no victim was found. I then walked over to check in with the crews and assist with picking up hose. The news crew continued with their preparations for the morning update with our fire scene in the background. As I walked to the motor home and observed the captain mucking out the interior of the vehicle, I noticed a tiny bathroom inside with the thin, cardboard style door closed on the bottom and burned away on the top from the fire. I asked the captain if crews had searched the bathroom. He said, "We checked it with the TIC." My heart immediately sank. I directed him to search again with his hands. After a few seconds, he said, "She's in there!"

The victim was a tiny elderly woman who had succumbed to the smoke and fallen off the toilet. The debris from the fire attack, including towels and remnants of the cardboard door, covered her. She had no heat signature visible to the TIC and was barely visible to the eye without moving the debris off her. Our paramedic firefighter pronounced her deceased at the scene. This was a terrible day and a lesson I would never forget.

First, while a great tool, the TIC is *not* a substitute for skilled searching by firefighters. The TIC is *not* X-ray vision. Rather, it recognizes the infrared spectrum, and that has inherent limitations. It cannot see through solid objects, including debris, drywall, lath and plaster, nor through windows. We must search voraciously, whether the VP is confirmed or low. Our job is to rescue civilians from fires, and technology will never outperform a well-trained firefighter who is bent on finding victims.

Second, mobile homes, motor homes, RVs, camper shells in the back of pickup trucks, and trailers can have victims just like a stick-framed permanent house. We must treat them as such and not be lulled into a less urgent mindset when searching these spaces. Kids can have sleep overs in campers and trailers (fig. 8–5), and many people are using these as alternative housing due to a host of issues, from the economy to green living to overflow (fig. 8–6). At the same time, if possible, do not get into check-box command and get a water supply before you search a motor home or trailer.

Search and rescue is the lifeblood of any fireground operation, especially in houses. We must have multiple tactical options available from conventional search, window-to-door removal, window-initiated search, and pincer search (chapter 6, page 131).

Fig. 8–5. Do not be complacent on travel trailer fires! If possible, skip water supply and get in and search ASAP.

Fig. 8–6. Mobile homes commonly have victims and the window shuts rapidly to make a rescue.

Tactical Supervision (Divisions/Groups)

Fire departments across America have an extremely disparate approach to commanding and organizing house fires. Some use no divisions or groups and feel that the span of control of the IC is manageable. This view lacks consideration for other benefits of ICS. Tactical supervisors bridge the tactical gap and prevent the NIOSH 5 from coming into alignment. This helps the IC remain ahead of the incident power curve, reducing radio traffic, and leverages division/group safety officers embedded into various locations around the incident.

Many FDs will go to the opposite extreme and utilize an operations position to supervise the fire and bridge the tactical gap. This places *all* the operational responsibilities onto the operations chief, and the above advantages vanish. Essentially, the IC gives the entire incident to operations, and that officer inherits all the problems the IC had. This is due to skipping the lower levels of the ICS hierarchy. *Divisions and groups should be utilized and employed well ahead of section chiefs like operations.* We will discuss this later in this chapter.

Another common phenomenon is the use of fire attack as a radio designator, rather than a true ICS tactical supervisor position like a fire attack supervisor, who supervises multiple crews performing the objective of fire attack and water supply. This can create confusion as fire attack may cause several companies to individually answer the radio, and they may or may *not* perform the critical objective of search due to the designation.

Commonly, an engine may be assigned to fire attack. Then that crew is called fire attack, even though several other companies may also be performing that task. Other companies may also be called the same, and confusion with accountability and radio traffic ensues. Transmissions from the IC such as, "E21, you are fire attack," can cause confusion. Is E21 now called fire attack? Are they simply performing the task of fire attack, or are they a group supervisor from whom other companies are supervised?

Details matter, and this disparity can cause confusion when lives truly hang in the balance. For this reason, the terms initiate versus assume, as discussed in chapter 5, are crucial to illuminate any confusion. Remember, initiate is for task-level assignment. Assume is for a tactical (division/group) supervisor assignment.

Another common practice is to assign the second chief officer arriving to a house fire to Safety, thereby widening the span of control as the safety officer does not have tactical responsibility nor do crews report to them. The safety officer is just one more person talking to command. Additional confusion is brought as the chief assigned to safety does a lap of the house. Crews often gravitate to this chief and unity of command (one boss) gets blurred. The crews seek direction and authority from the white helmet walking by who must resist the temptation to jump into the tactics. This happened during a house fire one night with a report of a victim. I was assigned safety officer. During my lap, I discovered the known would-be victim outside the house. I directed some crews accordingly and told command of the situation. During our debriefing, he scolded me for not staying in my lane as the safety officer. This exemplifies the check the box mindset of some incident commanders. But technically, he was right.

Alternatively, each tactical supervisor has the responsibility of safety for their respective areas of the incident, and each has the authority to make tactical decisions, account for crews, and communicate with them directly, all of which reduces the span of control.

> Having division and group supervisors gives the IC the benefit of multiple safety officers and allows them to get ahead of the incident power curve by remaining in the "then/what if" strategic mindset.

Later in this chapter, we will illustrate several alternatives for organizing house fires utilizing ICS.

Tactical Decision-Making Disclaimer—Science and Experience

Tactics have always been debated and will always be debated. Today, the two approaches often come from a science/research perspective and an experiential perspective. It was once around the coffee table only that stories of firefighting lore would be told by Jakes as they sipped coffee and spoke of great battles, tactics, tools, and tricks of the trade. We would sit, jaws agape, sopping up the sweet nectar of wisdom and experience. These firefighters had fought the beast, survived, and had the scars to prove it. Their tactics were proven, time and again, on the battlefield. As we aspired to their levels of saltiness, we too had experiences that became our toolbox from which to engage in battle. We adapted. We kept some things, modified some things, and tried new things.

There were no YouTube videos to watch on our phones, or UL FSRI studies to read. So, books and magazine articles were the records and means from which to pass along tactical wisdom. Icons and luminaries arose from the East and West Coasts who wrote books and shared their experiences, tactics, wins, and losses. Names like Dunn, Stapleton, Brennan, Norman, Brunacini, Anthony, Mittendorf, Castro, Halton, and countless others rose and wrote iconic books and articles. And thank God they did! This inspired the next generations to do the same. To dare speak tactics and place words on a written page, as if to say, “I know what I'm talking about, here.” This brought both criticism and praise. “Way to go!” one minute and “Who the hell do you think you are?” the next.

The advent of scientific studies appeared to polarize the fire service almost overnight. The two camps of experience and science seemed at odds. This was apparent at kitchen coffee tables across America, tailboard after-action reviews, and countless classrooms and conferences where some students would walk out in disgust if the instructor didn't match their ideologies. This is where it gets dicey. We never want to stop listening to each other. Debate is good, as it causes discussion to (hopefully) learn from each other to come up with best practices. Neither should be considered gospel; however, both should be considered excellent sources from which to discuss and choose the best practice for any given situation. And every situation is different.

Looking back, each generation reaches a tipping point where they start to think they know better than the previous generation. The problem is that the previous generation is still there, working in the firehouse, or more likely at headquarters. This happens every 20 years or so. That's why we seem to always go back to the same tools we did in generations past. The new generation finds a new way to use an old tool that was used previously. First, water was okay to put in windows, then not, then okay again. First, smoothbore nozzles were the only way, then fog was the new deal, then smooth bore again.

The point is that we are about to embark on some more tactical considerations in this textbook. This may cause you to wince, disagree, or agree wholeheartedly. It's based upon your experiences, training, SOGs, regional differences, riding position, rank, articles, books you've read, mentors you've had, and countless other factors. Just keep reading. We attempt to show both experiential and scientific approaches to a few of these topics as *both* have tremendous value.

As you enter the next sections and chapters, please maintain a humble heart and open mind. Sometimes there is even more to be learned through the shared experience or viewpoint of someone else than personally being present for the same event, because emotions are not a factor.

Fire Attack/Vent Coordination

Fire attack and ventilation have always required coordination. The UL FSRI studies have brought to light just how critical this coordination of tactics is from a scientific standpoint, and the next generation of firefighters have a clear understanding of this premise.

Often, crews on a roof performing vertical ventilation with a saw at high revolutions per minute (RPM) cannot hear the radio to coordinate with task-level crews inside the structure who are communicating through SCBA in low visibility. This is why a division supervisor in charge of the interior of the structure who is communicating for attack crews *and* a designated roof division supervisor who is communicating for vertical vent crews are *both* critical to the clear, effective, and expeditious process of communication and coordination of these vital tactics.

Division A, for example, often does not need to communicate with attack crews, as they can see the conditions and progress of the fire attack and can quickly confirm for the roof division that they can open the roof. As the division supervisor on countless house fires, I have often communicated to the roof division without having to confirm with my attack crews inside that ventilation is needed. I could see conditions that confirmed that water was on the fire and simply give the roof division a thumbs-up from the ground, standing in the front yard of the house without having to try to communicate with a crew at the end of a nozzle (fig. 8–7). Vertical ventilation is another subject of much debate. It works very well when it is done correctly and water is applied. One common question we face while teaching is whether fire venting the roof prior to getting on the roof is a cause to eliminate vertical ventilation as an option. It depends. “Fire venting itself” or “it’s already vented the roof” does not mean that the vent is doing what you want it to do, nor having the effects inside the structure. The location or size of the fire venting may be inadequate. For example, heavy smoke and heat may be in the living space from a fire that extended into the attic. The attic fire is minimal, just starting to breach the decking. The crews inside need relief. Or a strip mall may have fire over the involved unit, but trench cuts are required to prevent lateral spread over exposures B-1 and D-1.

A good truck officer will take note and identify that the fire has weakened the roof assembly to some degree if it has already vented. This is a risk/benefit size-up decision. Experienced truck officers who do vertical ventilation on a regular basis are more likely to go to the roof, even if it is venting upon arrival, if conditions inside need more ventilation. That said, *you must have adequate training and experience to perform this tactic.* And always get water on the fire.

In figure 8–8, fire vents after the hole is cut. This will commonly occur during an attic fire, or after the truck punches through the ceiling to relieve heat, smoke, and gases underneath in

Fig. 8–7. The close proximity of roof division and division A supervisors (orange helmets) makes for easier coordination of fire attack and vent.
Source. Photo courtesy of Jacob Salzman

Fig. 8–8. Fire is supposed to vent the hole in the roof, so get water on it from *underneath* ASAP. The small caliber line is for protection and prevention of ember casting. *Source.* Photo courtesy of Joe Lyneis

the living space. Note the small caliber hoseline. This is for protection of the crew as they cut. It was once taboo to take a hoseline to the roof, let alone place it into a vent hole. I have seen this have catastrophic results and injure firefighters below, as happened to two good friends on a duplex fire. That said, a protective hoseline for the crew is appropriate. In addition, my crews have occasionally used small caliber lines to cool down a very small attic fire that was hard to access from below, or from the gable or eaves. This must be done cautiously and judiciously. Again, training and experience are vital.

PPV is even more critical to coordinate, as the induction of fresh air into the box can have immediately catastrophic results. Fire conditions can quickly escalate and incur a flashover into what was an otherwise vent-controlled fire. This is precisely what occurred one morning at an apartment fire. I was the IC, and an eager captain turned a fan into a doorway without the attack team having water on the fire, nor an adequate exhaust opening. The result was a flashover. From the ICP, I saw a bedroom window go from black smoke to flashover in seconds. The search crew inside came out with their turnouts off-gassing. No coordination occurred and thank God we didn't have any injuries.

Garage Fires

Take a look at figures 8–9a and 8–9b, the two simulated images. The first (fig. 8–9a) has a well-involved attached garage. Note light smoke from the turbine vent on the roof. How do you expect the fire to grow, in what direction, by what avenue, and how soon?

Figure 8–9b shows extension into the house (likely the kitchen in an older home) and definitely the attic (two turbine vents). Would the fire grow in the time that you arrived, forced entry, and stretched a line? Would transitional attack on the way in the front door make any difference? How soon is the second-arriving engine on scene? What is your staffing? These all matter. If you have adequate staffing and/or resources, consider your first line in the front door to protect potential occupants and the uninvolved. First, cut off the fire (confine), then extinguish. Once it's in the attic, you should get inside the living space to check extension and occupants ASAP.

Attached garage fires are a common occurrence throughout North America. With the advent of McMansions and their "Garage Mahal" garages that often boast three or four bays, significant garage fires can easily burn down an entire home or ignite exposures, creating a tactical challenge upon arrival of the first-due company.

This occurred one morning when the first company arrived to find a well-involved attached garage fire. The company officer reported, "E201 on scene, we have a two-story house with heavy fire from the garage, and what appears to be bedrooms above. We are initiating fire

Fig. 8–9a. The fire in the garage appears to be extending into the attic.

Fig. 8–9b. The fire is now in the living space and extending into the attic. Get inside ASAP.

attack inside to push the fire out. Next in, give us a water supply. This is Main Street Command." The second-due company arrived and announced, "E202 on scene. We can hit it with our deck gun." This drove the E201 crew out of the house as they were getting pummeled by the heavy stream coming at them while they made it to the interior garage man-door. The result was that the fire extended into the home's living space and the house was a total loss.

Now, as you read this you are likely in one of two camps: those who believe that E201 was correct, and those who believe E202 was correct. Therein lies the notion of tactical confusion on garage fires. Who is correct?

As you can see, the garage is embedded into the home with what appears to be bedroom windows overhead. All incidents begin with a victim profile. Neither company identified a VP, nor announced one as they began their respective operations and gathered more information. In addition, with two three-person companies on scene, there are clearly enough personnel for several tactical options. They are not limited, for example, by two-person staffing, delayed companies, or distant water supplies. This further allows for multiple choices in this particular scenario.

Using our previous FPODP size-up process, consider the following from E201:

- Facts: Two-story house with fire in garage.
- Probabilities: Moderate VP, vertical and horizontal extension. Bedrooms overhead.
- Own Situation: First-arriving engine, with additional companies arriving directly.

- Decision: Life saving priority, offensive strategy.
- Plan: Establish command, go inside to confine the fire from the living space and search, secure a water supply.

Contrast this with E202's size-up:

- Facts: Two-story house with fire in garage.
- Probabilities: Moderate VP, vertical and horizontal extension. Bedrooms overhead.
- Own Situation: Second-arriving engine, with additional companies arriving directly.
- Decision: Life saving priority, offensive strategy.
- Plan: Attack the garage with a deck gun from the street without a water supply.

Which company is out of alignment? Given the circumstances, E202 is out of alignment, due to the VP and the fact that fire is potentially getting into the living space and there are bedrooms above. Now, some of you reading this would say, "but taking the heat out of the garage and resetting the fire is a good thing and can only improve conditions for potential occupants and firefighters."

Let's take a closer look at this. The UL/NIST studies have given us great information to consider when making tactical decisions. Transitional attack (hitting it hard from the yard) has become an option that was otherwise taboo in generations past. The advent of modern construction and the use of low-mass synthetics for interior contents and décor have certainly required new tactics, like transitional attack. Fires grow exponentially faster and reach flashover more rapidly than ever before. So, taking some heat out of the fire on the way into the structure has advantages.

Blasting the fire from the street with a deck gun is *not* transitional attack, especially when another company has already committed to interior operations. This is a defensive tactic. It is not even a combination tactic, as both companies are affecting the same space and opposing each other, to the detriment of one (E201). As we discussed in chapter 7, a combination strategy requires close coordination to prevent situations just like this one.

In addition, getting to potential victims as soon as possible is the priority here. Even if E201 was not in position behind the garage to get impacted by downstream master stream application, the deck gun is still not the best option with these conditions and staffing/resource levels. Any time taken away from getting into the house, confining the fire with interior hoselines, and searching behind and above the fire takes precious seconds away from potential victims who must be removed from the environment as soon as possible. Even when heat is removed, lethal doses of carbon monoxide and hydrogen cyanide are present and killing your victims.

From the inside, E201 can confine the fire from getting into the living space, take heat out of it, and allow the fire to exhaust steam and products of combustion out of the building and away from occupants. They are also able to get to any victims much faster, as they are *inside* the building with them.

The difference is discipline. *Tactical discipline* is the act of restraining oneself in the moment to consider alternative options that could make conditions more favorable, rather than reacting in the moment with adrenaline or emotion. Firefighters are born, bred, and wired to put water on the fire, break stuff, and go into danger. That is what we do, and it is what makes us who we are. That said, we sometimes act in the moment without taking a second to see the bigger picture and consider other options that may take a few seconds longer but have better overall benefits to the incident. That is why tactical supervisors are there, to see the bigger picture. Had E201 performed a quick transitional attack on the way into the home, that would have been appropriate. That is *not* what E202 did.

If E202 had arrived first with two personnel, an adequate water supply, and the expectation of the next company to have a prolonged arrival time, then a deck gun might be a better choice. Every factor must be taken into consideration (VP, construction, staffing, additional resources, timing, water supply, experience, and training of the crews, etc.), *not just what is in front of you*.

Generally speaking, when conditions and resources allow, getting into the house as soon as possible should be the first priority for the first company on an attached garage fire. This has several benefits:

- Firefighters get into the house to search for potential victims as soon as possible.
- In the likely event that the fire has breached the fire wall into the attic or living space, firefighters are positioned to confine the fire as soon as possible.
- If fire is confined to the garage, you have still checked the house for victims and can then consider alternate attack options, such as using a side door or front door, or front roll-up door. A charged line should remain inside in case conditions change.

Figures 8–10a and 8–10b show alternative methods of attacking an attached garage fire. Figure 8–10a shows indiscriminate use of a fog pattern from the open garage. This is *not*

Figs. 8–10a and 8–10b. Two methods of attacking the attached garage fire that is encroaching on the house.

transitional attack. This has potential to pressurize the attic and living space. Yes, you cannot push fire with water; however, you can introduce pressure and air with a misapplied attack and reverse the flow path of smoke, heat, steam, and other products of combustion. Contrast this with figure 8–10b, which shows the tactical discipline to check the inside for occupants and extension and put the line between the involved and uninvolved.

When the fire is clearly isolated to the garage (by the fire wall), as confirmed by an interior search for occupants and fire, another method is to attack from the person-door on the side of the house, if possible. These may be heavily secured (fig. 8–11). This may also be considered if you have a low VP, confined fire, and have limited staffing without the means to quickly open the garage roll-up door.

Fig. 8–11. A firefighter attacks an attached garage fire that is isolated to the garage from the side door.
Source. Photo courtesy of Rick McClure

Attic Fires

Attic fires in homes are common. Electrical wiring is a common culprit in many parts of the country. Fire extension into attic spaces and cocklofts from fires in the living space or on the exterior is likely, as the fire wants to travel upward due to convection. HVAC vents, kitchen ducting, and lighting fixtures in the ceiling make easy pathways for fire to extend into the spaces above firefighters. Garage fires commonly extend laterally into attached homes due to breached fire walls between the garage and attic over the living space.

Outside-in fires, where the origin is outside the house, often extend into the attic through soffit or rafter vents. Plastic garbage cans are a common culprit, as discarded ashes, fireworks, cigarettes, oily rags, and other ignition sources cause the fire to easily travel up the siding into vents. Other possibilities are intentional starts outside, vehicle fires, or vegetation fires, to name a few.

Like the garage fire, attic fires spark contentious debate among firefighters around North America. Most debate centers on two things: the location from which to attack the fire, and whether vertical ventilation is effective. The most important thing to remember before getting into these discussions is that, in most buildings, the garage and attic spaces are compartments separate from the living spaces and should be treated as such, with confinement and compartmentalization as the priority.

It is not uncommon for first-arriving firefighters to enter a house or apartment, encounter light smoke conditions in the living space, and report they have a working attic fire, immediately followed by the radio traffic, "we are going to start pulling ceiling." Once again, the attic is a separate compartment from the living space and the ceiling is the dividing wall. On far too many attic fire scenes, the living space of the dwelling and the occupant's belongings are in perfect condition upon arrival, and within 30 minutes, hundreds of square feet of drywall and cubic feet of insulation will be soaking wet on the floor and furniture through the work of firefighters.

An intact ceiling at a working attic fire is a very effective salvage cover and fire barrier. When it is recognized that a fire is in the attic (non-occupied) and not the living space (occupied), take advantage of the time this barrier provides to complete important objectives of primary and secondary search, followed by aggressive salvage operations to protect the most important property in the structure, the stuff below the roof line.

As with all tactics, each has a time and place, and attic fires are no exception. There is rarely one size that fits all in any aspect of fireground tactics. Attic fires can be quickly contained and extinguished by breaching the ceiling and attacking from underneath. The benefits are that the seat of the fire can be more directly hit by streams, and access is relatively simple. However, tall ceilings, lath and plaster, ducting, wiring, and hording conditions can make this challenging. In addition, heavy fire overhead with a tile roof should be a watch-out situation, as catastrophic collapse can occur quickly when lightweight web trusses and gusset plates are impinged by flame.

If you do attack the attic fire from underneath, here are some proven tactics that you may want to consider:

- A bathroom off a hallway is a good location from which to initiate an attic attack inside. These hallways are usually centrally located, so you can reach a lot of the attic space from that location. You can also access the ceiling easily without an attic ladder by stepping up onto the toilet, then the counter, using them as makeshift steps.
- Open a small hole with a hook. Insert the nozzle (1¾" line) with the *widest fog pattern possible*. Open the bale *fully* as it passes the plane of the ceiling into the attic. Shove the nozzle up into the attic space to the point that it starts moving around by itself, usually 3–5 feet. At that point (3–5 feet), hold the hose firmly while the nozzle tumbles around by itself in the attic. Move the hose in a circular motion to keep the nozzle moving and bouncing off the roof any joists, rafters, ducting, trusses, wiring, and so on. It's like inserting a giant sprinkler head into the attic space, yet easier than a cellar nozzle over your head (fig. 8–12).
- Keep the flow until conditions improve. This may take 1–2 minutes or more. The fog is converting to steam and the nozzle is acting like a giant sprinkler head. Again, depending upon the amount of fire, the British thermal units (Btus) may take a minute or two to succumb to the water and steam. Eventually, the gallons per minute (GPM) will outdo the Btus and prevail.
- If your FD performs vertical ventilation, then opening the roof further away from the fire (6–12 feet) will provide release of the heat, pressure, and energy out of the attic. The reason for opening further away from the fire and not directly over is for safety, and it allows the steam conversion so the water doesn't just flow right out of the hole without absorbing Btus.
- A common pitfall is not using a wide fog and spraying up with a straight stream pattern from a crouching position in the living space. This will only cause the straight stream to penetrate the attic and the stream will not absorb any Btus as it shoots out the roofline.
- Once an initial conversion or knockdown occurs, cut a scuttle and use a ladder to get into the attic to spray water directly onto the roof construction members and coat them with water as needed, or reach corners and distal areas. Be judicious with the water so you don't do more damage with the water than the fire did.

Fig. 8–12. Inserting the nozzle on widest fog several feet above the plane of the ceiling is crucial to knocking down an attic fire.

- Larger attics may require use of a ladder initially and more than one 1¾" hoseline. Consider attacking from other interior locations in large attics.
- Have a backup line in the living space below to protect firefighters from any change in conditions.

In 2012 UL FSRI completed a study titled *Study of Residential Attic Fire Mitigation Tactics and Exterior Fire Spread Hazards on Firefighter Safety.* The study included lab and field experiments involving attic fires with both internal and exterior starts and how traditional and non-traditional tactical approaches impacted them. The following are the tactical considerations listed from the study.

An overview of these tactical considerations includes:[2]

- **Increased use of plastics in exterior walls will change what you arrive to**—Changes in residential wall construction methods are playing an important role in how exterior fires are initiated, as well as how they spread and extend.
- **If the fire starts on the outside, start fighting it from the outside**—Rapid water application to knock down the exterior fire is a critical part of any attempt to control not only the fire's spread to adjacent structures but also the fire's migration into the interior of an exposed building. If the source of the fire is not suppressed, it will continue to supply heat energy to the fire developing on the interior, worsening conditions on the inside for occupants, and in many cases making it impossible for the interior crews to maintain or advance their positions.
- **Learn to anticipate where and how an exterior fire will migrate to the interior**—Exterior wall fires may easily spread to the interior at locations other than the eaves and soffits. Any penetrations—such as air vents, electrical receptacles, plumbing

penetrations to faucets and drains, and especially windows—provide the opportunity for fire spread into the interior of the structure. Leaving the interior fire barrier in place until the exterior fire can be controlled will limit the extension into the structure.

- **Attic fires are commonly ventilation-limited fires**—The openings provided for natural ventilation are not sufficient to maintain steady state burning and fuel-limited fire behavior. The size of the fire is limited by the available oxygen and will nearly always become ventilation-limited. Controlled openings created below the neutral plane (such as through the ceiling below the attic space) will not cause immediate growth and can provide access for suppression operations.
- **Closely time or limit vertical ventilation until water is in the attic**—A vented attic fire was more difficult to control with the indirect methods applied to the unvented attic test. The "open up above and then attack it from below" tactic can and has been successfully used at attic fires. However, it can create a large amount of property damage and puts both civilians and firefighters at high risk during the initial stages of the operation if not timed properly. Once initial water absorbs some energy, a vertical vent will assist the crews with suppression and overhaul because standard fireground ventilation tactics will be sufficient for exhausting the smoke and fire gases produced by the remaining fire. In the absence of suppression, the positive effect of a roof opening is a very short-lived phenomena. The accelerating fire can overwhelm all openings and push back into the occupied space. Increased visibility does not automatically mean a reduction in the size of the fire over your head.
- **Plastic ridge vents can affect size-up and fire dynamics**—As the vents heat, the plastic melts and collapses on the opening at the peak, creating a very effective seal. Once the ridge vent seals, the eaves will act as both the source of air as well as the exhaust and you may notice a pulsing of smoke out of the eaves.
- **Wetting sheathing with an eave attack slows attic fire growth**—If crews wet the sheathing, either as part of an offensive fire attack or defensively to slow fire spread to uninvolved sections of the structure, the major flame spread mechanism in the attic is eliminated until the moisture evaporates. Removing the soffit and flowing water along the eave line of these structures was the most effective way to gain the upper hand on a fire that was venting through the roof.
- **Attic construction affects hose stream penetration**—The most effective water application takes into consideration the construction of the attic, using the natural channels created by the rafters or trusses to direct the water onto the vast majority of the surfaces.
- **Consider flowing up instead of down with a master stream**—Consider using an aerial device or portable ladders and handlines to open up the eaves and flow water into the attic. This approach could result in controlling the fire enough to permit firefighting crews to transition back inside the structure to complete searches, suppression, and overhaul.
- **Knee wall fire dynamics**—During a structure fire, it is possible for fire to enter void spaces and surround crews conducting interior operations. Even though there is a delay between making the breach and the change in conditions, once initiated, the transition to untenable conditions in the area of operation occurs in seconds. Knee wall construction often provides the potential for ideal fire growth, with air entering low at the eave line and combustion gases exiting the peak through mushroom vents, ridge vents, or gable vents.
- **Apply water on a knee wall fire at the source and toward the direction of spread before committing to the attic**—Applying water utilizing the same path the fire took to enter the void space may be the most effective method at slowing fire growth. Water application to the knee wall will not be effective until the source below it is controlled with direct water application.

- **Interior operations on knee wall fires**—Tests have demonstrated that the most effective way to get a handle on knee wall fires is to control the source fire, cool the gases prior to making large breaches in the barrier, and then aggressively open the knee walls to complete extinguishment, focusing on wetting the underside of the roof decking.

Working Fire Example of Eave Attack with Coordinated Vertical Ventilation

Figures 8–13a and 8–13b are from a working attic fire that started on the exterior. Two lines were inside, with crews reporting very little progress on the fire by pulling ceiling and flowing. Command called for vertical vent and another line to try to get it at the gable end. Gable end siding and access with fence prevented water application and the line was moved to the A side, where soffit was pulled and ladder crew completed vent.

Heat trapped at the ridge auto ignited just out of the vent as the crew descended. With about one minute of flow along the Alpha-side eave, the fire was under control and all smoke was being drawn up and out of the vertical vent hole as the eave vents and interior ceiling breaches became low-point intakes with a high-point exhaust.

Notice smoke at the door and porch in one picture and not the next (figures 8–13c and 8–13d, respectively). This operation highlights the value of coordinated fire attack and ventilation operations and several of the UL FSRI attic fire tactical considerations.

Fig. 8–13a. Interior attack from underneath was not effective on this attic fire.

Fig. 8–13b. A coordinated eaves attack and vertical vent was ordered.

- **If the fire starts on the outside, start fighting it from the outside**—The attic fire was quickly mitigated by water application from the eaves. However, this was an adjustment by the IC, not the initial action. On the interior, water application was delayed due to decking added in the attic for storage. When there are enough members available to apply exterior water and extend an interior line simultaneously, take the opportunity.
- **Attic construction affects hose stream penetration**—The most effective water application takes into consideration the construction of the attic, using the natural channels created by the rafters or trusses to direct the water onto the vast majority of the surfaces.
- **Wetting sheathing with an eave attack slows attic fire growth**—If crews wet the sheathing, either as part of an offensive fire attack or defensively to slow fire spread to uninvolved sections of the structure, the major flame spread mechanism in the attic is eliminated until the moisture evaporates. Even if the eave line had not completely extinguished the fire, it would at least result in fire control.

Attic fires were common in Sacramento, and we had countless opportunities to fight them in different ways. An extremely effective way of attacking, as mentioned previously, was attacking from underneath, and coordinated vertical ventilation. The key is that the stream inserted into the attic must be at the correct location (as close to the fire as possible), and the stream

Fig. 8–13c. Gases venting from the hole auto-ignite as expected.

Fig. 8–13d. The water from the eaves combined with the vertical vent drew the fire and smoke up and out of the hole, improving interior conditions.

must be a very wide fog pattern. Essentially, you are inserting a giant sprinkler head into the attic to create massive steam conversion. This conversion will initially knock the fire.

Finally, it was once extremely taboo to take a hoseline and insert a stream into a vertical vent hole. It is still not a great option and has potential for bad results, as the weight of the water and impact of a misused straight stream can and has brought ceilings down onto firefighters, as was the case on a duplex fire we had that injured two firefighters.

This option may be considered in houses that are confirmed to be abandoned, as it may minimize risk to firefighters if they are not in the would-be living space, where other dangers may loom. I had crews take a booster line to a roof to access a stubborn attic fire that was localized, unable to be reached by crews below. The roof division surgically and judiciously used the line to knock down the fire and prevent spread until crews underneath could make adequate access. This should be an exception and *not* the rule and *must* be highly coordinated between division supervisors.

Exterior Control—Transitional Attack (NFPA 1700)

Transitional attack is a tactic executed in an *offensive* strategy. For some, transitional attack has become the magic silver bullet of tactics, where it is *always* used, despite other factors and conditions. The timing of the UL/NIST studies, combined with a new generation of firefighters who learn quickly through media and are much more adept at technology, YouTube, and the internet, has created a propagation of belief for some firefighters that transitional attack is a *must do* in *every* situation.

For example, I arrived on-scene at a house fire to find a firefighter performing transitional attack on a fully involved garage. The fire had burned most of the roof off the garage and the flames were 50 feet high. The 1¾" line was doing no good, as the Btus outmatched the GPM, and this first line should have been inside to confine the fire as the garage fire was venting away from the attached living space (fig. 8–14).

Contrast this with hitting a window on the way to the front door—two totally different approaches. One is simply an adrenaline-charged emotional reaction, the other is a well-thought tactical decision as part of an interior attack.

Like all tactics, transitional attack has a time and place on the fireground. When done properly, it has outstanding benefits. NFPA 1700 defines the objective of Exterior Control–Transitional Attack to improve occupant tenability and interior conditions for fire control (https://www.nfpa.org/codes-and-standards/1/7/0/nfpa-1700). This is our goal with tactics for life: to improve conditions for occupants and firefighters alike. While we cannot push fire with water, we can impact the flow path, pressure, and direction of heat flow. Even this is not as much of an issue, as long as the stream is not a fog pattern.

A deck gun is a defensive tactic. If a deck gun is required, it is by definition *not* a transitional attack. Even if it is for a few seconds, a deck gun into a building is a defensive tactic. Occasionally, this can be extremely effective with large amounts of fire, with exposures threatened, or with limited staffing and resources with a water supply. You may choose to switch to an offensive strategy; however, this will take longer than a true transitional attack performed on the way into a house.

The key to successful transitional attack is tactical discipline (see previous section). When flames are billowing out of a window, every firefighter who has ever lived has a primal instinct to put water on the fire. Tactical discipline answers the "why, when, where, and how" to the "what" of transitional attack.

Why: The objective is to cool the space to improve thermal conditions and allow for the imminent entry by the crew into the building.

When: An offensive strategy is declared, and the objective is to get into the living space quickly. Enough GPM must be flowing to accomplish this from the handline you are stretching. If Btus are beyond the GPM capability of hoseline used during the transitional attack (not reducing flames/Btus) then a transitional attack is not effective and, therefore, not an option. You must enter to be transitional. If you do not or cannot enter, you are not performing transitional attack.

Where: On the way into the building. Transitional attacks are most effective into windows and man-doors *on the way into* a house or other living space. Consider the layout of the building and assess locations of doors and windows. If a doorway is nearby and can be utilized for the attack, that may be the most direct path, and a nearby window may act as an exhaust to let any pressure out of the building.

How: The S's—*straight/solid stream*, at a *steep* angle to deflect the stream across the ceiling, held Steady to limit air entrainment, for a *sustained* period of time to maximize compartment cooling and surface coating to reduce the potential for fire regrowth. Early transitional attack methods were short 15- to 20-second flows, while visual condition changes appeared at the window. Later interior fire stream studies and experiments showed this was not enough to improve survivability.

Figure 8–15 shows an appropriately applied transitional attack. The building was fortified, including bars on the windows. The fire was seated at the front. Transitional attack was required prior to entry. Notice the upward straight stream pattern.

Fig. 8–14. This is *not* an example of transitional attack.

Fig. 8–15. This shows the appropriate time, place, and method of transitional attack.

ICS for House Fires

The alternative to the safety officer is to break the house fire into manageable and functional parts that can be logically and individually supervised, yet laterally coordinated. A division supervisor who leads the crews entering the home from the portal of entry maximizes accountability and safety while minimizing radio traffic.

This also works for two-story homes and above. Even if a home has multiple stories, the division supervisor at ground level should supervise the crews who entered their side of the structure, as that is the portal of entry. They should know if E20, for example, went to the second floor with an attack line and TR1 went to the second floor to search.

If the supervisor were to be on the second or third floor with a central open stairwell, they would not be in a position to truly supervise due to smoke and heat, nor to look at exterior conditions. While the supervisor may decide to temporarily enter the hot zone to check in with crews and look at conditions, this is *not* the ideal place to reside.

Let's look at several ICS templates for house fires.

Smaller Home

One template from which to start is a 4:1 model with the following components (fig. 8–16). A simple establishment of Division A to supervise crews inside the house has the following benefits:

- Significantly reduced radio traffic and fewer crews are transmitting on-air with inaudible radio transmissions.
- Reduced span of control for the IC.
- Greater tactical supervision and consolidation of resources.
- A safety officer is embedded into the incident.
- The Division A supervisor can have eyes-on, hands-on active accountability of crews in real time going in and out of the structure.
- The Division A supervisor can see and anticipate deteriorating conditions that neither the crews in the hot zone nor the IC can see, including performing laps, if possible, or delegating to another firefighter.
- Tactical coordination can occur with operations outside of the structure, such as ventilation, RIC, and medical.
- The IC can get into the "then/what if?" mindset and stay ahead of the incident power curve.

Division A

- Boundaries: The inside of the house.
- Resources: All crews working inside the house.
- Objectives:
  - ✓ Initially, fire attack and search.
  - ✓ After knockdown, transition into salvage and overhaul.
  - ✓ Positive pressure ventilation or horizontal ventilation would also be supervised under Division A to ensure maximum coordination under a single supervisor.

Fig. 8–16. ICS can be applied to a smaller home very quickly.
Source. Illustration courtesy of Sam Wallace

Roof Division

- Boundaries: The roof. If crews are on the roof, regardless of objective, a roof division alerts all personnel that rooftop operations are underway.
- Resources: Crews working on the roof.
- Objectives: If vertical ventilation is the tactic of choice with a crew(s) on the roof, a roof division supervisor is critical in maintaining situational awareness and tactical coordination with Division A, laterally. This can be through radio transmissions or hand signals. This eliminates the need to go through the IC and delay ventilation/fire attack coordination while unnecessarily increasing radio traffic.

Ideally, no less than three personnel should perform vertical ventilation on a roof: the officer to supervise and coordinate, one firefighter to sound, create a louver, and punch, and the other firefighter to cut. The combination of potentially being above active fire, along with falling off the roof, creates a significant need for tactical supervision of just one crew, let alone multiple crews.

Even if a three-person truck company goes to the roof without the expectation of a second truck, the company officer should be identified as roof division supervisor while the crew of two remains in the task level.

A crew of two was performing vertical ventilation on a house fire. The officer was task-saturated and attempted to louver a vent hole that was cut by his firefighter. While creating a louver in the hole, the officer did not see his firefighter fall through the roof. When he turned around, he saw fire from the hole. The result was significant injuries to the firefighter and a Mayday with RIC activation.

Unlike the inside of a burning house, where visibility is typically poor, a roof division supervisor is in the hot zone and able to see. This tactical advantage allows the first truck officer to the roof to assume roof division supervisor. Subsequent companies would report to that officer.

I once had a large two-story custom luxury home that was approximately 10,000 square feet and had significant fire upon arrival at night. The roof division was comprised of two truck companies, and two divisions (Division A and D) were needed at the ground level to supervise the two portals of entry.

If the roof division does not have the objective of vertical ventilation, other duties could include checking extension, or even confinement from an exposure. Again, if positive pressure or natural ventilation is the tactic of choice, it should be coordinated through Division A.

RIC Group

- Boundaries: The entire incident.
- Resources: All companies assigned to the function of rapid intervention.
- Objectives:
  - ✓ Coordinate with division supervisor(s) for accountability of crews.
  - ✓ Soften the structure without worsening fire conditions.
  - ✓ Secure utilities.
  - ✓ Develop a tactical plan for potential firefighter rescue based upon current and future conditions.
  - ✓ Establish a RIC cache of equipment readied in case of intervention.
  - ✓ Supervise rescue of downed firefighters if IWI occurs.

Medical

- Boundaries: The entire incident.
- Resources: All ambulances and other resources assigned to medical treatment and transport.
- Objectives: Treatment and transport of all civilians and firefighters. This includes establishing a medical area on the A side of the home with gurney and medical gear. Medical can often be a simple designation of one ambulance in the early stages of a house fire, or it can be comprised of a medical group with a supervisor and multiple ambulances and other resources based upon the needs of the incident as it escalates.

Figure 8–17 shows an organizational chart of 4:1 ICS template on a smaller house fire.

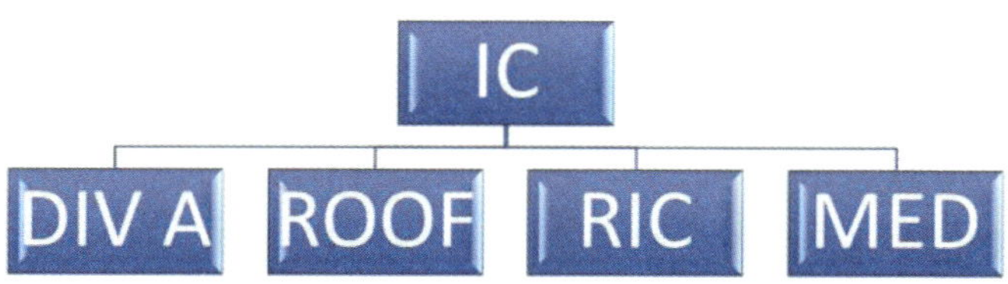

Fig. 8–17. ICS 4:1

Larger Home

An alternative template would be for a large home that requires more than one division. Extremely large homes or homes with difficult access may warrant more than one division at ground level (along with the potential for the previous divisions/groups as follows). The division supervisors should post up at the doorways on their respective sides (fig. 8–18).

Division A/B/C/D (fig. 8–19)

- Boundaries: The inside of the house as entered from the respective division. For example, crews entering to operate from the back of the home would work for Division C.

Fig. 8–18. ICS can be applied to a larger home that has more than one attack point.
Source. Illustration courtesy of Sam Wallace

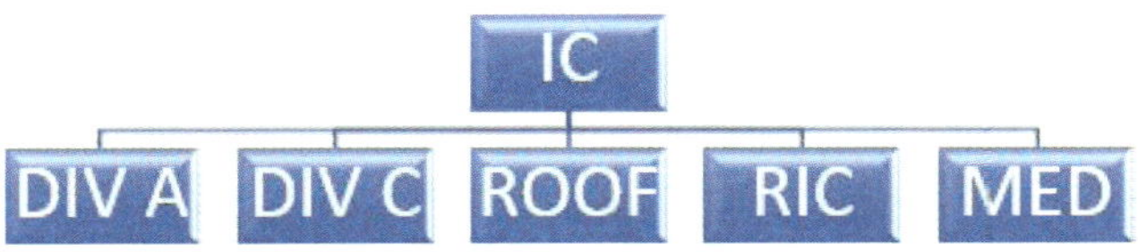

Fig. 8–19. ICS 5:1

- Resources: All crews entering from that side/division.
- Objectives: Crucial to coordinate with other divisions to prevent conflicting and redundant operations/effort. Initially, fire attack and search. After knockdown, transition into salvage and overhaul. Positive pressure ventilation or horizontal ventilation would also be supervised laterally under division supervisors to ensure maximum coordination under a single supervisor.

Confirmed Rescues in Homes

Another template would be for known rescues in homes with known victims (confirmed victim profile). Here, utilizing a fire attack group supervisor and a rescue group supervisor instead of division(s) to coordinate the critical objectives of fire attack and search divides the labor more clearly (fig. 8–20).

All things being equal, most firefighters naturally lean toward fire attack or search as a tactical priority. This is a derivative of experience, training, classes taken, articles read, UL/NIST studies, apparatus assignments, and so on. Obviously, conditions like the size and layout of the home, number and location of victims, size and location of fire, and number and type of resources responding all drive that decision. Still, as you read this, if you were responding to

Fig. 8–20. ICS 4:1 for known rescues

a house fire with victim(s) and no additional information, you would lean one way or another. And you would not be wrong either way. In general, both need to happen simultaneously and in coordinated fashion.

Therefore, it is easy for a single division supervisor to get overwhelmed if they were assigned to coordinate fire attack *and* search/rescue of multiple victims as Division A, for example. A more effective and efficient option is to break the crews inside the house fire into fire attack and rescue groups. A medical group would also be needed for treatment and transport of the victim. If vertical vent were performed, a roof division would also be appropriate. Otherwise, fire attack group could coordinate PPV, horizontal, or hydraulic ventilation.

The coordination of getting victims out and to the medical group is not easy and requires that the rescue group and medical group supervisors remain attached at the hip to minimize the TTT to get the victim(s) to ALS intervention. Remember, according to FRS data, the victim must be out and to definitive care within 6–8 minutes to survive.

This template works well for homes, duplexes, triplexes, fourplexes, condominiums, town homes, and garden apartments with relatively localized fire. If you have fire involving a lot of units, consider multiple divisions and a rescue group for the known victim(s).

Summary

House fires are the most common structure fire and where the most firefighter injuries and fatalities occur. Complacency kills. Considering a house fire a bread and butter job is a sign of disrespect that could cost you and/or your crew the ultimate price. At the very least, it is ignorant and dismissive of those firefighters who have lost their lives to these structures.

Countless home designs exist. The crucial aspect of incident command and tactical decision-making is knowing the home construction in *your* jurisdiction. McMansions are larger homes that are built relatively inexpensively for the amount of square footage they boast. Some of these building features, open spaces, and lightweight methods can have catastrophic results.

In addition to disrespecting house fires, additional pitfalls exist from a command and tactical standpoint. These include improper size-up, no victim profile, lack of tactical supervision (divisions/groups), delayed/missed search, lack of fire attack/vent coordination, garage fire tactical confusion, attic fire tactical confusion, and misuse of transitional attack. Both experience and science play an important role in determining the best tactic at the time. Every situation is different.

ICS templates for house fires, like all structure fires, are based upon building, conditions and resources. Most houses can be organized by either a Division A, Division A/C (or more based upon the above), or fire attack and rescue groups for known victims.

Never treat a house fire as routine or bread and butter and practice the varying ICS templates based upon your resource levels and building types.

Chapter Review

Review Questions

1. Explain some of the factors that contribute to complacency with regard to single-family dwelling fires.
2. Describe the differences between fire attack" as a radio designator and a group supervisor position.
3. Provide three tactical considerations for garage fires.
4. Provide three tactical considerations for attic fires.
5. Describe the application of exterior control/transitional attack.
6. Provide an example of how you would utilize three division/group supervisors given a first-alarm single-family dwelling fire.

FESHE Strategy and Tactics (C0279) Related Content

The content contained in chapter 8 provides detailed information to assist with instruction and education for all course objectives with a focus on the single-family dwelling. The chapter format supports the C0279 course outline by speaking specifically to basic division of tactics, rescue, exposures, confinement, ventilation and salvage for single-family dwelling fires to include fires in attic and garage spaces of these structures.

Access the Tactical Review Guide Appendix C #1: Three-Story Victorian Home with Basement Fire Extending for an objectives-based scenario exercise.

NFPA 1021 Job Performance Requirements

The information in this chapter can be utilized to support training and educational programs associated with the Emergency Services Delivery Fire Officer I JPR 4.6, 4.6.1, 4.6.2, Fire Officer II JPR 5.6, 5.6.1, and Fire Officer III 6.6 and 6.6.1.

Access the Tactical Review Guide Appendix C #1: Three-Story Victorian Home with Basement Fire Extending for 1021 Job Performance Requirement based scenario exercise for Fire Officer I, II, and III.

Endnotes

1. Marty Ahrens and Radhika Maheshwari, "Home Structure Fires," *NFPA Research*, November 2020, https://lowittalarms.com/wp-content/uploads/2021/10/oshomes.pdf
2 Steve Kerber and Robin Zevotek, *Study of Residential Attic Fire Mitigation Tactics and Exterior Fire Spread Hazards on Fire Fighter Safety* Underwriters Laboratories Fire Safety Research Institute [UL FSRI], November 26, 2014), https://fsri.org/research-update/study-residential-attic-fire-mitigation-tactics-and-exterior-fire-spread-hazards-0.

9

Multi-Family Dwellings, Hotel, and Motel Fires

Fig. 9–0. Oklahoma City FD fights a multi-alarm fire involving an apartment complex under construction.
Source: Photo courtesy of Oklahoma City (OK) Fire Department

Introduction

A *multi-family dwelling* (MFD) is a structure designed to be occupied by more than one family as the primary living space. *MFDs have potential for massive loss of life.* Many are often overloaded with considerably more occupants than designed or accounted for officially.

MFDs include smaller structures like duplexes, triplexes, fourplexes, condominiums, and townhomes. These often have an open-center stairwell; however, some are subdivided and have been remodeled without adherence to building codes. Larger MFDs include old and new law tenements, row frames/row houses, low-density housing, and center hall and garden-style apartments (figs. 9–1 and 9–2).

Some are specific to certain regions, like a brownstone in New York or New Jersey, while others are generic and can be found in many places throughout North America. Construction can be balloon frame, conventional platform, legacy, or lightweight. Again, the key is to know the buildings in your area. *The life loss potential is astronomical. Pre-planning and drilling in your MFDs is vital if you plan to save lives.*

Garden apartments are designed to have the door to each individual unit open to the outside. Multi-story garden apartments will typically have exterior open walkways and stairwells that serve one section/side of the building. Some have exterior partially enclosed hallways that lead to upper units (fig. 9–3).

Fig. 9–1. A small fourplex

Fig. 9–2. Brownstone row houses

Center hall apartments are designed to have the door to each individual unit open to an interior enclosed hallway. Stairwells to upper floors are enclosed and separated, feeding each floor (fig. 9–4).

Unlike a stand-alone apartment building, an *apartment complex* is a series of MFD buildings located in close proximity to each other, under one management company, with multiple access points and passageways into the area for parking and walking. To compound matters, apartment complexes are, by definition, *complex*. Confusion and lack of pre-planning can create a bottleneck of companies and delay access to the fire building (fig. 9–5).

Fig. 9–3. A typical garden apartment layout

Fig. 9–4. A center hall apartment building

Fig. 9–5. Apartment complexes are particularly challenging with access.

These facilities are often gated and may have hydrants located deep in the complex. Blocked fire lanes, parked cars, awnings, dumpsters, and narrow passages make apparatus access to fires in these areas challenging. Companies can commonly access the wrong driveway into the complex, and hose stretches can be long and complicated. The buildings in apartment complexes can be either center hall or garden style, or a combination of both. Pre-planning is vital to successful operations in apartment complexes.

In this chapter, we will also examine low-rise hotels and motels. Hotels and motels are designed to provide temporary living accommodations, yet can have long-term residents as well. A *hotel* is a center hall type of structure comprised typically of multiple stories with enclosed stairwells and interior hallways that access guest rooms (fig. 9–6). These can also have an atrium-style interior with an open center and doors to individual rooms that open into the inner atrium space, accessed by open interior hallways. The majority of atrium hotels are mid- to high-rise.

A *motel* is a garden-style building with doors opening to the outside atmosphere, often next to parking. Motels typically do not have the extravagance of amenities found in larger hotels. These are designed for travelers needing temporary overnight respite (fig. 9–7).

The term *motel* originated with the Motel Inn of San Luis Obispo, originally called the Milestone Mo-Tel, which was constructed in 1925 by Arthur Heineman. In conceiving of a name for

Fig. 9–6. A hotel can be mid-rise to high-rise.

Fig. 9–7. Motel rooms open to the outside.

his hotel, Heineman abbreviated motor hotel to mo-tel after he could not fit the words *Milestone Motor Hotel* on his rooftop.[1]

From a firefighting and command standpoint, center hall construction is the most lethal, as occupants must enter secondary and tertiary enclosed exit routes (enclosed halls and stairwells) to escape. These fires commonly extend beyond the unit of origin and can fill these exit routes with heavy smoke and fire. Quickly accessing the fire to contain it, open hallways, and maintain egress for occupants is vital.

Common pitfalls exist across North America regarding MFD fires:

- Lack of pre-planning and training
- Improper size-up/access problems
- Inadequate victim profile/underestimating the potential
- Lack of tactical supervision (divisions/groups)
- Use of two-out crews with confirmed VP (victim profile)
- Difficulty in containing attic fires
- Lack of sufficient medical capabilities

The following sections look at examples and, more importantly, solutions to these common pitfalls of MFDs.

CASE STUDIES IN COMMAND

Apartment Fire with Three Rescues

CHIEF ROBBY BERGERSON, WACO (TX) FIRE DEPARTMENT

On a March morning at 9:00 a.m., fire dispatch transmitted a commercial structure fire assignment for an apartment fire with people trapped at 3530 W. Waco Drive, sending five engines, two ladders, and two battalion chiefs, along with a private ambulance and EMS supervisor to the scene. Dispatch relayed that they were receiving multiple 9-1-1 calls reporting the fire. Knowing that the first-due battalion chief was on the far-west side of the city just clearing a large trash fire, I self-assigned from headquarters, recognizing that I would arrive with the first-due company Engine 6.

While en route, I confirmed with dispatch how many people were trapped, and in which apartment. Two occupants were reported trapped in apartment C. Engine 6 and I (Chief 2), arrived simultaneously in just under three minutes with heavy smoke showing from the first floor of a two-story garden-style apartment building. The living room window of the fire apartment had just self-vented, and flames were now visible.

I assumed Waco Drive Command and announced a "confirmed victim profile" and that we would be in an offensive strategy. Engine 6, already deploying a 1¾" hoseline, was assigned to initiate fire attack at the task level. This apartment complex, built in 1975, had eight units per building, with four units accessible by each exterior door leading to interior apartment entry doors, and an interior stairwell to the second-floor units. Upon exiting, the occupant of the fire apartment left the door to the apartment on fire open, which caused heavy smoke and heat to block the exit of two apartments upstairs, because the stairwell was filled with toxic smoke and heat.

Per our protocol with a rescue situation, next-arriving Engine 5 positioned to give their tank water to Engine 6 (as a booster backup), allowing all three members to be assigned to search the second-floor apartments immediately. Engines 8 and 9 arrived moments later and were also assigned to search, with the Captain of Engine 8 assigned as the rescue group (consisting of E5, E8, and E9) supervisor.

Battalion 1 arrived and was assigned to incident safety officer, and I requested a CAN report from Engine 6's lieutenant about their progress. The lieutenant advised me that they were making good headway on knockdown of the fire, so I instructed him to remain at the task level and that I would get them a fire attack group supervisor once more resources arrived.

Engine 4 arrived and was initially assigned to forward staging, but the radio crackled shortly after that assignment was given advising that the rescue group had removed three victims and were in the courtyard needing EMS. Not having communicated with any of the private ambulance personnel yet, I instructed fire dispatch to ask for two more ambulances and then assigned Engine 4 to start patient care. Battalion 2 arrived, and I assigned him to become the medical group supervisor with Engine 4 working for him. My immediate goal was to get our personnel to begin treatment, and for the medical group supervisor to determine which apartments the victims came from, and if we had gotten the people out of apartment C with the only known victims.

Because we were looking for two victims and found three, there was initial confusion over which occupants were from which units, and that needed to be cleared up quickly. Two of the victims were from apartment C directly above the fire apartment, and the previously unknown victim was found in apartment D across the smoke-choked stairwell also on the second floor.

The private ambulance personnel were assigned into the medical group and assisted with initial care. Truck 2 arrived and was assigned into the rescue group to assist with continuing the second-floor searches. And Truck 1 arrived and was assigned with Engine 6, with the Truck 1 lieutenant given the role of the fire attack group supervisor. The fire was held at a one-alarm, without a supply line laid, and all the victims rescued had minor injuries.

This fire was quickly contained because Engine 6, operating at the task level, began knocking down the fire and buying time, which allowed the incident commander to commit nearly all later-arriving resources toward the rescue group until the victims were located and removed. This was greatly aided by the second-arriving unit, Engine 5, making their tank water available to Engine 6 rather than taking the time to lay a supply line from a hydrant. This protocol, used in rescue situations, has effectively allowed our department to maximize the personnel that can go to work immediately for whichever tasks need be performed.

Simultaneous fire attack and primary search operations began almost immediately. In these situations, if needed, the third-arriving engine lays the supply line. Additionally, having a medical group to complete the rescue process with treatment is important for positive outcomes. This incident validates the data from Firefighter Rescue Survey, which says that apartment fires with a victim trapped are most likely to have three victims. Using a simple ICS (incident command system) for confirmed victim profile incidents using functional groups of fire attack, rescue, and medical quickly and effectively allowed for this incident of over 11 resources to be resolved with a manageable span of control of 4:1 (fig. 9–8).

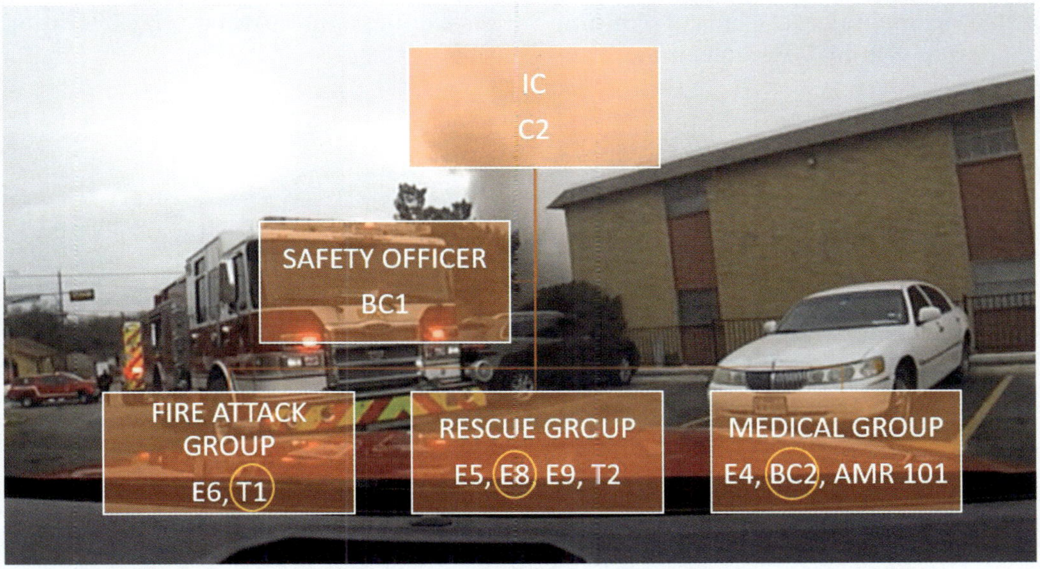

Fig. 9–8. An 11:1 span of control was brought down to 4:1.

Size-Up/Access

MFDs are not always easy to size up. Apartment complexes, in particular, create challenges, as described previously. Most civilian rescues happen at night, and the ability to see smoke coming from deep inside an apartment complex creates challenges. Often, 9-1-1 calls are made from apartment units adjacent to the involved unit and could have an entirely different street address.

Apartment complexes are commonly found on corners of city blocks and can have two different street names and associated addresses. In addition, the streets inside the complex could have their own names that are different from the main surface street names that are more commonly known by crews.

Water supply is often a problem, considering the following common issues:

- Engine companies sometimes race to be first in, and the first two companies go into the complex without having laid a stick of supply line. Limited access creates further problems, as the first truck may also dive into the complex. If no supply was laid, this can create significant delays in establishing a patent water supply.
- The first company goes into the complex with a wet lay, preventing subsequent companies from accessing the complex.

- The closest hydrant to the fire is accessed from a different driveway/entry point than where the first engine entered to attack the fire.
- Failure to bring multiple supply lines to support multiple apparatus can limit the availability of the additional handlines to address exposures or extension.
- If strategies change, the lack of planning for positioning apparatus for relay operations and larger volume flows may impact capabilities.

CASE STUDIES IN COMMAND

Apartments Under Construction

CHIEF DAN SHAW, FAIRFAX COUNTY (VA) FIRE

On a clear, crisp Sunday morning, I was off duty and at home when I received a phone call from one of our personnel working in the dispatch center to make me aware of a fire. This was not out of the ordinary and it was a step that ensured, despite us being a large fire department, that we maintained coverage if an incident depleted our resources.

Moments after the phone call, I clicked on a traffic camera that had a perspective of the area. The volume of fire I saw was momentarily incomprehensible. The call was received and processed swiftly, the first-due company was moments away and turned up quickly, so this volume of fire seemed impossible. Upon arrival of our personnel, an entire floor of a mid-rise wood-frame structure under construction was involved in fire. Despite a valiant effort in making an initial push and stop on the fire, the unimpeded path the fire had to the entire lumberyard of a structure was too much.

Fire had started in a trash chute and spread into the structure. With no sprinkler requirement to be in the building during various phases of construction, nor fire stopping or drywall, the fire had unlimited fuel and an unimpeded path. In my 30 years in the fire service, I have never seen fire travel so swiftly and consume absolutely everything in its path without hesitation, or appear to be undeterred by our efforts. An initial attack with hoselines, deck guns, and tower ladders did little to slow the fire travel.

The fire travelled faster than strategies and tactics could be developed, communicated, and implemented. The geographic area where this fire occurred consisted of several mid-rise residential apartments over commercial buildings and townhouses, as the area was undergoing redevelopment. As I mentioned earlier, without any firestopping, the fire spread from structure to structure. When I arrived, the fire was raining embers down onto the command post and other structures in this very urban area. It was seemingly creating its own weather, with a strong wind passing through the command post towards the ever-growing fire. The radiant heat made it impossible to walk down some streets that ran between the involved buildings and the exposures on the other side of the street. Telephone poles were alight, power lines were burning and dropping like strings, and windows were failing in the occupied exposure due to the intense heat, which allowed the fire to enter the threatened exposures (fig. 9–9).

In a phone conversation I had en route to the incident with our uniformed member in the dispatch office, I asked a question I thought I would never ask, having worked my career in a densely populated urban area. "Do we have any aerial assets available to drop water on this if our ground-based methods do not work?" Rightly so, this was met with, "what was that again, Chief?" Understandable, because we have never seen a foe like this that was so advanced and unphased by our firefighting efforts.

Fig. 9–9. The radiant heat from this fire created massive exposure problems.
Source. Photo courtesy of Fairfax County (VA) Fire

Eventually, from a change in the wind direction and a concerted fire attack effort, the fire was brought under control, but not before 5 five-story lightweight wood residential-over-commercial structures, 14 townhouses under construction, and 28 vehicles (in all, encompassing over 700,000 square feet of structures) were destroyed. We learned many lessons from this event and they continue to resonate today.

- We are a very successful "forward laying, one to two hoseline fire with a couple of rooms off in a residential structure" firefighting department. This is like many fire departments, given that every 64 seconds a residential structure fire occurs in the U.S. It is easy to be lulled into a level of security and success, but when your foe changes so drastically, you must recognize and meet the challenge head-on. The key element to even hoping to be successful in an event like this is that you must train for this event. How many of your personnel have a plan to tackle this fire? How often do you train on the rapid hook-up and implementation of your elevated master streams? Remember, the way we train will directly equate to how we will operate and perform on the fireground.
- Don't think a water supply group supervisor is only applicable in rural/non-hydrant fires! When a fire demands incredible amounts of water to suppress it, it is vital to dedicate one person, preferably a chief officer, to be consumed with assessing, planning, and executing a water supply plan in concert with the IC.

- When fire is consuming a large footprint, you need to adjust and get the aerial view. While it will depend on the assets you have available, you must be quick to utilize resources like putting one of your personnel in a police department helicopter to observe and report back, or use a drone, for example. This fire was so large that it showed up on the local news radar and looked like a thunderstorm in the area. There is no way you can begin to understand the scope of your problem if you cannot see it accurately.

Victim Profile/Underestimating the Potential

Since house fires are our most common structure fire for most of North America, they lull us into a rut, and we get into habits that are not as effective on other fires. They often create dangerous situations and near misses, or worse. We must recognize that MFDs have a much greater life hazard than a house fire. If a fire in an MFD has one known victim reported, it is likely that a total of *three* will be discovered, according to FRS.[2] Even a duplex, by definition, can have twice as many occupants as a single-family home. Apartments, row houses, tenements and the like all have massive potential for life loss.

Based upon your jurisdiction, SOGs, and resources, you may consider calling an additional alarm/resources prior to arriving to remain ahead of the incident power curve. You can always cancel them if not needed, but you will want them instantaneously if you *do* need them. Err on the side of the victims. Some triggers to consider include:

- Confirmed victim profile
- Visible column of smoke
- Multiple callers reporting the fire from inside or outside the building/complex

Most MFD fires should be considered a moderate to high victim profile at least. Apartment buildings, tenements, row houses and the like should be considered high victim profiles 24 hours a day. Economic pressures and trends in living conditions have created multiple families living in a 500-square-foot apartment. Many people are working from home, and many others have chosen not to work and are staying home during the day.

We were once dispatched to an apartment fire in a two-story garden-style apartment complex. The fire was dispatched at 10:30 a.m. on a weekday. Crews arrived to report a working apartment fire and initiated fire attack and search. As I arrived and assumed command, I backed into a parking space to stay out of the way. A sheriff deputy walked up to the window of my battalion chief (BC) buggy and casually said, “Yeah, he’s inside.” I responded, “Who’s inside?” Just then, M109 pulled out a severely burned, unconscious, non-breathing mid-20s male victim. Their medic unit was parked half a block away. Had I known we had a confirmed victim en route, I would have called an additional transporting medic unit.

Tactical Supervision (Divisions/Groups)

Another bad habit created from house fires is the lack of establishing divisions and groups early in the fire when confronting an MFD. This creates a tactical gap and causes the IC to get behind the incident power curve much more quickly than on a single-family house. As we have stated previously, the first company officer to an interior assignment is likely not the best choice to be a tactical supervisor due to task saturation; however, waiting too long will create more problems. Depending on the size and scope of the incident, you may need all your company officers to go to work at the task level with their companies. In that case, the second chief is a good choice to assume the hot division. They must be in full PPE and SCBA.

This occurred on an apartment fire early one morning involving a two-story garden-style apartment building. Heavy fire involved several units on the second floor and the attic above. As the first arriving chief on the second alarm, I was assigned accountability officer. The second chief on the first alarm was assigned safety officer, and no divisions or groups were established. The IC had a span of control of approximately 15:1, with more companies responding.

As the accountability officer, I went to all the apparatus and retrieved the FATS passports. One problem was that most of the passports were inaccurate and did not reflect the actual firefighters on the rigs that shift. Another problem was that it took a long time to determine where each company was located. Finally, the companies kept moving and the dynamics of their natural course of movement made any accountability obsolete. By the time I got back to the ICP with the passports, the effort was all but pointless.

If the IC had developed two divisions early, those supervisors would have had the role of accountability and safety officers as part of their tactical assignments. They can see where crews are moving in their areas, account for them in real time through eyes-on/hands-on active accountability, while looking at safety issues the entire time. As the second and third chiefs arrived, we could have filled in behind the initial two division supervisors (company officers) if needed, just as a chief takes over command from an initial IC.

Two-Out Crews

Two-out can be over- or underutilized depending on the respective fire department SOG and actual practice in the field. With this in mind, consider the greater potential for victims in MFDs. Fire departments sometimes wait for a two-out or back up crew before initiating a fire attack on an apartment fire, for example. Apartments, as stated above, should be considered occupied 24 hours a day. Even if the involved unit is determined to be unoccupied, it should be assumed that all exposure units are occupied until proven otherwise and be designated as high victim profiles. Therefore, two-out may need to be deferred based upon other conditions and considerations.

Containing Attic Fires

Like any fire, fires in MFDs will likely travel vertically through everything from balloon frame walls, utility chases, interior stairwells, and dumbwaiters, to heating registers, air ducts, interior windows, and airshafts. And during most fires, the first objective once a fire is located is to confine the fire from occupants, hallways, stairwells, and other means of egress. Once the fire hits the attic or cockloft area of an MFD, things change from a vertical confinement challenge to a lateral confinement challenge. Crews may think that they have the fire contained in the living space only to find that it has outflanked them and travelled unknowingly vertically into the attic. Attic fire access can be difficult, especially in older buildings.

The astute IC plans on vertical extension on most every fire, especially those in buildings with common attics. The key is to get ahead of the fire and create new tactical areas with supervisors who have clear control objectives to stop lateral spread. Without tactical supervision, individual crews will act at the task level and not think ahead to find a good TPO to make a stand, such as a firewall. Many times, fire walls are breached; however, they may still provide an anchor point to choke the fire down into a smaller pathway that is easier to contain. Roof division supervisors are particularly insightful in leveraging fire walls from their vantage point (fig. 9–10).

Fig. 9–10. Roof division on an apartment fire
Source. Photo courtesy of Sacramento (CA) Metropolitan Fire District

Medical Capabilities

The typical rule-of-thumb ratio for transporting medic units/ambulances to fire victims is 1:1. Unlike a typical medical call such as chest pain, fire victims often require much more care in the back of the medic unit, so doubling up patients may not be an option. Since an IC must be in the "then/what if" mindset, one additional medic unit/transporting ambulance should be available on scene at all times for unknown victims and firefighters. This creates the following formula for medic units/ambulances on the scene of a fire: *one transporting medic unit per fire victim plus one.* So, if you had two known fire victims on scene, you would ideally have ordered three transporting medic units.

Many regions in America do not have adequate resources for the one medic per fire victim plus one rule. When fires involve MFDs, the potential for fire victims escalates exponentially. Set up a medical group supervisor who can coordinate the treatment and transport of fire victims. This does not necessarily need to be a medic unit or even a paramedic. In fact, a typical ambulance crew member may not have the training and experience to fulfill a tactical medical group supervisor position. While a paramedic supervisor would be ideal, an emergency medical technician (EMT) company officer can fulfill the role with the right training. This allows their crew to assist with patient triage and treatment. A typical engine company should be able to fulfill this role. This keeps the often-minimal number of transport units available to transport.

On the center hall apartment fire described at the beginning of this chapter, I ordered three additional transporting medic units upon arrival, for a total of four, before I even knew how many victims we actually had. In the end, we had six transporting medic units on scene and all were utilized.

ICS and Communications for Multi-Family Dwellings

Duplex, Triplex, Fourplex, and Garden Apartments

Often, there is confusion on how to label or refer to the various units in these buildings. Whatever method you use, have common terminology in your agency/region to prevent confusion. Below are some considerations:

- First, it helps to refer to a unit and a building differently. For example, a *unit* is a separate living space in the same building that is adjoined to other units under the same roof, often connected by a common attic, wall, or floor. A *building* is a separate occupancy/structure that is not connected to other structures. For example, a duplex is a single building with two units. A strip mall is a single building with multiple units.
- Second, we can refer to the unit/building of origin where the fire started as the involved unit/building. The adjacent units/buildings are the exposed units/buildings or exposures. Exposed units can be above, below, or next to the involved unit.
- Third, the left or right side, in relation to the involved unit/building, can cause confusion. Regardless of where the fire starts, the involved fire unit of origin is typically called the A (Alpha) unit/building as a frame of reference. The exposed units/buildings are labeled as B for the left and D right. If the fire were to extend to another building to the right, that would become the D-1 exposure building, the next one over would be D-2, and so on. The building to the left would be the B-1 exposure building (fig. 9–11).

As with any structure fire, ICS is determined by the formula B+C+R = ICS. Often, a single division supervisor can manage a smaller duplex or fire held to one unit of an MFD. As the building gets larger or conditions become more complicated, more supervisors may be required (fig. 9–12).

If a lot of work needs to be done, then waiting for the second chief may be a better option. If the building, conditions, and resources (ICS formula) warrant, you may need to establish more than one division supervisor on the ground using Divisions A/B/C/D as needed.

Fig. 9–11. Because the fire started on the left, that is the involved/Alpha unit. The unit on the right is the exposed/Delta unit

Source. Illustration courtesy of Sam Wallace

Battalion 9 and crews arrived to find a fire in a duplex that had fire in the D-side unit on the right, in the attic overhead, and smoke in both units. As the second chief, I was assigned Division A with the objectives of fire attack and search of both units of the duplex (B- and D-side units). E54 and E62 were in the D-side unit attacking the fire and performing primary search and TR50 was on the roof as roof division performing vertical ventilation. E105 and TR65 were in the B-side exposure unit checking extension in the attic and searching.

As Division A, I communicated with crews face-to-face and roof division to coordinate the fire attack, search, and vertical ventilation. Since both doors opened to the A side, it was relatively easy to move back and forth between the two B and D units and talk directly to my crews. Figure 9–11 shows a diagram of this template.

A fourplex would likely necessitate the need for more than one ground-level division to coordinate tactical supervision. This was the case on a McKuen fourplex fire. Fire had started in the A side ground unit but traveled vertically through a utility chase into the common attic. This necessitated Divisions A and C to coordinate fire attack and search from both sides of the fourplex. I also established a roof division and RIC. Medical was assigned to stand by at the ICP.

Garden-style apartments and motels can be organized in similar fashion as all the units open to the outside. Even if you have two to three stories, you can still effectively supervise crews from the ground, as you can see the units and doors that flow out to the external walkways. You may need to utilize your portable radio more; however, most communications will be verbal or even visible hand gestures (fig. 9–13).

Townhomes, Condominiums, and Row Houses

Townhomes, condominiums, and row houses present a unique challenge in that they have a potential for vertical fire spread up multiple floors, and they have the potential for lateral fire spread to multiple adjacent buildings via a cockloft or common attic. Each building can be identified as a division with a supervisor and the companies assigned to that division working for the supervisor, regardless of the floor. Consideration should be given to setting up a Division C in these situations, if accessible. You may perform rescues off the back of the building, initiate fire attack from the rear, or want to watch exposures. Some cities have alleys behind MFDs, some do not. Again, know *your* area.

Fig. 9–12. This fire is held to one unit. Division A should be adequate for interior operations.

Source. Illustration courtesy of Sam Wallace

One additional challenge is understanding the layout of some MFDs, and which address goes with which door. Many MFDs can present and act like a row house yet be built or converted to be separate apartments. Many large homes are converted into multiplexes. Figures 9–14a, b, and c show two row houses that have been converted into three apartments each. Addresses and power meters may help size up the number of occupancies under one roof.

Figures 9–15a, b, c, d, and e show a four-story apartment building. Three doors to the upper floors empty into the A-side entry outside at the walk-up stairs, and one lower sub-level unit has a separate entrance to the right. Notice how each of the doors has a different address (fig. 9–15b). Each door leads to a separate floor, as seen by the internal stairs. How would you

Fig. 9–13. This three-story motel could be organized by Divisions A and C to get two sides of the building, especially if a fire is running the common attic.

a. b. c.

Fig. 9–14a, b, and c. Remodels can be identified by multiple addresses or meters on a former large single home or row house.

Fig. 9–15a, b, c, d, and e. (a) This may appear to be a single-family row house at first; (b) A closer look reveals multiple units; (c) The first-floor unit is at the entry door level; (d) Unit #1329 leads to the second floor; (e) Unit #1327 leads to the top floor.

organize this incident if you had a fire in Unit #1329? This is the second-floor unit (third-floor unit if you count the lower sub-floor unit). Using the template above, you could have Division A supervising all companies on all floors. The Division A supervisor would be busy and posted up at the bottom of the stairs at the sidewalk to see conditions on all floors simultaneously.

Division by floor number (Division 1, 2, 3) would handicap the supervisors, as they would be in the smoke and not able to see conditions. More radio traffic may be necessary between Division A and the task-level companies on the floors. The portal for each floor is the exterior doors, hence Division A would work effectively.

Townhomes, condominiums and row houses can be organized by a similar ICS template as the one described above. If fire were to extend laterally, you could grow your ICS to Divisions B or D as needed. Each division supervisor would have all the floors (usually two to three) in the occupancy. Since the portal of entry is the outside, divisions would make sense (fig. 9–16).

Fig. 9–16. The Division A supervisor accounts for all crews inside and on all floors. If the fire were to extend to the D-1 exposure, then Division D could be created.
Source. Illustration courtesy of Sam Wallace

We were dispatched to a fire in a townhome on a 115°F day at 2:00 in the afternoon on a summer weekday. E109 arrived first with the following arrival report:

E109 on scene. This is a two-story attached townhome with fire in the attic and smoke on the second floor. E109 is initiating fire attack. Passing command. Hydrant is 200 feet behind E109. This is a moderate victim profile.

E109 arrived first and initiated fire attack, passing water supply and command to the next engine. E109 still had a working command until the arrival of the next officer. HM109 (a 4-person truck company) arrived with E109 and split with two members inside for search and two to the roof for vertical ventilation. HM109 captain assumed roof division supervisor.

E108 arrived next and assumed IC from E109 captain as *working command.* His engineer assisted with securing a water supply and his firefighter pulled a second line to establish a 2-out crew, per the SOG.

I arrived, assumed command from E108, and announced the ICP location. After a brief transfer of command with E108 captain, I announced/confirmed an offensive strategy with low VP (low VP due to companies on scene with primary search underway, could have also been considered moderate), and that E108 captain would now assume Division A.

All units on scene, this is command. Battalion 7 is on scene. This is an offensive strategy. Low victim profile. E108 Captain is Division Alpha. We have a townhome involved. Division Alpha is E108. Everyone inside is working for Division Alpha. Objectives are primary and fire attack."

E108 captain assumed Division A supervisor, with E109, his crew (E108), half of HM109 (searching), and subsequent arriving units that would be assigned into his division. His objectives were fire attack and search of the involved townhome. E23, HM09 (Team 2), TR23 (Team 1), and E103 each arrived and was assigned to Division A as well.[3]

Division A: E108 Captain

- Boundaries: The inside of the involved two-story townhome, *both floors*.
- Resources: All crews working inside the involved townhome, both floors: E109, E108, HM109 (Team 2), E23, TR23 (Team 1), E103.

- Objectives:
  - ✓ Initially, fire attack and search of both floors.
  - ✓ Check for extension and evacuate the attached B-side exposure unit.
  - ✓ After knockdown, transition into salvage and overhaul.
  - ✓ Coordinate vertical ventilation with roof division.
  - ✓ After the fire is under control and roof division exits the roof, if positive pressure or natural ventilation is needed during salvage and overhaul, it should be coordinated through Division A.

Roof Division: HM109 Captain

- Boundaries: The roof.
- Resources: Crews working on the roof: HM109 (Team 1) and TR23 (Team 2). Total of four firefighters.
- Objectives: Vertical ventilation, including a strip cut to prevent extension into the B-side attached-exposure townhome. Coordinate with Division A.

RIC Group: E24 Captain

- Boundaries: The entire incident.
- Resources: All companies assigned to the function of rapid intervention: E24.
- Objectives:
  - ✓ Coordinate with Division A for accountability of crews.
  - ✓ Soften the structure without worsening fire conditions.
  - ✓ Secure utilities.
  - ✓ Develop a tactical plan for firefighter rescue based upon current and future conditions.
  - ✓ Establish a RIC cache of equipment readied in case there is need for intervention.
  - ✓ Supervise rescue of downed firefighters if an IWI occurs.

Medical: M65

- Boundaries: The entire incident.
- Resources: All ambulances and other resources assigned to medical treatment and transport.
- Objectives: Treatment and transport of all civilians and firefighters. This includes establishing a medical area on the A side of the townhome with gurney and medical gear. Medical can often be a simple designation of one ambulance in the early stages of a fire, or it can be composed of a medical group with a supervisor and multiple ambulances and other resources based upon the needs of the incident as it escalates. For this fire, heat exhaustion was a primary concern for crews.

If, in the above example, the B-side exposure unit were to become significantly involved in fire and Division A were to get overwhelmed with span of control, we could establish a Division B with the second-arriving chief. Otherwise, the second-arriving chief could assume Division A from E108, be available for another assignment, or be released back into the system. The second chief could also shadow E108 for mentoring to help develop the E108 captain since he was a newly promoted officer. He did a great job!

As with any confirmed rescue, you may have a hybrid ICS template of divisions and a rescue group. Division supervisors would coordinate laterally with the rescue group, and one supervisor per building may be a relatively simple template to manage.

Confirmed Rescues in MFDs

As with single-family dwellings, another template would be for known rescues (confirmed victim profile). Here, the primary objectives of fire attack and search are split between the functional fire attack and rescue groups. This would be utilized for smaller MFDs like duplexes, triplexes, fourplexes, townhomes, condominiums, low-density housing, and garden-style apartments that have limited fire but confirmed victim(s). Row houses in which the fire is confined to just one or two units with confirmed victims may also be organized with this template.

This was used on the El Camino Fire (described in the preface of this book) which involved a two-story garden-style apartment complex in which we had a child trapped. The incident was organized as follows: fire attack group, rescue group, medical group, and roof division.

If you recall from the preface, we had a commercial alarm already responding and on the road, headed to an extinguished sign fire at a Goodwill store. This pulled a very robust alarm with 5 engines, 3 trucks, 2 BCs, and a transporting fire medic unit.

While responding, I was notified by dispatch that another fire was being dispatched nearby for an apartment fire with child trapped. We literally had to pass the apartment fire with child trapped, so I directed that initial alarm to divert to the apartment fire, and within seconds we were on the scene of the apartment fire with smoke billowing out of the Charlie side of the complex. One engine continued to the Goodwill to find a man with a cut finger—case closed.

Meanwhile, the CAD system dispatched a separate alarm to the apartment fire, since our alarm was already shown as responding to the Goodwill fire. The net result was 9 engines, 6 trucks, 4 BCs, and 2 transporting medic units.

E106, TR106, and I (Battalion 7) arrived simultaneously. E106 *initiated* fire attack at the task level (Fig. 9–17). TR106 *initiated* search at the task level, announcing they would be all in for

Fig. 9–17. Heavy fire was blowing out the C-side apartment. Note the father pointing to the window for TR106.

rescue.[4] I assumed El Camino Command with the ICP across the street from the complex on the A side. This was an offensive strategy with confirmed VP (fig. 9–17).

Companies began to arrive in rapid fire, one after the other. There was a very brief tactical gap between E106, TR106, and I. E101 arrived quickly and I directed the captain to *assume* fire attack group (bridge the tactical gap). His objectives were water supply and to supervise all fire attack operations. He had E106 working for him and any other subsequent companies that were going to be assigned the objective of fire attack or supply as needed. The E101 engineer performed a reverse lay from E106 out to a hydrant several hundred feet away. He was accounted for in a non-IDLH assignment pumping back to the attack engine. E101 firefighter pulled a second attack line off E106 and worked under his captain in the fire attack group.

E110 arrived and I assigned that officer to *assume* rescue group supervisor with TR106 and his crew working for him. His objective was search and rescue operations for the whole incident, including known and unknown victims.

M30 arrived and was assigned to set up for treatment and transport on the A side by the ICP. M110 also arrived and joined M30, so we had one medic unit for the known victim and one for any other unknown victims or firefighters.

The next objective was vertical ventilation, as this was a second-story fire in the two-story garden-style apartment complex. Fire had extended into the attic and threatened three other apartments through a breezeway. I called the next-due truck company, HM109, to see what his ETA was in anticipation of assigning the officer roof division supervisor. His response was, "seven minutes." This was surprising, as I knew they were close by.

I then went directly to TR106 to see if he could split away two members of his crew to perform the vertical vent. Normally, if I needed to communicate with TR106, I would have gone through their boss, the rescue group supervisor. Herein lies a small detail. Rescue group supervisor, E110 officer, was not in position yet to communicate face-to-face with TR106, as he had just arrived on scene and was not back in the tactical space with TR106 on the C side of the complex. If I went through rescue group supervisor, it would have created much more radio traffic, as he would *still* have to contact TR106 directly on the radio.

I called TR106, "TR106, Command." A short response came, "Truck." I inquired, "Are you able to possibly split and give me vertical ventilation if we need it?" I received no response. TR106 was task-saturated with a VEIS operation and needed all four members to rescue the child from the second floor (fig. 9–18). They never responded.

Fig. 9–18. This image shows the VEIS in progress.

After a few seconds, HM109 announced, "HM109 will be on scene shortly, we can perform vertical vent." I then responded, "Copy HM109, you will be roof division upon arrival." Right then, another company arrived, "Command R20 (another 4-person truck) is on scene, do you have an assignment?" I then assigned them vertical ventilation since they were on scene before HM109.

The next company, E19 arrived. I assigned them directly to fire attack group (E101 officer) for an assignment.

The next two companies, Engines 105 and 20, were directed to come to the ICP for forward staging. This provided two companies that could be immediately assigned to any urgent request from rescue, fire attack, or medical. Also, if we had a Mayday (Heaven forbid) they could form a Rapid intervention group.

Just after Engines 105 and 20 arrived, the boy was run out to the waiting gurney (fig. 9–19), transported code 3.

The four-year-old boy was on a bed with a wool blanket that filtered the heat from burning his lungs, but not the carbon monoxide and hydrogen cyanide from rendering him unconscious with a lethal dose of CO in his system, at 23% (fig. 9–20).

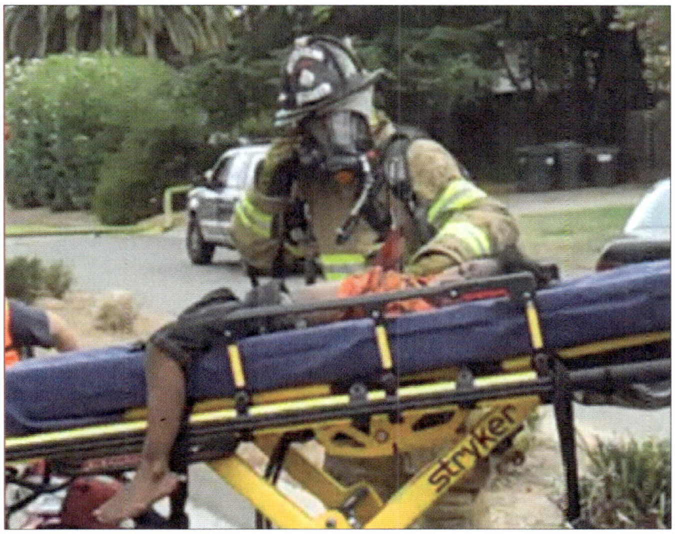

Fig. 9–19. A boy was rushed to the gurney in seconds.

Fig. 9–20. The victim was located under a wool blanket that shielded his airway from the intense heat.

This was a result of several miracles. First, we were already responding and passing the apartment fire, which saved two to three minutes. Second, the victim was in the back bedroom. If he had been in his own bedroom, which was fully involved on arrival, he would have died before we arrived on scene. Third, he was under a wool blanket in the summer, which filtered the heat from his lungs. Finally, the crews were trained and didn't hesitate. This was the right crew at the right time. At the time, VES (VEIS) was not a commonly known tactic in our region, yet Engine and Truck 106 were dedicated to training, knew VEIS, and worked flawlessly as a team that day to save a little four-year-old boy's life. Captain Kevin Wegener led the coordinated fire attack with his firefighter, Mike Lowry, while engineer J.P. Santos pumped the engine. TR106 performed VEIS. Firefighter Jeff Karges didn't hesitate. When he took the window for VEIS, he was blasted with hot, black smoke and dove into the window instantly. He felt the little boy under him in the dark and handed him out the window. Rather than close the open door that would have given him more refuge, he immediately handed the little boy out to TR106 Captain Steve Turner, who handed him to firefighter Walter Lewis, who handed him to engineer Brent Barsdale, who ran out to the gurney across the street in seconds, with the little boy over his shoulder. The M30 crew did an outstanding job of intubating the patient and regaining respirations on the way to the hospital. These men are my heroes (fig. 9–21).

I believe God got a call from the devil that morning. The devil told God he was going to take this little boy and use a fire to do it. I think God said, "You can try, but I am giving Engine and Truck 106 a two-minute head start."

All of the communications on this fire between the IC and companies were on a single tactical channel. I spoke to dispatch on a separate command channel to keep those transmissions off the TAC (tactical channel). At no time during the fire were there any radio transmissions regarding the VEIS, additional hoselines, or any other task-level communications. The division and group supervisors were trained, empowered, and entrusted to bridge the tactical gap and make those decisions as they deemed necessary without having to get permission or check with me, the IC. That would have only slowed down the operations, clogged the tactical channel with unnecessary radio traffic, and possibly cost the life of a child. Details matter. Remember, Training + Time = Trust.

Training + Time = Trust

Fig. 9–21. Engine and Truck 106 crew clockwise (from left) are Chief Kastros, Captain Kevin Wegener, Captain Steve Turner, firefighter Walter Lewis, firefighter Jeff Karges, our brave little fighter, and engineer Brent Barsdale.

Hotel/Motel

Occupant accountability for hotels and motels is critical and should be coordinated with management on scene. They can tell you what units are occupied and by whom. This will help with search-and-rescue efforts. While all units involved and exposed must be searched, a follow-up with on-site management can confirm what units had been assigned to potential occupants.

Hotels will likely be served by a division for the fire floor and any other floors requiring supervision of multiple crews. Like any structure fire in a larger building, you may choose to employ a rescue group for targeted areas of known victims. They would augment the divisions who have the objectives of fire attack and search as well. Remember, this is a lateral relationship between divisions and groups. Groups and divisions are both led by supervisors, meaning that they are at the same level of authority; therefore, groups work *with* divisions, *not for* divisions.

As with any MFD, think ahead regarding medical treatment and transport of victims of hotels and motels. Also, consider Red Cross for any displaced occupants of any MFD fire. Depending on the size and location of the fire, management of any MFD may have vacant units in which your displaced occupants may seek temporary overnight refuge.

Motels are likely to be organized by a Division A and Division C ICS template, as the doors open to the outside and even two or three floors can be supervised from the ground on either side of the motel.

Larger/Multi-Floor MFDs with More Fire

Once the MFD building becomes larger or center hall with multiple floors, enclosed stairwells, and the fire is larger, a combination of divisions and groups may be appropriate. Neither a single fire attack group supervisor nor a rescue group supervisor alone will be able to size up changing conditions effectively, account for crews on multiple floors, and communicate face-to-face as effectively if they are running up and down multiple stairwells to multiple floors of a large building.

The benefit of the enclosed stairwell is that the area provides the tactical division supervisor a tactical space/threshold against the hot zone from which to account for crews coming and going. Like the front door on an A side of a building, the enclosed stairwell door on an upper floor of a multi-story MFD can provide an excellent workspace for the tactical supervisor.

This was the template used in a three-story stand-alone center hall apartment building fire on a windy night that broke out at 2:30 a.m. Battalion 7, TR106, E101, E103, E105, E110, E19, R20, HM109, M105, and Battalion 3 were dispatched to an explosion with heavy fire and multiple people trapped at the Garden Plaza Apartments. The apartment building was a symmetrical 200' × 200' standalone center hall apartment building with no sprinklers, no standpipes, and carpeted hallways. Each side of four wings of the building had a mid-span lobby access point with interior stairwell that fed all three floors. Each floor was accessed by the interior stairwell and hallway. Approximately 100 apartments per floor contained one bedroom, with the exception of the corner units that had two bedrooms (fig. 9–22).

Fig. 9–22. This large center hall apartment building was the scene of a three-alarm fire.

Pre-planning gives you a tactical advantage that will pay enormous dividends on a dark and stormy night when you have multiple people/families trapped as you arrive first-due and wipe the sleep out of your eyes. *Nowhere is pre-planning more important than on MFDs.*

Conventional pre-planning or walk-through drills may not be enough. If you really want to prepare the troops for battle, consider hands-on drills in these buildings and complexes. This provides an exponential tactical advantage, which is what you will need from the arrival of the first companies on scene.

To say we trained there before this particular fire would be a gross understatement. We conducted multiple hands-on drills, pulling charged hoselines down hallways filled with theatrical smoke, performed VEIS with mannequins, and pumped engines from hydrants. To add to the tactical advantage, we built a full-scale tabletop model of the building and digital simulations.

Now, that may be extreme, but the fact was that we identified this massive life target hazard located directly behind the firehouse. To top it off, the normally first-due engine was temporarily closed due to staffing shortages that night. The winds were sustained at 25 mph, and a gas leak helped to spread the fire after the explosion. As the IC, I was first to arrive, followed by TR106, which had no water or pump. The disadvantages of the closed engine, lack of water, heavy winds, gas leak, and non-existent protection systems were outmatched by the tactical advantages of having trained so extensively in this structure—thank God!

Upon arrival, heavy fire was visible from multiple apartments. Unknown at the time, the fire was started by an Afghanistan War veteran who severed the gas line in his apartment on the third floor and blew the roof off and walls out. We had heavy fire showing from the interior side of the apartment building as well, high winds driving the fire through the center hallways due to the missing roof and wall, and multiple people trapped, including the severely injured vet (fig. 9–23).

All engines and the first truck went to the third floor for fire attack and search from two different stairwells. Due to the amount of labor and tasks to be done simultaneously, the Division 3 supervisor was not assigned until the second chief arrived. I needed to bridge the tactical gap with more direction to the incoming companies. Once Division 3 was established with the second chief, he posted up in the stairwell on the third floor, A side. His objectives were fire attack and search of the third floor. The supervisor must always be in full PPE with SCBA, as was the case on this fire (fig. 9–24).

The third chief officer was posted up in the D side stairwell, third floor, and was assigned to supervise the crews going into the third floor from the D side. He worked under the Division 3 supervisor. Another option here would have been to establish Branch 3, with Divisions A and

D, or just have Divisions A and D, who each had primary operations on the third floor and tertiary operations like salvage and evacuation on the lower floors. I chose a single Division 3 template initially to keep the span of control narrow and radio traffic minimal.

We had dozens of potential victims in addition to the ones that were confirmed. I established a rescue group with the objective of VEIS of windows to apartments on the A/D corner, third floor, nearest the fire. The rescue group supervisor coordinated laterally with the Division 3 supervisor to prevent redundant or conflicting operations (fig. 9–25).

Fig. 9–23. Collapse, gas leak, wind, and victims made this a challenging fire.
Source. Photo courtesy of Gary Lascelles

Fig. 9–24. The Division 3 chief on this fire was a battle-hardened veteran who leaned into the fray.
Source. Photo courtesy of Gary Lascelles

Fig. 9–25. The rescue group supervisor (chief) had companies perform VEIS.
Source. Photo courtesy of Gary Lascelles

Exterior-facing windows surrounded the perimeter of the building and were easily accessed by each respective parking lot. Interior-facing windows were accessible from an interior courtyard that was accessed by one of the four entry points on the four sides of the building. No windows had balconies for refuge of any occupants. Each unit had a window-mounted air conditioner located under the family room window to the apartment. Automatic fire doors were located at each stairwell, providing the TPO to isolate each floor into four quadrants and prevent lateral spread. Another TPO was the legacy construction that was composed of dimensional lumber and a narrow attic space.

Two hydrants were located on the A/B and C/D corners and the manager's office faced the A side main entrance, to the right of the lobby. Every single piece of information you just read was known prior to arrival by me and the majority of the officers and firefighters who responded that night. We had a significant tactical advantage due to pre-planning and realistic training prior to the fire. Figure 9–26 shows that the fire had extended past the interior hall to the outer units due to wind and a gas leak.

Nearly every officer responding who had trained at that apartment was several steps ahead of the fire as they leveraged their knowledge into tactical advantage, even before arriving. Had we not known the building so well, the results could have been catastrophic. Fire had already extended well into the hallways upon arrival, thus compounding the rescue challenge (fig. 9–27).

Fig. 9–26. Fire blowing out the windows, indicating the fire was on both sides of the hallway
Source. Photo courtesy of Gary Lascelles

Fig. 9–27. The fire owned the hallways of this center hall apartment fire.
Source. Photo courtesy of Gary Lascelles

Roof division was composed of two truck companies with the objective of vertical ventilation and preventing lateral spread past the two stairwells on the A and D sides.

Division 2 was established with the objectives of evacuation and salvage below the fire floor. RIC was established with the objective of RIC readiness and securing utilities (there was a gas leak on the third floor. *RIC was not established until the third alarm due to the number of confirmed victims.* Figure 9–28 shows the organizational chart for this fire.

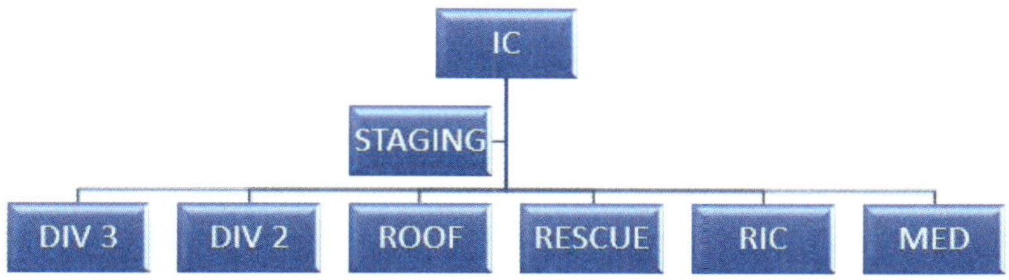

Fig. 9–28. ICS for multi-story apartment fire with confirmed rescues

Expanded Communications

Communications on a greater alarm fire will become more complex. While you ideally want to remain on one tactical channel, you have the potential for your tactical channel to get quickly clogged with multiple companies assigned to multiple divisions. This can become an obstacle to a crewmember transmitting a Mayday, victim found, or emergency traffic. In addition, the division supervisors and IC may speak on the channel, which will create even more traffic.

You may need more tactical channels as your incident grows. Consider giving each division/group a separate tactical channel, and/or grouping together divisions and groups onto separate channels. If your system does not have enough channels available, you may need to consolidate divisions based upon their working relationships to each other.

Looking at figure 9–29, the communications plan for the aforementioned center hall apartment fire had all the tactical supervisors and the IC on Command Channel A4 to facilitate lateral communication between tactical supervisors and vertical communications with me at the ICP. The only exception was roof division supervisor, as he did not have two radios. We spoke to each other on Tactical Channel A6 (fig. 9–29).

The tactical supervisors above communicated with the companies assigned to them on their respective tactical channels listed. Note that Division 3 (the hot division), roof division, and RIC group were all consolidated on the same tactical channel (A6). This was intentional, so that they could coordinate vertical ventilation and fire attack more efficiently. RIC was also on the primary/hot tactical channel to better track companies on the hot division. If we needed to launch RIC, they would already be on the appropriate tactical channel.

The timing of when you expand your communications plan is important so you do not lose companies or make things worse. This must be a strategic decision by the IC, and expanding

Division 3	Roof	Medical	Division 2	RIC	Rescue
BC3	HM109	M105	BC14	BC9	BC1
2402					
A4		A4	A4	A4	A4
A6	A6	A8	A9	A6	A7
TR106	HM109	AMR101	E109	R21	T2
E101	TR17	M109	E54	T6	E8
E103		M8	E24	T50	
E110		M64		E24	
E105		M19		E20	
E19					
TR23					
T2					

Fig. 9–29. Note that each tactical supervisor had a specific tactical channel assigned to them as follows:
✓Division 3—A6
✓Roof division—A6
✓Medical group—A8
✓Division 2—A9
✓RIC group—A6
✓Rescue group—A7

to additional channels too early or too late can each create complications. Adding channels too early can happen in two ways. First, your crews may be in the thick of the battle and have auditory exclusion while fighting fire in low or zero visibility. They may not hear the switch, hear what channel, or see their radios clearly enough to turn the knob or hit a key pad or screen. Secondly, if a Mayday occurs, switching to a separate frequency right away can have a similar effect. This technique is often taught in RIC training, and has merit; however, if done too soon, it can complicate the already tenuous situation. Switching or adding channels can also happen too late. This occurs when the IC calls additional resources and simply is behind the incident power curve, overwhelmed, and doesn't think of it. They are drowning in radio traffic and doesn't have the bandwidth to get help.

A good technique to hit the sweet spot of adding/expanding your communications plan and frequencies is when you call an additional alarm and/or expand your incident with more divisions/groups. For example, we had four-alarm fire one day, and as additional alarms were called I gave them to specific chiefs, each with a tactical channel. With four divisions, each division supervisor (chief in this case), had their own tactical channel and alarm or resources.

Depending upon your communications capabilities, you may request additional channels from dispatch and hold them until needed. As you expand tactical supervisors, give them a channel. Ensure you have a plan/command channel to communicate with your supervisors that is separate from their respective tactical channels. Again, you may be limited by the number of frequencies you have, or the range to which they reach.

CASE STUDIES IN COMMAND

Canton Apartments Incident

CHIEF MIKE WALKER, OKLAHOMA CITY (OK) FIRE DEPARTMENT

Crews from the Oklahoma City Fire Department were dispatched to check smoke coming from a building at 6:33 p.m., early February 2022. The Canton apartment building was a five-story mid-rise that was still under construction, which was composed of a combination of Type 1, Type 3 and Type 5 construction that totaled 374,507 square feet in size.

When the initial response of two engines (E-11, E-22) and a ladder company (L-22) arrived on the scene, crews observed a "haze of smoke" through the windows on the fifth floor on the southwest portion of the building (A/D). Upon this observation, the remainder of a commercial alarm was requested (two additional engines, one additional ladder, one heavy rescue, and two battalion chiefs). The on-scene crews then entered the building to locate the source of the smoke and, shortly thereafter, fire was observed coming from the roof in the same general location of the earlier observed smoke.

In all, the incident grew to a five-alarm fire consisting of 49 apparatus and 138 firefighters, as well as other support personnel. Despite the department's best efforts, the fire was a total loss. The dollar loss is still being decided at the time of the writing, but it is estimated to exceed 100 million dollars. Currently, the cause of the fire is believed to have been due to the neutral side of the electrical system being energized (fig. 9–30).

Fig. 9–30. Oklahoma City firefighters respond to a five-story luxury apartment building fire in February, 2022.
Source. Photo courtesy of Oklahoma City (OK) Fire Department

After a thorough investigation and multiple after-action reviews, the challenges firefighters faced during this incident included:

- Video footage showed smoke coming from the building for over an hour prior to the initial call to 9-1-1.
- The sheer size and layout of the building made it extremely difficult to adequately size up the situation.
- There were multiple areas of origin that were long distances from each other and on opposite sides of firewalls.
- Though the premises were compliant with applicable fire codes, the fire suppression systems in the building were not yet operational or had been turned off due to recent subfreezing weather.
- The wood truss design allowed for wooden void spaces that surrounded the firefighters (ceiling/floor space and walls), thus providing ample fuel for hidden fires to quickly grow.
- Fiberglass-reinforced Sheetrock (Firerock) posed a significant challenge when crews were opening up ceiling spaces.

Major lessons learned:

- Pre-plans should be completed and updated during the different phases of construction to include:
    - Temporary FDC location and markings.
    - Which standpipes are operational and which ones are not.
    - The operational status of the sprinkler system.
- Expect limited fire suppression system functionality when the building is under construction and plan on how to override that challenge.
- Develop specific policies and procedures for this type of construction to include while they are under construction because of the unique challenges they pose.

- Tools:
  - The department has invested in electric chainsaws to defeat fiberglass-reinforced sheetrock.
  - The department has purchased tethered drones for each battalion chief vehicle to aid in improved situational awareness.
- ICS:
  - Utilize a recon group to accomplish size-up within the time of relevance.
  - To effectively bridge the strategic/tactical gap, consider establishing branches for each of the affected sides/wings of the building, and division supervisors for the affected floors of the respective branches (i.e. Alpha Branch/Division 2) instead of divisions per side (i.e. Alpha Division/Charlie Division).

Summary

MFDs have some of the greatest potential for life loss. When a fire in an MFD has a confirmed VP, there is a probability that there will be three victims, on average. MFD occupancies include duplexes, triplexes, fourplexes, condominiums, and townhomes. Larger MFDs include old and new law tenements, row frames/row houses, low-density housing, and center hall and garden-style apartments. Some are subdivided and have been remodeled without adherence to building codes. Hotels and motels can also be considered MFDs from a command and tactical standpoint and can be equally lethal. Hotels are typically center hall while motels are generally garden-style.

Center hall construction is typically more deadly than garden-style, as exit passages are enclosed and require primary and secondary exit routes, while garden-style units open to the outside atmosphere.

Construction can be balloon frame, conventional platform, legacy, or lightweight. Again, the key is to know the buildings in your area.

The life loss potential is astronomical. Pre-planning and drilling in your MFDs is vital if you plan to save lives. Plan on a robust medical group to support operations through treatment and transport of multiple victims. A MCI may even occur. See chapter 15 for more on MCIs.

Common pitfalls include lack of pre-planning and training, improper size-up/access problems, inadequate victim profile/underestimating the potential, lack of tactical supervision, use of two-out crews, difficulty in containing attic fires, and lack of sufficient medical capabilities.

ICS can be similar to house fires for smaller buildings like duplexes, triplexes, and condominiums, while larger MFD buildings may require multiple divisions or a combination of divisions and groups when confirmed rescues are involved. For extremely large complexes, branches may be considered.

Chapter Review

Review Questions

1. Explain some of the challenges and tactical considerations associated with an apartment complex versus a stand-alone apartment building.

2. Describe water supply considerations for responses to multi-family dwellings.
3. The high life hazard of multi-family dwellings presents a potential challenge to the medical capabilities of a fire response. Explain how to proactively address this.
4. Explain the unique challenges associated with row house construction.
5. What are some ways to address the needs of expanded communication associated with working incidents at multi-family dwellings?
6. Provide an example of how you would utilize four or more division/group supervisors given a first-floor fire in a three-story multi-family dwelling fire.

FESHE Strategy and Tactics (C0279) Related Content

The content contained in chapter 9 provides detailed information to assist with instruction and education for all course objectives with a focus on multi-family dwellings. The chapter format supports the CO279 course outline by speaking specifically to basic division of tactics, rescue, exposures, confinement, ventilation, and salvage for multi-family dwelling fires to include apartments, row house, townhomes, condominiums, hotels, motels and duplex structures.

NFPA 1021 Job Performance Requirements

The information in this chapter can be utilized to support training and educational programs associated with the Emergency Services Delivery Fire Officer I JPR 4.6, 4.6.1, 4.6.2, Fire Officer II JPR 5.6, 5.6.1, and Fire Officer III 6.6 and 6.6.1.

Endnotes

1. Kristin Jackson, "The World's First Motel Rests Upon Its Memories," *The Seattle Times,* April 25, 1993, https://archive.seattletimes.com/archive/?date=19930425&slug=1697701.
2. Brian E. Brush, "2021 Fireground Civilian Rescue Research Project," Firefighter Rescue Survey, 2021, https://www.firefighterrescuesurvey.com/uploads/2/5/3/6/25366566/firegroundcivilianrescueresearchprojpost.pdf.
3. When splitting, truck companies are composed of two 2-person teams. Team 1 is the officer and saw firefighter. Team 2 is the engineer and irons firefighter. When arriving separately on scene, they may split opposite of each other so there is a company officer with each newly formed four-person crew.
4. "All In" is announced when a truck company of four members go all inside for search and/or forcible entry.

10

Commercial Fires

Fig. 10–0. Los Angeles City Fire Department works a commercial fire.
Source. Photo courtesy of Los Angeles City (CA) Fire Department

Introduction

Commercial buildings come in all kinds of shapes, sizes, configurations, construction types, uses, eras, hybrids, and purposes. Commercial structure fires require an entirely different mindset from the IC and the rest of the response team of officers. The most common pitfall is treating a commercial building like a residential fire. Commercial fires are much less forgiving of errors, and when the habits of house fires are applied to a commercial structure, we can end up behind the power curve very quickly.

While most firefighter injuries and fatalities occur in residential structures, there is a greater potential for larger numbers of firefighter injuries and fatalities to occur on a single commercial fire. Typically, the greatest life hazard on most commercial fires is us, the firefighters.

The IC must train and prepare the troops to come into commercial structure fires with tremendous sobriety regarding the potential risks involved. VP, tactics, risk, recon, resources, time, and ICS all have different characteristics. This chapter aims to give you clarity on the risks and requirements of commercial building fires.

Commercial Structure Fire Definition and Hazards

There are countless definitions of commercial structures. For example:

- *Any structure which is not a residential structure, including but not limited to shopping centers, offices, nursing homes, and similar structures.*[1]
- *A structure designed or used for human habitation, or human occupancy for industrial, educational, assembly, professional or commercial purposes.*
- *A commercial building definition is one in which at least 50% of its floor space is used for commercial activities, such as retail, the providing of services, or food service (restaurants and the like).*

Neither the NFPA nor the International Building Code (IBC) lists the term “commercial” in any of their occupancy classifications. *NFPA 101: The Life Safety Code, NFPA 5000: The Building Construction and Safety Code*, and the IBC use the occupancy classifications in figure 10–1.

For the purposes of this text, a *commercial structure* is defined as a building designed and used primarily for business, or a combination of business and residential. Examples include, but are not limited to, the following:

- Taxpayers (mixed use)
- Strip malls
- Restaurants
- Big-box
- Mega-box (regional malls)
- Retail stores
- Office buildings
- Warehouses
- Storage facilities

We will include and discuss assemblies later in this chapter, as they have many of the same hazards and challenges, including large life loss potential. As you can see from figure 10–1, assemblies are further characterized by the number of people allowed to legally occupy the building.

NFPA 101 and 5000 Occupancy Classification	IBC Occupancy Classification
Assembly	Assembly (divided into subcategories A-1, A-2, A-3, A-4, and A-5)
Ambulatory Health Care	Business
Educational	Educational
Day Care	Educational or Institutional
Health Care	Institutional (divided into subcategories I-1, I-2, I-3, and I-4)
Detention and Correctional	
Residential Board and Care	Institutional or Residential
One- and Two-Family Dwelling	Residential (divided into subcategories R-1, R-2, R-3, and R-4)
Lodging or Rooming House	
Hotels and Dormitory	
Apartment	
Mercantile	Mercantile
Business	Business
Industrial	Factory and Industrial (divided into subcategories F-1 and F-2)
Storage	Storage (divided into subcategories S-1 and S-2)
No equivalent occupancy classification (see surrounding paragraphs for additional information)	High Hazard (divided into subcategories H-1, H-2, H-3, H-4, and H-5)
No equivalent occupancy classification (see surrounding paragraphs for additional information)	Utility and Miscellaneous

Fig. 10–1. NFPA and IBC building classifications[3]

Construction of these buildings can be any type, era, size, and combination of materials and uses. As mentioned in previous chapters, pre-planning can save firefighter and civilian lives alike.

While most firefighter and civilian injuries and deaths are in residences, commercial buildings create potential for extreme life loss per fire. The Ghost Ship Fire on December 2, 2016 killed 36 civilians in Oakland, CA. The occupancy was in a commercial warehouse that was zoned for industrial purposes, but a concert was being hosted there. Residential and entertainment purposes were illegal. The warehouse had been illegally converted into an artist collective with living spaces. This was the deadliest building fire in America since the Station Nightclub Fire in 2003 that killed 100 people.

The Sofa Superstore Fire in Charleston, SC claimed the lives of 9 firefighters on June 18, 2007. This was the deadliest loss of firefighters since the 9/11 attacks. On May 16, 2020, the Boyd Street Incident in Los Angeles, CA severely injured 12 firefighters. The stakes are high.

As the first-due officer, company officer acting as the IC, or the chief who assumes command for the duration, you must respect commercial fires and teach your members to do the same.

These buildings are wrought with hazards, including but not limited to:

- Excessive fire loading
- Excess flammable liquids beyond permitting
- Hazardous materials without proper placarding and labeling
- Illegal remodeling against code
- Illegal storage of materials

- High-piled storage racks
- Security gates, bars, and doors
- Roll up doors
- Older and dilapidated construction
- Lightweight and truss roof assemblies
- Excess dead loading on the roof
- Unsupported mansards and facades
- Loading docks
- Basements, chases, and freight elevators
- Access and egress issues
- Boarded-up doors and windows
- Sudden collapse of roofs and walls
- Sudden fire increase, including deflagration
- Lack of adequate water supply
- Lack of fixed protection systems or systems that are shut down

CASE STUDIES IN COMMAND

Boyd Street Incident IC Perspective and Overview

CHIEF LUIS ALDANA, LOS ANGELES CITY (CA) FIRE DEPARTMENT

The LAFD's Boyd Street Incident was a major emergency structure fire on Saturday, May 16, 2020. The incident occurred in the heart of downtown LA's industrial/commercial district, also adjacent to the skid row area of Los Angeles.

For the last few years, the downtown LA area has seen a marked increase in structure fire activity, primarily due to the large homeless population living in and around commercial buildings. A typical scenario is a rubbish fire or tent fire adjacent to a commercial building extending into the building. Depending on the time of day or night, a fire might be very well developed before fire resources arrive.

The fire occurred on a Saturday afternoon and was our third fire of the day (for Battalion 1). As a result, the typical weekend training routine had been interrupted because of this busier-than-normal fire tempo combined with staffing and other routine duties. I had been assigned to Battalion 1 for two months before the incident, having been transferred from my first assignment as a newly promoted battalion chief in Battalion 10 in the San Fernando Valley. The training plan was to visit each station in the battalion and conduct Mayday drills beginning with my single-engine companies and eventually getting to my task force houses (two engines and a truck). This plan had been in place since the previous weekend, when I had conducted a Mayday drill with Engine 4 and Rescue Ambulances 4 and 804 (advanced life support [ALS] and basic life support [BLS] ambulances, respectively).

At 18:26 hours, our Metro Fire Communications Center (MFC) dispatched a category B assignment to 357 E. Boyd Street. The first units arrived on scene within 2 minutes and 27 seconds, and the Battalion 1 command team (myself and my emergency incident technician [EIT]) arrived 30 seconds thereafter. Initial companies noted smoke in the area but no obvious loom up, so they dispersed in different directions to see if they could find the source.

Truck 9 was the first to find the fire building and took an appropriate outside spot directly across from the address. The truck captain assumed command, gave an arrival report, and deployed his resources offensively. He sent his truck company to the roof for ventilation and his two engine companies on fire attack and backup fire attack. As he was doing this, we arrived on scene. We selected our command post location, just east of the address and across the street from the intersection (fig. 10–2).

We transitioned command and continued placing resources in the offensive mode. At the same time, my EIT proceeded to the rear of our rig to set up for a potential extended operation. The initial impression was that we had a fairly incipient fire inside this one-story commercial building due to the current time of day and light-to-moderate lazy smoke in an area known for storing textiles and other consumer goods.

This appeared to be "just another one of the many commercial fires we were all used to." The first thing I noticed that seemed odd was a thunderous and resounding rumbling noise that sounded akin to a military jet on full after-burner (I used to work on a military airfield). The sound was very noticeable, but I initially did not associate it with the relatively mild to moderate smoke conditions I saw and the fairly routine fire communications on the tactical channel. As resources continued to arrive on scene, I began surrounding the fire by placing companies on all four exposure buildings.

At approximately five minutes, all the resources from the first alarm arrived on scene and received an assignment. I asked the fire attack team to provide me with a CAN report by asking them to tell me, "What was the occupancy doing business as?" and "What was the fire load like inside?" Unfortunately, I was never given an answer and did not follow up because the conditions quickly deteriorated.

I ordered additional resources for a rapid intervention component and to have additional companies in staging. The loud sound kept growing to the point of interfering

Fig. 10–2. Truck 9 sustained major thermal damage on the Boyd Street Incident.
Source. Photo courtesy of Mike Meadows

with communications and vibrating the entire street. Finally, around the 6½ minute mark, an enormous after-burner-like, clean-burning, bright orange fireball engulfed the whole street. The fireball completely enveloped the aerial ladder and the truck itself for a prolonged period. It also mushroomed towards the sky in what I would estimate was at least 60' up and out of the building.

As the fireball engulfed my crews, I broadcasted a Mayday tone, ordering everyone out of the building in preparation to transition to a defensive mode (fig. 10–3). Per our SOGs, my EIT ordered additional resources. At the same time, I began requesting a personnel accountability report and company accountability report (PAR/CAR) from one of the captains that was able to make it outside of the interior of the building and was asking for help.

From the vantage of my command post, I could see the crews completely engulfed in flames but had no idea who had made it out of the building and who was unaccounted for. I remember thinking, "They are certainly dead, and we are probably next." The captain talking to me on the tactical channel gave me a body count, trying to determine which company the injured members were from. Ultimately, he was able to account for both engine companies and two ambulance crew members who were inside the building when it went up in flames, but was not able to give me any information on the truck company.

While sounding the tones, I remember Command 22 (our assistant chief) rounding the corner. He tapped me on the shoulder and said, "You run the Mayday, and I'll take the fire." As the next couple of resources arrived on scene, I remember thinking that I needed to build out both a medical unit for our injured members and a rapid intervention group in case anybody was trapped. When the next company checked in, I had to decide the priority assignment. I knew I had at least eight injured firefighters outside and a truck company that was unaccounted for. I prioritized the injured firefighters outside first to avoid delaying their treatment and transportation. Fortunately, all the firefighters were able to exit the building. Still, it weighs heavily that I might have had

Fig. 10–3. Crews wore personal protective equipment (PPE).
Source. Photo courtesy of Mike Meadows

to delay a trapped firefighter rescue to provide immediate care for the burned firefighters outside the building.

I remained in charge of operations for the remainder of the fire until a knockdown was declared, at which point I was relieved and I went to the hospital to check on my guys. It was approximately 0130 hours the next morning.

Tactical lessons learned:

- The main fire building was a long, skinny building nestled between two taller buildings (a two-story and a six-story), limiting roof access and, more importantly, the ventilation team's egress options to just the building's Alpha side.
- Two rolling steel doors occupied the entirety of the Alpha side of the building, eliminating any ideal spot to throw a ground ladder or the aerial ladder (creating an access/egress hazard).
- Having a well-thought-out and rehearsed Mayday procedure with your command team is vital.
- If you are working with an overtime member, reviewing your Mayday procedures with them should be your priority. I always make it a point to challenge my captain II's and captain I's on how they will run a Mayday. I have them show me everything from how they will set up their radios to how they will document and track resources, what they will order additional, and how they will deploy their resources as they arrive on scene.

Incident command lessons learned:

- The importance of an officer, especially a chief officer, thoroughly understanding the roles and responsibilities within the incident command system is imperative.
- During extreme stress, this intimate knowledge will help you identify and prioritize your most significant problems and what group or division you will build to address them.
- Not to sound cliché, but regularly practicing broadcasting the Mayday tones and drilling with other companies (teaching) was the most important thing I did to prepare for this unplanned event.
- When everything was quickly unfolding and time seemed compressed, I remember vividly thinking, "This is not happening," and seeing in my mind's eye my hands going through the motions of sounding the tones from the previous Saturday's company drill. Then, when the pressure was on, I fell back to muscle memory from my recent training.
- Our command rigs have a plasticized Mayday script. Many of our members memorize the script and assume that, if faced with a Mayday, they will simply regurgitate the script and everything will be alright. Unfortunately, the incident I encountered did not match any of the talking points in the script. These scripts are prompts for things that will most likely occur and that you should think of.
- It is more important to prepare by quickly recognizing the challenges, prioritizing, and addressing the problem(s) in front of you, and giving the person(s) in

need the proper leadership, guidance, and hope to hang on while the rescue team is on their way.

Techniques I applied:

- When sounding emergency tones or broadcasting vital communications, I have always practiced keying both microphones (tactical and command channels) and communicating on two channels simultaneously.

For more details and photos, see the Boyd Street Incident Green Sheet in Appendix B.

Commercial Structure Fire Watch Out Situations

Several warning signs on commercial fires should be considered commercial structure fire watch out situations that are mirrored from the Wildland 18 Watch Out Situations. These are in addition to the LAFD structure fire watch out situations in chapter 5 (fig. 10–4).

Fig. 10–4. Los Angeles (CA) Fire Department at the scene of a fire. *Source.* Photo courtesy of Zak Holman

Each is listed below with an explanation:

1. **Inadequate size-up (360° when possible) and recon not performed.** This will be evident by the lack of a sufficient arrival report. If the first-arriving company officer/IC and subsequent chief assuming command do not take note verbally of the size, construction, and potential use of the building, there is likelihood that this is being treated like a house fire without adequate pause to repeat to all responding units that this is a commercial structure with more potential risks. If the fire is not clearly visible, reconnaissance is appropriate before over-committing members inside the structure.
2. **Night operations.** Overhead powerlines, construction features, remodels, modifications, placards, and hazards of all kinds are missed during night operations. The inability to get apparatus around the structure to provide lighting will further complicate visibility. Combined with heavy smoke conditions, night operations at commercial fires are always wrought with greater hazard and accident potential.
3. **Boarded up and/or heavy security bars and doors.** Commercial building egress is challenging enough without the addition of boarded up doors, windows, and security bars. The front, A side of a commercial building may seem open and accessible. The C side rear often tells a completely different story. For this reason, RICs must access all sides of a commercial building to soften the structure and gain access to hard-to-reach areas.
4. **Unfamiliar with building, have not pre-planned.** Because some fire departments do not perform pre-planning, and commercial occupancies change, this can be a moving target. That said, get out and walk your commercial buildings as often as possible. You will notice changes and trends that may make a huge difference.
5. **Command has not announced a clear priority, strategy, or tactical objectives and/or there is a lack of incident alignment between them.** Remember, the incident is successful from the top down. You must be in the right priority/strategy before you can maintain alignment in the tactics and tasks. Often, commercial fires are in the wrong priority and strategy because everyone is in groupthink, whereby no one challenges what the team is doing. Look up and ask yourself, "What is our priority here? Are we in the right strategy to support it? What is my risk versus gain? Is a combination strategy a possibility? Should we switch to a defensive strategy?"
6. **Communications overload on the tactical channel.** Like setting up divisions and groups or calling additional alarms *before* you get behind the curve, the same applies to utilizing additional tactical channels. If you have the capability and the SOGs to do so, consider adding additional tactical channels for your divisions/groups if the primary tactical frequency is overloaded with radio traffic. Countless Maydays have been missed because of this issue.
7. **Unable to contact crews inside the building due to a repeater issue or construction type.** Heavy concrete and reinforced buildings, basements, and security rooms can make repeaters ineffective. You may have to switch to a simplex or direct channel to regain adequate communications with your interior crews. This was the case in a Kaiser Permanente Hospital that was less than a mile from my battalion office. Our pre-plan for that building was to switch to a direct channel upon arrival, thus bypassing the repeater system. As the IC, I spoke to my crews inside on the direct channel and to incoming units on the repeated tactical channel.
8. **Water supply not established or available.** Insufficient water from the system can have a significant impact on your strategy and tactics. Dead-end hydrants and some water systems may not be able to adequately supply your fire, depending upon the size and flow requirements.

9. **Limited access around the perimeter.** Overhangs, awnings, fencing, storage, equipment, and a host of other challenges can prevent getting around the building and access to certain areas.
10. **Attacking the fire from a downwind position.** This may seem obvious, but it has happened. Like a house fire, crews often go to the front door of a commercial building even if it is downwind. This is an example of a residential habit on a commercial fire. Unlike a house, attacking a commercial fire at the closest point to the seat of the fire is likely the most effective and expeditious, rather than a long interior stretch where conditions may deteriorate rapidly during the stretch.
11. **Little to no visibility without knowing where the fire is located.** Commercial fires are unforgiving and do not have the readily escapable compartments that residences do. If you do not see visible flames upon arrival, proceed with caution. Even light smoke from the outside of a large commercial building can have a well-seated fire deep inside the structure. This may lull you into a false sense of security as you advance into a structure because the heat may be concealed in the overhead, behind closed doors, or below you.
12. **Excess hazardous materials, flammable liquids, high-piled storage, and/or excess fire load.** In some commercial buildings, these are normal contents, such as textiles plants, paper processing facilities, manufacturing facilities, automobile repair shops, swimming pool supply stores, and paint suppliers, to name a few. That is the point. These commercial buildings are not offices, small retailers, compartmentalized strip malls, restaurants, and so on. These present greater hazards and challenges. Beware of getting ambushed (such as with the Boyd Street Incident).
13. **Primary access is under an unsupported mansard.** This was the case on the 65th Street Fire in Sacramento. The mansard collapsed just after firefighters evacuated, pinning their hoselines in place.
14. **Conditions worsening during interior attack.** Conditions in residences typically improve relatively quickly, as access can be made easily when compared to commercial buildings, and required fire flow is much lower than in a commercial structure. In many cases, firefighters at the task level do not recognize worsening conditions, especially above and below them. Mentally, they want the same knockdown that they have experienced many times and most likely recently from their last house fire. When they don't get it, they get emotionally involved and stay in the fight too long despite worsening conditions. This was the case on an old Main Street USA strip mall that we were losing. I was Division C and when the IC switched to a defensive strategy, the crews were furious. I was able to convince command that a combination strategy would minimize risk while saving the rest of the building and narrowly avoided a mutiny of the IC by the crews who made the fire personal.
15. **Resources are inadequate to prolong the fight.** If you are unable to muster enough resources to meet your strategic goal of going into the building, and corresponding tactical objectives, then you must switch to a sustainable strategy and appropriate objectives.
16. **Vacant, abandoned, or low victim profile.** As stated in this book previously, the VP drives the rest of the incident. A low VP or obviously vacant and/or abandoned building should drive an incident stabilizing priority from the onset. Unless there is a confirmed VP, minimize risk to your troops. They may want to go in and fight fire. Your role as the IC is to take out the emotion and adrenaline and make logical decisions.
17. **Obvious remodels and modifications look out of code.** Notice the different rooflines, boarded-up windows and doors, and construction features. This building has obviously been tacked upon and remodeled multiple times without proper permitting. This can be a trap for firefighters.

18. **Crews are fatigued due to weather conditions such as extreme heat or cold.** Firefighters are accustomed to working intensely for short durations of time in normal conditions for their area. While the average ambient temperature for different parts of North America vary, firefighters are acclimated to their particular area. Commercial fires typically are for longer duration than firefighters are accustomed to, and extreme weather conditions can add to their fatigue.

Commercial versus Residential

Commercial fires are not as common as single-family and multi-family dwelling fires (fig. 10–5). Therefore, the habits born from the more common dwelling fires are often repeated in these larger/more complex structure fires. These habits are often known as performing residential tactics at commercial fires and include, but are not limited to:

- Not performing recon when appropriate or possible
- Going at the same speed as a dwelling with higher victim profiles
- Using small caliber hoselines when larger 2½" lines or master streams are needed
- Companies not checking the rear when possible
- Remaining on the roof too long
- Remaining in the building too long
- Pushing forward when the fire is not easily located or unable to find the seat of the fire

As an example, you are dispatched to a large commercial warehouse on a Saturday evening in the summer (fig. 10–6).

Signs of inadequate size-up and recon occurring would be the following example transmissions:

Engine 1 on scene. We have heavy smoke visible. Initiating attack with a 1¾" line. Next engine bring a water supply.

Engine 2 on scene. Assuming command. Water supply is complete to Engine 1. We're going to pull a second 1¾" line.

Chief 1 on scene. This will be Main Street Command. The ICP is on the A/D corner in the parking lot.

Fig. 10–5. Commercial fires like this are not as common as house fires.
Source. Photo courtesy of Zak Holman

Fig. 10–6. Simulated fire at commercial warehouse

Did you note the following issues?

1. No one acknowledges or describes the type and size of the building.
2. The first two lines pulled are 1¾" on a large warehouse with heavy smoke. Would recon and forcible entry be a higher priority before committing to a small caliber hoseline? Would a 2½" line be a better choice once forcible entry is made on the D side?
3. The chief/IC never announces the incident priority, strategy, or tactical objectives.

All of these issues can cause all incoming companies to automatically go into a residential mindset without even realizing it. Contrast this with the following:

> *Engine 1 on scene. We have a large 200' by 200' one-story warehouse. Heavy smoke from several roll up doors on the D-Delta side. Light smoke from the A-Alpha side. This is a low-moderate victim profile. We will be in an offensive strategy. Engine 1 is going to pull a 2½" line and begin cutting roll up doors on the D-Delta side. Next in bring a water supply. First truck, assist with forcible entry on the D-Delta side roll ups.*

Size-Up and Recon

Once again, lack of or improper risk assessment (size-up) is number one on the NIOSH 5. Commercial fires require much more than a simple three-sided view, and sometimes more than a 360°. Obviously, some commercial buildings are much too large to perform a 360° lap on your feet. And countless obstacles can stand in your way of getting a 360° like fences, gates, equipment, stock, zero-clearance property lines, security measures, and adjacent buildings (fig. 10–7). If possible, have other companies respond from a different direction to see the back of the building or complex. Or, size-up from the roof is a great way to get information.

Fig. 10–7. Forcible entry challenges may require an entire truck company or more on commercial fires.
Source. Photo courtesy of Greg Doyle

From the street, the usual FPODP size-up provides an excellent starting point. Look for the following as you arrive on scene:

- Building (era, use, type, size)
- Business signs or other indicators of business type and contents (fig. 10–8 and fig. 10–9)
- Layout, access, and exposures
- Overhead hazards
- Placarding
- Smoke (volume, velocity, density, color) to determine if a contents or structural fire
- Visible flames and location (seven sides)

Notice that looking at the smoke and flames was listed last. This is due to tunnel vision. When you first arrive on scene, if you first focus on smoke and/or flames, it is easy to lose sight of the other previous factors that can be missed or obscured. A good rule of thumb is to look at any structure fire first as though it was not burning. Then you can see construction features and other clues you would otherwise miss when experiencing tunnel vision on the flames. I once missed an entire attached exposure and fire wall because I was focused on the 75-foot-tall flames coming out of the roof of the involved commercial building. Don't let this happen to you!

As discussed previously (chapter 5), recon is a very effective process to gain additional information that a windshield size-up in the street or parking lot would not offer. Remember, the objective of recon is to find the enemy's location, size, and direction of travel,as well as any TPOs that could serve to provide tactical advantage. It should be swift, and you must be ready for changing conditions and know when to retreat. Recon is most appropriate when you have no visible smoke/flames, light smoke, or a large volume of smoke that does not appear to be coming from any particular area. Stay cautious and ready for conditions to change.

Fig. 10–8. Can you predict what's inside by looking at the overhead signs?
Source. Photo courtesy of David Blair

Fig. 10–9. High rack storage is a deadly hazard.
Source. Photo courtesy of David Blair

As you perform recon, ensure that you maintain situational awareness as to your entry point, how far you have entered, and what doors or other obstacles you have passed. Knowing the distance you have traveled in the structure will be very important if you happen upon fire. You can tell the IC and engine companies the portal of entry, location of the fire, and distance you have traveled. Depending upon the information and distance, they may or may not follow your path of travel. Possibly, a closer location to the fire provides an alternative location to enter and stretch to attack.

Try to minimize flow path and compartmentalize the fire by closing doors. Identify the size of the fire, what is burning (if possible), and the avenue and direction of fire spread. Identify TPOs as you go. Are there fire walls, partitions, fire doors, breezeways, separate attic spaces, and so on?

Another means of size-up is from the roof. Even if vertical ventilation is not a tactic appropriate for the fire, or not used by your fire department, a crew on the roof may be able to further identify a heat signature with a TIC, see excess smoke from HVAC systems or vents, or simply identify a burned belt on an AC unit.

If a truck is assigned to the roof for vertical ventilation, a *roof report* (construction, conditions, load) should be given ASAP prior to committing to the vent (fig. 10–10). This will help

Fig. 10–10. Even if the roof division does not vent vertically, they can perform recon and give a roof report.
Source. Photo courtesy of James Hedlund-Kaiser

the IC determine if vertical ventilation is in alignment with the IAP. Between the companies on the ground gaining entry and giving CAN reports and confirming a VP, and the roof division giving a roof report, the IC can verify if an offensive strategy can be executed, or if a combination or defensive strategy should be employed.

Some commercial facilities have 24-hour on-site engineering or maintenance personnel who may know the situation or have called 9-1-1 prior to your arrival. They often have radios to communicate with each other. During pre-planning, you may find the location to contact such personnel or have access to one of their radios.

If you arrive and assume command on a commercial structure fire assignment with either no signs of fire, or light smoke, and your primary objective is recon, then setting up the incident early for several permutations by remaining flexible will pay huge dividends downstream.

During recon, do not overcommit your companies. Rather than having all your resources dive into the building and start hunting down the situation, be judicious with your companies. One engine and one truck (or equivalent) should be sent in for recon. One engine can be staged at the primary hydrant, or if your SOG requires that commercial fires be pumped from the hydrant, then stage a second engine at the primary attack pumper to be ready to lay out and pump back, if needed. Other considerations would be a company to the rear to check conditions and a company to the roof to check for signs of fire. If you do have a working fire, you can then assign additional companies based on the report back from recon. This way, you do not over-commit resources to one area then have to pull them out to re-deploy to another area. Restraint at the beginning will pay off.

If the company in the rear finds smoke or fire conditions, they may be the best choice to attack the fire, and the engine standing by on the front can then supply the engine in the rear. Or, if the recon crews determine there is fire closer to the rear of the building and it is more accessible from the back, the engine in the rear can attack the fire from that closer, more tactically advantageous position.

Cold smoke conditions also provide an opportunity for recon to help you truly size up what you have going on inside the structure. *Cold smoke conditions* occur when a large structure has a large volume of smoke that is lazy (like a fog), often without velocity, that is likely the result of a sprinkler system keeping a fire in check. The smoke will not have the typical characteristics of structure-fire smoke with pressure or turbulence. *This can conceal a larger fire, so always be ready to be ambushed in a commercial fire.* Cold smoke doesn't typically rise, since there is no convective heat or current to provide lift, so the roof is likely not an effective location to garner additional intel in cold smoke scenarios.

Flowing water, audible alarms, water gongs, and water issuing from under doors are all common signs of a likely sprinkler activation or broken pipe or sprinkler head.

Victim Profile and Search

Generally speaking, the greatest life hazard at a commercial structure fire is the firefighters (fig. 10–11). Unlike a residence, civilians do not hesitate to evacuate a business on fire, nor do they run back into commercial buildings for pictures, pills, and pets. Early warning systems can cause a large number of civilians to evacuate the fire in the incipient phases. Unlike MFDs and assemblies, commercial fires generally have a low-to-moderate VP. That said, you can always have a surprise like the aforementioned Ghost Ship Fire that was in a building classified as industrial. The Station Nightclub Fire was in an assembly classification building. These were two of the most lethal fires in recent American history.

If the fire is after business hours, you may have the benefit of closed occupancies; however, employees can be present. Attempt to make contact with an employee when possible to get the VP. Look for appropriate attire. Fast food restaurants and other retailers will have people in uniform. Look for nametags, aprons, or hats that indicate an employee.

In office buildings and doctors' offices, look for business attire and medical scrubs, respectively. Ask if anyone knows where the manager or owner is. If people are lingering outside in the parking lot of an office building, on cell phones with computers and briefcases, they are likely employees of one of the offices. Get intel from them.

Crews arrived at a large commercial building to find heavy fire in the overhead of a lightweight roof assembly. A bogus report of a possible occupant still inside delayed the IC from announcing a defensive strategy and evacuating crews. Shortly thereafter, the roof collapsed, nearly killing eight firefighters. By a miracle, none had significant injuries.

A fire occurred in a Tuesday Morning store. Bogus reports of occupants in massage booths of an adjacent parlor caused crews to attempt a search. They were driven out by heavy fire. No one was inside (fig. 10–12). The next day, the manager came to the scene to survey the total loss. When he asked the IC about the fire, he was told that crews were inside the building for a brief period of time to attack the fire. The manager's response was incredibly enlightening. "Why would you go in there? We are going to write it off anyway, and the last thing we need is publicity that a firefighter was killed at a Tuesday Morning store." Go figure.

Large commercial buildings like grocery stores, furniture stores, and big-box stores are not realistically going to be searched with heavy fire and smoke inside the structure. They are more likely going to be *evacuated*. There is a big difference. Evacuations occur when firefighters assist occupants in leaving the structure under their own power, and with minimal risk to firefighters. An example would be a confined fire with light smoke in the building.

Fig. 10–11. LAFD Boyd Street Incident in which multiple firefighters were severely injured
Source. Photo courtesy of Rick McClure

Fig. 10–12. Reports of occupants inside a building were made before this photo was taken.
Source. Photo courtesy of Sacramento (CA) Metropolitan Fire District

Search in commercial buildings is either for known victims in targeted areas, or when conditions allow for a search. An example would be search in a cold smoke condition, or an occupant who is trapped by fire calling 9-1-1 from a cell phone inside a bathroom. Absent a confirmed VP by a credible source, heavy fire and smoke in a large commercial building is likely a defensive strategy.

Fire Attack and Water Supply

Commercial fires often require fire attack from multiple locations. The fire flow required for these larger fires requires multiple access points to prevent bottlenecking of companies into one area. As additional large caliber hoselines are placed in service, the access point can become clogged (fig. 10–13).

And, as fires grow, they can go beyond the reach of the initial attack lines and alternative access points can get you closer to the fire or help you take advantage of TPOs and protect uninvolved areas. If you or the crews in your division must stretch a line further than 200 feet into a commercial building to get to the fire, you should consider another portal of entry from which to make the attack with another company. If conditions deteriorate, will crews be able to retreat quickly? Always be ready for rapidly changing conditions.

In residences and MFDs, our first objective is to confine a fire to protect occupants and means of egress. This often requires attacking the fire from the unburned area, likely from the front door. Even if we perform transitional attack (offensive attack from a defensive position) on a house fire, we are still attempting to get inside for victims as soon as possible. This requires coordination.

Commercial fires are often best attacked at the *closest location to the fire as possible* (fig. 10–14). Commercial structure fires typically require longer, more complex hose stretches with larger caliber hoselines and more firefighters. In addition, access issues and forcible entry challenges further complicate and delay the time it takes to locate and get water on the fire.

We arrived to find a well-involved tractor/trailer in the loading dock area on the C side of a large retail store one morning. The fire was starting to get into the business. Initial companies on scene went to the C side and quickly knocked down the fire. Subsequent companies went in the A side main entry and had to stretch a 2½" line over 200 feet into the building before they reached the rear. By the time they wove their way through racks of clothing, multiple doorways, around corners, and down halls, they had almost 300 feet of hose on the ground. In that time, how much can the fire grow? This is not to criticize or even imply that the line in the front door was not appropriate. The intent was correct to protect the majority of the uninvolved business.

The initial attack and knockdown on the main body of fire was able to keep it from further encroaching on the interior of the building than it already had. Damage inside was relatively minimal (fig. 10–15). There was a person-door immediately adjacent to the rear roll up door where the fire was encroaching.

After knockdown, would that person-door have been a better location from which to attack the fire inside? Would it have been a much faster stretch with less hose and exposure to personnel and prevented crews from potentially getting too far into the building from the A side? Perhaps (fig. 10–16).

Fig. 10–13. Large caliber hoselines will quickly limit access for incoming companies.
Source. Photo courtesy of Mike Meadows

Fig. 10–14. The two doors provide quick access to this fire upon arrival.
Source. Photo courtesy of Sacramento (CA) Metropolitan Fire District

Fig. 10–15. Minimal extension into the C side of the structure.

Fig. 10–16. The person-door could allow quick access.

Homes and MFDs usually have a window for most every compartment inside the building. Commercial buildings do not. Hallways, corridors, high-piled stock, storage rooms, machinery, basements, remodels, shafts, coolers, and a host of other hazards lie waiting inside commercial occupancies, often with no windows or exits from which a firefighter can make a quick escape if conditions deteriorate.

Fire flow on commercial building fires will require the use of multiple hydrants. Supply engines should pump from the hydrant to the attack engines and trucks with elevated master streams (fig. 10–17 and fig. 10–18). This will bolster pressure and flow for the attack pumpers. Often, engines will lay into a commercial fire several hundred feet and begin to attack the fire without an engine at the hydrant to pump. This is another example of using residential tactics on commercial fires.

FDCs and sprinkler connections should be pumped by a separate engine than a primary attack engine. This requires a lot of resources and can tax the water system. Err on the side of supplying the attack engine(s) if you have limited resources and cannot support the sprinkler system and the fire flow needs of your crews inside.

Consider the presence of fire pumps, on-site building engineers, and contacting the water district if you need more pressure in the system. A water supply group supervisor may be able to coordinate multiple companies, including water tenders, and supervision of relay pumping operations. A separate tactical frequency for the water supply group should be considered in these situations.

Fig. 10–17. Pump the hydrant on commercial fires.
Source. Photo courtesy of Mike Meadows

Fig. 10–18. Even a short distance to the attack pumper warrants pumping the hydrant.

As with all fires, your IAP starts with the VP. You must balance the risk to your firefighters with the gain of civilian lives. Your firefighters will want to get in and fight fire, which is what they are trained to do. Realize that you and your tactical supervisors may need to either redirect them or pull them out of a commercial building because of deteriorating conditions. You are staying in the "then/what it" mindset. The "what if?" part of that mindset is what will save your firefighters.

Ventilation

The IC and tactical supervisors have several options for ventilation on commercial fires that can be more complex and require more coordination than on smaller buildings. These include:

- Vertical ventilation (offensive and defensive)
- Horizontal ventilation
- PPV
- HVAC systems
- Hydraulic ventilation

A combination of the above tactics may be utilized, or you may begin with one and end with another as conditions change.

As with recon on a commercial structure, you may send a truck to the roof. *Vertical ventilation* is the tactic of opening the roof and/or roof assembly to exhaust heat, gases, and products of combustion up and out of the building. The first priority is to get a roof report (construction, conditions, load) to help you determine if vertical ventilation is an appropriate option or if the roof is untenable. This can help greatly with overall recon in the early stages of the fire to ensure you are in the correct strategic alignment.

If vertical vent is the tactic of choice, then the truck officer should be assigned the roof division supervisor. Depending upon the size of the roof, you may need more than one truck on the roof. The roof division supervisor will coordinate and supervise these companies for you. They must step back from the noise of the saws cutting and see the bigger picture to remain in the tactical "next" mindset. The roof division supervisor is your safety and accountability officer on the roof and this must be an experienced and well-trained officer.

I was fortunate to have extremely experienced, well-trained, and dedicated truck captains for many years. The level of trust I had in them was marrow deep. Roof operations require an even deeper level of understanding regarding construction and the hazards/tactics associated with each type.

Like any other supervisor, a roof division supervisor can really drive the incident tactics and potentially the strategy, because they see things you do not see as the IC, especially from the roof. A roof division supervisor can see construction features and the size of the building all at once. They will also likely see exposures unseen from the ground and other hazards like HVAC units, arching power lines, antennas, and other dead loads. The roof can tell you what the fire is doing and the smoke conditions and whether the fire is so advanced that you must switch to a combination or defensive strategy. TPOs like parapet fire walls can also be identified, which can become significant landmarks and anchor points from which to save a large portion of the building or draw a line from which to mount a combination strategy.

Units responded to a reported fire in a strip mall on a Sunday evening. Moderate smoke was visible from the A side upon arrival with heavy fire from the C side. This was a low-moderate VP, as all the businesses were obviously closed. As crews opened the doors to the occupancies, smoke quickly dissipated, there remained heavy smoke from the roof. I assigned the first truck officer to roof division. This was an outside-in fire that began outside the C side wall and went up into the roof assembly.

A *rain roof* is a newer roof built over an older existing one, often creating a void space between the two. The fire was burning in the space between the old roof and the rain roof. The challenge for engine companies was accessing the space between the roofs due to the old roofing in place that needed to be breached to get to the void space above. This caused a significant delay to getting water on the fire. Meanwhile, roof division began opening the roof.

The delay in fire attack caused the fire in the roof assembly to increase, since the ventilation continued. This was the result of a lack of coordination between Division C and roof division. Division C did not communicate the challenges with accessing the space for fire attack, and roof division was not communicating with anyone, despite the fact that he had two truck companies on the roof working for him. He did not step back from the work into the tactical space.

Meanwhile, Division C saw the fire through the roof and attempted to contact roof division. There was no response from roof division despite multiple attempts. I could see them working and the fire was burning cleanly without any smoke in the living space below. I suspected a rain roof and knew the trucks were extremely experienced. The problem was not that the officer filling the roof division supervisor role was not experienced. He was a very experienced and capable truck officer. The problem was that he was working overtime that day and had not been trained in the role of a roof division supervisor. He certainly knew what a division supervisor was since he had basic ICS training (ICS 100 and 200); however, this is *not* the same as hands-on training in the role of being tactical on the roof and stepping back from the work and noise to see what is really happening, coordinating laterally with divisions below you, and vertically with the IC. Most truck officers take a lot of pride in being autonomous and not talking on the radio. We could not afford that in this case.

When performing vertical ventilation on commercial fires, the roof division will either execute offensive cuts, defensive cuts, or a combination of both. For example, offensive cuts would be utilized to remove heat and products of combustion to support fire attack and search below (heat cuts). Defensive cutting would be to protect an uninvolved area of the building from a fire that is spreading laterally (strip/trench cuts). Both require coordination with the IC and division supervisors below.

Trench and strip cuts often require the roof division to write off a portion of the building to get ahead of the fire by removing roofing materials and creating a fire break in the roof from which the fire can escape and be held by hoselines. If the fire is fast-moving and the TPO of a fire wall is present, cutting on the leeward side of the fire wall gives the crews an escape route

Fig. 10–19. A truck company prepares to go to the roof of a large commercial building. The officer will become roof division supervisor.
Source. Photo courtesy of Greg Doyle

and buys them time to support the fire wall, as it slows the fire's progress while engines are below doing the same with hoselines (fig. 10–20).

Horizontal ventilation is the tactic of opening doors and windows to provide fresh air into the building. This tactic is extremely common and can be accomplished quickly and easily. Simply opening a front door to initiate fire attack is horizontal ventilation. *The challenge is that it must also be coordinated and can negatively impact the fire if not communicated.* This can be an assignment given to a company within a division without having to establish a separate ventilation group. For example, Division 2 on the fire floor or Division A in the front may want ventilation and simply directs a company/crew to open up the building accordingly. The division supervisor is aware and that is really all that matters.

PPV is the tactic of using fans to pressurize a space with outside air to exhaust smoke, gases, and products of combustion out of the building. PPV can also be coordinated through a company already assigned to the division. *While the key is to have the fire knocked down significantly and an exhaust in place prior to executing the vent, that is easier said than done.* PPV has evolved into a tactic that is likely to be completed after the fire is extinguished and/or during cold-smoke situations. The potential to "blow up the fire" with the induction of air is too risky and has caused many firefighter injuries and contributed to firefighter line-of-duty deaths.

HVAC systems may be utilized to move/exhaust/isolate air traveling through a building. Utilizing HVAC must be done in coordination with building engineers and division supervisors to prevent introducing smoke to unexposed parts of the building. If the fire is extinguished and you have a large volume of smoke to evacuate, you may establish a ventilation group to coordinate PPV, natural vent, and HVAC systems, or any combination of the three. *Again, the ventilation group supervisor and division supervisor whose area will be affected must coordinate and communicate.* This is the benefit of tactical supervisors. They are not too task-saturated to communicate and can ensure the safety and accountability is in place much more effectively than an IC trying to do so from the ICP over the radio alone.

Hydraulic ventilation is the use of water through fog streams to evacuate smoke, heat, and gases from a room. Hydraulic venting can occur in concert with fire attack as crews move through the different compartments and find windows from which to exhaust. This is likely more common in single- and multi-family dwellings, but is always a viable option, especially with limited resources.

Fig. 10–20. Defensive trench cuts can prevent the fire from spreading laterally.

Assemblies

The IBC has five Assembly Subcategories: A-1 through A-5. For the purposes of this text, an *assembly building* is defined as an occupancy designed for large numbers of people to congregate (fig. 10–21). Examples include, but are not limited to, the following:

- Churches
- Schools
- Libraries
- Night clubs
- Conference centers
- Bowling alleys
- Movie theaters
- Stadiums

Fig. 10–21. Los Angeles City (CA) Fire Department works a church fire.
Source. Photo courtesy of David Ortiz

CASE STUDIES IN COMMAND

The Ghost Ship Fire

CHIEF JAMIE BOWRON, OAKLAND (CA) FIRE DEPARTMENT

In the fire service, departments are often defined and known by major incidents that have occurred over the course of their existence. The Oakland Fire Department, over the years, has had many historic and notable incidents that have shaped the fire service, from the Loma Prieta earthquake of 1989, the Oakland Hills Fire of 1991, and the Oscar Grant incident in 2009, to the Occupy movement of 2011.

It was 2330 hours on December 2, 2016, and I was wrapping up paperwork in the battalion office when the crackling of the speaker broke the silence of the fire station. A still alarm came over the air for a warehouse fire with multiple confirming stills on 31st Ave. The voice of a 9-1-1 caller was a young girl describing a warehouse fire with a sense of urgency that you could hear in her voice.

I rushed to the BC buggy, putting on my turnout bunkers and radio strap. I responded on Oakland Channel 2 and traveled south on High Street to Foothill, moving as quickly as I could to get into the Fruitvale district. Prior to my arrival, a significant column of smoke could be seen. I heard E13's arrival report of a single-story warehouse with heavy smoke showing. They had secured their own water supply and were going to initiate fire attack.

Upon my arrival, I came down 31st Ave, where I did a quick three-point turn, positioning my buggy where I could see the Alpha and Bravo sides of the building. My initial assessment of the building was that I had what appeared to be a multi-floor cinder-block warehouse with heavy smoke showing from all the windows, front door, and through some of the cracks in the grout lines. Post-incident data collection confirmed that it was built in 1930 with a total square footage of around 9,880 square feet. The roof was constructed of wood gables built upon steel trusses with rolled composition roofing material.

I immediately struck a second alarm, which would quickly double my resources. The second- and third-due engines on the first alarm, Engine 4 and Engine 17, grabbed hydrants and brought in water to Engine 13. I ran up to get the quick 360° and make a face-to-face with the engine crews. I told the officer from E17 that I wanted a 2½" in there due to extensive smoke conditions and the size of the structure. E17 officer would be assigned Division A supervisor.

The Alpha side of the building appeared to be the only access and egress to the building and was heavily covered by various art creations. The Bravo side of the building had a large amount of debris, cars, trailers, pallets, and a fortified chain-link fence. There were three attached business on the D side of the building, which stretched the span of the block. These were identified as the D1, D2, and D3 exposures.

Division A was established with E17 as the supervisor and E13 and E4 advancing a 2½", as well as a 1¾", line through the front door of the building. While returning back to the BC buggy to update my status board for accountability, a civilian told me that he had two friends missing. Upon reassessing the survivability profile in such a large building with heavy, pressurized smoke conditions, I determined that a low likelihood of survivability existed. At that point I had three crews committed to the entry with two hose streams in place. I was already considering a defensive strategy, and if someone was going to be found, one of those crews would have a high probability of coming across them.

The crews were incurring heavy smoke conditions and immense amounts of contents, similar to hoarder conditions, including artwork, pianos, and a labyrinth of contents. It was discovered later that this was an artist collective with heavy contents, and that a dance party was occurring in an upper mezzanine level, in violation of occupancy codes.

The fire was found to have started on the first floor, likely due to an overloaded electrical system installed for the audio and visual demands of the event. Makeshift stairs to the upper mezzanine were constructed out of pallets, leading to where the occupants were located. This was an old dairy and the mezzanine was open to the smoke, gases, heat, and products of combustion below. 33 people succumbed to the smoke and gases, and crews were unable to reach them due to fire/smoke and hoarding conditions.

There was a large crowd in the street and adjacent fast-food restaurant. This mass of people was calm and quiet, watching in a state of complete shock. No yelling, no screaming, no crying—just silence. This was the time that another individual who appeared to have made it out of the building told me that there was easily 50–70 people inside the building. This information solidified why every civilian standing around the scene was in a state of shock. However, operationally I knew there was no way any more companies were going to be able to squeeze into that single man door on the Alpha side of the building. This was a tremendous weight to carry as the IC—the knowledge of potentially so many civilians trapped, and my crews who were clearly not making enough progress in this large building to reach them.

The Division A supervisor reported that he thought they were getting good knockdown. However, I advised them that their knockdown was isolated to their specific area where they were located, and that there was still large volumes of smoke and fire on the B side of the building, starting to extend to the second floor. At this time, second-alarm units started to arrive. E23 was assigned to assume Division B with the tactical objective to gain access into the Bravo side and ensure that all the debris and material on the B side did not light off, and to keep the fire from further extending to the second floor.

A roof division was established with Truck 6 as the supervisor and Truck 2 working for him. They were advised to work together and give me a roof report and report on conditions once they got the roof opened up. The sounds of multiple saws ripped across the incident and crews advised that they had a significant amount of pressurized smoke coming from two large holes towards the Bravo/Charlie and mid-Charlie sides of the building.

Division A advised that they were having a hard time making progress and that the building was extremely compartmentalized. This information confirmed what I was seeing from the exterior, with the small portions of the building looking like they were getting some knockdown. A third alarm was struck due to fire growth in the Bravo/Charlie side of the building, hose advancement diminishing due to hording and compartmentalization in the building, and the time spent by crews inside.

However, no sooner than the third being struck when I saw the night sky light up. It appeared that conditions changed, and a flash of fire came out the front window. That glow and noise will live with me forever. For a brief second, I thought that the fire had flashed over on Engine 4, 13, and 17.

I prepared to request a fourth alarm and activate a RIC (E16) as I attempted to contact Division A. The first attempt resulted in negative contact. I transmitted a second time, requesting an update and hoping to receive a response. My mic opened and they advised that they were okay and all members were accounted for. That 15–20 seconds of that incident will live with me forever. I advised them of the change in conditions I had experienced for the exterior and, fortunately for them, it must have been isolated in one of the various compartments of the building.

Seeing conditions deteriorate in the Bravo/Charlie corner of the building, I made contact with Division B (E23) who advised that they had found a small man-door on the B side. I knew that, once the fire had become established on the second floor, our ability to extinguish this fire from an offensive attack was finished. Which also meant that any possibility of finding any victims was done as well. From the command post I could see the smoke from the vent holes increase. I asked for a current roof report and the officer advised that the fire had become very established on the second floor.

My strategic decision point had been hit. The fire progression was too much for offensive operations. I advised the roof division to grab their crews and get off the roof, and that we were transferring to defensive operations. At that time, I made an announcement on the radio and advised all members off the roof and out of the structure. I did a check with all divisions and fire attack to ensure that all members were out of the building. After 31 minutes of aggressive offensive operations, I received an all clear from all divisions and groups that defensive operations were initiated, and divisions were made for all sides of the building.

This was the correct decision for all the members on the fireground—to ensure their safety. It was the decision that confirmed that we would not be successful in finding any of the possible 70 people who might still be inside. Five hours later, the fire was brought under control, but it took us 6 days to respectfully extricate the 36 civilians that lost their life in the Ghost Ship Fire.

I wish I would have had more companies available for RIC by calling additional alarms sooner. If we had a Mayday, only E16 was available, which is not adequate. This is the kind of fire that lives with you forever. I hope none of you reading this will ever have this type of experience.

The biggest hazard to assemblies is loss of civilian life. This is an exception to the rule that the biggest life hazard at commercial fires is firefighters. As discussed earlier in this chapter, two of the deadliest fires in recent U.S. history are the Ghost Ship Fire on December 2, 2016 that killed 36 civilians in Oakland, CA, and the Station Nightclub Fire in 2003 that killed 100 people. Both were being utilized as assemblies, although illegally.

You could arrive to find dozens of people trapped and/or evacuating. Unlike many other commercial-type buildings that have large egress pathways, roll up doors, and multiple exits, a small assembly can easily have many people who are unable to escape and become clogged in the exits. One of the most horrifying images in the history of fires is the people stuck in the doorway of the Station Nightclub. They clogged the doorway, half their torsos sticking out, with countless people behind them who burned alive.

Overcrowded nightclubs and illegally converted occupancies can become incredibly chaotic scenes of people screaming and dozens or more dying. These spaces are often filled with low-mass synthetic furniture, wall coverings, and décor that make fires grow rapidly. Pre-planning assemblies can save many lives, and being aware of special events in your jurisdiction or first-due area will give you a tactical advantage.

A fire in an assembly can rapidly become an MCI in which a separate medical group, or even medical branch, may be required, with separate tactical channels. Depending upon the scope of the MCI, you may need separate channels for triage, treatment, transport, and medical communications with the hospital. This will be discussed in much more depth in chapter 15.

CASE STUDIES IN COMMAND

Bronx Lebanon and Montefiore Fires

CHIEF DANNY SHERIDAN, FIRE DEPARTMENT OF NEW YORK CITY (FDNY)

I'd like to share stories of two different hospital fires. They were both different, but somehow eventually became very similar. The first was Bronx Lebanon. I was acting battalion chief. We received a Class 3 alarm for Bronx Lebanon Hospital at the change of tours, 18:00 hours. I will admit that the last thing on my mind was that this would be an actual fire. We get inundated with these types of alarms all day. It becomes very difficult to not become complacent. To say what happened next was a complete surprise would be an understatement. I heard the following transmission from the officer of Engine 92: "Engine 92 to nozzle, start a line." "Ladder 27 to 27 OV take the windows in the front of store."

There was a luncheonette on the first floor, Bravo side. The fire had started in the kitchen and spread to the dining area. We had a raging fire on the first floor of a fully functioning 11-story Type 1 hospital. My first mistake was that I transmitted a signal 10-75, that's equivalent to a first alarm. The FDNY has a particular code, 10-76-2nd alarm, for fires in commercial high-rises. This would give us the adequate resources to handle all that would be entailed in dealing with these types of fires.

The SOP in a high-rise commercial would be:

- Determine the fire floor
- Verify the fire floor
- Control evacuation
- Gain control of the buildings systems
- Elevators
- HVAC
- Communications
- Fire pumps
- Confine and extinguish the fire

I hadn't quite made the adjustment in my mind that we were dealing with a high-rise fire. Instead, I was treating it like it was any other fire in a Type 1 building. Fortunately, the fire was on the first floor and not one of the upper floors. Had it been an upper floor, I would have had done the following in concert with the fire safety director.

- Ask if the fire floor had definitely been determined.
- Find out what is the extent of the evacuation that has been implemented.
- Were there any reports of severe life hazards?
- Inquire as to the status of the elevators and the HVAC system.

- Determine if there are any access stairs in the vicinity of the reported fire floor.
- Lastly, what communications have been established between the fire floor and the fire command station?

In the midst of the fire attack, the fire safety director approached me with a major decision. There were surgeries being performed. Did I want to stop all surgery procedures? I made a decision to keep all surgeries going and we would shelter-in-place. It worked out well. The fire was extinguished pretty quickly.

The magnitude of what had just transpired didn't fully hit me until much later. I will admit that I had, in my mind, associated the signal 10-76 with buildings in Manhattan. It had not occurred to me that it was the building that dictated the code, not the location.

The second hospital fire was Montefiore Hospital in the North Bronx. It was a quiet summer day, around noon. We got called to a street box with the only information being "smoke in the area." Again, my guard was down, I was expecting this to be a nonsense call. While responding, the dispatcher notified me that "EMS is now reporting smoke in the Emergency Room."

This got my attention, I was surprised that we hadn't received a class 3 alarm, but nevertheless I was now at least expecting that something was happening. When we arrived on scene, there was a heavy smoke condition with hundreds of people pouring out of the building. I radioed the dispatcher, advising him we needed a 10-75 and that we were using all hands, but I had no idea what we had. This time the dispatcher was 100% on his A Game. He transmitted a 10-76-2nd alarm. I didn't know that he had done this.

I did two things right at this one. I normally listen to what is said between the units themselves, internally. I heard the following transmission, but I went with my gut and discounted it. "OV to Command, it's just a rubbish fire." My experience told me that this was much more than a simple rubbish fire and I continued to run the operation. The transmission that was the game changer is when I heard the officer of the ladder company give this transmission, "Hey Mike, bring the search rope to the basement." Hearing this was an indication that there was a heavy smoke condition in the basement.

It turned out that the diesel generators in the basement had a leak and the diesel fuel had spilled out and ignited. We had a pretty good fire condition in the generator room that was semi-contained, and we still had a heavy smoke condition permeating the building. To compound the issue even further, the power was completely knocked out in the entire building. We had people on life support that now had to be moved, which was very labor-intensive.

I ordered a foam hoseline to be stretched to extinguish the fire, and then focused on the arduous task of getting the patients out of harm's way. In total, we moved 165 patients, and some of them required up to four persons to move them.

Again, I was drawn into the idea that this building was in the Bronx and not Manhattan. I never thought about the 10-76 code. The takeaway from both these operations, which both ended very successfully, is to consider the building and the fire problem and try to match the resources, no matter what the cost. I've seen it many times that once you are behind the power curve, it is very difficult, if not impossible, to catch up.

Strip Malls

CASE STUDIES IN COMMAND

Webster Avenue Laundry Incident

CAPTAIN DOUG MITCHELL, FIRE DEPARTMENT OF NEW YORK CITY (FDNY)

It was just before noon on a crisp day in early March. It was shaping up to be a busy morning and we had been running around, with a variety of responses since our 09:00 change of tours. We were just released from a second-due "odor of smoke," which had turned out to be food left on the stove. As we were taking up from the block, we heard an engine company come over the air and say: "Bronx, we are getting a verbal (FDNY lingo for a civilian who reports an incident directly to the members) reporting smoke in a laundromat on Webster Avenue."

The dispatcher gathered a few additional informational tidbits and quickly transmitted the alarm over the air. We were assigned as the first-due truck. This was, however, not our normally assigned first-due box (the first-due truck was still operating in the apartment from the previous food on the stove). The engine that gave the initial verbal promptly transmitted a 10-75, the signal for a working fire.

As we made our turn onto Webster Avenue, we noted smoke pushing under pressure from a large one-story commercial occupancy. We arrived as the first-due truck company and went right to work, the inside team members grabbing ten-foot hooks, the outside team making the roof and heading to the rear, while I tied off our search rope

Fig. 10–22. Fire companies operating in a defensive strategy.
Source. Photo courtesy of Fire Department of New York City (FDNY)

to a street sign in front of the main entrance. Initially, the owner of the store stated that all his employees were accounted for.

As the engine was in the process of stretching and flaking out their 2½" hoseline, we entered the store. Knowing that, many times, fires in laundry facilities start in and or around the clothes dryers, and noting heat with the thermal imager in that direction, we made our way toward the rear of the store. This was not before poking a hole in the ceiling with the ten-foot hook just inside the main entrance. We wanted to ensure there was no fire above us in the cockloft (the space between the ceiling and the underside of the roof boards, which can be as large as six feet in some commercial buildings).

Moving toward the rear, looking and listening for fire and victims, we encountered light heat but blacked-out smoke conditions. We needed to get a path, the right path, for the engine's advance, and until we found the seat of the fire. This would prove difficult due to the rows and rows of washing machines.. As we continued to search for the seat of the fire and any potentially trapped civilians, things began to rapidly change.

Our roof firefighter informed both the IC and I that there was a significant fire in the rear of the cockloft and it was moving with great speed toward the front of the store (most cocklofts are larger in the front than the rear to create a slight pitch to the roof, allowing for water runoff). They had cut a hole in the flat tar paper roof towards the middle of the building and fire was venting through it. Simultaneously, fire began to now show itself in the rear of the store. I met the engine officer and brought him to a position where he could apply water to the fire. The hoseline was charged in the street and brought into our position. With the line in place, we began to start expanding our search off our search rope when conditions once again changed, with significant heat now over our heads.

In the street, members also noted smoke conditions changing, darkening, and increasing in pressure and density. Incident commanders, with fire now venting through the roof nearly three stories into the air and threatening exposures, backed all members out of the fire building. While an aggressive offensive fire attack was initially employed, with advanced fire throughout the cockloft and rollover/flashover type conditions emanating from the store entrance, commanders made the decision to withdraw operating forces and utilize large caliber streams from a position of safety. This fire eventually went to a fourth alarm (fig. 10–23).

Every fire provides learning lessons. This fire proved to provide several for all of us:

1. 'First-due' knowledge: This cannot be stressed enough. While only a few blocks outside of our first-due area, our lack of information regarding this large (100' × 150') one-story commercial building's layout, the type of construction used, and the presence or absence of a cellar provided additional stressors for the team.

2. The response: With several apparatus responding from a previous incident and not from quarters, the potential to arrive out of standard assignment sequence is great. Also, apparatus chauffeurs and officers must be alert for the response routes and locations of other apparatus, as they may be much closer and take a different route than usual. As units arrive at the fire location, the chief or IC must ensure a firm understanding of which companies are operating and where they are operating.

Fig. 10–23. Companies on the A side of the commercial building.
Source. Photo courtesy of Fire Department of New York City (FDNY)

3. Apparatus Positioning: Our strong FDNY SOPs ensure that rear-mount and aerial ladder trucks are positioned away from the front of the fire building, open for potential placement of tower ladders and/or large-caliber streams (LCS). Sound tactical SOPs allowed for a simple transition from offensive to defensive operation when deemed warranted by the IC.

4. Operations: Ensure team members that are operating remotely are checked in with at timely intervals. There is also a time-space disconnect from the IC to the companies operating in the IDLH zone (tactical gap). Remember that time does not pass at the same rate for the incident commander in the street as it does for the task-focused company officer. Five minutes in a store searching for fire seems to go rather quickly. Conversely, five minutes for an incident commander seems like an eternity. Have trust in your officers and commanders, ensure communication is maintained, and let each other know the conditions as you see them.

5. Reflex Time: Changes in the modes of attack require forethought and reflex time to put into motion. Ensure that all members are aware of any tactical changes that have been made on the fireground and have positioned themselves accordingly, prior to starting LCS, and so on.

6. Be Critical: There is always room for improvement in all our operations, both on a personal and operational level. Critique yourself and share your experiences: what you were thinking, what did you see and hear, and share what went well and what you would do differently the next time.

As firefighters and fire officers, we routinely operate in hostile, rapidly deteriorating environments. Much of our initial tactical decision-making must occur quickly, often

with limited information. We must remain adaptable, resilient, and mission-focused. SOPs and SOGs, building knowledge, and clear, concise communication will assist us in our operational successes.

As discussed previously in chapters 8 and 9, attic fires are a common occurrence. In the commercial arena, this is especially true in strip malls. *Strip malls* consist of several businesses connected with common walls and a common roof assembly. They are all over North America in practically every town on main street or in commercial areas. They can be legacy or modern lightweight construction, depending on the era built. They have common attics, cocklofts, and often have facades that traverse the entire front of the building (fig. 10–24). These are all common avenues for lateral fire spread (fig. 10–25). The tactical advantage is that even if there is a common attic, party walls will separate the business occupancies. These provide good separation from which to protect the exposed business living spaces from the primary unit burning.

Obviously, if a fire wall is found in the attic of a larger strip mall, this can also be leveraged to provide tactical advantage by providing trench cuts to prevent lateral spread in the attic space. Heat cuts may be needed to relieve pressure to attack teams and buy time to then cut trenches. This must be coordinated with hose teams underneath. Here, your ground-level division supervisors and roof division must coordinate.

Fig. 10–24. Fire through the roof of a multi unit stripmall.
Source. Photo courtesy of Rick McClure

Fig. 10–25. This fire ran the common facade and attic.

Upon arrival of a fire in a strip mall, you will find one of three common conditions. First, the fire could be isolated in one business and not appear to have extended into the common attic and/or exposure units. This can be knocked down quickly with the proper-sized hoseline. If the fire is confined to a bathroom, kitchen, or rear office, for example, a 1¾" line may suffice. If the fire is significant, a 2½" line may be the best option, or if you have limited staffing, a quick deck gun blast may buy you time and keep the fire from spreading into the roof assembly. As with all fires, consider the VP first and ensure a water supply is nearby.

A second scenario, and perhaps the most common, is heavy fire in the involved business and fire is clearly in the attic/cockloft and spreading laterally. This is common, as these buildings often have lower ceilings with heavy fire load in the form of stock, office furniture, and low-mass synthetics. Hazardous materials in businesses like dry cleaners, paint stores, and mom-and-pop hardware stores can also propagate the fire problem. In this condition, consider getting ahead of the fire by placing lines in the B and D side exposure units to cut the fire off from spreading further in the common attic. This will take tactical discipline, as most firefighters will want to immediately put water on the fire they see coming from the involved unit. With heavy fire, that unit should be written off if there is a low VP and no refuge in the back for potential victims.

The first lines should go, again, into the exposures on the side that has the most to lose. If you do not have the resources to place lines in both exposure units, the first line should go to the side that has the greatest exposure potential for loss.

Figures 10–26 and 10–27 show fire that occurred in a strip mall with a large, supported mansard over the front entries. Note that the fire originated in one unit from the end (fig. 10–26). If the main origin unit is fully involved and getting into the mansard and clearly in the attic, where would you stretch your first line? The B-1 exposure to the left is the ideal choice here because there are several more attached businesses that could be lost. There is more to save to the B-side/left of the involved unit. So, your initial line should go there to prevent lateral spread in the attic (fig. 10–27). Note also that the D-1 exposure unit to the right is the last occupancy on end *and is vacant.* This further illustrates the need to go to the B-1 exposure first in this situation.

The third scenario would be fire that has engulfed more than one unit upon arrival and is through the roof and clearly has significant structural involvement and potential for collapse. Here, you should consider getting one or more businesses ahead of the fire as the potential for collapse and the amount of fire will require a significant fire flow. This will take time and resources. By making a stand ahead of the fire, you will give yourself time to get resources in place inside the exposures to truly stop the fire. In the meantime, if you have a patent water

Fig. 10–26. The fire was next to the end D-1 unit, which was also vacant.

supply, a deck gun attack at the fire's edge between the involved unit and exposure units can further slow the fire's progress. Work your way back after you knock down the fire at the edge for the exposure and involved unit.

Remember to check the C side and potentially set up a Division C with the objectives of forcible entry and fire attack with a separate water supply. Ensure Divisions A and C coordinate to not have opposing streams. Expect reinforced doors and windows in the rear (fig. 10–28).

The first example would be an offensive strategy. The second and third examples would be combination strategies, with the involved units being in the defensive space, and the exposures being in the offensive space.

The same considerations can be made in standalone businesses of size when a fire wall is in place. If the building is large enough and has fire walls that can be identified, these can be leveraged to prevent spread by using a coordinated attack with vertical ventilation. The challenge in these buildings is that you may not have the party wall separation needed that separates occupancies as you would in a strip mall. You and your tactical supervisors must be suspicious and watch out for changing conditions.

Remember, once the VP is low and/or the building is confirmed evacuated, vacant, or all clear, you really must reconsider what you are saving. Chances are the contents of the business are a loss from the smoke and water, let alone heat and flame. Consider retreating to a combination strategy, or just switch to a defensive strategy if you have any doubts about containing the fire in a timely manner. Your tactical supervisors will be your eyes and ears and you should notify them that you are considering a strategic change if conditions do not improve.

Fig. 10–27. Note the B-side exposures had much more to lose. This is where the first exposure line should go.

Fig. 10–28. An example of reinforced windows and doors *Source.* Photo courtesy of Rick McClure

This was the case in a leather garment factory shop where machinery and raw leather materials were used to make fashion accessories. The building had no fire walls identifiable. The Division A supervisor had excellent accountability. The fire was knocked down in the work space; however, a stubborn attic fire was burning in the attic and was difficult to access from below. Division A had declared an all clear of the building as secondary searches were complete. This meant that we were in an incident stabilization priority. The life risk now was the multiple companies working inside and on top to access the attic fire. I notified Division A and roof division that I was considering switching to a defensive strategy. If I didn't see significant improvement in conditions, I was going to evacuate and go defensive. The tactical supervisors agreed. Just about then, conditions improved dramatically, and we were able to continue the offensive strategy, put the fire out, and perform salvage and overhaul.

Taxpayers and Mixed-Use Occupancies

CASE STUDIES IN COMMAND

Westlake Taxpayer Incident

CHIEF KEN COOK, LOS ANGELES CITY (CA) FIRE DEPARTMENT

I had only been a chief officer for over a year, but the fire activity had been abundant in the city. The truth was that some of our citizens experiencing homelessness had been lending to our professional development when it came to fires. The probability was force-multiplied when I picked up an overtime day downtown, as the area had seen a significant increase in fires over the past few years.

That same evening in late January, a two-story pre-33[2] center hallway apartment building erupted into flames in the Westlake district of Los Angeles (fig. 10–29).

Around dinner time, more than 100 firefighters responded to the 2800 block of 7th St. in 11's first-in district. We were the second Battalion Command Team on the ticket, and when we arrived, the fire was already well-advanced on the second floor and into the attic.

The Los Angeles Police Department (LAPD) had initially warned the fire department of a barricaded suspect at the location. After an intense standoff, the suspect started a fire in a room on the second floor and narrowly escaped from an adjacent window. LAPD made a valiant effort to evacuate the building, but it was feared there were still people trapped inside. As fire department resources arrived, LAPD advised that the suspect was in custody, and with permission from law enforcement, the firefight began.

The building was constructed in 1921, situated on the corner of a bustling intersection, with most businesses closing shop for the evening. It had multiple retrofits with the famous kings' row and tie plates giving away its decade. The establishment was a pre-33 taxpayer with residential units and businesses mixed-use on the top floor and commercial occupancies lining the first. In the Westlake district, these businesses were all most people had. A transom window constructed above a door on the Alpha/Bravo corner marked the location of the stairwell leading to a typical T-shaped hallway where initial resources made their attack.

At the top of the stairwell, firefighters encountered intense heat and heavy smoke, banked down to the floor. The kind of heat that burns when making minor adjustments to your limbs in your coat. They attempted to advance with little to no progress, even cooling the hallway for a small gain. Indeed, if someone was overcome by smoke in the hallway, they were already dead, but that didn't mean there weren't still lives to save. It was communicated to the IC that there was a lack of suppression efforts or forward movement on the second floor. The IC acknowledged this, but the companies would have to wait to engage without ventilation.

As additional resources arrived, an engine company assigned to the first floor to ensure that the fire hadn't started in a basement or a wall at ground level provided the IC with some additional communication that better access may be on the Charlie side—the opposite side of where fire attack was initiating their assault. The immediate concern was opposing hoselines, but this would be the best and only opportunity with no advancement and fire attack ahead of ventilation. The IC directed the fire attack team to back out and ordered the next company to engage from the Charlie side on the second floor. It was the most thoughtful approach, involved to uninvolved, at least in the hallway, to cool and suppress the environment—giving occupants the greatest opportunity to survive.

A few minutes later, ventilation was initiated from the Charlie side to support the suppression efforts. When the roof team opened the 100-year-old conventional roof, the vent-limited fire found its oxygen and immediately ignited in the attic. Now with the hallway, two rooms, and the entire attic ablaze, a request for additional resources was made.

Fig. 10–29. Los Angeles City FD Engine 10 at the Westlake Incident. *Source.* Photo courtesy of Los Angeles City (CA) Fire Department

The incident was organized into divisions and groups (i.e., roof division, Division 1, Division 2, medical group, rapid intervention group, and eventually search group). Multiple companies were deployed to the second floor and began the intense firefight. As the roof team continued to cut, they knew they were no match for the extensive involvement and asked for an additional truck company to the roof. Then, the coordinated attack ensued.

As we pulled up to the incident, about 20 minutes in, 30-foot flames raged above the parapet. Fire could be seen from two windows on the Bravo side, as heavy smoke poured out of the adjacent windows. By this time, I knew the IC was probably leaning in a direction to change his strategy, and I knew we only had a few minutes to help him to decide otherwise, if it made sense.

As we made our way toward the building, I was impressed with the sight of the Los Angeles City Fire Department going to work. Aerial and ground ladders were strategically placed, multiple engines tactically maneuvered in the street to provide the most advantageous hoselines, and ambulances were staged, ready to race to the hospital with the possibility of occupants being trapped.

But there wasn't enough time to get to the ICP, let alone find it. I momentarily stood in front of the structure, taking in every detail, keyed the mic on command, and said, "IC from Battalion 1, I'm standing in front of the building; I'll take Division 2 if you concur?" He obliged and rattled off my companies. I immediately went to the engineers to see what companies were operating in each portal with access to the second floor. I chose the Charlie side for my ascent, remembering the communication from the later-arriving engine company. As I ascended the staircase, I saw the fire raging in the attic. Smoke was about head-high, and companies were just starting to get a knockdown in two rooms; however, the attic was still fully involved.

Most companies were directing their hoselines vertically in the attic from the burned-out rooms, but it wouldn't be effective unless we could get our suppression efforts horizontal. As I stepped foot on the second floor, I immediately heard the sounds of multiple warbling whistles indicating our people had already exhausted their air. I transitioned command over the radio with the engine company managing the division and started to develop a plan for suppression and search of each unit. My priorities were accountability of the assigned companies, their portal of attack, additional support, and determining which companies needed to replenish their air. I ordered one additional task force and methodically selected each company to rehab based on their remaining air and usage rate.

Next, I assessed the risk associated with our offensive attack. Balancing my decision against the resources assigned, the time elapsed, the conditions on the floor and within the attic, and the construction. I chose to stay offensive, communicating our plan to the IC. My general concern was the roof collapsing onto the ceiling joists or heavy equipment in the attic falling through the ceiling. After evaluating the joists in the hallway and surveying the attic, I knew we could safely stay offensive and successfully achieve a knockdown.

> We began to cut the ceiling joists with chainsaws and strategically placed our inside ladders to develop hoselines in the attic. We coordinated with the ventilation team and started working the fire from uninvolved to involved. Eventually, I notified the IC that roof division had accomplished its mission and communicated we could operate without them on the roof.
>
> We set up horizontal ventilation with another truck company. A thorough search followed, methodically checking every square inch of the eight units. Remarkably, the closed doors had kept most of the rooms from becoming engulfed by the flames. After an extensive search, we determined that all occupants were out.
>
> At 18:05 hours, just 50 minutes from the time of the alarm, we declared a knockdown. Every business on the first floor was saved, as salvage efforts played a huge part. At least four units on the second floor were protected from the heat and flames, most of them being businesses that depended on their contents to stay afloat.
>
> The incident was a success; however, without the company's tenacity and willingness to work past the typical work/rest ratio, we wouldn't have been able to complete final extinguishment. The ventilation on the roof was extraordinary and assisted us with providing an environment to make critical decisions about risk and tactical operations to stay offensive. It was assumed we were initially in the rescue profile, and I felt comfortable about that. After about 30 minutes of fighting the fire, we reduced our life profile to a property profile.
>
> Later the next day, we accompanied four business owners in retrieving their belongings from the second floor, which was especially rewarding. That experience alone encourages me to continue to be highly critical about my decision-making and determining the level of acceptable risk when deciding between life and property, and the safety of our firefighters.

A *taxpayer (mixed-use)* is defined as a building with a retail-type business(es) on the ground floor and residences on the upper floors. Sometimes the upper floors may have offices. The challenge with these buildings is fire load, occupancy, and access. Another term is residential over commercial for these buildings.

The fire load on the ground floor may be extreme due to retail stock, such as low-mass synthetics or the possibility of hazardous materials. If the upper floors consist of apartments (most likely), the life loss potential from a fire on the first floor is significant. If a fire were to occur after business hours, at a time when tenants are sleeping in upper floor units, the results could be catastrophic.

Access can be challenging, as many times the A side door can be difficult to distinguish from the retail business or restaurant access. The upper floors are normally center hall configuration, which can add to the challenges regarding evacuation, search, and hose stretches.

As an IC, you should be thinking about a high VP at the minimum in these buildings, as they are likely occupied 24 hours a day, either in the business below, or the dwellings above, or both. Plan on rescues of entrapped victims and the need for a lot of medical resources.

Many of these buildings were built in the early 1900s, comprised of unreinforced masonry that has been reinforced and retrofitted. Most will have a basement and common attics or cocklofts.

Concrete Tilt-Up

Concrete tilt-up buildings are characterized by pre-poured concrete walls that are set and connected to the roof assembly for structural stability. Older tilt-ups from before 1960 are more likely to have conventional roof construction, such as bowstring truss, while post-1960 tilt-up buildings will likely have a more lightweight construction. Panelized or lightweight truss roofs are extremely common on the West Coast and in newer developments (fig. 10–30).

The contents and businesses inside these buildings vary dramatically, from furniture stores and carpet and textiles retailers to indoor sporting venues and multi-unit warehouses. The relative ease, speed, and less-expensive manner in which concrete tilt-up buildings are built makes them more common every day.

The primary hazard associated with this construction is collapse of the walls. The roof essentially holds the tilted-up walls in vertical plumb position. The concrete walls in tilt-up construction notoriously fall outward when the roof assembly (that holds them together) burns away (fig. 10–31).

As the Division C supervisor on a large piano factory fire, I was supervising multiple engine and truck companies. This was at night in the early hours of the morning. The fire had already blown through one fire wall on its way to the next. The roof was burned off 60% of the massive commercial building and this was tilt-up construction. The fire was in an incident stabilization/combination strategy.

Fig. 10–30. Concrete placement of walls and buildings is fast and easy and is more common every day.

Fig. 10–31. Note the walls separating at the corner, where you can see fire in the seam.

Once I arrived in the rear to assume Division C, I could see fire between the seams of the walls and at the pilasters. This told me that collapse was highly probably. Multiple companies were working in the collapse zone, utilizing 1¾" handlines, which had absolutely no effect on the fire. In addition, a truck company was operating an aerial master stream, but they were set up extremely close to the structure in the collapse zone.

I directed the engines to move 100 feet down to the offensive space, and the truck to break down and move back out of the collapse zone. We then employed offensive tactics of interior handlines and a defensive strip cut overhead to support the next fire wall and prevent the fire from blowing through it as it had done before. The defensive area was written off and the offensive area was saved. This required a lot of direction and coordination, as mission creep was constantly occurring as crews would migrate into or near the collapse zone.

Brick and Unreinforced Masonry

Brick and unreinforced masonry (URM) buildings consist of brick, masonry, or concrete that is not reinforced with metal, such as rebar. The Long Beach (CA) Earthquake of 1933 resulted in significant changes to building codes and URM buildings were becoming less common, hence the term "pre-33" building. Signs of these buildings include reinforcing tie plates, arched/recessed windows (fig. 10–32), white lime mortar, and king rows in the brick (bricks laid with end showing every four to seven rows [fig. 10–33]).

Collapse is the primary concern in these buildings, as the collapse zone can stretch significantly further than the height of the walls. These buildings are often retroactively reinforced, remodeled, converted, and may have a mixture in internal construction features. Often, reinforcement plates with through-bolts will be seen (fig. 10–34).

As with any fire, an IC with a fire in these structures should focus on the VP; however, the unpredictable nature of these structures, combined with age and collapse potential, should trigger going to a defensive or combination strategy sooner than later.

Sometimes, all you can do upon arrival is begin protecting exposures. This was the case one night at a massive vacant warehouse fire at approximately 02:00. The entire first alarm went to the burning warehouse that already had 100' flames coming through the roof upon arrival. The building was approximately 200' wide and 300' long and was fully engulfed in fire.

As the first BC on the second alarm, I arrived to find elevated master streams, deck guns, ground monitors, and 2½" lines operating. This was clearly a defensive strategy, as

Fig. 10–32. Arched and recessed windows are a sign of URM buildings.

Fig. 10–33. White lime mortar and king rows

Fig. 10–34. Reinforcing plates

appropriate. I was assigned to assume Division A by the IC. My primary objective was to ensure that no personnel or apparatus was in the collapse zone of the URM building.

As I walked over to the B side, I immediately noticed something that I should have seen driving into the scene, but missed because I was looking at the flames. The B side business that was clearly operational, but closed at the time of the fire, was attached by a parapet fire wall. This was a clear TPO that could be leveraged to save this yet-uninvolved business.

I immediately contacted the IC and recommended that I be reassigned to assume the Division B supervisor role and that he give me the entire second and third alarms to save the business. It was approximately 100' wide and 300' long. He concurred, as the involved building was a complete loss.

I directed companies to open the roll up doors and noticed clear conditions inside the party supply warehouse. This was obviously a low VP, but there were helium tanks and massive amounts of flammable stock such as balloons and other decorations, not to mention giant clowns!

The old fire wall of URM brick and mortar was holding, but not for long. Small flames began licking through the mortar cracks. I directed companies to stretch 2½" lines inside with the objective of protecting the fire wall. My intent was to have crews on the roof perform a defensive trench cut to support the engines inside. The net result for the IC would be to switch from a defensive strategy to a combination strategy. He agreed.

The problem was that I only had two engines at the time, and it would be considerably more time before the balance of the second alarm would arrive, let alone the trucks get to the roof in a timely fashion and begin this significant trench cut. As companies stretched lines, conditions changed. I looked to the rear C side of the building and saw active fire. I then saw a silhouette of a firefighter running toward me. It was the captain off of Rescue 20, Johnny Brenner. He notified me that the fire wall to the rear had failed and actually cut off his egress back out the C side from where he came. I immediately notified his Division C Supervisor that he was accounted for.

As I turned around to exit back out of the A side roll up doors, smoke from the fire caused by the collapse of the C side wall (300' away) beat me out of the doors. By the time I reached the doors, we had black smoke ten feet off the floor in the entire party warehouse. The moment had passed, the window of opportunity shut. We had a low VP at 02:15 in the party warehouse, a failed URM wall, and nowhere near the resources to save the exposure. By the time I called the IC with an update, smoke was banking further down and more of the mortar was showing flames in the wall.

Lessons learned:

- If I had immediately identified the parapet wall between the involved unit of origin and the B side attached exposure, we *may have* had time to redirect units earlier and save the business.
- If the first alarm had seen the exposure, we likely *would have* had the time, because the first alarm could have gone directly to the exposure and write off what was lost.
- Always look at the big picture as you come into a fire. Discipline yourself *not* to have tunnel vision on the flames first. Once you do, you are more likely to miss exposures, other TPOs, or hazards that you would have otherwise seen.

CASE STUDIES IN COMMAND

Certified Bakery Fire—Union City, New Jersey

CHIEF ANTHONY AVILLO, NORTH HUDSON (NJ) REGIONAL FIRE

The Building, Exposures, and Arrival Conditions

This fire occurred at rush hour in heavily congested Union City, New Jersey. The fire building was a U-shaped one-story of ordinary construction. The building was located on Park Avenue between 36th and 37th street. On Park Avenue, adjacent to the bakery (fire occupancy) was a beauty salon, a furniture store, and a hardware store, designated as exposures B, B1, and B2 respectively. The bakery occupied both the corner section of the building and the rear part of the U. The sales section for the bakery was located on Park Avenue. Behind the sales section and located on the 37th street side was the delivery section. The production area occupied the part of the U at the rear of the building. On the Bravo/Charlie side of the fire building was a one-story homeless shelter connected to a three-story homeless shelter, both of ordinary construction, which were attached to the rear of the production area. The three-story part of exposure B/C was originally an embroidery factory with many large windows facing above the roof at the rear of the fire building. It was a very congested layout, which would prove extremely challenging at this fire. In between the two legs of the U was a service alley that was about five feet wide. Part of it was covered with plywood. There was a fire wall between the production area and the delivery area of the bakery. The doors between these areas were open at the time of the fire. There was also a power line that ran from the utility pole on 37th St. to the service connection for the building, which was located in the service alley. It ran just about head high across the roof of the delivery area. Utility lines also ran along the 37th St. side of the structure. There were two large doorway openings to the delivery area on the 37th St. side.

The fire originated in the production section of the bakery. The North Hudson Regional Fire Rescue response to this fire was four engines, two ladders, a heavy rescue, myself as the second battalion chief, a safety officer, and the deputy chief who would take command when he arrived. As I was leaving my quarters about a half mile away, I could see a large column of black smoke spiraling up from the area. As the employees tried to fight the fire themselves, it got a really good head start. The fire started in the area of the ovens and spread into the ceiling, feeding on the byproducts of the forty-year-old baking business. It appeared there may have been a dust explosion or at least a dust-fed traveling flashover prior to our arrival. When I arrived and established command, I saw fire showing through the roof in the area of the chimney (in the production area) and a heavy smoke condition in the delivery area. The homeless shelters were severely threatened by the fire venting through the roof. I immediately struck a second alarm, bringing another two engines and ladder, as well as another battalion chief to the scene.

Strategic Action

The strategy was to attempt to keep the fire from extending to the homeless shelters on the Bravo/Charlie side and from spreading into the stores on Park Avenue. Our SOPs assigned the first-arriving ladder to primary search and, as per SOP, the first two engine companies teamed up to stretch a 2½" line to the delivery area. A second 2½" was being stretched by the third-arriving engine. The second ladder company laddered the roof on the Park Avenue side. The companies on the interior reported some good news and some bad news. The good news was that the primary search was all clear and the fire had not yet passed the fire wall between the production area and the delivery area. However, the bad news was that the roof in the production area was collapsing due to the HVAC units on the roof. The report from the roof also notified me as to the presence of the service alley. Because of the collapses in the production area, I ordered the lines to stay on the safe side of the fire wall in the delivery area and direct the streams into the burning production area. The plan now was to keep the fire out of the delivery area that led to the bakery sales area and the adjacent stores.

Exposure Protection

The deputy chief arrived on scene and assumed command. I went to recon the Park Avenue side and as I looked down the Bravo side, I could see the fire licking up against the exposed second- and third-floor windows of the homeless shelter. Several of the only trees left in Union City were also on fire. There was also a two-and-a-half-story frame dwelling about 15 feet from the furthest store in the row. Luckily, the fourth-arriving engine happened to be a Tele-Squirt, so I had them secure their own water supply and utilize their boom to protect the exposures on the bravo and Charlie sides. I also ordered three more 2½" lines via the Bravo side of the Park Avenue stores. These lines were stretched to the roof of the Park Avenue stores and operated on the safe side of the service ally to assist in exposure protection and to darken down the fire issuing through the roof. These lines were staffed by second alarm companies. Not so luckily, at around this time, the chimney in the production area collapsed, opening up the roof and spewing even more angry fire.

Command ordered a third alarm and directed all responding companies to 36th street to protect the interior of the exposed three-story homeless shelter. The third-alarm battalion chief was made the division supervisor of that operation. As the fire had already entered that building, multiple lines were stretched inside and the building was searched and evacuated. Once the fire was controlled, those lines were also used defensively from the windows onto the burning production area.

Confinement Operations

I was then assigned to the roof division. While I was concerned about fire spread via the service alley, I was more concerned with the fire spread from the production area into the delivery area and into the stores on Park Avenue. The roof crews told me that the cockloft was about 10 feet deep. We decided to utilize a trench cut just behind the skylight on the roof to keep the fire out of the delivery area and the Park Avenue stores. Once we opened the skylight, we set about cutting the trench cut and I ordered all remaining roof personnel to stay on the safe side of the trench cut, which was above the sales area. This was because the collapses in the production area made the delivery area more of a threat, so the lines in that area were backed out to the exterior sidewalk and operated from outside the two large doors. The fire did eventually pass the fire wall, but we were able to stop its progress with the trench cut and lines from Park Avenue inside the sales section of the bakery.

Another division commander was set up on the alpha side to supervise the Park Avenue store operation. A fourth alarm was struck due to the depletion of the tactical reserve. We also stretched lines into all of the Park Avenue stores, as each of them had access through a back door to the service alley. The plywood over the delta section of the service alley was also burning. We were concerned that the fire would spread into the drop ceiling space in the stores and run across the entire open area above the space. Companies in the stores pulled the ceilings to ensure the fire was not running past them.

Results

This fire was contained to the area of origin, the production area of the bakery. There was only minor extension into the delivery area and the homeless shelter. The service area of the bakery and the adjacent stores on Park Avenue did not become involved.

Several key factors allowed us to keep the fire from spreading not only to the adjacent interior exposures but also to the homeless shelters. The first was the decentralization of the command structure, assigning chief officers in each major area of operation. The second was the willingness to request additional alarms early in the operation. If you don't have enough people, you will never put out any fire. The third was the execution of several strategies at the same time in different areas of the fire operation. This allowed a complex plan to proceed and be successful. Lastly, the willingness to modify the action plan and strategy based on reports generated from the division commanders made this operation a success instead of a conflagration.

Fast-Food Restaurants

As with most commercial buildings, fast-food restaurants are likely to be evacuated upon your arrival, or evacuation will be in progress. Most will have hood extinguishing systems; however, they can be easily overwhelmed with a large enough fire or arson.

On February 14, 2000, two Houston firefighters were killed in a vacant McDonald's restaurant. On June 5, 2018, a McDonald's restaurant burned down in Seaside, CA when a drunk driver's car struck the building. The restaurant was rebuilt within months. Large chain restaurants have the means to rebuild. Figure 10–35 shows a McDonalds that burned down. It was rebuilt in the same location soon after.

Many times, fire in the roof assembly can present with no fire visible in the common area. This was the case on a fire in which I was the IC at a McDonald's fire one night. Fire was visible from the roof. I had a report of a firefighter who fell off the roof. Miraculously, he sustained no injuries. Again, the building was vacant with no fire or smoke in the common area.

Fast-food restaurants can lull us into a false sense of security because they are relatively small in size, yet they have countless hazards lurking inside. Grease, oil, buildup in hood system, tile floors over basements, heavy HVAC systems, heavy wiring and ducting in the roof assemblies, truss construction, and many other hazards are common. Standalone McDonald's restaurants often have a basement, for example.

Any time you respond to a fire in a fast-food restaurant, you should remain extra vigilant and always err on the side of caution. Your instinct to switch to a defensive strategy should be heightened. Remember, they can rebuild, and putting these fires out is not easy once they are structural.

Fig. 10–35. Companies operating in a defensive strategy at a stand alone McDonald's
Source. Photo courtesy of Demetrius Kastros

Big-Box Stores

Big-box stores are extremely large retail outlets that are between 100,000 and 200,000 square feet. Examples include Home Depot, Walmart, Target, and countless others (fig. 10–36).

They are commonly tilt-up construction. These are piled high with stock, hazardous materials, and low-mass synthetics, and can generate flame lengths in the hundreds of feet once they breach the roof (fig. 10–37).

It may be easy to think a big-box would not burn due to the sprinkler systems in these huge buildings. Not true. Plenty have burned down. Most of the mega outlets have the plan and means to replace a store within a short time.

The key is evacuation of occupants. Unfortunately, some citizens do not evacuate quickly or do not see the fire as an actual life hazard until conditions literally drive them out of the store. If you have a fire that is localized and can be quickly confined and extinguished, you still must consider the time lag to get a line stretched and on the flames, while evacuating occupants.

Be ready for crews to experience worsening conditions and to be outmatched. Division supervisors must correlate what they see inside with what they see outside. Often, this will create a reality check that triggers a strategic change. Ensuring firefighters do not overcommit too deep into the building when conditions are tenuous is critical.

Fig. 10–36. Home Depot is a big-box store.

Fig. 10–37. Typical high-rack storage of inventory in a big-box.

Two-Out Crews and Backup Lines

It may require several engine companies to get just one 2½" attack line stretched, in place, flowing, and supplied to be effective in a commercial building. In addition, backup lines will likely be necessary to provide sufficient fire flow for the Btus that are generated from these fires.

A ready two-out crew should be standing by at every portal of fire attack to watch the backs of the crews inside. This would include vigilance for worsening conditions, potential collapse, and support at the door with the stretch. The division supervisor should ensure that at least one or two crews are standing by at the portal of entry on an additional 2½" line. They can assist with evacuations at the door while watching conditions and be ready to flow their line if necessary to assist the primary crews with retreat.

All of the above could require at least four engine companies or more, depending upon staffing. If your FD does not have the resources to support this, or they will not be there in time, you will likely find yourself quickly switching to a defensive strategy. In suburban and rural areas, a dozen different fire departments may respond to a commercial fire. The response time alone may warrant a defensive strategy.

Whether the IC, a tactical supervisor, a company officer supervising a crew, or a firefighter, when you are fighting fire inside a commercial building, always ask yourself, "What are we risking and what are we really saving?" If the VP is low and you do not have the resources needed to support the plan as described above, change the plan!

Rapid Intervention and Safety for Commercial Fires

RIC teams are especially critical on commercial fires. Rather than a single engine company that would be utilized on a house fire, commercial fires require a greater level of preparedness for your troops. Consider an engine and truck company *as a minimum* to start.

Assign one of the officers as the RIC group supervisor. As with all templates, your tactical division and group supervisors also act as the safety and accountability officers for their respective areas/functions. That said, consider a specific safety officer on large, complex commercial fires if you have the extra overhead. They can also assume the RIC group supervisor role and take over for the company officer should you have a Mayday (Heaven forbid). If resources are available, consider a heavy rescue company and transporting ALS unit with a chief officer as the RIC Group Supervisor. You may also need to post up multiple companies around the commercial structure, as one RIC company on the A side does very little good for the companies on the C side of a big-box store, for example. See more on rapid intervention in chapter 12.

Remember, hazards are lurking all around commercial structure fires well beyond the fire itself. Unsupported mansards/facades (fig. 10–38), high-piled stock, hazardous materials, machinery, excess inventory, collapse, getting lost, and countless other possibilities exist.

Tactical supervisors are your safety officers and will see these hazards and potentials unfold. For example, they may see crews standing under an unsupported mansard with fire overhead and be able to remove them. Also, listen to your tactical supervisors about switching strategies from offensive to combination or defensive. They see things that you and your crews do not see.

Fig. 10–38. Unsupported mansards/fascades collapse suddenly and quite frequently on commercial buildings.
Source. Photo courtesy of Mike Meadows

ICS and Communications for Commercial Fires

First, we will discuss the nuances and differences between residential and commercial tactical supervision and then present several ICS templates for common commercial fires.

Tactical Supervision (Divisions/Groups) and Communications

Tactical supervision is vital on commercial fires. This is another area where residential habits may be unintentionally duplicated on larger fires involving commercial buildings. A delay in setting up divisions/groups can easily occur. Meanwhile, a second alarm (additional resources) are called, the IC gets behind the incident power curve, and communications get overwhelming.

The second or third officer to an area will likely be your tactical supervisor. For example, E1 arrives to a fire in a large 200' × 200' warehouse on a Thursday morning, evacuation in progress. E1 *initiates* fire attack with a 2½" line on the A side reporting light to moderate smoke. E2 establishes a water supply and assumes command from E1. E3 arrives and is assigned to stretch a backup line for E1.

You arrive as the first chief and assume command, declare an offensive strategy with high VP, and call a second alarm. Who is the best choice for Division A? In this case, E2 is the likely candidate, because the E2 officer is already in command, has a water supply being established, and has E3 to assist with a backup line. In other words, there are enough companies to perform the initial tasks on the A side. The E2 officer is already in command and no doubt has accountability of E1, E2, and E3 crews. By leveraging this *tactical advantage* of the mindset of E2 (IC) being in the "next," you can then assume command and assign the E2 officer to *assume* Division A, with E1 and E3 working for them.

Conversely, you may be on a fire where a lot of work needs to be done simultaneously and not enough companies are on scene yet. These task-intensive fires are challenging. As the IC, you want to bridge the tactical gap; however, if all of your initial companies are working, none of the officers may be available to step back and assume a division/group role. As discussed in chapter 7, consider this

the tactical gap from the *strategic level to the task level.* As the IC, you may be required to get in the weeds a bit more by directing each incoming unit and giving them more specific details regarding their assignments. This is counterintuitive to the notion of decentralization of command and tactical decision-making; however, you may have to shoulder the burden of tactical decisions by directing crews as they arrive during a more tenuous commercial fire situation, then bridge the gap when the second chief officer arrives.

Let's look at another situation with the same commercial fire in a 200' × 200' warehouse. E1 has initiated attack on the A side. E2 is getting a water supply and has assumed command from E1. When you arrive as the first chief to assume command from E2, a lot of work still needs to be done and not enough hands are on scene to do it (E3 is not on scene). Here, you may choose to have E2 rejoin the crew and remain task-oriented because a second line is needed immediately. Then, TR1 arrives and assists with forcible entry and search from the A side. You keep TR1 at the task level. E3 arrives and you assign them to the rear on the C side to give a CAN report since this is an extremely large building. Your span of control is 4:1.

The next resources will drive your ICS (remember ICS formula B+C+R = ICS). If you need vertical ventilation and the next company is a truck (TR2), for example, you will likely assign them to the objective of vertical ventilation, with the company officer *assuming* roof division supervisor. Your span of control is now 5:1. Then, E3 reports from the rear that there is heavy fire in the loading dock on C side. This appears to be the origin of the fire and it is getting into the structure. E3 requests another engine to bring them a separate water supply and assist with stretching a 2½" line into the structure.

E4 arrives with Battalion 2. How would you assign them? You clearly need E4 on the C side to assist E3 and bring water. And, you have multiple companies operating on the A side with no tactical supervisor, evacuations in progress. Size up your situation. The VP is high, as evacuation is in progress and employees are exiting the A side. So, you should continue your offensive strategy there. Meanwhile, you have heavy fire on the C side that is the likely origin in a loading dock. Your span of control is widening, and you have called a second alarm. The roof report from the roof division supervisor is that this is a panelized roof. There is fire on the C side and light-to-moderate smoke from the vents, as well as four large HVAC units on the roof. They recommend beginning vertical ventilation to support the crews inside and limit the fire's extension and requests another truck to assist.

In this situation, Battalion 2 can assume Division A with the objectives of fire attack and evacuation/search from the A side. Division A has E1, E2, and TR1 assigned. E4 can assume Division C with E4 crew and E3 assigned with the objectives of establishing a water supply, fire attack, and any evacuation/search that can be made from the C side. Your span of control is now 3:1 with Division A, Division C, and roof division. This means you also have three safety and accountability officers embedded into the incident at each of those locations. Your three supervisors are expected to monitor conditions and crew locations as part of their job description. As additional resources arrive, you feed the system. The next truck (TR3) would go to roof division, for example.

You have eight units on scene (E1, E2, E3, E4, TR1, TR2, TR3, BC2), yet your span of control is now 3:1 with three safety/accountability officers in place as division supervisors. Division A and C have the primary objectives for getting search, evacuation, and fire attack completed. Roof Division's primary objective is to coordinate vertical vent with Divisions A and C.

What is your "then/what if?" Medical and RIC should be at the top of your list as you get enough units on scene. Staging should be considered for your additional alarms. Contacting the owner/on site manager, and any support like law enforcement, air units, and utility companies should be considered.

You should remain in an offensive strategy until an all clear is declared (after secondary searches) by Divisions A and C, or if conditions deteriorate to the point of switching to a defensive strategy, the primary clear may be as good as it gets before going defensive. You and your division supervisors will continually monitor conditions and communicate about this.

Remember, one of the benefits of utilizing divisions and groups is that your supervisors will do most of the communicating with their crews face-to-face. In addition, the fact that they stand in the tactical gap allows them to see, which reduces their tendency to talk on the radio. Both reduce radio traffic exponentially.

Smaller Commercial Buildings

One template from which to start is a 4:1 model (fig. 10–39).

The establishment of Division A to supervise crews inside entering the A side has the following benefits:

- Significantly reduced radio traffic and fewer crews are transmitting on air with inaudible radio transmissions.
- Reduced span of control for the IC.
- Greater tactical supervision and consolidation of resources.
- A safety officer is embedded into the incident.
- The Division A supervisor can have eyes-on, hands-on active accountability of crews in real time going in and out of the structure.
- The Division A supervisor can see and anticipate deteriorating conditions that neither the crews in the hot zone nor the IC can see.
- Tactical coordination can occur with operations outside of the structure, such as ventilation, RIC, and medical.
- The IC can get into the "then/what if?" mindset and stay ahead of the incident power curve.
  On a commercial building, even if Division A is capable of supervising all companies, they must be able to perform a 360° lap, or the IC should assign a company to monitor the rear (C side).

Division A

- Boundaries: The inside of the commercial building.
- Resources: All crews working inside the building.
- Objectives:
  - ✓ Initially, fire attack and search.
  - ✓ After knockdown, transition into salvage and overhaul.
  - ✓ Positive pressure ventilation or horizontal ventilation would also be supervised under Division A to ensure maximum coordination under a single supervisor.

Roof Division

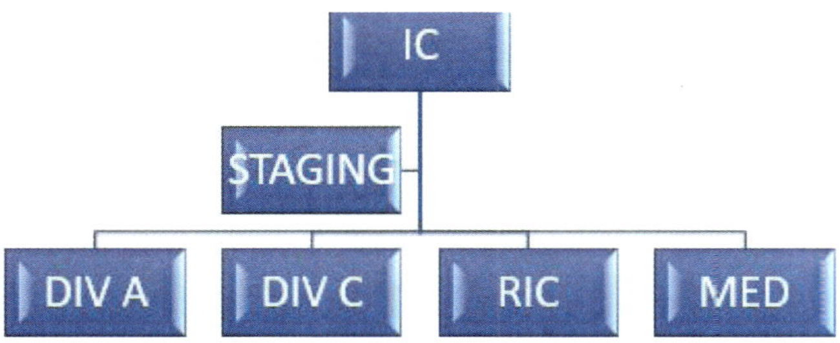

Fig. 10–39. A 4:1 is a good foundation from which to build.

- Boundaries: The roof. If crews are on the roof, regardless of objective, a roof division alerts all personnel that rooftop operations are underway.
- Resources: Crews working on the roof.
- Objectives: *Give a roof report ASAP!* If vertical ventilation is the tactic of choice with a crew(s) on the roof, a roof division supervisor is critical in maintaining situational awareness and tactical coordination with Division A, laterally. This can be through radio transmissions or hand signals. This eliminates the need to go through the IC and delay ventilation/fire attack coordination while unnecessarily increasing radio traffic.

Ideally, no less than three personnel should perform vertical ventilation on a commercial roof: the officer to supervise and coordinate, one firefighter to sound, create a louver, and punch, and the other firefighter to cut. The combination of potentially being above active fire, along with falling off the roof creates a significant need for tactical supervision of just one crew, let alone multiple crews.

Even if a three-person truck company goes to the roof without the expectation of a second truck, the company officer should be identified as roof division supervisor while the crew of two remains in the task level.

Unlike the inside of a burning house, where visibility is typically poor, a roof division supervisor is on the hot zone and able to see. This tactical advantage allows the first truck officer to the roof to assume roof division supervisor. Subsequent companies would report to that officer.

If the roof division does not have the objective of vertical ventilation, other duties could include checking extension, or even confinement from an exposure.

Again, if positive pressure or natural ventilation is the tactic of choice, it should be coordinated through Division A.

RIC Group

- Boundaries: The entire incident.
- Resources: All companies assigned to the function of rapid intervention.
- Objectives:
  - ✓ Coordinate with division supervisor(s) for accountability of crews.
  - ✓ Lap and soften the structure without worsening fire conditions.
  - ✓ Secure utilities.
  - ✓ Develop a tactical plan for potential firefighter rescue based upon current and future conditions.
  - ✓ Establish a RIC cache of equipment readied in case of intervention.
  - ✓ Supervise rescue of downed firefighters if IWI occurs.

Medical

- Boundaries: The entire incident.
- Resources: All ambulances and other resources assigned to medical treatment and/transport.
- Objectives: Treatment and transport of all civilians and firefighters. This includes establishing a medical area on the A side of the home with gurney and medical gear. Medical can often be a simple designation of one ambulance in the early stages of a house fire, or it can be comprised of a medical group with a supervisor and multiple ambulances and other resources based upon the needs of the incident as it escalates.

Larger Commercial Buildings—Single Story—Division A/B/C/D

An alternative template would be for a larger commercial building that requires more than one division. Extremely large buildings or those with difficult access may warrant more than one division at ground level (A/B/C/D based on side of entry), along with the potential for additional divisions/groups (roof, RIC, medical, etc.) (fig. 10–40).

- Boundaries: The inside of the building as entered from the respective division. For example, crews entering to operate from the side of the building would work for Division B.
- Resources: All crews entering from that side/division.
- Objectives: Crucial to coordinate with other divisions to prevent conflicting and redundant operations/effort. Initially, fire attack and search. After knockdown, transition into salvage and overhaul. PPV or horizontal ventilation would also be supervised laterally under division supervisors to ensure maximum coordination under a single supervisor.

Strip Malls

As with all ICS templates, you will base your organization upon the ICS formula of B+C+R = ICS. A strip mall can initially begin with Division A as described previously, with companies entering more than one business from the A side to attack or confine the fire. For example, you have fire in a center unit with extension in the attic above exposures B-1 and D-1. A Division A supervisor should be capable of managing companies in both sides from the A side. As described previously, a key point is for roof division to coordinate with Division A for heat and strip cuts to stop lateral spread (fig. 10–41).

If the fire has more complex conditions, or the building is too large, you can establish more divisions as needed to make a stand (A/B/C/D). Divisions B and D, for example, would have the control objectives of protecting all exposures on each of their respective sides.

Again, even with only a Division A template, consider a Division C to get companies to the rear of the strip mall to soften the structure and monitor conditions.

Fig. 10–40. Expand divisions based upon sides of entry requiring supervision.
Source. Illustration courtesy of Sam Wallace

Fig. 10–41. A strip mall can be organized with a similar template until it grows.
Source. Illustration courtesy of Sam Wallace

Confirmed Rescues in Commercial Buildings

Another template would be for known rescues in commercial buildings (confirmed victim profile). Here, utilizing geographic divisions for the objectives of fire attack and search from portals of entry around the building (as described above) *and* a rescue group supervisor to supervise crews performing *just* the rescue may be warranted.

Remember, a lot of resources are required to mount an offensive strategy in a large commercial building. While all divisions are likely performing interior attack and primary search, an additional rescue group could focus on getting a SIGNAL and execute the rescue in support of the divisions along with the potential for additional divisions/groups (roof, RIC, medical, etc.).

The coordination of getting victims out and to the medical group is not easy and requires that the rescue group and medical group supervisors remain attached at the hip to minimize the time-to-task (TTT) to get the victim(s) to ALS intervention. Remember, according to FRS, the victim must be out and to definitive care within 6–8 minutes to survive.

Multiple Floors (Low- to Mid-Rise) in Commercial Buildings—Division 1/2/3/4, and So On

Once the commercial building becomes larger (multiple floors, enclosed stairwells) and the fire is larger, a combination of divisions and groups may be appropriate. The benefit of the enclosed stairwell is that the area provides the tactical division supervisor a tactical space/threshold against the hot zone from which to account for crews coming and going. Like the front door on an A side of a building, the enclosed stairwell door on an upper floor of a multi-story commercial or mixed-use building can provide an excellent workspace for the tactical supervisor.

As with other templates, if a known victim(s) are trapped, you may add a rescue group to support the divisions.

Depending upon conditions, you may want a Division C in the rear access of the building, or the rescue group may be pulling victims from windows. In multi-floor buildings, these should still be organized by floor number if there is an enclosed stairwell.

Summary

Commercial fires are much more complicated events to command and coordinate than single-family dwellings and even MFDs. Both allow compartmentalization, whereas commercial fires are wrought with additional hazards, including but not limited to hazardous materials, void spaces, shafts, rack storage, and often mysterious contents and processes. The Boyd Street Fire is an example of firefighters getting ambushed by unknown contents and rapid fire deflagration. The Ghost Ship Fire is an example of the lethality of commercial buildings being used against code unbeknownst to firefighters. We have introduced 18 commercial watch out situations that should be considered.

VP and search is more likely to take the form of evacuations. Tactics such as fire attack, ventilation, and water supply are vastly different, requiring more effort to remain ahead of the incident power curve, including additional alarms, water supply considerations, RIC, and overhead.

Each type of classification and building type includes assemblies, strip malls, taxpayers (commercial over residential/mixed use) tilt-up, URM, fast-food, big-box, and more. Each has unique fire behavior, construction, and tactical considerations.

ICS and communications must be planned for well ahead of time, including bridging the tactical gap from the IC down to the crews when a lot of tasks need to be done simultaneously. The good news is that the same systems you have been learning throughout this book will serve you well on a commercial fire: size-up, VP, priorities, strategy, tactics, decentralized command through ICS, SAW-CSS-RECEO-VSS, and remaining ahead of the incident power curve.

Chapter Review

Review Questions

1. Define a commercial structure and provide some examples.
2. What are some of the LAFD structure fire watch out situations?
3. Explain some of the differences in tactical considerations between residential and commercial occupancies.
4. What is a roof report?
5. Describe cold smoke.
6. Provide tactical considerations for ventilation methods with regard to commercial occupancies.
7. What are some of the strategic and tactical considerations for strip mall fires?
8. What is a taxpayer?
9. Provide an example of how you would utilize four or more division/group supervisors given a center-unit fire in a five-unit strip mall fire.

FESHE Strategy and Tactics (C0279) Related Content

The content contained in chapter 10 provides detailed information to assist with instruction and education for all course objectives with a focus on commercial structures. The chapter format supports the C0279 course outline by speaking specifically to basic division of tactics, rescue, exposures, confinement, ventilation, and salvage for commercial fires to include assemblies, strip malls, taxpayers (commercial over residential/mixed-use) tilt-up, URM, fast-food, big-box, and more.

Access the Tactical Review Guide Appendix C#2: Two-Story Taxpayer, Fire from A-Side First-Floor Restaurant and C# 3: Strip Mall–Large Regional, Fire Running Mansard for objectives-based scenario exercises.

NFPA 1021 Job Performance Requirements

The information in this chapter can be utilized to support training and educational programs associated with the Emergency Services Delivery Fire Officer I JPR 4.6, 4.6.1, 4.6.2, Fire Officer II JPR 5.6, 5.6.1, and Fire Officer III 6.6 and 6.6.1. The content in the chapter regarding the watch out situations and incident reviews also aligns with post-incident analysis-related job performance requirements of Fire Officer I 4.6.3, Fire Officer II 5.6.2, and Fire Officer III 6.6.2.

Access the Tactical Review Guide Appendix C#2: Two-Story Taxpayer, Fire from A-Side First-Floor Restaurant and C# 3: Strip Mall–Large Regional, Fire Running Mansard for 1021 Job Performance Requirement-based scenario exercises for Fire Officer I, II, and III.

Endnotes

1. *NFPA 5000: The Building Construction and Safety Code*
2. Term for buildings constructed before the 1933 Long Beach Earthquake that resulted in changes in building codes. Pre-33 buildings are typically unreinforced masonry.
3. Valerie Ziavras, "Occupancy Classifications in Codes," *National Fire Protection Association (NFPA)*, May 7, 2021, https://www.nfpa.org/news-blogs-and-articles/blogs/2021/05/07/occupancy-classifications-and-model-codes.

11

High-Rise Fires

Fig. 11–0. High rise fires can happen anytime and are lethal, complex incidents.
Source. Photo courtesy of Steve Gentry.

High-Rise Fire Definition and Hazards

A *high-rise* is a building over 75 feet in height. A high-rise fire may be the most challenging and potentially lethal structure fire you will experience. The immense occupancy loads, tremendous resource requirements, and labor-intensive operations all make these incidents staggering. A working high-rise fire can be like battling multiple commercial fires at once that are stacked on top of each other. Civilians located many floors above the fire can be killed by smoke and fire gases.

Everything moves slower on a high-rise fire, except the smoke and flames. If crews must ascend a stairwell from the lobby to the fire floor, it may take 60–90 seconds per floor for companies to get to the fire floor from arrival on scene and have water flowing. Meanwhile, the fire is growing, and smoke is quickly rising through countless vertical avenues.

The good news is that the building blocks of incident command you have learned thus far will provide an excellent foundation from which to build your knowledge, skills, and abilities, so you can ensure that you save savable civilian lives and your crews go home safe to their families.

Pre-planning and training, including interface with the personnel at your high-rise buildings, is paramount. Knowledge of occupancy trends, fire suppression systems, elevators, HVAC systems, internal communications, stairwells, fire control rooms, FDCs, standpipes, and countless other factors will come into play on a significant high-rise fire. Never has the adage, "Every battle is won or lost before it is ever fought" been truer than in a high-rise fire. Knowing your enemy, battlefield, and tactical points of opportunity, combined with training, and knowing the time it takes to get water flowing on upper floors, will drive your incident command decision-making.

Consider the following from the NFPA:[1]

From 2009–2013, U.S. fire departments responded to an average of 14,500 structure fires in high-rise buildings per year.

These fires caused an annual average of:

- *40 civilian fire deaths*
- *520 civilian fire injuries*
- *$154 million in direct property damage*

Five property use groups account for almost three-quarters (73%) of high-rise fires:

- *Apartments or other multi-family housing (62% of all high-rise fires)*
- *Hotels (4% of high-rise fires)*
- *Dormitories (4% of high-rise fires)*
- *Offices (2% of high-rise fires)*
- *Facilities that care for the sick (2% of high-rise fires)*

Most of the remaining fires occurred in mixed-use residential or office buildings (such as restaurants, stores, and parking garages) or probable miscodes of properties that cannot be high-rise (such as one- or two-family homes and sheds).

The fire death rate per 1,000 fires and the average loss per fire are generally lower in high-rise buildings than in other buildings of the same property use. High-rise buildings have lower percentages of fires with flame damage beyond room of origin, providing further evidence of impact from fire protection systems and features:

- *Apartments (4% of high-rise fires vs. 10% in shorter buildings)*
- *Hotels (4% of high-rise fires vs. 11% in shorter buildings)*
- *Dormitories (2% of high-rise vs. 1% in shorter buildings)*
- *Offices (10% of high-rise fires vs. 21% in shorter buildings)*
- *Facilities that care for the sick (4% of high-rise fires vs. 9% in shorter buildings)*

High-rise buildings are more likely than shorter building to have fire-resistive construction and wet pipe sprinklers. These types of protection help prevent fire spread.[2]

CASE STUDIES IN COMMAND

Granite Tower

CHIEF DAVE MCGRAIL, DENVER (CO) FIRE DEPARTMENT

I responded with companies to investigate an automatic fire alarm from a downtown Denver high-rise building. There were no second-source 9-1-1 calls. I had been to this building (Granite Tower on 18th Street) many times before, mostly on elevator rescues, automatic alarms, and other emergencies. In fact, I've been to thousands of automatic alarms during my 42-year career, and will go to many more before I retire from active duty. Like most firefighters and fire officers across our great fire service, most of my automatic alarm responses have turned out to be something less than a significant emergency or working fire. And therein lies the crux of the problem—these high-frequency, low-acuity calls can cause a false sense of security among us as human beings. We can slowly drift into complacency, and that could be catastrophic; not only for us, but for those we've sworn to protect.

In terms of high-rise command and coordination, here's my two cents. We are professional firefighters, and we must always act like it. Every time you respond to a high-rise building (or anywhere for that matter), take it seriously. High-rise buildings are large, complicated, and dangerous buildings. If you have drifted into complacency, you must get it together or find another job. Anyone around you, and especially under your command, who has drifted into complacency must be rescued today! It could be the most important grab you ever make.

Item number one for the high-rise incident commander is call for help! You're going to need a significantly greater number of resources (specifically boots-on-the-ground firefighters) at a high-rise fire than for most other ground-based operations. The very best and most experienced big city fire departments across our fire service have automatic, built-in upgrades in response upon confirmation of a working fire in a high-rise building. For any large fire department, it should be at least an automatic second alarm. For small and medium-sized fire departments, figure out what your resource options are, both internal and external, well ahead of time. Furthermore, determine who they are and where they're coming from, such as automatic and mutual aid resources. Ideally, train with those resources, design and operate under the same procedures, protocols, guidelines, and so on, and ensure you're using the same equipment packages. A good baseline number to shoot for is 100 firefighters as soon as possible.

When I arrived at Granite Tower just past midnight on that summer night, Tower 4 and Engine 6 were already on scene. There was nothing showing from the two sides I could immediately see of this large 33-story commercial high-rise building. Keep in mind, nothing showing really means nothing, especially at a large high-rise building. The first communication I received from Tower 4 was twofold: first, a very frightened building security guard met them at the lobby and reported a *"big fire"* (his words) in the elevator machine room at the 33rd-floor level. Further, Tower 4 reported that the elevators were malfunctioning and that Tower 4 and Engine 6 were proceeding to climb up and go to work. That's 33 flights in a commercial building, with an extra 75 to 100 lbs. of equipment, and doing it one mile above sea level. That requires top physical condition, including cardiovascular/muscular endurance. Denver firefighters are some of the most physically fit firefighters in the world. For me as the incident commander, this opens up a whole lot of operational concerns for the welfare of the members under my command. Managing the risk just became much more difficult.

Now, when I arrive at an automatic alarm with companies, I will usually go into the building and establish my CP at the fire command center (FCC). From this position, I can wear multiple hats and allow the fire companies, and specifically the company officer who would have had to stay at the FCC, to join, lead their crew, and focus on tactical and task-level operations. For those automatic alarms that escalate and become a significant emergency, including when an actual working fire is found upon investigation and recon, I will move back outside and establish my ICP, ideally with a two-sided view of the fire building. I will subsequently assign an incoming resource, ideally a chief officer, to the FCC as a systems control officer. Again, remember, the key to success here is that you must call for help early, from the outset, and call for lots of it.

The primary focus must be to stop the fire and complete a search of the primary fire area. Yes, there is still plenty of strategic, tactical, and task-level work to complete, but stopping the fire and completing a primary search must be the focus. I emphasize this because I have found that too many fire departments are focused primarily on building a fancy command structure (checking boxes on tactical worksheet) than they are on stopping the fire and getting a search.

All of these can and should occur simultaneously, but the vast majority of resources you have from the outset must be doing tactical- and task-level work.

At the Granite Tower fire, I initially assigned the company officer from Tower 4 to be the fire floor division supervisor. Under their supervision was Tower 4, Engine 6, and Engine 3. Subsequently, I assigned my second-arriving district chief to assume the fire floor division supervisor position, with the Tower 4 company officer assigned back to his crew to work at that tactical and task level.

We utilize divisions, especially in a high-rise environment, to divide up the geography and maintain a reasonable span of control. However, most of us don't have enough command resources to establish a division supervisor on every floor of a typical high-rise building. In addition, when things get really big and complicated, we tend to move toward section chiefs, and specifically establishing an operations section chief. Don't forget about branches. My recommendation is before we go for the finance, logistics, operations, and planning (FLOP), consider the much-underutilized branch component: specifically, for an escalating situation, a fire suppression

branch, responsible for the fire floor, floor above, and floor below. Just three floors, but there's a lot going on in this area. The upper branch is responsible for all the floors above the floor above the fire floor (recon is critical here, with search and vent groups typically established and utilized). The interior support branch is responsible for everything below the floor below the fire floor (including staging, rehab, stairway support, medical support, etc.). And last but not least, the exterior support branch is specifically responsible for everything necessary to support the firefight from outside the building (including a water supply group, base, etc.).

At the Granite Tower fire, early reports from my first companies to reach the fire floor was that there was a working fire in the elevator machine room. What had started as an electrical fire was now very large and involved not only the elevator machinery but had communicated to the combustible (Class A) contents of this large machine room. The engine companies had stretched an attack line, but we needed to control and shut off the electrical power before a fire attack using water could be made. Further, truck companies needed to initiate and complete a search for life. At the ICP, I'm thinking what if an operating member or members sustain an electrical shock? What if the shock stops their heart? How do I get them down? My third-due truck company was assigned as a RIT from the outset. They made the climb up and established a position on the floor below. They were equipped with an automatic external defibrillator (AED).

During the operation, second-alarm resources, including my systems control group, were able to establish emergency power to one low-rise bank elevator, giving us access from the main floor lobby (street level) to the 16th floor sky lobby. At this position (16th floor), I established a medical group with paramedics with ALS equipment.

Needless to say, there were significant logistical challenges at this high-rise fire due to the lack of elevators. Multiple carbon dioxide (CO_2) fire extinguishers were requested by the fire floor division. These are very heavy, cumbersome, torpedo-like pieces of equipment. Getting several from the street to the fire floor without elevators was a demanding task. I assigned four engine companies to assemble a stairway support team. One member, every other floor gave us a connection from the lobby to the 32nd floor. This made the call for and transport of additional equipment much easier to accomplish.

We were able to get the local utility company to quickly respond and control/shutoff electrical power to the fire floor. Companies were able to extinguish the fire and complete searches, which were negative for life.

My message to you is this: take every automatic alarm seriously! Most will turn out to be benign, unremarkable events. However, some will be the real deal, challenging, high-risk operations. High-rise fires are very labor-intensive and logistically demanding events. Always be prepared mentally and physically. Design, implement, and frequently train on the ICS/NIMS components necessary to command and coordinate the high-risk operations of a high-rise firefight efficiently and effectively.

SAW-CSS-RECEO-VSS for High-Rise Fires

The acronym SAW-CSS-RECEO-VSS is the tactical template for all types of incidents. This was first introduced in this text in chapter 7 for structure fires; however, it will be modified to apply to other incidents throughout the rest of this book, beginning with high-rise fires.

High-rise fires are far more complex than any other structure fire. You can consider each floor involved and above the equivalent of a commercial structure fire and more. Most fire departments with a significant high-rise building presence will have a detailed high-rise policy and/or SOG. The FD document dictates agency-specific actions based upon their respective staffing and resource levels, history, fire and building codes, and interpretation of high-rise tactics (fig. 11–1).

That said, there are universally accepted practices regarding high-rise fires that should be considered as a minimum and provide an excellent foundation for incident commanders to plan for, train upon, and execute high-rise fire operations. Each component has key points that should be considered. These are benchmarks that the IC and tactical supervisors should share and communicate. Again, these are minimums. Your agency's SOGs will likely be much more comprehensive.

The foundation of high-rise operational and command considerations lies in a classic acronym of ALS-BASE: attack, lobby, staging, BASE (where apparatus are parked). These components comprise the initial considerations during the early stages of a high-rise fire. Simply stated, the first alarm should typically size up the fire floor location, initiate attack with water supply, control the lobby and elevators, set up staging two floors below the fire floor, and identify the location of a base for subsequent alarms.

The following SAW-CSS-RECEO-VSS components go into more detail regarding the totality of high-rise tactical and command considerations. Key points are in *italics* and correspond with the tactical worksheet presented in this chapter. Again, refer to your own agency policy and SOGs.

Size-Up

Unlike a more common structure fire in a residential or commercial building, high-rises present immense size-up and recon challenges. In addition to any sign of smoke or flames from the street, reconnaissance will be required in the lobby, control room, elevators, shafts, and on

Fig. 11–1. High-Rise fires present with both interior and exterior access challenges.
Source. Photo courtesy of West Metro Fire (CO)

multiple floors. *Remember, size-up is a mental, ongoing process. You must think exponentially regarding the size-up of a high-rise fire.*

- Based upon training and *pre-planning*, you may have information on the building that you can convey to the rest of the alarm while en route or upon arrival before you enter. High-rise buildings have different construction features, systems, retrofitting, and based on the era they were built, may or may not have adequate fire protection systems functioning.
- En route, try to determine if this is an office building or a residential apartment building and consider occupancy loads based upon the time of day you are responding.
- Start with your *FPODP or FIRST* algorithm from the street. If you have visible smoke or flames from the street, identify the estimated floor(s) involved and call additional alarms per your SOG.
- Locate an on-site *RP or building engineer* to get information and reconnaissance from the building systems and personnel.
- Control the *lobby* and get information from the *annunciator panel* and from any occupants or on-site personnel, engineers, security, and so on.
- Recall and control the *elevators* per your SOG.
- Identify the location of smoke and fire and the *occupancy level with potential VP*. Is this an office building or residential apartment building, or mixed use? Do you have confirmed victims trapped or multiple reports of fire, or is this an alarm only with no 9-1-1 calls from occupants?
- If you have a working fire and decide to set up fire attack, *identify the fire attack stairwell versus the rescue/support stairwells* and communicate ASAP. This will alert subsequent companies and get them in proper positions to support your operation.
- Your initial size-up will conclude with an arrival report to incoming units; however, remember that size-up is continuous. As you talk to the RP, bystanders, occupants, employees, and check system status, you will garner more critical information. The IOCAN (see chapter 5) system is designed to give a quick arrival report. Your arrival report is likely dictated by your SOGs.
- Remember, size-up is continuous and high-rise fires will present uniquely challenging factors that cause you to be relentlessly vigilant as you ascend the building. Be sure to maintain clear communications as the incident escalates and conditions are confirmed.
- As you ascend, check lower floors to get oriented to the possible layout of the fire floor(s).

Apparatus Placement

The first engine will likely stage at the *lobby entrance*. The second engine will likely stage at the FDC for the standpipe of the attack stairwell while the crew assists in the lobby. If still in recon mode with no signs of fire, consider staging the balance of the alarm a block away with the exception of the first two engines and truck. Again, your SOG will dictate.

- Apparatus, hoselines, equipment, and personnel are all susceptible to falling glass and *debris* during a high-rise fire. Consider identifying safe access passages that are well-marked for members to move to and from the building.
- *Staging* for a working high-rise fire is typically two floors below the fire floor where subsequent personnel and equipment are located to support operations above. Based upon conditions and SOGs, a minimum of one full alarm should be considered minimum in staging at all times in a high-rise fire. A staging area manager should be identified ASAP.
- Unlike a ground-level structure fire, *equipment*, including SCBA cylinders, hose packs, forcible entry tools, long-handled tools, search ropes, portable radio batteries, EMS

equipment, spanner wrenches, lights, water cans, and a host of other equipment should be taken to staging when initially entering the building.

- *Base* is the apparatus parking area for high-rise fires and should be located at least 200 feet away from the building. A dedicated base manager should be identified (fig. 11–2).
- *Traffic flow* into base should be identified early, as multiple alarms will likely be called to a working high-rise fire. Companies should avoid crossing into any falling debris area, the front of the building, or where hoselines are in place to support operations.
- Consider law enforcement early to assist with traffic coming into base.

Water Supply

Once the *fire attack companies and stairwell* are identified and confirmed, a water supply should be established as soon as possible to get water into the standpipe. Consult pre-plans or on-site building engineers for the presence and effectiveness of a fire pump to augment supply.

- Who is securing the water supply? This will be dictated by your SOG. Is this being accomplished with one or more companies? Hydrants are usually right next to the FDC; however, this is no guarantee that the hydrant is functional. Will this require a relay pumping operation and the establishment of a water supply group supervisor?
- Will there be a separate company pumping a sprinkler connection or is the standpipe a combination *FDC* for both sprinklers and firefighting standpipes (fig. 11–3)?
- Consider utilizing the fire control room or a building engineer to assist with management of *water supply systems and pumps* when applicable. Depending upon the size of the

Fig.11–2. Ambulances in base at a high-rise fire
Source. Photo courtesy of Steve Gentry

Fig.11–3. Some FDCs are combination.
Source. Photo courtesy of Rick McClure

fire, a lobby control group may be adequate for lobby and systems (fire, HVAC, elevator, communications) management. If not, you may need both lobby and systems control groups.
- Once a patent water supply is established, it should be announced on all of the channels.

Command

Command must be established and announced ASAP. *There can only be one IC at a time, and the IC must be on scene to assume command and must name the incident.* While your SOGs will determine specific parameters, some fundamental best practices exist. There will be a lot on the shoulders of the first-arriving company officer who must assume command on a high-rise fire (even if it will be passed quickly to the next officer).

- Unlike a more common structure fire, a first-arriving company officer must contend with access to the lobby, annunciator panel, and interaction with multiple occupants and employees.
- Depending upon your SOGs, you may also be required to identify the fire attack stairwell and ascend to initiate fire attack on the fire floor with your crew(s).
- Command should be passed as soon as possible to the next officer, or first chief officer.
- Once a chief officer or permanent IC has arrived, the *ICP location* should be announced. The ICP is typically not best suited in the lobby of the involved building.
- The ICP should be away from the building, if possible, and away from the incident base (fig. 11–4). The IC should be able to see the conditions of the fire and flow of personnel into and out of the building, ideally. Consider the lobby of a nearby building with a good view.
- Call *additional alarms* early and confirm the base location.

Fig.11–4. The incident commander should be able to see the conditions of fire and flow of personnel in and out of the building.
Source. Photo courtesy of Steve Gentry

- Set up your *ICS* early; however, do not forsake resources needed to perform the immense number of tasks required to get water on the fire, and evacuations/searches underway.
- Develop a *communications plan* to prevent overtaxing a single tactical and/or command channel. Add additional command and tactical frequencies ahead of the incident. If possible, utilize the building public address (PA) system, fire phones, and obtain the lobby phone number.

Strategy

In a high-rise fire, size-up and initial condition reports on the fire floor will be delayed and likely change as companies ascend. You may find everything from fire showing from the street to a deep-seated working fire that grows exponentially as companies ascend to the fire floor, a room-and-contents fire that was contained by the sprinkler system, a burned-out light ballast, or something you never expected. Plan for the worst.

- *VP* can change dramatically as well. As with any incident, you must determine the incident priority/strategy: life saving, incident stabilization, property/environment priority; and *offensive, combination, or defensive strategy.*
- If you have a working fire with a well-developed high-rise SOG that dictates what each company should do, then follow it accordingly.
- Keep it simple. Keep it calm. High-rise fires will likely begin as life priorities and offensive strategies that will likely be sustained longer than a more common structure fire. That said, using your routine terminology and confirming incident priority, strategy, tactical objectives, and location of the ICP will exude confidence in your crews that you are handling business. Do not let your voice pitch, tone, or volume change from what it would be on a house fire. This calm command presence will extend to all your companies and members.

Safety

Depending upon occupancy level and resources available, consider a rapid intervention plan early. The lag time to get companies to the fire floor and flowing water is significant. Plan ahead and get companies assigned to rapid intervention early to be in place as they get into the IDLH zone. Depending upon conditions, however, this may not be until later alarms. Just plan ahead.

- *RIC:* Unlike a more common structure fire, rapid intervention preparation will require at least two, if not more, companies. Consider an RIC group supervisor early located in staging on the fire floor with crews. When resources allow, bolster your RIC group.
- *The lobby* determines who comes and goes to and from the building and ascertains civilian occupancy levels. This may easily require two companies or more. As mentioned above, the lobby will also potentially be responsible for control of building systems via the *control room.*
- This includes *controlling elevators* to prevent crews from ascending too far in one bank. Crews should only use elevators if there are split banks to ascend to the fire floor.
- Separate from a medical group, the *medical unit* is focused solely on fire and emergency personnel injuries, treatment, and transport. Plan early for hydration and rotation of crews.
- 2-out and/or a backup line is critical on a high-rise fire. The fire flow required, and potentially advanced stages of a working high-rise fire, will necessitate multiple lines and potentially from multiple stairwells.
- If the fire is being attacked from two stairwells, the division supervisor of the fire floor should have a second officer on the opposing stairwell to assist with crew accountability. The IC would only communicate with the designated division supervisor. The other officer would be the division supervisor's eyes and ears on the other stairwell and communicate with the floor division supervisor.

- A separate *ISO* should be established and, depending upon the size and scope of the incident, may require assistants. The ISO can provide a level of redundancy and a second set of eyes to the IAP, and factors such as accountability, structural integrity, changing conditions, and so on.
- Additional considerations for high-rise fire safety include, but are not limited to, the following:
    - External weather, wind, and flow path severely impacting fire conditions on fire attack crews
    - Stack effect in stairwells creating lethal conditions above the fire floor
    - Reverse stack effect creating lethal conditions below the fire floor
    - Inability to control elevators
    - Fire rapidly ascending through chases, elevator shafts, and lapping of compromised windows
    - Excessive fire load from contents inside the structure
    - Falling glass and debris in the street
    - Crews becoming disoriented and running out of air
    - Apparatus in the hot zone (see apparatus placement mentioned previously)
    - Fatigued and dehydrated firefighters
    - Extremely inexperienced firefighters or officers
    - Highly emotional scene due to firefighter Mayday or excess victims
    - Building's structural components compromised if exposed to prolonged flame impingement

Rescue

Remember, the victim profile drives *everything*. This is still true of high-rise fires. You must think globally regarding VP. High-rise buildings can have thousands of people inside, depending upon the building type, size, use, and time of day.

- Life is the first priority and rescue the first tactical consideration. That said, *you will have to get attack lines in place to save the most people, and that can take a lot of time.*
- For high-rise fires, search-and-rescue operations take on an entirely different scope.
- The potential for civilian injuries and fatalities is extraordinary. High-rise fires like the Twin Parks Fire in The Bronx, New York, 2022, are tragic examples.
- *Primary and secondary searches* will be much different and require expedient tactics. Searches on fire floors will likely be supervised by the division supervisor of the respective floor.
- If the building is residential, the potential for life loss could be immense. Lobby may be able to assist in ascertaining actual occupancy levels and locations for greater efficiency.
- A *Rapid Ascent Team (RAT)* is utilized to go to the upper floors above a fire and begin search and rescue as soon as possible due to the stack effect in stairwells. This may occur in the fire attack stairwell or a *separate evacuation stairwell*, or both, depending upon conditions. Consult your SOGs.
- Any RAT companies would report to the evacuation group supervisor, once established.
- *Evacuation* is typically for civilians away from and/or above/below the fire. They should be ambulatory and able to assist in their evacuation. *That said, the stack effect can be lethal when deadly smoke and gases ascend to upper floors, as was the case in the Twin Parks Fire.*
- If you have minimal smoke and fire conditions and thousands of occupants, you may consider shelter-in-place based upon your SOGs.
- *Medical treatment and transport* functions should be established early under the direction of a medical group supervisor (MGS). Ensure you have advanced life support and

transport capability on scene at all times, including ALS treatment in staging. Think MCI until proven otherwise (fig. 11.5).

Exposures

Interior stairwells, shafts, chases, HVAC systems, and floors above all have potential for smoke and fire extension (fig. 11–6).

- The potential for smoke and gases to kill occupants above the fire due to the stack effect and flow path is exponentially greater in high-rise fires.
- Take nothing for granted and get above the fire ASAP. Extreme wind conditions, air operations, and *lapping of flames* from broken windows should all be factors that must be controlled to prevent extension as well.

Fig.11–5. Any high-rise fire, no matter how small, can become an MCI due to stack effect on the floors and stairwells above. Plan ahead!
Source. Photo courtesy of Steve Gentry

Fig.11–6. Smoke is the biggest killer in high rise fires.
Source. Photo courtesy of Miami (FL) Fire-Rescue

Confinement

As presented previously, confining a fire in a high-rise is far more challenging.

- The objective for confinement should be established and communicated early. Will the fire be contained to *one unit, one floor, or multiple floors*?
- Typically, we don't think of confining smoke or gases. Smoke and fire gases are far more difficult to confine in high-rise fires and are more lethal than the flames (Twin Parks).
- Utilize HVAC systems and pressurized stairwells to confine gases and smoke when possible.
- Fire attack lines on the floor above may be required to confine the fire in addition to attack lines on the fire floor.

Extinguishment

Unlike supply, extinguishment is about the *size and number of attack lines.*

- What is the *fire flow required* for the floor(s) involved?
- The rule of thumb for high-rise fires is 2½" attack lines, although some fire departments may use 2" attack lines.
- *Backup teams and crew rotations* make for a large demand on resources. Plan on attack teams, backup teams, and crews ready to rotate into the fight.
- The ability to sustain the fight to the point of extinguishment is vital, yet also dependent upon your staffing, training, response levels, water supply challenges, and systems (or lack thereof) in the building.

Overhaul

Do *not* rely on the TIC alone. Otherwise, you are asking for a rekindle.

- *Who* is responsible for overhaul and *where* is the priority? Ultimately, the division supervisor(s) of the fire floor(s) will be responsible to confirm total and complete extinguishment.
- Be aware of hidden fire in shafts, chases, and any other vertical avenues of fire spread in *floors above*, all the way to the top of the building.

Ventilation

Proper pressurization of the stairwells is absolutely critical in high-rise operations.

- *Stack effect* is the movement of air in a high-rise building due to external and internal temperature fluctuations whereby lethal smoke and gases can settle far above the fire floor and other locations, putting occupants in peril.
- Remember, wind and weather can have an adverse effect on fire and smoke conditions on the *fire floor and throughout the building.* This can put both civilians and firefighters in peril.
- A combination of ventilation/pressurization tactics will be required to effectively control and remove smoke and gases from the building, and they must be *coordinated* through a ventilation group supervisor.
- Pressurization of stairwells and fire floors must be *coordinated* with the IC, ventilation group supervisor, and Division supervisors on the affected *fire floors.*
- *HVAC* systems must also be controlled and coordinated through the lobby or systems group supervisor.

Salvage

Salvage is a tertiary priority compared to fire attack, search, evacuation, ventilation, RIC, and medical. Floors below the fire floor are obviously most susceptible to water damage. Division supervisors will likely coordinate salvage operations in the later stages of the fire. What is the salvage plan? *Who* is conducting the salvage, *where* is the priority, and contact any *RP* if the fire is confined to one or two units.

Support

High-rise fires will require an immense amount of support. As a minimum, consider the following:

- Building engineers, site specialists, managers
- Lobby/systems control groups and control room access
- Dedicated dispatchers and additional command and tactical channels and phones
- Law enforcement (le)
- Incident safety officer (ISO)
- Public information officer (PIO)
- Liaison officer (LO)
- Logistics section (medical, communications, food/rehab, ground support, supply)
- Planning section (resource status, situation status, documentation, technical specialists, demobilization)
- Utility companies (gas, electricity, and water)
- Red Cross
- Rehab and medical unit for members
- Air/light units
- Air operations
- Hazmat team
- Investigator

Tactical Worksheet—High-Rise Fire

S	Size-Up	☐ Pre-Planning ☐ FPODP ☐ RP/Engineer	☐ Lobby/Panel/Elevators ☐ VP	☐ ID Stairwells ☐ IOCAN
A	Apparatus	☐ Lobby ☐ Debris	☐ Staging ☐ Equipment	☐ Base ☐ Traffic Flow
W	Water Supply	☐ FA Co. ☐ Attack Stairwell	☐ FDC/Supply Co.	☐ Pumps/Systems
C	Command	☐ Name ☐ ICP	☐ Additional Alarms ☐ ICS	☐ Comms
S	Strategy	☐ Confirm VP ☐ O	☐ C	☐ D
S	Safety	☐ RIC ☐ Lobby	☐ Control Room ☐ Elevator Control	☐ Med Unit ☐ ISO
R	Rescue	☐ 1st ☐ 2nd	☐ RAT Stairwell ☐ Evacuation	☐ Medical T/T
E	Exposures	☐ Interior Stairwells/Floors Above		☐ Exterior Lapping
C	Confinement	☐ Unit	☐ Floor	☐ Multiple Floors
E	Extinguishment	☐ Number of lines	☐ Flow Required	☐ Back-Up/Crew Rotations
O	Overhaul	☐ Who	☐ Where	☐ Floors above
V	Ventilation	☐ Stairwell Pressurization	☐ Fire Floor/Above ☐ Coordination	☐ HVAC
S	Salvage	☐ Who	☐ Where	☐ RP
S	Support	☐ Building Engineer ☐ Control Room ☐ Comms ☐ Law ☐ Safety	☐ PIO ☐ Liaison ☐ Logs/Plans ☐ Util ☐ Red Cross	☐ Rehab/Med for FF ☐ Light/Air Units ☐ Air Resources ☐ Hazmat ☐ Invest

ICS and Communications for High-Rise Fires

ICS is always incident-driven, and templating can help you get ahead of the incident power curve (fig. 11–7). That said, high-rise fires are extremely dynamic events that have the capacity to quickly become overwhelming. There is a balance between organizing your incident with ICS to stay ahead and not getting overwhelmed, while not draining your limited resources from accomplishing the tasks required. As with any structure fire, the ICS Formula (B + C + R = ICS) applies. Consider the following:

Building

- Is this an office or residential building, or a combination of both?
- Is this a modern high-rise with sophisticated fire suppression systems, or an old building without functioning sprinklers?
- Are there multiple stairwells on the corners or a single stairwell in the center?

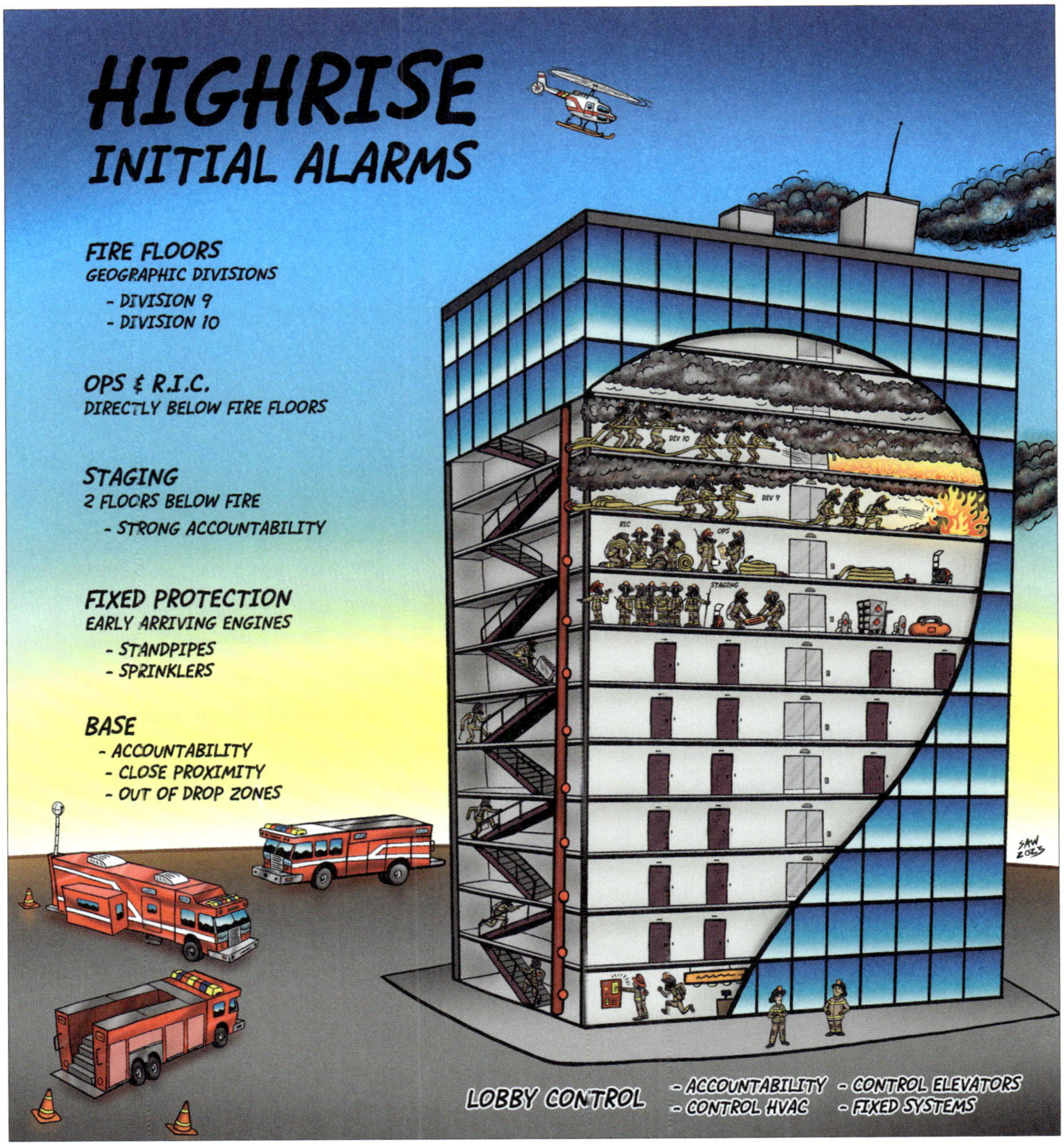

Fig.11–7. A foundational template for high-rise fires. Use the parts you need for your incident.
Source. Illustration courtesy of Sam Wallace

Conditions

- Is this an alarm with no signs/reports of fire?
- Is there a confirmed report of smoke and flames in the building?
- Is there smoke and fire visible from the street?
- What is the estimated occupancy load and are there confirmed victims trapped?

Resources

- Do you have the capacity to fill multiple alarms?
- Will you rely on automatic/mutual aid, and will response times be delayed?
- How much overhead do you have to support your ICS?
- Do you have air resources and how would they be utilized?

Consider the following templates as a starting point for given conditions. Remain pessimistic and anticipate the incident escalating rapidly. These templates are just considerations. The actual incident and your SOGs and staffing will drive your operations.

High-Rise Fire with Alarm Sounding Only

On these more common alarms, the opportunity to practice effective use of ICS and clear communications will provide the foundation for more significant incidents later.

- As crews ascend and reach the fire floor, that floor will become your *first division*. The supervisor will likely be one of the first company officers to the fire floor; however, they will need to be backfilled with a chief officer if a working fire is confirmed.
- Division supervisors of fire floors will have the tactical objectives of fire attack *and* primary search on their respective floors.
- *Lobby/systems group* will be required to gain control of the lobby, read the annunciator panel, control access and elevators, and work with on-site personnel to better ascertain conditions above.
- Staging may not be needed if the alarm is false, or if there is an extinguished food-on-the-stove situation, for example. That said, all resources should continue as if there is a working fire until proven otherwise. This includes setting up staging and getting companies coming up, per your SOG, until conditions are confirmed to be stabilized (fig. 11–8).

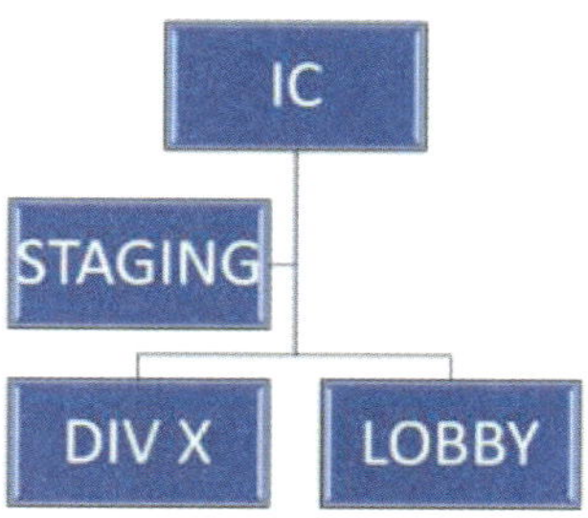

Fig.11–8. A basic initial ICS

High-Rise Fire with Confirmed Fire Confined to One Unit

Any confirmed fire in a high-rise has immense potential to be far more complex than is reported or seen upon arrival. Consider time-to-task completion and lag time to get resources on scene and to the fire floor. Call additional alarms upon confirmation of a high-rise fire (fig. 11–9).

- As crews ascend and reach the fire floor, that floor will become your *first division*. The supervisor will likely be one of the first company officers to the fire floor; however, they will need to be backfilled with a chief officer if a working fire is confirmed.
- Division supervisors of fire floors will have the tactical objectives of fire attack *and* primary search on their respective floors.
- *Lobby/systems group* will be required to gain control of the lobby, control access and elevators, and work with on-site personnel to better ascertain conditions above. In a confirmed fire, this will include utilizing any fire protection systems in the building including water pumps, PA, HVAC control, and so on.
- Staging will be established as companies muster toward the fire floor. *Identify a staging area manager.*
- Base will be established as the second alarm arrives. *Identify a base manager.*
- Consider an evacuation group supervisor for rapid ascent teams and additional division supervisor and companies above the fire floor to check extension. Depending upon your resource levels and SOGs, this may not occur until your second alarm, as the first alarm will be focused on getting water on the fire floor and controlling the lobby.
- Consider a medical group for potential civilian victims.
- Note the span of control is quickly maximized at the limit of 7:1. What could the IC do prior to this point to keep from getting overwhelmed?

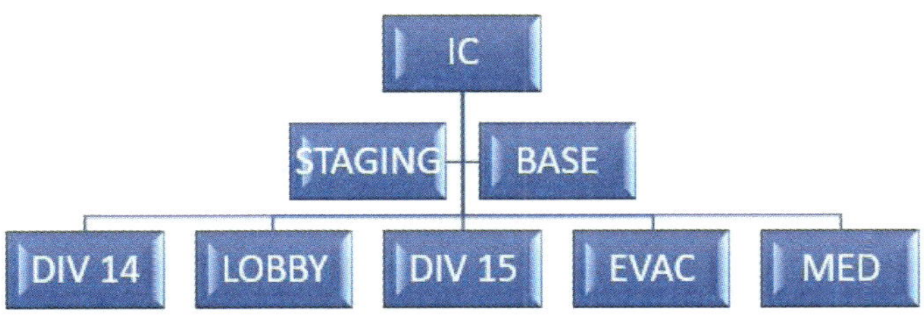

Fig.11–9. ICS for a smaller working fire

High-Rise Fire with Smoke/Fire Visible from the Street—Multiple Units/One Floor

Visible smoke and flames from the street means that the fire has ventilation and will grow fast. Sprinklers are either not effective, not functioning, or not present. This means a well-seated fire that will be much larger and more complex by the time crews get first water on the flames.

Conditions for civilians above the fire will only deteriorate as lethal smoke and gases rise into the building. Additional alarms, overhead, command, and tactical channels should be ordered upon arrival. This includes additional ALS capabilities and ambulances with appropriate supervision. You can get behind the power curve very quickly (fig. 11–10).

- As crews ascend and reach the fire floor, that floor will become your *first division*. The supervisor will likely be one of the first company officers to the fire floor; however, they will need to be backfilled with a chief officer if a working fire is confirmed.
- Division supervisors of fire floors will have the tactical objectives of fire attack *and* primary search on their respective floors.
- *Lobby/systems group* will be required to gain control of the lobby, control access and elevators, and work with on-site personnel to better ascertain conditions above. In a confirmed fire, this will include utilizing any fire protection systems in the building including water pumps, PA, HVAC control, and so on.
- Staging will be established as companies muster toward the fire floor. *Identify a staging area manager.* The first three alarms will likely go directly to work.
- Base will be established as the second alarm arrives. *Identify a base manager* and traffic pattern for subsequent alarms as they arrive.
- Consider an evacuation group supervisor for rapid ascent teams and additional division supervisor and companies above the fire floor to check extension. Depending upon your resource levels and SOGs, this may not occur until your second alarm as the first alarm will be focused on getting water on the fire floor and controlling the lobby.
- *Expand your communications plan carefully.* Do not add too many channels too fast; however, waiting too long for radio traffic to be too congested can be too late.
- Consider a ventilation group supervisor to control pressurization of stairwells and coordination of HVAC with lobby/systems group.
- If the fire grows beyond the involved floor, consider an RIC group, which may not be properly staffed until your third alarm or later, based upon conditions and VP.
- Consider a medical group for potential civilian victims.
- Note the IC is beyond the span of control at 9:1. If the IC does not rearrange the ICS, the span of control will only grow and become completely unmanageable. Radio traffic and accountability is now chaotic. How could this have been prevented? Could expanding the ICS prior to this point prevented confusion and chaos? Yes! See the solution below in the next template.

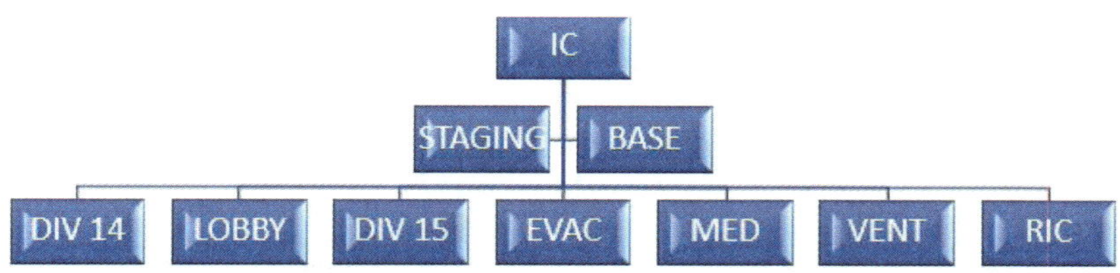

Fig.11–10. The IC is well beyond the span of control.

High-Rise Fire with Smoke/Fire Visible from the Street—Multiple Floors

Visible smoke and flames from multiple floors means a structural high-rise fire that could compromise the integrity of the building if not brought under control within a few hours.

Sprinklers are either not effective, not functioning, or not present. Conditions for civilians above *are* lethal because smoke and gases *have risen* into the building. Additional alarms, overhead, command, and tactical channels must be ordered upon arrival. This includes additional ALS capabilities and ambulances with appropriate supervision. Any automatic/mutual aid resources must be called if your FD cannot sustain this level of resource demand. Think ahead (fig. 11–11).

- As crews ascend and reach the fire floor, that floor will become your *first division*. The supervisor will likely be one of the first company officers to the fire floor; however, they will need to be backfilled with a chief officer if a working fire is confirmed.
- A *second division* will be established for the second fire floor, and a *third additional division* above to confine the fire.
- Division supervisors of fire floors will have the tactical objectives of fire attack *and* primary search on their respective floors.
- A second stairwell should be identified for evacuation and support.
- *Lobby/systems group* will be required to gain control of the lobby, control access and elevators, and work with on-site personnel.
- With more than one floor involved, this will include utilizing any fire protection systems in the building including water pumps, PA, HVAC control, and so on.
- Staging will be established as companies muster toward the fire floor. *Identify a staging area manager.* The first four alarms will likely go directly to work.
- Base will be established as the second alarm arrives. *Identify a base manager* and traffic pattern for subsequent alarms as they arrive.
- Utilize law enforcement early to control streets in several directions.
- Establish an evacuation group supervisor for rapid ascent teams to quickly access upper floors and stairwells.
- Consider a ventilation group supervisor to control pressurization of stairwells and coordination of HVAC with systems/lobby group.

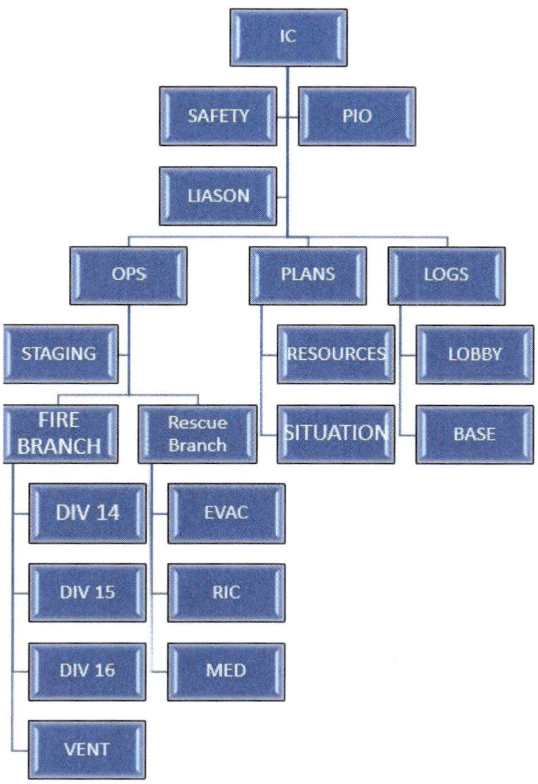

Fig.11–11. A more robust ICS for a larger high-rise fire.

- A medical group supervisor should be established for civilian victims with separate triage and treatment areas. They will operate on their own tactical channel if necessary.
- Confirm that there will be an adequate supply to sustain the fight. You may have to contact the water company.
- Establish an RIC group, which may not be properly staffed until your fourth alarm or later, based upon conditions and VP.
- Consider a structural engineer if the fire lasts more than a few hours. Sooner may be required depending upon the era, construction, and fire resistance in the building.
- *Expand your communications plan carefully.* Do not add too many channels too fast; however, waiting too long for radio traffic to be too congested can be too late.
- Get the phone number to the lobby group to reduce radio traffic.
- Consider establishing an ICP in the lobby of a neighboring building.
- Augment ICP support for radio management, situation status, and accountability.
- As overhead personnel arrive and the tactical supervisors are in place, establish a command staff (safety, information, liaison officers).
- As you can see, the span of control for the IC could become impossible. The IC would no longer be effective and radio frequencies will likely become so clogged that communication is impossible. This situation is ripe for a sentinel event. If a Mayday were to occur, it would likely not be heard. If a civilian were to be found, the transmission would likely not get out from the company to the rescue group, let alone to the medical group.
- *Splitting the fire into two sections, operations and logistics, will dramatically reduce the span of control and make accountability and communications more effective and less chaotic.*
- You will now have room for safety, public information, and liaison officers. Your span of control would then be 5:1: operations, logistics, safety, PIO, and liaison.
- If the battle will last several hours or into the next operational period (shift), consider establishing a planning section for support as described in SAW-CSS-RECSO-VSS. This would bring the span of control of the IC to 6:1.
- Note that the operations section span of control was also reduced by the insertion of two branches: fire branch and rescue branch. Fire branch is responsible for all fire attack and coordinated ventilation operations. Rescue branch is responsible for all human life, whether firefighters (RIC), or civilians (EVAC). Evacuation group will be focusing on the stairwells and floors above. Division supervisors will be focusing on their floors.
- Either firefighters down or civilians found can land in medical. All civilian and firefighter victims land in the medical group, whether they come from evacuation group, RIC group, or a division.

Roof Division

You may have operations on the roof of the building for ventilation, HVAC, evacuations, air operations, and so on. Consider a roof division supervisor if you have the capacity to be the eyes and ears on the roof.

Summary

High-rise fires can be the deadliest events for civilians and firefighters alike. A small fire on a lower floor can cause smoke to stack on floors above, killing dozens of civilians who never felt the heat of the fire.

Pre-planning and training well before an incident are paramount to success. Resource demands will be immense, members will be taxed, and you must remain calm. Call additional alarms and resources early. Utilize ICS and think globally to remain ahead of the incident power curve, maintain a manageable span of control, and keep the NIOSH 5 from coming into alignment. This can happen exponentially faster on high-rise fires.

SAW-CSS-RECSO-VSS tactical considerations for high-rise fires, the tenants of incident command that you have learned, and proactive application of ICS, will allow your members to be aggressive, effective, efficient, and properly supported so they save lives and go home when the work is done.

Consider training upon the ICS templates in this chapter and adjust as needed when your turn comes to be the IC on a high-rise fire. Plan ahead!

Chapter Review

Review Questions

1. Explain the differences between staging and base with regard to a high-rise incident.
2. Define a rapid ascent team.
3. What are some tactical considerations for ventilation of high-rise fires?
4. Describe the role and responsibilities of a lobby/systems group.
5. Explain the use of branches for ICS at a high-rise fire incident.
6. Create an ICS chart for a high-rise fire with smoke visible from the street for a single apartment on the 12th floor of a 15-story residential high-rise building.

FESHE Strategy and Tactics (C0279) Related Content

The content contained in chapter 11 provides detailed information to assist with instruction and education for all course objectives with a focus on high-rise structures. The chapter format supports the C0279 course outline by speaking specifically to basic division of tactics, rescue, exposures, confinement, ventilation, and salvage for high-rise fires.

Access the Tactical Review Guide Appendix C#5: High-Rise Fire Upper Floors for objectives-based scenario exercise.

NFPA 1021 Job Performance Requirements

The information in this chapter can be utilized to support training and educational programs associated with the Emergency Services Delivery Fire Officer I JPR 4.6, 4.6.1, 4.6.2, Fire Officer II JPR 5.6, 5.6.1, and Fire Officer III 6.6 and 6.6.1. For working high-rise incidents and pre-planning responses, the content in this chapter also supports training and education for Fire Officer III 6.6.3. Develop a plan for the organization given an unmet need for resources that exceed what is available in the organization.

Access the Tactical Review Guide Appendix C#5: High-Rise Fire Upper Floors for 1021 Job Performance Requirement-based scenario exercise for Fire Officer I, II, and III.

Endnote

1. Marty Ahrens, "High-Rise Building Fires," *NFPA Research,* November 2016, https://www.nfpa.org/education-and-research/building-and-life-safety/high-rise-buildings

12

Rapid Intervention and Maydays

Fig. 12–0. Los Angeles City (CA) Firefighter's personal protective equipment after the Boyd Street Incident
Source. Photo courtesy of Rick McClure

Introduction

Firefighters have been trapped and Maydays have been called during all phases of a fire (fig. 12–0). Many statistics have been captured by such enterprises as Project Mayday that give compelling data regarding downed firefighters. The bottom line is that it can happen anytime, anywhere, and to anyone.

As a firefighter of 32 years, I (Anthony Kastros) have sustained burns to the face, had my SCBA mask come off during an interior attack and had to evacuate, fallen through a roof, been the primary caregiver of a downed firefighter who was a friend, and been the IC of a Mayday on a house fire. As my dear friend Bobby Athanas of FDNY Rescue 3 once said, "This job gets real, really quick."

Alan Brunacini had a quote that fits: "The IC should be the biggest pessimist on the fireground." You cannot and *must* not rest until all your firefighters are home safely back at their fire stations after a fire. Firefighters have been trapped or killed in the first few minutes of arriving on scene of a fire, during overhaul, and everywhere in between. As the IC or tactical supervisor, you must *never* let your guard down, nor let your firefighters do so. Consider the following facts:

- During a structure fire, a building is under demolition.
- The longer the building burns, the more unstable it becomes.
- Gravity and weight of water also put stress upon the building and continue to do so after the fire is extinguished.
- Buildings are not designed to remain intact when flames, forcible entry, and water impact their structural components.
- Lightweight construction may fail rapidly within a few minutes because it is not designed nor intended to withstand the impact of direct flame impingement.
- Conditions in structure fires can change in seconds.

An Ounce of Prevention is Worth a Pound of Cure

The old adage, "An ounce of prevention is worth a pound of cure" is never more appropriate than for preventing a Mayday. As we have described and reinforced in this text, the work done *before* the fire may prevent the Mayday in the first place.

How so? Well, it all starts with you! Every firefighter, driver, and officer *must* ensure the following at the beginning of each and every shift. Remember, every company and chief officer sets the example:

- Your SCBA cylinder is full and all straps on your pack and mask are in working order.
- Your portable radio battery is full and not left on during your shift.
- Your PPE is in working order, free from rips, tears, defects, or hazardous materials that would cause it to not protect you.

When responding to and arriving at a fire or other emergency scene, ensure you:

- Have your portable radio on the correct tactical channel when responding to a fire.
- Know the IAP, priority, strategy, and tactics on your incident. This will give you the risk/gain threshold.
- Take a deep breath and size up the situation. Look at the big picture before you experience tunnel vision on the fire.
- Wear all your PPE, including hood, gloves and SCBA (including waist strap) when in the IDLH atmosphere, or over the IDLH (roof operations).
- Get a lap of the building before entering, when possible.
- Know where the fire is and the layout of the building before entering, when possible.
- Know the VP.
- Know who your boss is.
- Know what your job is.

From the IC standpoint, you should do all of the above and have reinforced your expectations through your example. Chiefs, you are first a firefighter. Ensure your rig, radio, and PPE is ready at the beginning of each shift. The astute IC (any officer) begins each shift with a sober reminder that either a Mayday, a civilian rescue, or *both* could happen any day.

Once on scene, setting up divisions or groups early can prevent a Mayday in the first place in the following ways:

- Tactical supervisors are in the tactical gap to see risk that could be missed by crews and the IC, thereby preventing a Mayday. This was the case on a strip mall fire. The Division A supervisor prevented crews from going underneath an unsupported mansard. Seconds later, it collapsed.
- Radio traffic is minimized, and face-to-face communication is maximized. This will also allow critical traffic to be transmitted during a Mayday and may prevent it with safety transmissions.
- Accountability is eyes-on and hands-on in real time. This maintains alignment with the overall strategy and risk management plan and keeps crews from freelancing or getting into untenable conditions.

The philosophy of proactive and decentralized incident command is, in itself, a key component in either preventing a Mayday outright, or managing one much more calmly and effectively.

When division and group supervisors are in place, they have bridged the tactical gap and can see bad things unfolding (ounce of prevention). They can stop crews from going into a space that is going to collapse or have a significant fire event. They can grab firefighters by the straps and say, “What are you doing? Don’t go in there. The floor is going to give way!”

Conversely, relying solely on the rapid intervention crews is completely reactive (pound of cure). According to Project Mayday by Don Abbott, the crew assigned to rapid intervention only rescues downed firefighters 6% of the time (fig. 12–1).[1]

The vast majority of downed firefighters are rescued by other nearby crews inside the structure. This means less radio traffic is required through tactical supervision of companies.

Fig. 12–1. Los Angeles City (CA) firefighters tend to one of their own at a commercial structure fire.
Source. Photo courtesy of John McCoy

Remember, as the safety officer of their respective division or group, the tactical supervisor is in place to forecast bad things and prevent task-focused firefighters from jumping into a bad situation. As a division supervisor, I have pulled firefighters out of buildings that were about to collapse, held them back from walls that were about to fall, and stopped them from going into areas they had no business going into. Most of the time, the IC never knew.

Number one on the NIOSH 5 is inadequate risk assessment. This is a key factor in every Mayday and firefighter LODD that has ever happened. Having a guardian angel (tactical supervisor) over your shoulder keeps you from unnecessary risk. Good firefighters are aggressive, great firefighters are aggressive *and* smart. Yet, we all miss the bigger picture when the fire is lulling us into its grasp.

CASE STUDIES IN COMMAND

Christmas Eve Mayday

CHIEF KILEY KEELEY, SACRAMENTO (CA) METRO FIRE AND
CHIEF RUSTY VAN VUREN, SACRAMENTO CITY (CA) FIRE DEPARTMENT

Battalion Chief Kiley Keeley: Christmas Eve night was a cool evening with calm winds and light intermittent rain in Sacramento, CA. Late in the evening (2246 hours dispatch time), a fire would break out in a home in North Sacramento, within the City of Sacramento's borders. This is an older part of the city, with small homes. These large-lot property homes often have multiple non-permit-approved additions that have been constructed over the years. This house fire was one of these types of buildings.

Sacramento City Fire and Metro Fire share a border, and throughout the years our agencies have developed a great relationship, where we respond on automatic aid without respect to jurisdiction. We share common SOGs, and the battalion chiefs commonly train together for larger-scale incidents to be familiar with our expectations of other responding BC duties upon arrival, including full turnouts, SCBAs, and two handheld radios.

Our common residential response levels include two BCs, four engines, two trucks, and one medic. A second alarm receives the same type and kind of resources. BC3 (City) and BC5 (Metro) were the responding BCs on the first alarm. Unknown to the crews upon dispatch, three people were trapped inside. However, BC3 was updated en route and it was immediately passed on to the responding units on their TAC (tactical) channel. An additional medic was also immediately requested by BC3. Given the serious rescue anticipations and difficult access, along with an initially incorrect street and heavy fire conditions faced by first arriving crews, BC3 (as the IC) quickly ordered a second alarm, including additional ambulances.

I was on duty at Sacramento Metro Fire Battalion 7 and responded as part of the second-alarm assignment. As I got in the buggy and began driving to the scene, the TAC channel was overwhelmed with crews dragging victims out of the house, crews fighting the fire, coordinating vent, protecting exposures, and other pertinent radio traffic that the IC relies upon to maintain his/her plan for the incident.

Upon my arrival, I checked in with the IC at his command post. He let me know that BC5 was Division C and he assigned me to run Division A. The objectives for Division A were to complete a primary search, fire attack, and coordinated vent with the truck on the roof. I made my way forward through the front yard strewn with cars and waist-high household debris to find the last of the victims being removed from the house and placed onto the gurney for transport to the hospital. Crews were outside and I was able to have face-to-face updates as to the status of the search and where the seat of the fire was.

One of the engines in my division was finishing a bottle change when a muffled voice came over the radio "I'm lost, I'm out of air, I NEED HELP!" I looked at the engine with fresh bottles and said, "That was a Mayday get inside and start searching. Listen to the radio for the name, unit, conditions, actions, needs (NUCAN)." Crews went to work to locate the firefighter with the Mayday. He was quickly located and removed from the IDLH zone with no injuries, thanks to the quick efforts of the men and women on scene.

Battalion Chief Rusty Van Vuren: I responded from quarters with the units that would be first on scene with situational awareness of E20 having an acting captain as well as a first-day probie, along with multiple crews affected by the Christmas holiday. So, a "then/what if" mindset was queued up well before this particular dispatch.

For BCs, the ability to hear multiple radio channels is incredibly important and can only be honed by having your radios on all day, regardless of whether one is on the incident or not.

The update for confirmed victim profile from dispatch happened while units responding were finishing an automatic TAC channel roll call. This allowed me to acknowledge dispatch and request an immediate supplemental ambulance on the primary city channel before immediately advising units en route. This gave them the chance to make that next mental prep within their rigs and avoid surprise.

That same "then/what if" is what has one ready to make important decisions that ultimately are validated when the "what if" occurs. In this case, additional ambulances and the second-alarm units were both quickly available for the additional victims.

The incident power curve continued to rise and fall throughout the incident due to the rescues, Mayday, multiple roofs, a gas meter that was hidden inside the fire building, and interior access issues. We BCs decided that we would stand down the aggressive suppression efforts after a confirmed all clear was announced after secondary searches.

Takeaways from this incident:

In order for command to live in the "then/what if," the IC should establish a division/group supervisor to make tactical decisions based on information gathered from a forward position. Oftentimes, from the command post all you can see is a glow and some smoke rising above the apparatus.

Division/group supervisors provide great intel and will help to limit radio traffic by having face-to-face communications with the task-level crews (this was a critical point accomplished by both BC5 and BC7). Without divisions on this fire, the Mayday could have been walked on, or not heard, due to the need for radio transmissions to command. With this being a Christmas Eve fire, three victims already pulled out and two under cardiopulmonary resuscitation (CPR), the scene was highly charged and needed forward coordination.

Communication between BCs that fill these roles begins well before the incident. The manner and reason this was achieved is/was because the BCs meet nearly every rotation either in the form of training or coffee sit-down, where we learn each other's preferences, personalities, strengths, and weaknesses. This creates a relationship of confidence and trust that eliminates ego and territorialism. This naturally trickles through to the crews as well, who eventually were doing inter-departmental station drills on their own to create a borderless feeling of a team.

Personal note from the incident commander: The sense of gratitude I had for my division supervisors (BC7 and BC5) after an incident like this was overwhelming. Their work that day, being hostile, agile, and mobile, filling the tactical gap, so that I could multi-task as a focused, proactive IC at the strategic level was critical. The decision-making pace needed to stay ahead of the incident power curve and is really what made the incident come to a safe and effective conclusion.

Two-Out Crews and Back-Up Lines versus Rapid Intervention

Perhaps one of the greatest disparities in American firefighting is the interpretation, terminology, procedures, and use of lines to fulfill the Title 29 Code of Federal Regulations (CFR) 1910.134 (g)(3) and (g)(4) "two-in/two-out" regulation. Some terms used are "two-out line," "back-up line," "secondary line," "initial rapid intervention team (IRIT) line," and countless others. In addition, some agencies will wait until a RIC is established to stretch a two-out line. In many jurisdictions, the RIC company is a later-arriving crew, well after interior firefighting is underway. Some agencies treat every building as occupied to elude the regulation. Others go to the opposite extreme and will wait to establish a two-out crew with line before entering any structure, regardless of high or confirmed victim profile.

Firefighters are smart and want to know the "why" behind any policy or SOG. Like any rule, firefighters will challenge, push the limits, find loopholes, and skirt policies and SOGs that they think make no sense, or that have not been given a proper "why" when implemented. If the only reason a fire department has a two-out policy is to stay out of trouble or prevent paying fines with OSHA, that is *not* good enough, and firefighters will likely blow it off.

Properly placed, staffed, and utilized two-out lines (or whatever your FD calls it) have saved firefighter lives and prevented catastrophic outcomes. Simultaneously, crews have waited too long to enter fires because of fear of getting in trouble, potentially costing civilian lives. Therein lies the disparity.

Another problem is misunderstanding the difference between two-out regulations and rapid intervention crews/teams. First came the two-out regulation. There is no mention of rapid intervention teams/crews or otherwise in CFR 1910.134. The fire service put its spin on it and created rapid intervention crews/teams after the fact.

Two-out is the minimum under the law. Rapid intervention can be subjectively applied based upon conditions, perceived need, staffing, response levels, traditions, SOGs, state regulations, and a host of other factors, including the IC.

Let's look at the *actual* OSHA standard:[2]

1910.134(g)(3)
Procedures for IDLH atmospheres. *For all IDLH atmospheres, the employer shall ensure that:*

> *1910.134(g)(3)(i)*
> *One employee or, when needed, more than one employee is located outside the IDLH atmosphere;*
>
> *1910.134(g)(3)(ii)*
> *Visual, voice, or signal line communication is maintained between the employee(s) in the IDLH atmosphere and the employee(s) located outside the IDLH atmosphere;*
>
> *1910.134(g)(3)(iii)*
> *The employee(s) located outside the IDLH atmosphere are trained and equipped to provide effective emergency rescue;*
>
> *1910.134(g)(3)(iv)*
> *The employer or designee is notified before the employee(s) located outside the IDLH atmosphere enter the IDLH atmosphere to provide emergency rescue;*
>
> *1910.134(g)(3)(v)*
> *The employer or designee authorized to do so by the employer, once notified, provides necessary assistance appropriate to the situation;*
>
> *1910.134(g)(3)(vi)*
> *Employee(s) located outside the IDLH atmospheres are equipped with:*
>
> > *1910.134(g)(3)(vi)(A)*
> > *Pressure demand or other positive pressure SCBAs, or a pressure demand or other positive pressure supplied-air respirator with auxiliary SCBA; and either*
> >
> > *1910.134(g)(3)(vi)(B)*
> > *Appropriate retrieval equipment for removing the employee(s) who enter(s) these hazardous atmospheres where retrieval equipment would contribute to the rescue of the employee(s) and would not increase the overall risk resulting from entry; or*

> *1910.134(g)(3)(vi)(C)*
> *Equivalent means for rescue where retrieval equipment is not required under paragraph (g)(3)(vi)(B).*

1910.134(g)(4)
Procedures for interior structural firefighting. *In addition to the requirements set forth under paragraph (g)(3), in interior structural fires, the employer shall ensure that:*

> *1910.134(g)(4)(i)*
> *At least two employees enter the IDLH atmosphere and remain in visual or voice contact with one another at all times;*
>
> *1910.134(g)(4)(ii)*
> *At least two employees are located outside the IDLH atmosphere; and*
>
> *1910.134(g)(4)(iii)*
> *All employees engaged in interior structural firefighting use SCBAs.*

Note 1 to paragraph (g): *One of the two individuals located outside the IDLH atmosphere may be assigned to an additional role, such as incident commander in charge of the emergency or safety officer, so long as this individual is able to perform assistance or rescue activities without jeopardizing the safety or health of any firefighter working at the incident.*

Note 2 to paragraph (g): *Nothing in this section is meant to preclude firefighters from performing emergency rescue activities before an entire team has assembled.*

Note that a hoseline and the term "rapid intervention crew" are not mentioned above. Neither is a list of required rescue/extrication equipment. The AHJ is the local fire department and each one determines how they will adhere to the above. In reality, what is done in practice may differ from what is written in policy or SOG. Fire departments all over America have interpreted this regulation in countless ways. The key is that you know *your* policy, how to interpret it, and train your officers to properly engage each fire as thinking leaders who understand the intent of 1910.134 (g)(3) and (g)(4). *The intent is to balance the protection of firefighters' lives with rescue of civilian lives.* This will be a judgement call, and the VP is a very effective means of gathering information to make an informed decision when seconds count and lives hang in the balance.

Note that

> *One of the two individuals located outside the IDLH atmosphere* may be assigned to an additional role, such as incident commander *in charge of the emergency or safety officer, so long as this individual is able to perform assistance or rescue activities without jeopardizing the safety or health of any firefighter working at the incident.*

This means that as an IC or division/group supervisor, you may perform either of those roles and still be part of the two individuals located outside the IDLH atmosphere (two-out). This should be liberating and allow you many options on an incident.

In the early stages of a working house fire with low or moderate VP, for example, two three-person engines should be able to initiate interior fire attack, stretch a second two-out line for protection, secure a water supply, and have one officer in command.

A RIC, also known as a (RIT is composed of at least a three-person crew, placed to be ready for immediate rescue of firefighters inside a structure. An RIC will have equipment cached in advance of a firefighter down to expedite the response during a Mayday. Minimum equipment should include:

- Spare SCBA or RIC pack with umbilical air
- Spare SCBA cylinders
- Forcible entry tools such as a flathead axe, Halligan tool, pry bar, officer tool, and so on
- Powered circular and chain saws
- Rescue rope bag and accessories
- EMS equipment or a medical group nearby
- Ladder appropriate to size of structure
- Portable radios for all members
- Portable lighting

In addition, a separate size-up of the incident, including situation status and resource status and location of crews working inside and around the building, should be assessed. This should be done in coordination with division supervisors and command. This will help verify that active accountability is accurate and will reduce reflex time if a Mayday were to occur.

As the IC, you must weigh risk versus gain, as we have illustrated throughout the text. Based upon the VP, your staffing and response levels, and your SOGs, you will determine when to establish a rapid intervention plan and to what extent it will be staffed and supported. For example, you may have confirmed victims trapped and not enough firefighters on scene to perform fire attack, rescue, vent, medical, and have enough personnel left over to adequately staff a RIC Group. The priority is the civilian victim.

As the IC of a three-alarm apartment fire, I did not have rapid intervention crews in place until the third alarm, as we had several trapped occupants upon arrival.

CASE STUDIES IN COMMAND

Lullwater Incident

CHIEF DAVID RHODES, ATLANTA (GA) FIRE RESCUE AND EDITOR-IN-CHIEF, *FIRE ENGINEERING* MAGAZINE

Hearing the rumble and feeling the ground shake from a structural collapse is not something you want to experience as a chief officer managing crews that are interior. As the dust settles and the reality hits you square in the face, it is a nauseating feeling. There is, however, no time to waste in gathering your thoughts and moving forward with first making sense and orienting yourself to what has occurred. Then, requesting and assigning the appropriate resources to a strategy to mitigate the hazard, all while thinking of contingency plans if that strategy doesn't work.

I was assigned to Battalion 3 operating as the C (Charlie) side division supervisor on a very large mansion fire in Atlanta, Georgia. I was part of a three-chief command team that was managing this fire that also included Battalion 5 as IC, and the shift commander assistant chief who was the A (Alpha) side division supervisor. The alarm came in at 22:10. The structure was an early-1920s-era, 6000-square-foot, two-story, L-shaped residence (fig. 12–2).

The initial alarm included three engines, three trucks, two battalion chiefs, and the heavy rescue. There was heavy fire upon arrival and we operated in the offensive mode, rescue phase for the initial strategy. A second alarm was requested due to the size of the structure and the amount of fire involvement. This brought another three engines and three trucks and the addition of the shift commander.

Interior crews made progress and had the majority of the first-floor fire under control within the first 10 minutes. The heavy rescue was able to get an all-clear on the second-floor primary search. At the 20-minute mark, some early signs of roof collapse began to present on the A (Alpha) side and the decision was made to switch to a defensive mode, since the primary search was completed. Exterior lines and a ladder pipe operated for approximately 45 minutes before the fire was under control. The roof on the main part of the house burned through on both the A and C (Charlie) sides.

Now we were faced with a large overhaul operation and numerous pockets of fire. An assessment of the structure was completed and a decision was made to send crews back interior for overhaul. I was managing an engine and truck company working to get into the attic of the second floor in the smaller part of the L directly over the garage. The house had metal casement windows and the windows on the C side extension had not been removed. The crew squeezed their way in the window frame in place (fig. 12–3).

I did not like this and called for an additional truck company to come and cut the window frames out. Crews remained inside and line was also advanced through the window frame. The interior crews started working to access the attic by pulling ceiling in the wire-and-lathe ceiling covering.

I went inside the first floor directly under where crews were working to double check the structure and make sure there was no additional fire that had rekindled. Upon exiting the first floor, I returned to check the status of the window removal. A piece of the cornice fell across the ground ladder that was used to enter the second floor. The truck company assigned to remove the window had still not arrived. I had a worse feeling about the window still being in place and decided to pull the crews out until we could get the window removed. All crews acknowledged the order to exit and began the process.

Then, I heard a low-decibel rumble and felt vibration from the ground, and a burst of dust occurred. I knew I had four members working on the second floor. A large section

Fig. 12–2. Alpha-side view of Lullwater incident.

of the roof had collapsed, pinning one firefighter against the wall and one other under the debris. Two others were able to escape the area and immediately started working to rescue the trapped members. I notified command that we had a collapse and firefighters trapped, declared a Mayday, and advised I would no longer be the C (Charlie) division supervisor and would be managing the Mayday.

The assistant chief responded to the rear of the structure and assumed the C (Charlie) division supervisor role and requested two additional engines, one truck, and a medical supervisor. An RIT was deployed consisting of an engine and a truck company. The firefighter pinned against the wall was removed quickly and continued assisting the rescue of the other firefighter. Crews continued attempting to cut the metal casement window, but the standard department abrasive blades were just disintegrating with little impact. Additional saws were requested and the captain of Truck 15 was placed in charge of the rescue.

The heavy rescue had left the scene for another fire once we were under control. They were requested back in case the airbags and other lifting equipment was needed. Chainsaws and pry bars were used interior to cut away rafters, ceiling joist, and decking. Handlines were deployed on either side of the rescue, as fire was now flaring up in the debris with the influx of oxygen and dislodged building materials.

Sixteen minutes into the Mayday, the firefighter was freed. He was pinned in a supine position and was unable to move any part of his body. He was in full PPE and on air when the collapse happened. As the rescue crews removed a critical piece of the roof, he was able to reach his regulator and disconnect. He had just run out of air the instant before the pressure and weight on his arm was lifted.

The rescued firefighter was able to get out of the structure and down the ladder on his own. He appeared to be okay and looked at me and requested another assignment. Of course, he was done for the day and we insisted he be transported to the hospital and evaluated. He and the firefighter that was pinned against the wall were both sent for evaluation. The firefighter pinned against the wall suffered torn tendons and ligaments in his foot that required surgery to repair. His recovery was nearly six months. The

Fig. 12–3. A C side view of the Lullwater Incident.

trapped firefighter under the collapse sustained second and third degree burns on his legs, back, and buttocks. This occurred from the compression of his body on embers that were smoldering on the floor. He was out for nearly a year and received several skin grafts before returning to full duty.

Lessons Learned

This house had a full slate roof, which added a considerable amount of weight that was not initially considered in the risk assessment. Our department had a culture of using manpower for overhaul instead of waiting on resources from other department with heavy equipment.

Always, always trust your intuition! I didn't like the fact that firefighters had to squeeze through that metal casement window, but instead of stopping them, I let them go with a plan to remove the window behind them. Had I not ordered them out when I did, the trapped firefighter would have been in an area that would have been even more difficulty to remove. Since he was moving towards the window, he was only a few feet inside, which made access a little bit easier.

ICS and Communications for Maydays

The first thing to do if a Mayday occurs is to remain absolutely calm. As with all transmissions, whatever the IC does, the rest of the alarm will do. Not all Maydays are equal. You may have a quick Mayday that is minor and resolved before you even have a chance to respond on the radio. Or, you may have several members trapped, with an extensive rescue operation requiring additional alarms and support.

Do not overreact and make the response more complicated. Wait a few seconds and see what you have. For example, dispatch may suggest adding more radio channels, additional chiefs may arrive, and companies may assign themselves to the incident. You must control this chaos and ensure you know what the extent of the problem is. Everyone will want to help, but not everyone can.

A Mayday can also be extremely labor-intensive. The Southwest Supermarket Fire in Phoenix, AZ showed that it takes 12 firefighters to get 1 out. This 12:1 ratio may tax your system immediately. Or, you may have adequate resources. Additional Maydays could occur and you may find multiple firefighters in cardiac arrest. This is why a medical group should *always* be standing by on all working fires.

When a Mayday occurs, there are several steps that should occur by the IC and division/group supervisors, in unison and often simultaneously, to maximize efficiency and minimize time-to-task completion. *Remember, the majority of downed firefighters are rescued by nearby crews operating close by.* Allow them to do so. This will congeal the task-level response as crews muster around their fallen comrade(s).

As soon as the Mayday occurs, firefighters will do whatever they can at their levels and in their locations to rescue their brothers and sisters. Your job, as the IC, is to coordinate, support, bridge the tactical gap, and maintain alignment of response to maximize effectiveness. Remember, remain in the "then/what if" and let your tactical- and task-level bosses get the work done.

Remember, the majority of downed firefighters are rescued by nearby crews operating close by.

Consider the following:

- The IC should stop all radio traffic upon a Mayday and acknowledge the Mayday directly by conducting a *LUNAR* report with the downed firefighter, if possible. The LUNAR should be repeated clearly by the IC to verify accuracy and ensure that all members on scene have heard the LUNAR.
- Simultaneously, the division supervisor of the firefighter/company down and the RIC group supervisor should meet face-to-face, confirm the LUNAR, and develop a tactical plan. This may require the division supervisor giving crews to the RIC group supervisor to augment the Mayday response.
- The IC and tactical supervisors should confirm the tactical plan so the IC can support it. This includes notifying the IC of any movement of crews from division to RIC group.
- As needed, the IC will call additional alarm(s) and ambulances for the Mayday, based upon the LUNAR report. Plan ahead accordingly.
- The RIC group supervisor's objective is to rescue the downed firefighter(s).
- The division supervisor's objective is to continue the firefight to ensure that conditions are tenable. This may include additional attack lines and/or ventilation.
- Based upon the VP and fire conditions at the time of the Mayday, companies on scene can be reassigned to the RIC Group. In addition to the above example, you may have a company on the roof performing vertical ventilation that is no longer needed. They can be reassigned to the RIC Group.
- If you have a Mayday *and* a confirmed rescue of civilians simultaneously, the IC must support both. ICS will help you organize and decentralize both operations at once. You may have the respective division supervisor lead the civilian rescue, or establish a separate rescue group, depending upon the nature and needs for the civilian rescue(s).
- The medical group should be augmented and the medical group supervisor should be working face-to-face with the RIC group supervisor to receive downed firefighters.

We responded to a house fire with explosion and gas leak at 10:30 on a weekday. Upon arrival, E101 reported a one-story house with heavy fire from the rear and heavy black smoke from the front. This was a low VP, as the sole occupant was out and he stated that the fire started at his propane barbecue in the back of the house. E101 initiated fire attack. E20 arrived shortly after and assumed command from E101. E20 secured a water supply and stretched a two-out line, per the SOG, to protect E101.

I arrived and assumed command from E20. It was clear that this fire required more water and E20 requested to take their two-out line inside as a secondary attack line. I agreed. E103 then arrived with TR20. E103 was assigned to assume Division A with the objectives of fire attack and search inside the house. Division A had E101, E20, TR20, and his crew (E103) working for him. The E103 crew pulled a second two-out line.

TR106 arrived and was assigned to roof division with the objective of vertical ventilation. The TR106 captain was roof division supervisor. E105 arrived and was assigned to RIC. M19 arrived and was assigned to medical and to stand by at the ICP.

Within three-and-a-half minutes of my arrival, the rear patio cover collapsed on E101 captain and firefighter. Per the SOG, E20 immediately called a Mayday on their behalf. As I asked for a LUNAR report, E20 confirmed that E101 was not entrapped and was fine. This was obviously a huge relief! Had this been an extended Mayday with entrapment of two firefighters, we were set up for success. At the time of the Mayday, the radio was dead silent because tactical supervisors were in place to communicate with their crews and eliminate unnecessary radio traffic. As the Mayday was being called, the Division A Supervisor immediately responded to the front door and made face-to-face contact with his crew who were staffing the two-out line. They were in an excellent position to continue firefighting and protecting the area around E101 if needed. This would have been critical, as E101 was down and E20 was assisting them, leaving both of their hoselines unstaffed. Simultaneously, RIC (E105) was ready to proceed with augmenting any firefighter rescue that E20 was obviously conducting with TR20. Medical was at the ICP, ready to receive the firefighters. If needed, TR106 could have been called off the roof to further augment RIC operations until the arrival of subsequent alarms, had they been called.

As the IC, you must be thinking in the "then/what if" all the time.

CASE STUDIES IN COMMAND

Carr Fire

CHIEF CHRIS STAVROS, GLENDALE (CA) FIRE DEPARTMENT

Your engine is driving back from your maintenance facility and you see a bit of smoke coming from a residential area on the border of your city. You notify your dispatch center, and as your engine proceeds with a smoke investigation, you arrive on scene, in station attire, to a working apartment house fire, with immediate rescue needs on the second floor. Twenty-five minutes later, you and one of your firefighters fall into a burning basement.

The Carr Incident was a structure fire, intentionally started with a propellant, in the basement of a two story pre-33 center-hall apartment building in Glendale, California. Pre-33 refers to the construction date and construction standards prior to the Long Beach, California earthquake of 1933. Pre-33 apartment buildings are typically URM and can house many people from lower income levels, sometimes with six or seven people living in a one-bedroom apartment, so there is a large loss-of-life possibility at all hours.

The incident began with rescue of several civilians from upper floors and ended in a double Mayday of two firefighters who fell through the first floor into the burning basement. There were rescued by the assigned rapid intervention group, which consisted of a Glendale (CA) truck company and a Burbank (CA) engine company.

Some incident commanders can struggle with when to assign an RIC. If there is a confirmed rescue or a high victim profile, the RIC assignment can be delayed, but what is your *personal* risk tolerance for making that decision? How will you, or your companies, feel if RIC is assigned early, and that possibly results in a missed rescue and a civilian fatality? How will you, or your crews, feel if RIC is delayed, and a firefighter is killed who may have been rescued had a dedicated RIC been assigned? These are situations that should be discussed with your firefighters and officers. Regardless of what trigger you have for making the RIC assignment, you need to plan for many different scenarios, and you need to practice simulations to prepare yourself and your crews for the actual event. You should expect the unexpected (fig. 12–4).

Fig. 12–4. Division supervisors interacting face-to-face at the Carr Incident
Source. Photo courtesy of Glendale (CA) Fire Department

From a command perspective, I use the acronym EDWARD-D if I have a Mayday.

E Emergency declaration, clear all radio traffic on the incident, and announce a Mayday in progress. Attempt to get a LUNAR.

D Deploy RIC. Begin a Mayday timer.

W Get a warble tone from the dispatch center to announce a Mayday in progress, the tactical objectives, and tactical frequencies.

A Add an alarm (additional companies to the staging area).

R Roll a PAR (conduct a personal accountability report).

D Double the resources to and from (if not already formed) the RIC.

D Deal with the incident.

On the Carr Incident, several factors came into alignment that may have contributed to the double Mayday. The first was that the first-in engine company was out in district and responded to a "smoke in the area," and were dressed in their work uniforms. The standard routine that a company officer would have of being properly dressed, giving a solid initial report on conditions, getting a 360°, or at least a three-side view, was disrupted. Upon arrival, they were faced with immediate rescue. The second factor was that the BC had a delayed response because he did not get paged that there was an assignment. That one-minute delay prevented them from possibly getting a drive-by of the building, something that is very useful. The third factor was that the first-in engine captain was a well-respected officer, so subsequent arriving officers did not reverify the origin of the fire, which was in the basement, but reported as being on the first floor (fig. 12–5).

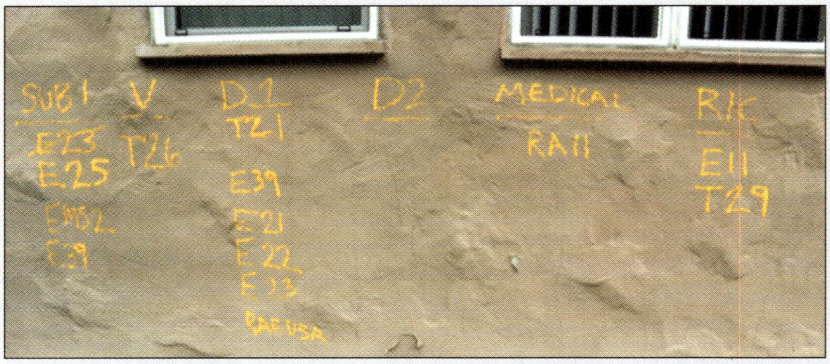

Fig. 12–5. RIT group supervisor accountability tracked on the side of the building. *Source.* Photo courtesy of Glendale (CA) Fire Department

Lessons Learned

1. Company officers and divisions/groups need to be brief on the radio. Long, rambling transmissions may not be received by the IC, and make the situation more chaotic.

2. Divisions/groups are responsible and accountable for the operations in their areas of responsibility. They need training to ensure their operational readiness for that responsibility.

3. When a company officer's routine is disrupted, it is important for the officer to reset the incident. After the initial disruption, the officer needs to start over and go through the routine that they practiced (update the report on conditions, complete a 360°, etc.).

4. In a complex incident, the IC, in addition to stating objectives, should remind officers and divisions/groups of vital information. For example, a reminder in clear text, "All units operating on the Carr Incident, the fire has been burning in the basement for 30 minutes, use all precautions, including sounding floors prior to operating above the fire."

5. Always assign an engaged, quality officer and crew as the rapid intervention company or group supervisor. Know your lineup in the morning and think about how you may use them that shift (fig. 12–6).

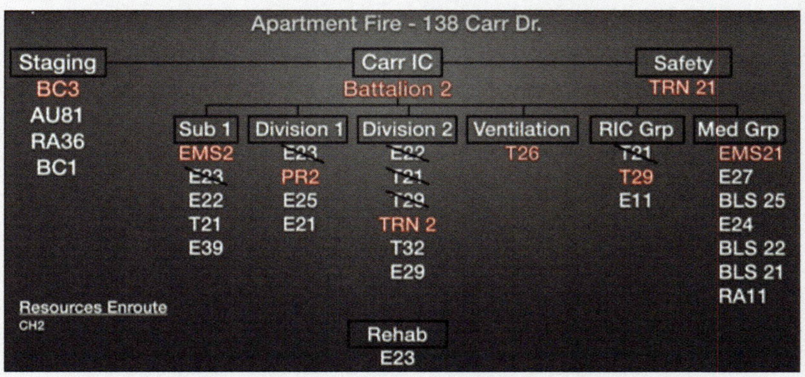

Fig. 12–6. Carr Incident Command Board[3] *Source.* Photo courtesy of Glendale (CA) Fire Department

Summary

A Mayday can happen to anyone, anytime, in any place. As the IC, your first priority is to get all your troops back to the station after every call, and home, safe to their families after each shift. The safety of our troops must be balanced with the need to save civilian lives.

Each incident starts with a VP. From there, you will determine the level of gain versus risk. What do you have to gain (civilians) and how much risk are you willing to take with your firefighters? You may have multiple confirmed victims and delay RIC implementation until much later in the fire or until enough resources are on scene. Conversely, you may have a low VP and have a two-out and an RIC group established. The timing and level of rapid intervention is largely subjective, based upon the FD SOGs, staffing, response levels, conditions on scene, VP, and even the IC. You must plan ahead and always be in the "then/what if" as the IC and remain the pessimist until you clear the scene.

Tactical supervisors are a tremendous help in preventing the Mayday in the first place. They may see risks such as worsening fire conditions and deteriorating building conditions, thereby preventing firefighters from going unnecessarily into high risk/no gain situations.

Set up your incident for success by creating divisions and groups to keep the span of control manageable, radio traffic minimal, and risk assessment exponentially more effective and decentralized. Once a Mayday does occur, you will have tactical supervisors in place to handle the immediate rescue of the firefighters, while you remain ahead and able to support the operation, whether you have heavy fire, civilian victims, or otherwise.

Chapter Review

Review Questions

1. Explain how setting up division/groups early can help prevent a Mayday.
2. What is the number one NIOSH 5 contributing factor?
3. Describe the foundation of two-in-two-out.
4. What does LUNAR stand for?
5. Explain the functions of a RIT/RIC crew and some of the tactical considerations.

FESHE Strategy and Tactics (C0279) Related Content

The content contained in chapter 12 provides detailed information specific to rapid intervention and Mayday incidents. While the FESHE C0279 course does not specifically reference rapid intervention of Mayday incidents, they will occur within a structure fire incident and require adjustment to tactics and ICS/NIMS. The chapter format supports the C0279 course outline by detailing information on fireground communications associated with these events and ICS/NIMS roles and responsibilities to support and manage them.

Access the Tactical Review Guide Appendix C#4: Mayday—Walmart. Fire in B Loading Dock for an objectives-based scenario exercise.

NFPA 1021 Job Performance Requirements

The information in this chapter can be utilized to support training and educational programs associated with the Emergency Services Delivery Fire Officer I JPR 4.6, 4.6.1, 4.6.2, Fire

Officer II JPR 5.6, 5.6.1, and Fire Officer III 6.6 and 6.6.1. For active rapid intervention or Mayday incidents and pre-planning responses, the content in this chapter also supports training and education for Fire Officer III 6.6.3 "Develop a plan for the organization given an unmet need for resources that exceed what is available in the organization."

Access the Tactical Review Guide Appendix TRG #4: Mayday-Walmart. Fire in B Loading Dock for a 1021 Job Performance Requirement-based scenario exercise for Fire Officer I, II, and III.

Endnotes

1. Don Abbott, Project Mayday Download!, December 25, 2018, https://www.firefighterclose-calls.com/project-mayday-download/.
2. Occupational Health and Safety Administration, "Standard 1910.134: Respiratory Protection," OSHA, last modified September 26, 2019, https://www.osha.gov/laws-regs/regulations/standardnumber/1910/1910.134.
3. Glendale Fire Training Videos, "Carr Incident After Action Review," YouTube, May 23, 2020, video, 26:57, https://youtu.be/hvMNnsue6MM?si=i1L3riuW0lL1Yz4_.

Section III

Commanding All-Risk Incidents

13

Vegetation Fires

Fig. 13–0. Los Angeles City Fire units initiating operations on a vegetation fire off of a roadway.
Source. Photo courtesy of Mike Meadows

Introduction

Units responded to a reported grass fire endangering structures and livestock. This was a fast-moving fire in light, flashy fuels on a red-flag day. This fire was in Battalion 5, adjacent and directly north of Battalion 7. I was assigned the hot division, ahead of the fire where the head was going. Multiple resource requests were being made by command and the column of smoke from the head could be seen from several miles away.

The fire jumped across several roads and was moving north. As I drove to get ahead of the fire, recon the values at risk, and account for the companies assigned to my division, I found a Type 1 Engine (typical structure engine) at the end of a driveway. The captain and his crew were in full structural turnouts. The captain had decided to deploy a 1¾" attack line into the green, unburned fuel. He unknowingly dropped his portable radio on the ground at the engine when he got out. We had no communications. The head of the fire was rapidly moving toward the captain and his firefighter. I used the PA system on my BC buggy to hail him and ordered him to evacuate. They were approximately 200 feet away. He began to take up hose with his firefighter. I told him to drop the hose and run to me. As they did, the fire front swept over the area and they narrowly escaped.

This was an example of the NIOSH 5 coming into alignment on a grass fire. The fire was large enough to require units from far away that were not accustomed to this type of firefighting. This can be an added risk to your operations if personnel are not trained, equipped, or experienced in these types of fires.

Like all incidents in this textbook, we will focus on the Type 4 and 5 scope of vegetation fires, also known as initial attack. These can easily grow to include larger vegetation fires from first to greater alarm, known as reinforced response, and sometimes extended attack where the fire may transition to a Type 3 (operational area/county-level). Type 3 would be indicated by a size that encompasses multiple jurisdictions and/or multiple operational periods wherein a Type 3 IMT would respond for additional operational, logistical, and planning support.

Like most fires, vegetation fires start small, and most are caught before they reach devastating levels. That said, trends have shown that these fires have reached unprecedented levels of death, destruction, and frequency compared to decades past.

The Sacramento Metropolitan Fire District (Metro Fire) is located directly north of the San Joaquin Valley of Central California. Metro Fire has vegetation fires throughout the summer that are affected by high heat, Delta breezes, and low humidities. Battalion 7 is directly adjacent to all the battalions at Metro Fire. This means a lot of fires in the summers with neighboring battalions. Some are roadside grass fires, some are thousands of acres and burned structures within our jurisdiction, and many are everywhere in between, from one to hundreds of acres on any given summer day. Typical tactics consist of pump-and-roll mobile attack, progressive hose lays, use of hand tools, dozers, and support of helicopter operations (fig. 13–1). On the north and east ends of the Metro Fire District boundaries are state and mutual threat zones that cause automatic and mutual aid responses with CAL FIRE on a regular basis throughout the summer fire season.

Simultaneously, Metro Fire crews respond to statewide wildfires as part of the California State Master Mutual Aid system. This program, managed through the California Office of Emergency Services (CAL OES) has been in place for decades and is the means by which hundreds of resources and thousands of firefighters respond all over the state to support each other in the cities and communities throughout California. In addition, like many agencies, several Metro Fire members respond as part of IMTs and/or part of the mutual aid system to fill individual overhead positions as needed.

Fig. 13–1. A Sacramento Metro Copter working a fire.
Source. Photo courtesy of Sacramento (CA) Metropolitan Fire District

Definitions

Terminology in the vegetation/wildland fire arena is relatively universal. That said, there are several authorities that have similar definitions and we have referenced several different sources to expose you to various entities. We hope this expands your awareness.

Vegetation Fires involve dry grasses, shrubs, and light flashy organic fuels. These can range from roadside grass fires to large vegetation fires that engulf thousands of acres and structures with lives threatened. *Wildfires* are typically in larger forested areas and can involve national/state parks, national forests, private land, logging areas, and may also encroach on populated areas. In recent years, wildfires have devastated entire communities with catastrophic life loss. The WUI is the area in which communities, neighborhoods, homes, and other buildings exist in regions of heavy vegetation growth and more mountainous topography. Vegetation fires, wildfires, and those in the WUI are more lethal than ever. The potential for these fires to grow rapidly and exponentially has caused unprecedented death and destruction. That said, WUI can happen on smaller Type 5 and Type 4 fires in communities and neighborhoods requiring access through properties, fences, yards, and driveways (fig. 13–2).

Glossary

Anchor point: The location from which a fire attack is initiated.
Black: Fuel already burned.
Control objectives: Geographic boundaries from which to stop the fire.
Direct attack: Also known as an offensive strategy, this occurs when firefighters remove the flames from the fuel by directly extinguishing the fire with water, usually with a mobile attack or progressive hose lay.
Escape routes: A path that firefighters utilize to get to a safety zone or away from the fire.
Extended attack: That phase of wildland fire management when initial attack resources are no longer capable of achieving the objectives (usually control) set for the incident, and

Fig. 13–2. Grass fires can threaten structures in first-due areas.
Source. Photo courtesy of Sacramento (CA) Metropolitan Fire District

additional resources and overhead are committed, generally extending incident activities into another operational period.[1]

Firing operations: A defensive tactic whereby fire is placed on the ground by qualified firefighters to remove fuel ahead of the fire.

Flanks: Outer left and right edges of the fire line.

Fuel: Any material that burns (grass, brush, trees, or manmade materials).

Green: Fuel not yet burned.

Head: Leading fire edge.

Heel: Area of fire origin, where the base of the fire is located.

Indirect attack: Also known as a defensive strategy, this occurs when the fuel is removed from the fire. Examples would be cutting lines or firing operations.

Initial response/attack: Resources initially committed to an incident.[2]

Kind: A classification of resources in the incident command system which refers to function; e.g., hand crew, helicopter, engine, dozer.[3]

LCES: Lookouts, communications, escape routes, safety zones

Mobile attack: Also known as pump-and-roll, this occurs when apparatus is capable of driving and pumping water simultaneously.

Progressive hose lay: A wildland firefighting tactic whereby hose is deployed and extended continuously in successive sections, from the nozzle, for as long as necessary to contain the fire (fig. 13–3).

Rate of spread (ROS): The speed in which a fire grows away from the area of origin. The rate is usually measured in chains per hour. One chain equals 66 feet. There are several means to measure ROS. In the field, one simple method from the Southwest Fire Science Consortium is to estimate by an average person walking or running. A slow ROS would be slower than a walk. A moderate ROS would the average person walking. A rapid ROS would be a fast walk. A critical ROS would be an average person running, also characterized by spotting.

Red-flag conditions: Weather conditions in which there is a greater potential for vegetation fires igniting and growing rapidly due to high heat, low humidity, and high winds.

Reinforced response: Those resources requested in addition to the initial response.[4]

Safety zones: A location to which firefighters can escape if conditions worsen rapidly.

Spotting: Embers casting out ahead of the fire, creating new fires, causing rapid fire growth.

Strike team: A team of five of the same kind and type of resource with a leader that acts as one resource.

Tanker: A fixed-wing aircraft designed to drop water on wildland fires.

Fig. 13–3. Crews working off a progressive hose lay.
Source. Photo courtesy of Rick McClure

Task force: A team of different kinds and types of resources with a leader, assembled for a specific task/mission/objective that acts as one resource.

Type: A classification of resources in the incident command system which refers to capability. Type 1 is generally considered to be more capable than Types 2, 3, or 4, respectively, due to size, power, capacity, or in the case of incident management teams, experience, and qualifications.[5]

Values at risk: Any item of value that could be threatened by a fire, including but not limited to structures, outbuildings, vehicles, utility infrastructure, fencing, stored materials, livestock, and so on.

Water tender: A mobile ground apparatus comprised of a large tank of water with the objective of providing remote water supply to other resources.

CASE STUDIES IN COMMAND

The Sand Fire

CHIEF MARK BRUNTON, CAL FIRE

The Sand Fire started in late July 2016 in the area of Sand Ridge Road and Hwy 49 in El Dorado County, California at 17:10 hours. The weather at the time of the dispatch was 101 degrees Fahrenheit, 12% relative humidity, with winds from the South at 7 mph. The response to this area from CAL FIRE called for six Type 3 engines, four Type 1 fire hand crews, two Type 2 dozers, two battalion chiefs, one air tactical group supervisor (ATGS), two Type 1 airtankers, and one Type 2 helicopter with crew. Additionally, four local government engine companies, three water tenders, a medic unit, and a chief officer were part of the initial response. In California, CAL FIRE has the jurisdictional responsibility for vegetation fires on State Responsibility Area (SRA) land. This is essentially any non-federal land.

Upon arrival of the first engine company, they reported a fire of approximately 10 acres in grass/oak-woodland fuels moving uphill at a moderate-to-rapid rate of spread (fig. 13–4).

At this point, there were no structures threatened. The engine company officer took IC and personnel started a progressive hose lay. The engine company officer kept the IC role until the first-in CAL FIRE chief officer arrived a few minutes later. The update

Fig. 13–4. The Sand Fire
Source. Photo courtesy of CAL FIRE

from the chief officer after a transfer of command was that the fire was at approximately 20 acres, continuing to move uphill with the same rate of spread.

I was dispatched as the second CAL FIRE chief officer. Additionally, more CAL FIRE battalion chiefs as well as a CAL FIRE assistant chief responded. Due to the location of this fire, fuel, and weather conditions, it was anticipated that this fire would grow significantly and go beyond initial attack and into the next burning period.

Upon arrival I was assigned as operations. An augmented resource request of ground resources was ordered by the IC. Additional air resources were ordered by the ATGS upon seeing the view from over 20 air miles away of a significant growing column (fig. 13–5).

A unified command structure was established between CAL FIRE, the local government fire district in which the fire was burning, and the El Dorado County Sheriff's Office (EDSO), who is responsible for evacuations and coordination of road closures with the California Highway Patrol (CHP).

At approximately 17:30 hours the fire had jumped the Consumnes River, which is the county line between El Dorado and Amador Counties. This prompted the involvement of additional resources from Amador County, as well as the Amador County Sheriff's Office (ASO).

The observed fire behavior was erratic and increased in intensity and rate of spread as it moved from the original fuel type into brush and light timber. Significant long-range spotting, as well as group torching of timber, attributed to the spread (fig. 13–6).

The IC and I established the initial incident organization, which started as 3 Divisions and then grew eventually to a two-branch, eight-division organization. Control and management objectives were established. Evacuation and road closures were decided upon and initiated by the law enforcement agencies.

Due to the difficult and limited access points along with the fast-moving fire, assignment of resources resulted in long distances that needed to be traveled in order to get to their assigned areas. At approximately 18:45 hours the first report from the ATGS

Fig. 13–5. CAL FIRE Air Base in Sacramento, CA

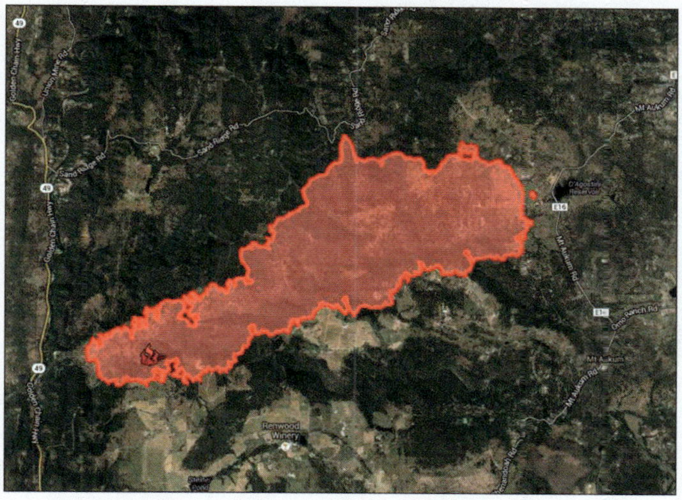

Fig. 13–6. Outline of Sand Fireline
Source. Photo courtesy of CAL FIRE

was that a structure was now involved. The area initially that was rural and had no WUI now moved into an area dotted by residential structures as well as various outbuildings, and eventually threatened numerous wineries in the Amador County side of the fire.

At approximately 19:30 hours, the blaze was being battled by 6 air tankers, 100 engines, and more than 10 hand crews, as well as 5 dozers (fig. 13–7). From 19:00 hours until 19:30 hours, the fire went from 600 acres to 1,300 acres, more than doubling in size in 29 minutes. Plans were being set for the fire to burn into the next operational period. Additional resource and logistical requests were made for that next day.

Suppression continued to take place throughout the evening. An update of acreage and containment at 23:00 hours was at approximately 2000 acres, 20% contained, and 2 structures destroyed. The fire at that point was in a large river canyon and still in two counties. Weather conditions improved only slightly, and active fire was observed throughout the nighttime and early morning hours. Ground resources continued to battle the fire and air resources had ceased operation due to nighttime restrictions at approximately 20:00 hours (fig. 13–8).

Fig. 13–7. An air tanker laying down a retardant drop
Source. Photo courtesy of CAL FIRE

Fig. 13–8. Night operations at the Sand Fire
Source. Photo courtesy of CAL FIRE

July 27 initially saw limited growth of the fire in the overnight. Predicted weather on that day was calling for high temperatures with rising humidity, and thunderstorms east of the fire, which were predicted to cause erratic winds. As the fire started into the second operational period that day, fire intensity increased, which challenged the control lines. By midafternoon the fire had taken off again. Air resources continued to fly and reinforce lines and attempted to catch multiple spot fires. Reports of additional structures involved were reported and ground resources were sent from staging areas or as soon as they checked in to the incident.

By midday, it was decided to request a CAL FIRE Type 1 IMT. It would take approximately 24 hours for the team to arrive and be briefed. In the meantime, the local regional Type 3 IMT integrated with the CAL FIRE unit incident structure to manage the fire.

With the fire progression, a firing plan was devised to take advantage of the change in topography and fuel type to slow the forward progression of the fire. A deputy operations chief was given a CAL FIRE Type 3 engine strike team, a dozer, crew, and supported by aircraft to conduct the operation as the fire advanced towards County Road 16. The plan was approved by me as the operations section chief and resources briefed and the operation conducted. The firing operation was successful and greatly reduced the forward progress of the fire before it breached the control objective of crossing County Road 16.

On July 28, a transition was completed with the CAL FIRE Type 1 IMT. Most active fire progression had been stopped, yet greater planning and logistical support was required to fully contain the fire. On July 28, the fire was contained at 75%. It had burned 3800 acres and destroyed 13 structures and 38 outbuildings. No lives were lost.

Some challenges of this fire from the initial attack were a fire that had prime burning conditions in terms of weather, fuel, and topographical alignment. Poor access, as well as natural barriers such as a river where the fire jumped, made it difficult to effectively deploy resources. The fire, which also burned in two separate counties, created a need for a larger unified command structure. Rapid fire growth over two operational periods with limited resources to staff the incident caused many initial attack resources to be double-shifted and therefore reduced the effectiveness of those resources.

Successes of the incident can be attributed to experienced company officers who were aggressive and tactically proficient. Experience within the chief officers assigned to the incident accounted for the development of a quickly designed and executed strategy and deployment of resources at critical points of the incident, and allowed for the ability to re-adjust objectives as needed.

Fundamentals of Vegetation Fires

Like any incident, the key is to stay ahead of the incident power curve. Fuel (light/flashy, medium, heavy), weather (relative humidity, temperature, and wind), and topography are the three factors that drive the size and complexity of a vegetation fire (fig. 13–9).

Preplanning, including access points and weather predictions, are vital to being prepared on any given shift. Depending upon the AHJ, a red flag warning may be in place, which could

Fig. 13–9. Crews performing pump-and-roll

Source. Photo courtesy of Sacramento (CA) Metropolitan Fire District

enhance your response levels. All these should be part of the incident commander's mindset and situational awareness at the beginning of each shift.

Initial attack response resources to vegetation fires in local jurisdictions are varied. You may have many of the kind and type of resources you need, or you may have to make structural resources work (Type 1 engines, trucks, squads, medics) (fig. 13–10). The response can be limited or mixed, with many resources from mutual/automatic aid agencies. This requires you, the incident commander, to order the appropriate resources ASAP, by specific kind and type.

Calls from reporting parties can be specific (at the corner of A and Z Streets), general (smoke seen from miles away), or come from a combination of the two. You may have homeowners calling because they see smoke, others who have direct flame impingement, and still others who got a call from a loved one across town.

Fig. 13–10. Be specific when ordering resources. *Source.* Photo courtesy of Mike Meadows

Like a structure fire, a working vegetation fire will likely produce a visible column of smoke that could be obscured by wind; however, unlike a structure fire, a vegetation fire may have grown to several acres before you arrive. This creates a challenging situation to size up the fire and determine the size, rate and direction of spread, and values at risk. Consider units responding from different directions to give you a report from their vantage point. The fire may be better anchored and fought from a different location than which you arrived on scene or were dispatched to.

Vegetation fires in local jurisdictions usually start as Type 5 or 4 (initial attack/reinforced response) before getting to a Type 3 or larger status and are described in terms of size, rate of spread, fuel type, potential, and values at risk. Some considerations for each include, but are not limited to the following:

- **Size:** Can be described by acreage. For rough estimates, a football field is roughly 1.32 acres. Other descriptions can include "roadside grass fire" or "100 × 100-foot grass fire," and so on.
- **ROS:** Slow, moderate, or rapid. If you have spotting (embers igniting ahead of the fire), then consider the fire moving at a critical ROS.
- **Fuel type:** There are thirteen fuel models in four fuel groups (grasslands, shrublands, timber, and slash) for more sophisticated fuel modeling and fire behavior predicting.

These are typically for larger Type 1–3 incidents. Type 4 or 5 incidents can be described as light/flashy, medium, or heavy fuels.

- Light/flashy fuel (dry grass) is considered the most dangerous to first responders, as the fuels burn very rapidly and can create enough heat to cause major injuries, death, and damage to equipment.
- Medium fuel would be ¼" to 3" diameter and more densely spaced. Examples would be Manzanita and heavy brush.
- Heavy fuel would be timber with laddering fuels from the ground to the crown, for example.

- **Potential:** Potential is your estimate as to the ultimate size that the fire can grow before it is contained.
- **Values at risk:** Homes, other structures, livestock, vehicles, power distribution, and any other assets that the fire could engulf or damage. Do not forget, with the right conditions, modern fires can consume entire communities and kill scores of civilians.

While performing your FPODP size-up, ask yourself, "Where is the fire going, and what am I saving?" The primary overarching objective for any vegetation fire is to put a box around the fire by containing it as soon as possible before it gets into values at risk like homes. Putting a box around the fire means identifying the perimeter of the fire and establishing control objectives as soon as possible.

For example, a fire may start at the intersection of Gerber and Waterman Roads and move into a vacant field with light/flashy fuel, moving at a rapid ROS (fig. 13–11). By the time the first units arrive on scene, the fire may be threatening the next road to the north (Rondeau Way), heading directly toward homes. The box would be the four sides of the fire from which crews will contain it: Keep the fire north of Gerber, south of Rondeau, west of Waterman and east of Souffle.

This can be accomplished by anchoring the fire at the heel and beginning a direct attack upon the flames, by either mobile attack or progressive hose lay. There will be a left flank, a right flank, and the head of the fire. One of the two flanks will be moving faster and

Fig. 13–11. An example of using roads to put a box around the fire

considered the hot flank that is downwind of the black and moving in the direction of the wind. The head (front) of the fire is often obscured by smoke on larger fires and difficult to see.

Tandem attack is composed of two mobile attack engines working in line, one behind the other, usually beginning on the hot flank. This allows the first engine to move rapidly toward the head of the fire, while the second back-up engine ensures complete containment and that no flare-ups occur. This is just like a primary fire attack line being followed up with a back up line in a structure fire. The next two engines would initiate attack on the cold flank that is upwind of the black.

Another option is to split the head of the fire when the fire is moving at a rapid ROS and about to involve homes or other significant values at risk. In this situation, the first two mobile pumping units would drive through the black directly to the head of the fire, anchor off each other, split the head, and work their way back the left and right flanks toward the heel. This would be supported by other units performing structure defense from the street side of the homes

Vegetation fires commanded in the urban arena will often follow best practices from the National Wildfire Coordinating Group (NWCG). Other state-specific resources like FIRESCOPE provide excellent field operations guides (FOGs) and direction, such as the ICS-420-1 FOG.

Once a fire goes to a Type 3 or above, more specific *wildfire* terminology, tactics, and resources are utilized. However, many of these components can and are utilized by urban fire departments that have this threat when fires are in the initial attack (Type 5) or reinforced response (Type 4) phase of growth.

Safety

General safety for vegetation/wildland fires includes the acronym LCES (Lookouts, Communications, Escape Routes, and Safety Zones). The incident commander and responding troops must be cognizant of LCES at all times, even in seemingly benign fires.

Whereas a lookout on a large wildland fire would be a more formal assignment, a lookout on an initial attack vegetation fire can be the incident commander, division supervisor, safety officer, airborne resources, company officer, or all of the above. The key is to expect the unexpected and respect these initial attack fires. That is why communications are so critical to get established early.

An escape route to a safety zone on an initial attack fire may be a dirt road, or simply getting back into the black (burned area) during a mobile attack. When the fire is extending into a more complex incident, the incident commander should ensure that the division/group supervisors on the fire have identified escape routes and safety zones within their respective areas of operation.

The NWCG has developed fire orders and watch-out situations for vegetation and/or wildfires. These are not just for large Type 1–3 incidents. These are for Type 4 and 5 initial attack, reinforced response, and extended attack fires as well. Ensure your troops are well trained in the following areas.

10 Standard Fire Orders

1. Keep informed on fire weather conditions and forecasts.
2. Know what your fire is doing at all times.
3. Base all actions on the current and expected behavior of the fire.

4. Identify escape routes and safety zones and make them known.
5. Post lookouts when there is possible danger.
6. Be alert. Keep calm. Think clearly. Act decisively.
7. Maintain prompt communications with your forces, your supervisor, and adjoining forces.
8. Give clear instructions and be sure they are understood.
9. Maintain control of your forces at all times.
10. Fight fire aggressively, having provided for safety first.

18 Watch Out Situations

1. Fire not scouted and sized-up
2. In country, not seen in daylight
3. Safety zones and escape routes not identified
4. Unfamiliar with weather and local factors influencing fire behavior
5. Uninformed regarding strategy, tactics, and hazards
6. Instructions and assignments are not clear
7. No communication with your company or supervisor
8. Constructing line without a safe anchor point
9. Building fire line downhill with fire below
10. Attempting a frontal assault on the fire
11. Unburned fuel between you and the fire
12. Cannot see the main fire and not in communication with someone who can
13. On a hillside where rolling material can ignite material below
14. Weather is getting hotter and drier
15. Wind increasing or changing direction
16. Getting frequent spot fires across the fire line
17. Terrain and fuels make escape to safety zone difficult
18. Taking a nap near the fire line

Structure Defense Watch-Out Situations

Often, structures will be at risk in the WUI where metropolitan fire departments work. These require a distinct set of safety and tactical considerations. The WUI has a separate and distinct set of watch-out situations to augment the 18 WUI watch-out situations:

- Poor access and narrow, one-way roads
- Bridge load limits
- Wooden construction and wood shake roofs
- Power lines, propane tanks, and hazmat threats
- Inadequate water supply
- Natural fuels 30 feet or closer to structures
- Structures in chimneys, box canyons, narrow canyons, or on steep slopes (grade 30% or more)
- Extreme fire behavior
- Strong winds
- Evacuation of public (panic)
- Underground utilities threat
- Structural collapse zone when structures are exposed to fire
- Smoke byproducts often laced with chemical compounds not found in pure wildland fires

According to the FIRESCOPE Field Operations Guide (ICS 420-1, 2022 ed.), the following structure defense tactics are options for consideration as structures become threatened:

- **Tactical evacuation**—in areas and situations with imminent life threats. As a result, fire personnel will switch from suppression to evacuation and assist law enforcement.
- **Check and go**—a rapid evaluation to check for occupants requiring removal or rescue:
  - Structure triage category—threatened non-defensible.
  - This tactic is most appropriate when there is no safety zone or temporary refuge area (TRA) present and the forecasted fire spread, intensity, and the projected impact time of the fire front prohibit resources from taking preparation action to protect the structure.
  - Complete a rapid evaluation to check for occupants at a structure, evaluate life threat, and to assist in evacuation.
  - Used when fire spread, intensity, lack of time, or inadequate defensible space prohibit firefighting resources from safely taking action to protect the home when the fire front arrives.
  - Evaluate the structure for follow-up action when additional resources become available, the fire front passes, or fire behavior intensity is reduced.
- **Prep and go**—implies that some preparation of the structure may be safely completed prior to resources leaving the area:
  - Structure triage category—threatened non-defensible.
  - A tactic used when a safety zone and TRA are not present and when fire spread and intensity are too dangerous to stay in the area when the fire front arrives, but there is adequate time to prepare a structure for defense ahead of the fire front.
  - Utilized for structures where potential fire intensity makes it too dangerous for fire resources to stay when the fire front arrives.
  - There is some time to prepare a structure ahead of the fire; resources should engage in rapid, prioritized fire protection preparations and foam the structure prior to leaving.
  - Resources should leave with adequate time to avoid the loss of escape routes.
  - Advise residents to leave and notify supervisors of any residents who choose to stay so that you can follow up on their welfare after the fire front passes.
  - As with check and go, prep and go is well suited for engine strike teams and task forces.
- **Prep and defend**—a tactic used when a safety zone and TRA are present and adequate time exists to safely prepare a structure for defense prior to the arrival of the fire front:
  - Structure triage category—threatened defensible.
  - An ideal multiple-resource tactic, especially in common neighborhoods where efforts may be coordinated over a wide area. A tactic used when it is possible for fire resources to stay when the fire front arrives. Fire behavior *must* be such that it is safe for firefighters to remain and engage the fire.
  - Adequate escape routes to a safety zone must be identified. A safety zone or TRA must exist on site.
  - Firefighters must be vigilant to sudden changes in fire intensity and be prepared to move to the TRA or withdraw along the escape route to the safety zone.
  - Adequate time must exist to safely prepare the structure for defense prior to the arrival of the fire front.
- **Fire front following**—a follow-up tactic employed when check and go, prep and go, or bump and run tactics are initially used:
  - A tactic used to come in behind the fire front.
  - This action is taken when there is insufficient time to safely set up ahead of the fire or the intensity of the fire would likely cause injury to personnel located in front of the fire.

  - The goal of fire front following is to search for victims, effect perimeter control, extinguish spot fires around structures, control hot spots, and reduce ember production.
- **Bump and run**—a tactic where resources typically move ahead of the fire front in the spotting zone to extinguish spot fires and hot spots, and to defend as many structures as possible:
  - Bump and run may be effective in the early stages of an incident when the resource commitment is light and structure defense is the priority.
  - Bump and run may also be used on fast-moving incidents when there are adequate resources available, but where an effort must be made to control or steer the head and shoulders of the fire to a desired end point.
  - Perimeter control and structure defense preparation are secondary considerations with the bump and run tactic.
  - Resources must remain mobile during bump and run and must constantly identify escape routes to safety zones and TRAs as they move with the fire front.
  - Bump and run is a defensive tactic when fire front impact in the WUI is imminent and there are not enough resources to effectively take perimeter control action. It is an offensive tactic when resources are steering the head of the fire to a desirable end point.
  - The tactic is useful when terrain and fuels are suitable for mobile attack.
  - Fire line supervisors and strike team/task force leaders must realize that bump and run places resources in front of the advancing fire front and that extreme caution should be exercised.
  - Control lines in front of the fire should be identified and prepared with dozers and fire crews, enabling the bump and run resources to direct the fire to a logical end point. This is a frontal attack strategy and a watch-out situation. Control lines in front of the main fire must be reinforced with retardant drops, coordinated firing operations, and engine support.
- **Anchor and hold**—anchor and hold is a tactic utilizing control lines and large water streams from fixed water supplies in an attempt to stop fire spread. The goal is to extinguish structure fires, protect exposures, and reduce ember production.
  - Anchor and hold can be referred to as taking a stand to stop the progression of the fire.
  - Anchor and hold tactics are more effective in urban neighborhoods where the fire is spreading from house to house.
  - Establishing an anchor and hold line requires considerable planning and effort and utilizes both fixed and mobile resources:
    - Fixed engines should be spotted in safe areas where they can safely withstand any fire situation.
    - Mobile engines or task forces can engage in individual structure defense actions or perimeter control and re-supply from a fixed water source.
    - Mobile engines should be prepared to redeploy to other areas should the fire escape the anchor and hold line.
    - Ground resources, such as engine crews and fire crews, should staff hoselines and be prepared to extinguish hotspots, fire perimeter, and structures.
    - Hand-crew strike teams should be deployed to construct fire control lines wherever needed and conduct firing operations.
- **Tactical patrol**—a tactic where the key element is mobility and continuous monitoring of an assigned area:
  - Tactical patrol can be initiated
    - After the main fire front has passed and flames have subsided, but when the threat to structures still remains;

  - In areas where the fire has passed but the risk to structures remains from fire brands smoldering in void spaces, on roofs, and in rain gutters and stored material near buildings; or,
  - In neighborhoods away from the interface where there is predicted to be significant ember wash and accumulated ornamental vegetation.
- The goal is to patrol areas downwind of potential ember showers.
- This tactic should be used to extinguish hot spots or secondary structure ignitions and address safety issues such as power lines, weakened trees, and other hazards.
- Vigilance, situational awareness, and active suppression actions are a must.[6]

SAW-CSS-RECEO-VSS for Vegetation Fires

The SAW-CSS-RECEO-VSS system has been used nationwide for years. You will notice that *italicized words correspond with key points on the tactical worksheet* that follows.

Size-Up

Vegetation fires have unique size-up considerations that are very different from structure fires. The good news is that you can still use your FPODP system for mental size-up when you arrive on scene. At the beginning of your shift, look at weather patterns and expected wind events that could create red flag conditions. Remain vigilant about weather changes.

- *Facts* identify the size of the fire, type of fuel, rate of spread, and direction the fire is moving. (Three-acre grass fire in light flashy fuels, moving at a rapid rate of spread to the north.)
- *Probabilities* identifies the potential size to which you estimate the fire could grow, and any potential values at risk, such as structures, livestock, electrical distribution, vehicles, and so on. (Potential for 30 acres, houses threatened to the north off Main Street.)
- *Own situation* determines the number, kind, and type of resources you have on scene and responding. (Two Type 3 engines on scene, one Type 1 engine, and one water tender responding.)
- *Decision* determines your priority and strategy, just like any other incident (life saving, incident stabilization, or property/environment priority; offensive, combination, or defensive strategy). See section on strategy for more information below.
- *Plan*, as with all incidents, determines your tactical and control objectives. (Initiate direct attack on the left and right flanks. Keep the fire south of Main Street, east of 1st Street, and west of 2nd Street.)
- *Four sides* refers to knowing the left, right flanks, base, and head of the fire. The *box* is the 4 sides or control objectives in which you intend to confine the fire.
- *Values at risk*, as described above, must be identified early so that subsequent resources can position ahead of the fire to defense.
- *Fuel* is either light (dry grasses), moderate (brush and heavier vegetation), or heavy (densely spaced fuels including heavy brush, ladder fuels, and trees).
- *Weather* has three key factors that drive a vegetation fire: relative humidity (RH), temperature, and wind. RH, temperature, and sustained wind speed and direction should be

identified at the beginning of shift and monitored throughout the day, so you will know it upon dispatch.

- *Topography* is the amount of grade, or level of incline and steepness of the land. The steeper the topography, the faster the fire will travel.
- End your mental size-up with an *IOCAN* arrival report. For example, "E303 is on scene at the corner of 1st and Pearl Streets. Three-acre grass fire in light flashy fuels, rapid rate of spread to the north. Potential for 30 acres, houses threatened to the north off Main Street. E303 is initiating direct attack on the right flank. E303 has Pearl Command. Command is mobile. E305 come behind E303 for tandem attack to catch the head of the fire. E106 go to Main Street for structure defense." Notice that we didn't say "moving with the wind" because that gives the rest of the alarm no frame of reference from where you are. Announce the direction in terms of the compass: "Moving to the north. . . "
- Order additional resources as soon as possible. With this arrival report, how many additional resources would you order and what type? Would you have any go directly to Main Street for structure defense?

Apparatus Placement

Unlike a structure fire at a single address, a vegetation fire can be spread over several miles. Once the fire is anchored by first-in crews, subsequent resources may arrive and be assigned to initiate attack from several different locations, based upon the size of the fire, values at risk, and conditions.

- The hot zone on a vegetation fire is different than on a structure fire. Resources working on the fire must be cognizant of being in the fuel, wind shifts, and falling trees and powerlines.
- Remain in the black or burned-out area as much as possible. On a grass fire as a captain, my firefighter (on overtime) and I were attacking the fire with a hoseline, and the wind shifted. The smoke enveloped us, blinding us for several seconds. The sound of the flames intensified. He started to panic and yell for me. I told him that we were safe because we were in the black and to "stay calm and stay put!" Then the wind shifted and we were okay.
- Keep apparatus and your attack position upwind when possible.
- Consider law enforcement early to assist with traffic coming into the scene.
- If needed, identify a *staging* area for additional apparatus. This may not be needed, as resources will likely go right to work during initial attack (fig. 13–12).

Water Supply

Water supply will likely be a water tender, but it could be a nearby hydrant(s), or even a body of water that you must draft. Tank transfers from large Type 1 engines are also an excellent resource if you have Type 3 or smaller engines that can nurse from them.

- Who is securing the water supply? Where is/are it/they located? What is it (hydrant, water tender, etc.)?
- Once a patent water supply is established, it should be announced on all channels.
- Consider a water supply group supervisor if you have a complicated water supply plan and, depending upon the amount of radio traffic, you may consider giving them a separate tactical channel.

Command

Command must be established and announced ASAP. *There can only be one IC at a time, and the IC must be on scene to assume command and must name the incident.* While your SOGs will determine specific parameters, some fundamental considerations exist.

Fig. 13–12. For more apparatus at the scene, identify a staging area. *Source.* Photo courtesy of Mike Meadows

- Command may be passed to the next company officer, or first chief officer, depending upon your SOGs and the size and scope of the fire.
- The benefit of the first-arriving company officer establishing and retaining command (until the first chief arrives) is that they can develop an IAP since it may be easier to see more of the fire and what values are at risk early. And, subsequent units can be assigned to different locations, if needed. Also, additional resources can be called ASAP.
- As with any incident, the first-due company officer will contend with tasks. This could include initiating a progressive hose lay, mobile attack, or structure defense.
- Once a chief officer or permanent IC has arrived, the *ICP location* should be announced; however, it may be mobile or move during the incident.
- Call *additional alarms/resources* early and confirm staging, or response locations that may be different from the original location.
- Set up your *ICS* early. Unlike interior structural firefighting, the first officer to an area can assume the division supervisor. The visibility and situational awareness regarding the best tactics and assignments must be made early and the front of a flank typically has the best view of where the fire is going and what is threatened.
- Develop a *communications plan* to prevent overtaxing a single tactical or command channel. Add additional command and tactical frequencies ahead of the incident.

Strategy

On a vegetation fire, you still have the same three choices for priority and strategy: life saving, incident stabilization, property/environment priority; and *offensive, combination, or defensive strategy.*

- An offensive strategy would be indicated when crews are performing direct attack on the fire with water. The objective is perimeter control by extinguishing the flames (removing the fire from the fuel).
- A defensive strategy would be indicated when crews are performing an indirect attack with containment lines (removing the fuel from the fire—back-burning, firing out, dozer lines, hand-crews, air drops) as seen in figure 13–13. Or the primary tactic is structure defense, as described above in the FIRESCOPE Structure Defense Guide for the WUI.
- A combination strategy would utilize both offensive and defensive tactics simultaneously. I have utilized a combination strategy many times due to threatened structures, rapid rate of spread, and resource types. For example, we had a fast-moving grass fire in light, flashy fuels with structures threatened in front of the head. While mobile attack units performed direct attack on the flanks, structural units, including truck companies, performed structure defense ahead of the fire. In an urban setting with limited wildland engines, this can be a very effective combination. Air support is a major tactical advantage (fig. 13–14).

Fig. 13–13. Crews performing firing operations. Do not attempt unless trained and approved by the IC.
Source. Photo courtesy of Sacramento (CA) Metropolitan Fire District

Fig. 13–14. Air support is a tactical advantage.
Source. Photo courtesy of Mike Meadows

Safety

Risks for vegetation fires in the wildland and WUI are unique and must be trained upon and taught to all firefighters who may respond to such incidents.

- *LCES:* lookouts, communications, escape routes, and safety zones. As described above, LCES is the minimum and usually the initial set of safety considerations.
- The 10 Standard Fire Orders are universal and should be applied to any incident. Additional safety considerations like the 18 Watch Out Situations or WUI Watch-Out Situations will be incident-specific and change as the incident grows and becomes for complex.
- *Power lines* are a common hazard. Often, power poles, lines, and other distribution components become involved. In addition, lines can fall and remain energized. Smoke can also cause arcing of power lines.

- *Snags* are burned-out trees that can fall suddenly.
- Remain vigilant with head up watching out for lines, snags, smoke, and wind shifts, and *spotting* ahead of the fire. Spotting indicates the fire is growing rapidly ahead of you and may not be caught with a direct initial attack. Consider structure defense or containment lines ahead of the fire.

Rescue

Remember, the victim profile drives *everything*. This is still true of vegetation fires; however, you will likely have more discretionary time during the early stages of initial attack, and reinforced response phases to evacuate and perform structure defense.

- As stated previously, large wildfires (Type 1 and 2 incidents) can claim many lives.
- For the scope of this book, the Type 4 and 5 incident level, the victim profile on a vegetation fire is usually low to moderate; however, you may have occupied dwellings ahead in the path of the fire during initial attack, especially at night.
- *Evacuation* of entire communities is an excellent tactic that can be assigned to law enforcement, when available, during a fast-moving fire.
- Structure defense automatically implies that evacuation of the defended dwellings is occurring; however, do not take this for granted. Crews may overlook evacuation if the home seems abandoned, or the occupant may refuse to leave.
- *Medical treatment and transport* functions should be established early under the direction of a medical group supervisor (MGS). Ensure you have advanced life support and transport capability on scene at all times, including ALS treatment in staging.
- Firefighters can get heat exhaustion during prolonged vegetation fires in extreme heat. Ensure your crews begin hydrating early in the day. I had an engineer who routinely "hydrated" with soda or coffee in the morning. He went down on two separate fires. Not a coincidence.

Exposures

Exposures on a vegetation fire can be any values at risk. Values can come in all shapes, sizes, and quantities that may or may not seem valuable to you. Some of these items can be extremely valuable to the owners, including, but not limited to:

- Structures, including homes, outbuildings, sheds, barns, and businesses. This can include vehicles, trailers, campers, and semi-trucks and trailers.
- Livestock, including cows, horses, sheep, goats, and any combination of farm animals.
- Infrastructure, including power poles/lines, transformers, distribution stations, phone, cable, water tanks, pump stations, propane tanks, gas lines, fencing, rail trussells, and bridges.
- Other values can include pallets, baled haystacks, agricultural/farm equipment, old, rusted vehicles, and junkyards.
- Roadways which may be impacted by fire or smoke conditions must also be considered as exposures and requests for traffic control should be made early.

Confinement

Unlike a structure fire, containing a vegetation fire takes more specific planning, communication, coordination, and support. Even Type 4 and 5 vegetation fires can be hundreds of acres in size or more. *Control objectives* must be identified by the IC in conjunction with the division and group supervisors to put a box around the fire.

- Roadways, rivers, creek beds, dozer lines, and handlines can all be utilized for control objective locations.

Extinguishment

The fire may be contained; however, it may not be extinguished. Smoldering fuels along the perimeter can easily re-ignite, or wind can change, causing rekindle and spotting.

- Extinguishment means the perimeter is cold. One rule of thumb to prevent rekindle is to have the fire cold with no smoking fuels (smokers) *100 feet in from the perimeter.* If winds are erratic or expected, this distance should be increased.

Overhaul

Overhaul is dependent upon the weather conditions and surrounding values at risk.

- *Mopping up* a vegetation fire is the act of overhaul and extinguishing additional hot spots and smoking fuels inside the burned area, or black (fig. 13–15).
- If weather conditions are not favorable, with erratic wind behavior, the fire may need to be made cold with overhaul throughout the entire burned area to prevent rekindle or spotting.
- The IC has the sole responsibility to declare the fire contained, and then extinguished before leaving the incident. Do *not* take this responsibility lightly. I would always drive the entire perimeter of every grass fire I ever commanded. Depending upon the weather, I would either drive the center or assign crews to patrol, especially at night, for flare ups.

Ventilation

While fuel, weather, and topography all influence fire behavior, weather is the most influential and most dynamic. At beginning of shift, you should already know the fuel types and topography in your area. Weather predictions can change. Temperature, RH, and wind all contribute to fire behavior, with the most influential being wind. Always watch for weather changes, especially wind.

Fig. 13–15. Crews mopping up
Source. Photo courtesy of Sacramento (CA) Metropolitan Fire District

Salvage

Salvage may happen on a vegetation fire, although it is not common. You may have a building or storage unit that could have contents to be salvaged.

Support

Vegetation fires will require an immense amount of support. As a minimum, consider the following:

- Law enforcement (LE)
- Public information Officer (PIO)
- Incident safety officer (ISO)
- Public information officer (PIO)
- Liaison officer (LO)
- Red Cross
- Utility company
- Rehab and medical unit for members
- Air operations (fig. 13–16)
- Bulldozers
- Hand crews
- Investigator

Fig. 13–16. LAFD Fire 4 launching for an incident.
Source. Photo courtesy of Mike Meadows

Tactical Worksheet—Vegetation Fire

S	Size-up	☐ FPODP ☐ 4 Sides ☐ Values at Risk	☐ Fuel ☐ Weather (RH, Temp, Wind)	☐ Topography ☐ IOCAN
A	Apparatus Placement	☐ 3-Sides	☐ Hot Zone (in the black, upwind)	☐ Staging
W	Water Supply	☐ Who	☐ Where	☐ What
C	Command	☐ Name ☐ ICP	☐ Add Res ☐ ICS	☐ Comms

S	Strategy	☐ O	☐ C	☐ D
S	Safety	☐ LCES ☐ 18/10	☐ Lines ☐ Snags	☐ Spotting
R	Rescue	☐ Evacuation	☐ Medical T/T	
E	Exposures	☐ Structure Defense	☐ Livestock	☐ Infrastructure
C	Confinement	☐ Control Objectives/Box (roads riverbeds, dozer lines)		
E	Extinguishment	☐ Perimeter control (100 feet in)		
O	Overhaul	☐ Mop-up		
V	Ventilation	☐ Weather changes		
S	Salvage	☐ Who	☐ Where	☐ RP
S	Support	☐ Law ☐ PIO ☐ Safety ☐ Liaison	☐ Red Cross ☐ Util ☐ Rehab ☐ Air Ops	☐ Dozers ☐ Crews ☐ Invest

ICS and Communications for Vegetation Fires

The potential of even Type 4 and 5 fires to grow in complexity with structures threatened necessitates that the IC remain ahead of the incident power curve from the onset. This means resource ordering, decentralized command with ICS, and potential for multiple radio channels from which to communicate and coordinate. This does not mean you have to order the world and five extra frequencies for every roadside or vacant lot fire. This means you must think ahead at all times.

Communication en route while responding to the scene amongst the responding units is paramount to an adequate size-up. I have arrived on countless vegetation fires to the dispatched area, only to find the fire had long since moved well away into another area entirely. The smoke visible was vast enough that it was often difficult to assess the best means of access. Air support helps tremendously in these circumstances and Metro Copter 1 and 2 saved our hides more than a few times.

An example of an FPODP for a 10-acre vegetation fire in dry grass:

- Facts: 10-acre vegetation fire in light flashy fuel, moving at a rapid rate of spread with spotting to the north.
- Probabilities: Potential for 150 acres and structures threatened ahead of the fire.
- Own situation: I am the first-arriving Type 3 engine with two more Type 3 engines, a water tender, and a chief responding.
- Decision: Life saving priority. combination strategy.
- Plan: Anchor the right (hot flank). Have E32 tandem behind me to attempt to catch the head. Have E33 anchor the left flank. Call five Type 1 engines for structure defense, four more Type 3 engines for mobile attack, a second chief, three more water tenders, and air attack. Retaining command until the chief arrives.

Fig. 13–17. Sample ICS template for vegetation fires
Source. Illustration courtesy of Sam Wallace

An example of an IOCAN arrival report:

- "E31 on scene at the corner of This and That Roads. 10-acre vegetation fire in light flashy fuels, moving at a rapid rate of spread to the north toward homes. E31 is assuming Veg Command, anchoring the right flank. Next Type 3 engine tandem behind me. Third Type 3 engine anchor and attack the left flank. Calling an additional five Type 1 engines, four Type 3 engines, three water tenders, air attack and a second chief."
- Unlike structure fires, calling a second alarm may or may *not* get you what you need. Be sure to refer to your response matrix. When ordering engines, you may not have the exact *type and kind* in your area, so requesting pump-and-roll capability may cast a wider net to get what you need sooner from a closer firehouse.

Fortunately, ICS for Type 4 and 5 vegetation fires is relatively easy and consistent, compared to structure fires. The primary objectives for each division are perimeter control and defend values at risk/structure protection. Essentially, "put out the fire and protect any buildings or other stuff you can." Divisions quickly and easily organize a vegetation fire, even if it grows into a more significant incident involving structures.

In the past, a structure protection group would be established for structure defense. The challenge was that a fire could be so large (hundreds/thousands of acres), with multiple communities or groupings of structures threatened, that a single structure protection group supervisor could not properly manage all the areas. And what about cars, fencing, utilities, and all the other values we described above? This does not mean that a structure protection group or branch is wrong. You may find it useful for a concentrated area of a fire. A branch may also be appropriate with divisions within it.

If a particular division has structures threatened, they must call resources commensurate with that situation. This could be structural Type 1 engines, trucks, or a combination of smaller-type engines. The trucks primarily evacuate and provide staffing to assist the engines with hoselines, and can utilize hand tools, or prep the building. This would include removing fences, excess fuel, and debris.

One way to organize your fire is to anchor the fire on the left and right flanks at the heel. The left flank would become Division B, the right flank would become Division X. This allows both divisions to grow if necessary to the left and right and create Divisions A and Y, respectively. Division O would be the head of the fire and you may assign structure fire resources there ahead of the fire. This way, you literally put a box around the fire (fig. 13–18).

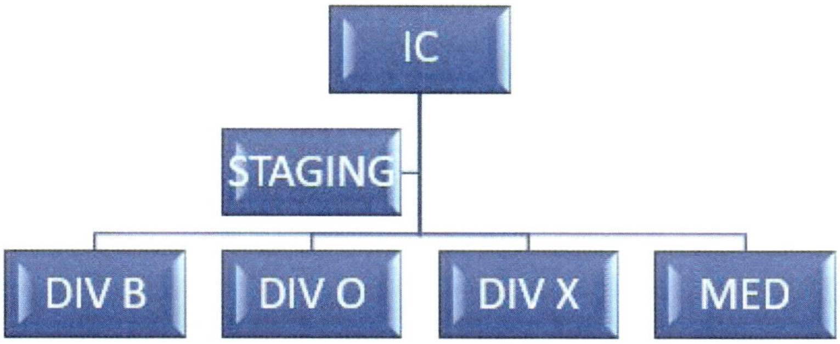

Fig. 13–18. Organize your fire by assigning resources so that divisions can grow.

Summary

Vegetation fires in the Type 5 and Type 4 arena will most often be contained quickly to under an acre in size. That said, red-flag conditions can easily create a fire that quickly overwhelms resources. In recent years, catastrophic fires with multiple fatalities have become more frequent.

Like all fires, training, pre-planning, size-up, and many of the other factors you have learned in this book apply. Remain ahead of the incident power curve. Prevent the NIOSH 5 from coming into alignment and decentralize early to bridge the tactical gap and organize the incident with ICS. Bedrock safety principles like the 10 Standard Fire Orders, 18 Watch Out Situations, and LCES can apply to a fire of any size. In addition, SAW-CSS-RECEO-VSS can be applied to this arena very effectively.

Nomenclature and tactics for vegetation fires are significantly different, as are the risks involved in these incidents. Mobile attack, progressive hose lays, hand tools, dozers, and air resources may be utilized in these fires, in addition to firing operations. Depending upon the threat to populated areas, evacuations may be required with support from law enforcement.

The key tactical objectives are perimeter control, known as putting a box around the fire, and structure defense. Several structure defense tactics are available. As with any fire or tactic, do not get in over your head. Only attempt tactics in which you are proficiently trained and have been approved and properly supported by the incident commander.

Use of ICS was founded in large-scale wildfires and works extremely effectively in smaller Type 5 and 4 vegetation fires. Anchoring and attacking the left and right flanks can be supported by setting up divisions while moving toward the head and achieving full containment.

Chapter Review

Review Questions

1. Describe vegetation fires, wildland fires, and the wildland urban interface.
2. What are the three factors that drive the size and complexity of a vegetation fire?
3. Describe LCES.
4. What are the 10 Standard Fire Orders?
5. Explain three of the tactical options for structure defense.
6. Explain why initial establishment of divisions on a vegetation fire would not follow traditional alphabetical order.

FESHE Strategy and Tactics (C0279) Related Content

The content contained in chapter 13 provides detailed information specific to vegetation fires. While the FESHE CO279 course does not specifically reference vegetation fires, they involve the same processes and principles of structural firefighting requiring size-up, determination of incident priorities, selection of appropriate strategy, and coordination of tactics and resources.

The FPODP method is applied to vegitation fires in the ICS and Communications section of this chapter. The chapter format supports the CO279 course outline by detailing information on fireground operation, communications, and coordination of these events and ICS/NIMS roles and responsibilities to support and manage them.

Access the Tactical Review Guide Appendix C#6 Vegetation Fire—Terrain with Structures Threatened for an objectives-based scenario exercise.

NFPA 1021 Job Performance Requirements

The information in this chapter can be utilized to support training and educational programs associated with the Emergency Services Delivery Fire Officer I JPR 4.6, 4.6.1, 4.6.2, Fire Officer II JPR 5.6, 5.6.1, and Fire Officer III 6.6 and 6.6.1. For larger-scale incidents and pre-planning wildland urban interface responses, the content in this chapter also supports training and education for Fire Officer III 6.6.3 "Develop a plan for the organization given an unmet need for resources that exceed what is available in the organization."

Access the Tactical Review Guide Appendix C#6: Vegetation Fire—Terrain with Structures Threatened for a 1021 Job Performance Requirement-based scenario exercise for Fire Officer I, II, and III.

Endnotes

1. National Interagency Fire Center, *Chapter 2: Wildland Fire Management Terminology,* 2004, https://gacc.nifc.gov/swcc/management_admin/Agency_Administrator/AA_Guidelines/pdf_files/ch2.pdf.
2. FIRESCOPE, *Field Operations Guide—ICS 420-1,* ICS Publication, 2022.
3. National Interagency Fire Center, *Chapter 2.*
4. FIRESCOPE, *Field Operations Guide.*
5. National Interagency Fire Center, *Chapter 2.*
6. FIRESCOPE California, *Wildland Urban Interface (WUI) Structure Defense,* October 21, 2013, https://firescope.caloes.ca.gov/ICS%20Documents/WUI-SD.pdf.

14

Hazardous Materials Incidents

Fig. 14–0. Sacramento Metropolitan Fire crews address a hazardous materials condition that resulted from a motor vehicle accident response.
Source. Photo courtesy of Sacramento (CA) Metropolitan Fire District

Introduction

Hazardous materials (hazmat) incidents have more potential than ever. Secondary devices, lithium-ion batteries, dirty bombs, white powder substances, vehicle accidents, and unintentional leaks/releases all have the potential for a lot of chaos. Today's structure fires are wrought with hazardous materials. Even if you do not have a formal hazmat response, your structure fires can be laden with hazmat. Look at the average home garage and you need look no further. Gasoline, oils, solvents, pesticides, paints, canisters, and countless other hazmat are commonplace. Refer to the Boyd Street case study in chapter 10. See also Appendix B, Boyd Street Incident. Below is an excerpt from the Boyd Street Incident green sheet:

> *The Boyd event was an extremely violent, hot, and rapidly expanding fire due to the abundant fuel supply in the form of smoke (carbon monoxide/CO), plastics, butane, and nitrous oxide (N_2O). Of particular concern was the extremely large quantity of the N_2O.*
>
> *Although N_2O is a non-combustible gas, N_2O, when heated, will decompose, creating an exothermic reaction. During the decomposition phase, its compounds split, releasing oxygen (0_2), and nitrogen (N_2). If N_2O is mixed with a fuel source, it will also create a rapidly expanding pressure wave of carbon dioxide (CO_2) and water vapor (H_2O).*
>
> *The small N2O cylinders found on scene of the Boyd Incident contained approximately 8 grams each of net product and have been known to release their contents in temperatures as low as 106° Fahrenheit. As the fire intensity and pre-heating increased, the release of the N_2O increased. The temperature around the N_2O storage was approximately 400° to over 900° Fahrenheit early into the event.*
>
> *According to industry experts, when large quantities of product contained in the cylinders are stacked in quantity and exposed to heat, they can release or explode from the outer edges of the storage configuration, working inward until a critical temperature is reached. This can cause a simultaneous release of a large portion of the product to occur, with violent results.*

Fortunately, we have taught you to use the same process for all incidents: Size-up (including victim profile), priority, strategy, tactics... Your ICS will flow from there and help with your incident action plan, resource ordering, and so on. Hazmat is no different. That said, there are specific objectives and considerations to make you more safe, effective, and efficient as the incident commander of a hazmat event. Reference 29 CFR 1910.120 for the Code of Federal Regulations regarding hazmat operations and response for further information.

Again, like an MCI, a hazmat must first be recognized to be properly commanded and resolved. Once the incident goes beyond the capability of the first company on scene, the incident commander must think big. When in doubt, make calls. You may have a department or county hazmat team that will advise you prior to their arrival.

Making proper notifications early is crucial to resolving the issue in a timely manner. In addition to the hazmat response team (HMRT), county hazmat, the county office of emergency services (OES), National Guard Civil Support Teams (fig. 14–1), and law enforcement may all be required to be notified. If a restaurant, cafeteria, grocery store, or any food processing facility is involved, county health must be notified.

Fig. 14–1. National Guard Civil Support Team operating with a local agency. Jurisdiction, authority, and expertise with regard to hazardous materials often lead to interagency management.

CASE STUDIES IN COMMAND

Lithium-Ion Battery Fire

CHEF JOEL MENDOZA, NORTH COUNTY FIRE DISTRICT, MONTEREY, CA

The North County Fire District in Monterey County is responsible for providing fire protection and emergency services to several communities, including the typically peaceful Moss Landing, California. Moss Landing is a small coastal community located in the northern part of Monterey County, known for its picturesque harbor, diverse wildlife, and agricultural activities.

At 01:00 on September 20, 2022, Fire Station No. 1 received an alarm indicating an unknown fire at the intersection of CA-Highway 1 and Dolan Rd. Initially perceived as a nuisance incident, it was quickly escalated as the dispatcher upgraded the response to a structure fire at the Elkhorn Battery Energy Storage Facility.

On that day, I served as the on-duty chief officer, having previously held the same role during a battery incident in September of the previous year at a neighboring facility.

The Elkhorn Battery Energy Storage Facility was commissioned in April of 2022. The facility is owned and operated by Pacific Gas and Electric (PG&E). Situated next to the federally protected Monterey Bay National Marine Sanctuary, this facility plays a crucial role in storing and releasing excess electricity. The facility supports the integration of renewable energy sources while ensuring a reliable power supply. With a capacity of 182.5 megawatts (MW), the Elkhorn Battery Energy Storage Facility houses 256 Tesla

Megapack battery packs. These lithium-ion battery arrays are specifically designed to store the surplus renewable energy that solar installations generate. During periods of low demand, the facility stores this energy and releases it during peak-demand periods.

Upon the arrival of the first fire engine at the scene, the company officer reported witnessing a glow from within the secured premises of the PG&E facility. When I reached the facility, I met up with the company officer, who had been on the scene for approximately five minutes. The company officer confirmed what we had heard on the radio: a Megapack was engulfed in flames. At that moment, I realized we were facing a challenging incident due to our knowledge of lithium-ion battery fires.

Lithium-ion batteries pose a significant threat to firefighters when they go into thermal runaway and ignite. The combustion of these batteries can release intense heat and toxic fumes, putting firefighters at risk of severe burns and injuries. Extinguishing such fires can be challenging and sometimes impossible. In particular, the Megapack cannot be extinguished once it goes into thermal runaway, and it may take anywhere from four to six hours to burn down.

After confirming the safety of all plant personnel, my immediate priority was to consult the site's emergency pre-plan (pre-plan) to guide my actions. As I opened it, memories of the planning process flooded back. The fire district and PG&E had worked together diligently to create the pre-plan, meticulously outlining necessary actions to safeguard the well-being of first responders during an emergency event.

I then assigned two engine companies to set up two portable water monitors to safeguard adjacent Megapacks not involved in the fire, as stipulated by the pre-plan. The on-site water system provided ample water volume and pressure, allowing us to avoid deploying fire engines into the hot zone. Once the monitors were in place, we relocated the engine companies to the warm zone, away from toxic fumes or smoke. We knew that as the fire evolved into a hazmat incident, we needed to assume the necessary roles and establish a unified incident command structure. To do so, I requested assistance from law enforcement for evacuation or shelter-in-place orders. When the representatives from the Monterey County Environmental Health Department (EH) and the Monterey County Office of Emergency Services (OES) arrived, they became technical advisors, providing valuable information and guidance throughout the incident.

In anticipation of the event generating significant publicity, I called Cal OES to inform them about the burning Megapack emitting toxic fumes and gases into the atmosphere. As the fire continued to burn and the winds picked up, the safety of the surrounding community became an even greater concern. Command made the decision to issue shelter-in-place orders for the immediate area surrounding the facility, which also included the closure of a small section of California's CA-1 Highway. These measures were taken to ensure the well-being of residents, commuters, and others in the vicinity of the facility. The Megapack continued to burn for approximately six hours, causing the shelter-in-place order to remain in effect for most of the day. The safety of the public was our utmost priority, and we wanted to ensure that there was no further threat to their well-being. To confirm the absence of any lingering hazards, proper air sampling was conducted. Once the results verified that there was no longer a threat to the public, the shelter-in-place order was lifted, allowing residents and commuters to resume their normal activities with peace of mind.

Collaboration between the incident commanders and PG&E was pivotal in mitigating the incident. Working closely with PG&E, our team coordinated efforts to prevent additional Megapacks from being affected. PG&E's extensive knowledge of the facility's layout and operational intricacies proved invaluable in strategizing containment measures. Together, we devised a comprehensive plan that included targeted suppression efforts and protective measures to safeguard non-involved critical equipment.

Through close collaboration with environmental agencies, we documented potential ecological harm and implemented effective measures to prevent the spread of harmful substances within the facility. Equipped with a fire suppression water system, onsite water monitors, and several holding ponds for fire suppression water, the facility ensured safety and actively safeguarded the surrounding ecosystem.

The preparedness and knowledge gained through the pre-planning process proved crucial during the actual emergency. By understanding the risks associated with lithium-ion battery fires and implementing appropriate safety measures, we were able to protect lives, property, and the environment. It cannot be overstated how important it is to work with developers early on and provide ongoing training. The importance of these measures became evident in the successful outcome of the incident.

In conclusion, it is essential to raise awareness about the risks associated with lithium-ion battery fires and ensure the implementation of proper safety measures to protect firefighters, the public, and the environment. Extensive planning with the facility, onsite training, and close collaboration with PG&E management ensured that the incident resulted in no injuries to site employees, the public, or firefighters. This incident highlights the dedication and professionalism of the individuals involved in emergency response, who fulfill their duties as part of their commitment to the communities they serve.

Glossary

Assistant safety officer (ASO): Separate from the incident safety officer, the ASO must have the same training as the entry team and focuses solely on hazmat operations safety.

Backup team: The technical hazmat team that is standing by if the entry team requires assistance or extrication.

Contamination reduction zone (CRZ): The area between the exclusion zone and the support zone in which decontamination occurs (fig. 14–2).

Entry team: The technical hazmat team that enters the exclusion zone to mitigate a spill or leak .

Exclusion zone: The area immediately around a hazardous material leak or spill in which specialized PPE must be utilized by technical-level responders.

Gross decontamination: The act of decontaminating large numbers of people who may be exposed to hazardous materials. This may be done with water alone by firefighters and hoselines.

Hazardous materials group supervisor/branch director: The individual in charge of the tactical and technical hazmat operations on a scene.

Site access control: The act of preventing persons from going into and out of the exclusion zone to prevent cross contamination.

Support zone: The area furthest away from the exclusion zone in which support functions such as command and medical are staged and working.

Fig. 14–2. Entry team working through the contamination reduction zone.

Technical decontamination: The act of decontaminating the entry and backup team members with multiple steps and solutions appropriate to the hazardous material. If an entry is required, a backup team must be ready. This will also require technical decontamination for the team as they exit the area.

Technical reference specialist: The individual charged with identifying the characteristics, dangers, operating procedures, mitigation measures, and overall safe and effective handling of the hazardous material product.

Tactical Considerations for Hazardous Materials Response

Identify, Isolate, and Deny Entry

For an active incident, the first tactical objective is to identify the material, isolate, and deny entry to the area to prevent further exposure. This could be anything from simply keeping people back with tape, to mass evacuations, closing rail lines, or freeways. Do not be afraid to think big. Wind and weather are a major consideration. Refer to your SOGs and the U.S. Department of Transportation (DOT) Hazardous Materials Emergency Response Guidebook (ERG) for details, including removing those in immediate danger (fig. 14–3).

The Safety Data Sheet (SDS) lists the hazardous ingredients of a product, its physical and chemical characteristics (e.g., flammability, explosive properties), its effect on human health, the chemicals with which it can adversely react, handling precautions, the types of measures that can be used to control exposure, emergency and first aid procedures, and methods to contain a spill. This will contain critical information regarding the material you are dealing with,

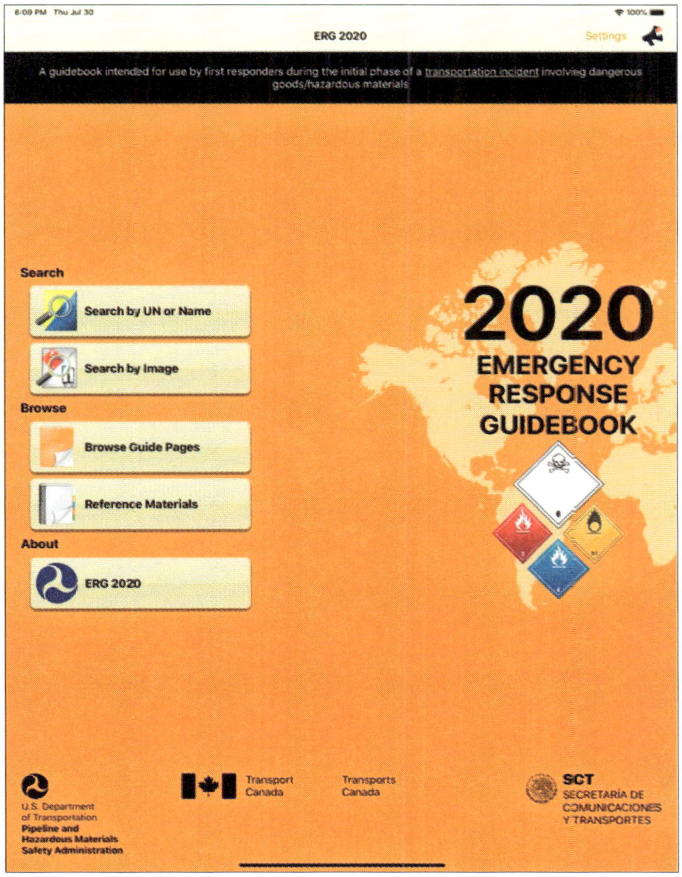

Fig. 14–3. 2020 Emergency Response Guidebook from U.S. Department of Transportation.

including but not limited to hazard classification, flammability, toxicity, health hazards, and potential medical response information. Refer to your agency and county hazmat response policy. Your HMRT will provide the expertise for the actual tactical considerations and procedures regarding entry, mitigation, and decontamination. Your job is to support and organize the incident.

Entry and Mitigation

As with all incidents, determine the VP ASAP. This may be a worker or bystander incapacitated in the immediate area, employees exposed, or civilians evacuating the building or area. An MCI may be generated as well. Your VP drives your priority, strategy, tactics, tasks, and resource needs. This will also help you determine if you have discretionary time or not to mitigate the hazard

Remember, the first priority is to identify, isolate the hazmat, and deny entry. That said, refer to your ERG, hazmat team, SOGs, MSDS, policy, and on-scene experts if you are faced with a confirmed VP with a person down in the immediate area of a spill or release.

The hazmat group supervisor (HMGS), or branch director with multiple hazmat groups is responsible for the tactical and technical operations. This includes technical reference, site access, entry, back up, and decontamination. The entry team must have a backup team place along with an ASO to approve the entry.

Decontamination and Medical

Once mitigated, all personnel that entered must go through technical decontamination through the CRZ (fig. 14–4). This includes all equipment. Non-technical companies can assist and support this under the direction of the HMGS.

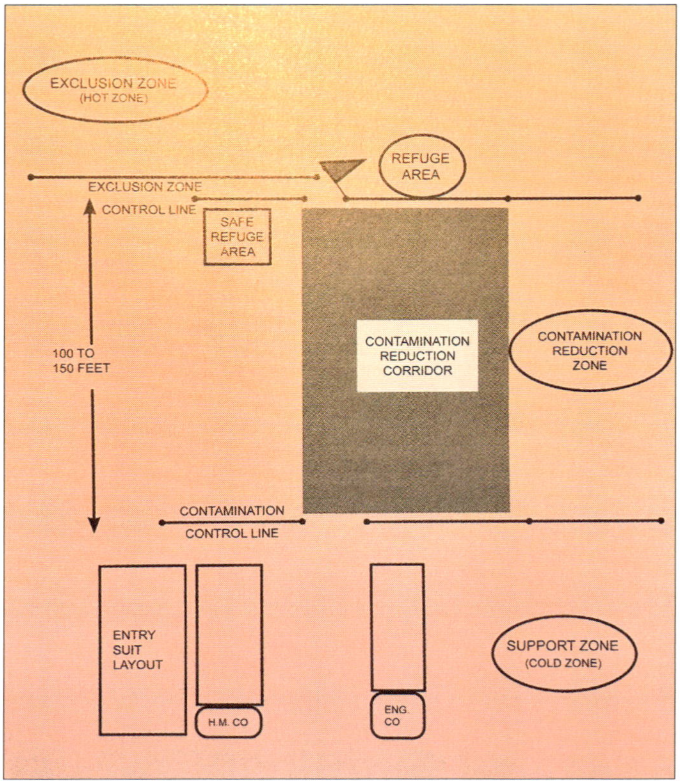

Fig. 14–4. Control zone layout[2]

If gross decontamination is indicated, engine companies can perform this task early in the incident prior to the arrival of a hazmat team. Refer to your SOGs.

Another primary objective for the incident commander is medical. Whether just one ambulance, or a medical group/branch due to the size and complexity, you must have medical standing by. The minimum is a standard for your hazmat team to be monitored and provide potential treatment. In addition, depending upon the victim profile, you will have additional medics for treatment and transport of civilians. Decontamination must be considered prior to treatment and transport so you do not transmit contaminants from the scene to local hospitals.

Also, consider the weather and duration of the incident. Your hazmat teams will likely need rehab and shelter, so two teams will likely be required any time an entry is planned.

WISDOM FROM THE MASTERS

Technical Rescue Command Considerations

CHIEF GREGG AVERY, LOS ANGELES CITY (CA) FIRE DEPARTMENT

Technical rescue incidents can have a component of hazmat. For example, a confined space entry may have a leak, or a rescue may involve a spill. To that point, we have added this contribution from Chief Avery.

When faced with a technical rescue incident, your role as the incident commander is challenging, due to the complex and specialized nature of the incident. More importantly, your workforce will/should possess unique skill sets and certifications and will challenge you to build an appropriate strategy that meets the needs of the incident and satisfies all legal requirements. As incident commanders, we must consider *NFPA Standard 1710: Standard for Organization and Deployment for Suppression, EMS, and Special Operations* as well as *NFPA 1670:* (now bundled into 2500), *Standard on Operations and Training for Technical Search and Rescue Incidents*.[1] I always utilize a subject matter expert, generally the US&R Task Force commander, to assist with this vital function.

When commanding technical rescue incidents, the standards of command always apply and are essential for an organized, efficient, and safe operation. As with any incident, the commander should implement and utilize the following:

- **Assume command**: In my organization we assume command by announcing to the current IC (captain) and providing Metro Fire Communications (MFC) a comprehensive size-up. I operate under the assumption that I have incomplete situational awareness when I assume command and as I gain further clarity I give updated size-ups.
- **Gain situational awareness**: This is accomplished by combining the information you have obtained while en route and with your personal observations as well as information from your assigned officers. I always say that situational awareness is ever-changing and ever-growing and I am fully committed to pivoting/changing my strategy and tactics accordingly.
- **Communications**: My command style utilizes a conversational style with my officers and division and group supervisors. It is incumbent on the IC to ensure that everyone on the scene is briefed on the communication plan.
- **Develop objectives and strategy**: This is done in a rapid fashion based on situational awareness, utilizing standard operating procedures, and always with the mindset that your strategy should and will be adjusted as needed. Your strategy and tactics should be driven by your department's operational risk management policy.
- **Resource deployment**: Initial deployment will be single resources and driven by standard operating procedure.
- **Incident organization**: Building out your organizational chart utilizing divisions and groups. I always consider carefully my leadership choices, and I try to match skillsets to assignments. In a technical rescue this is an absolute must. Allied agencies will be a part of your organization, so consider them early.
- **Evaluate and revise**: Looking at your plan continuously is essential so that revisions occur as needed and in a timely fashion.
- **Transition of command**: I always anticipate transitioning my incidents, and we routinely utilize the first battalion chief in the operations role.

I have a background in US&R operations and have had the opportunity to command several incidents. One of those incidents was a confined space/trench incident at a construction site in the Hollywood Hills. A worker had stepped into a recently drilled 15-foot-deep caisson hole. When the first-arriving engine company determined the type of incident, they called for a full technical rescue dispatch. I was assigned and responded, arriving after most of the resources were on scene. What I discovered was a chaotic scene with several officers attempting to implement their own plans. After assuming command, I placed the Captain II from the US&R Task Force (TFC) in charge of the rescue group and directed him to develop a plan for the rescue. I simultaneously ordered all personnel away from the hole. This simple act had the effect I was looking for: the scene immediately became calm and orderly. I gathered the officers together and briefed them on the site safety plan and on the plan to extricate the victim in conjunction with the TFC from the US&R TF.

Technical Rescue (Trench)

First Responders

- ☐ Approach from head of trench
- ☐ Ladders
- ☐ Spoils pile 2' from lip <3' high
- ☐ Edge protection
- ☐ Depth of trench (prior to cave in) _______
- ☐ Victim(s) location
- ☐ Victim Condition ___________-
- ☐ Keep all first responders out of trench
- ☐ Secure Scene
- ☐ Shut Down Machinery (300')if possible
- ☐ Limit unnecessary people at trench lip
- ☐ Lock Out Tag Out hazards (LOTO)

Resources

- ☐ Two US&R TF + UR88 (min)
- ☐ Trench Truck Out Of 88's
- ☐ HR56
- ☐ Vacuum trucks (Two) EMPTY
- ☐ Competent Trench Person – US&R 88
- ☐ Certified Trench safety officer- US&R 88
- ☐ DOT
- ☐ LAPD- Crime Scene
- ☐ CAL OSHA
- ☐ Building and Safety

Considerations: Leave spot for UR88 and HR56.
Air monitoring/ventilation for the trench

Note: As per new Trench regulations and state OSHA a Certified Trench Competent Person and Trench Safety officer shall be Part of the operation. UR 88 normally has members certified assigned on a daily basis. Operations and the trench permit can start prior to these members arriving

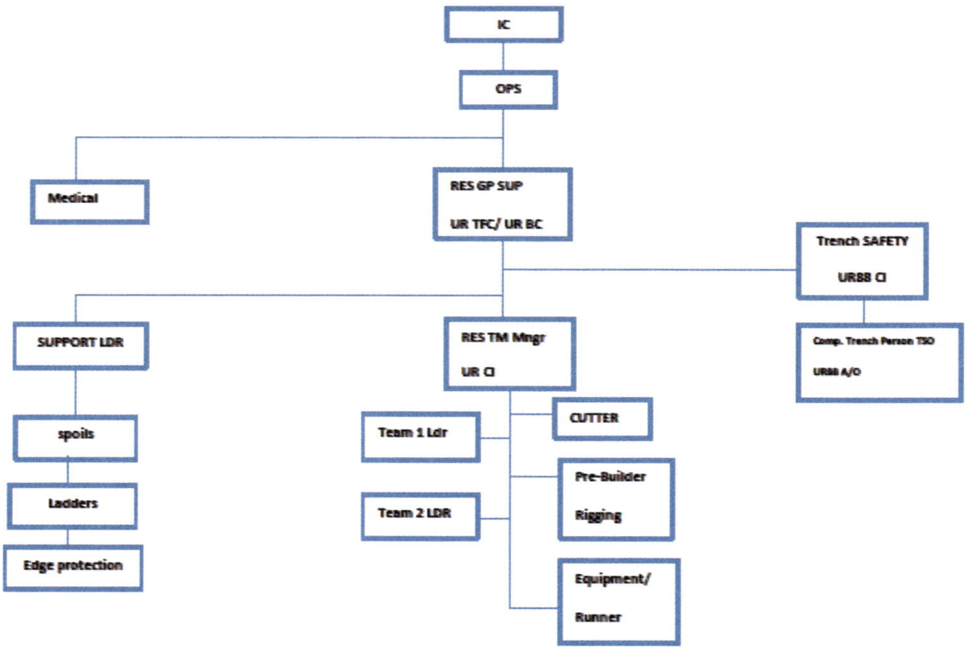

Additional considerations and actions include notifications to OSHA, city attorney, local council office, and command staff.

I am attaching two organizational charts for technical incidents. I normally utilize templates on specialized incidents so that I don't overlook vital positions.

Technical Rescue (Confined Space)

First Responders

- ☐ Upgrade INC. to Confined Space Dispatch if not already typed
- ☐ Size Up
- ☐ Perimeter Control (Deny Access into Space)
- ☐ Respiratory Protection
- ☐ Monitor Atmosphere surrounding portal
- ☐ Space use and Type
  - o Any product
  - o MSDS
- ☐ Locate knowledgeable Site Person
- ☐ Entry Exit portals
- ☐ Any Con space permit on Site
- ☐ Victim(s) location
- ☐ Victim Condition ____________-
- ☐ Limit unnecessary people at Site
- ☐ Lock Out Tag Out hazards (LOTO)
- ☐ Ventilation (if possible) Non combustion Blowers

Resources

- ☐ UR BC
- ☐ UR88
- ☐ US&R TF with UR Apparatus
- ☐ UR RA For Rescue Ops
- ☐ HR56
- ☐ HAZ Mat TF
- ☐ EMS Captain
- ☐ DOT
- ☐ LAPD- Crime Scene
- ☐ CAL OSHA
- ☐ DWP

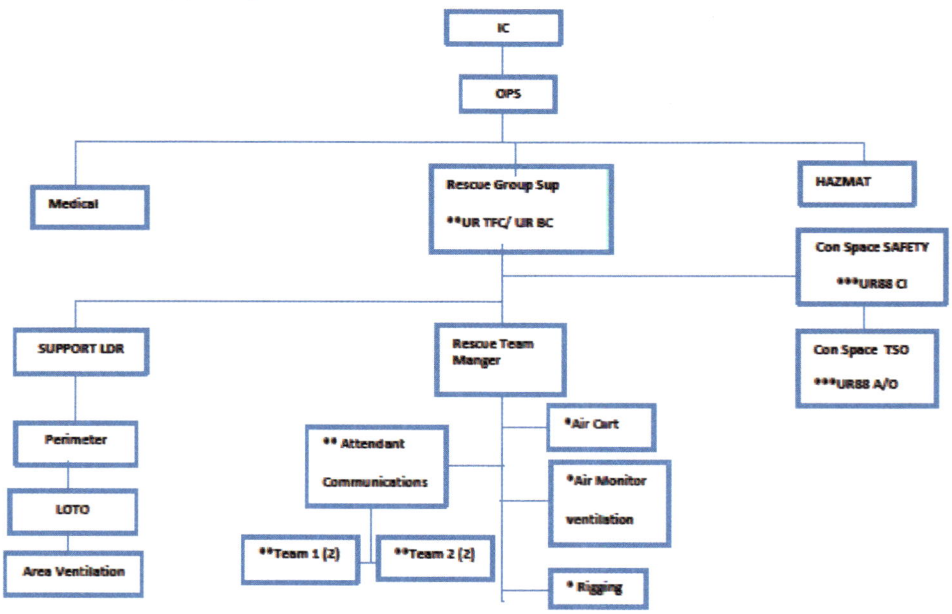

*** =Mandatory certified Confined Space technician (Filled By US&R 88)

** = Mandatory Certified Confined Space Technician (Filled By Members From US&R TF)

* = Filled by US&R TF Member

SAW-CSS-RECEO-VSS

Depending upon the nature of the hazmat, you may enter into unified command with law enforcement or another AHJ, such as environmental health agencies. In addition to an overall safety officer and an assistant safety officer (ASO), liaison and public information officers can prove vital. You may have to notify communities to evacuate, close rails, freeways, or just evacuate a building. The staff officers will be a huge help!

The SAW-CSS-RECEO-VSS system has been used nationwide for years and is easily applied to hazmat incidents. You will notice that *italicized words correspond with key points on the tactical worksheet* that follows.

Size-Up

Remember, size-up is a mental, ongoing process.

- Start with your *FPODP or FIRST* algorithm.
- As you respond to the scene, take note of the *weather*, including wind direction. This will prevent you from approaching downwind and becoming contaminated before you ever arrive on scene.
- As with all incidents, get the *VP* ASAP. This will determine your discretionary time, priority, strategy, tactics, resource needs, and ICS.
- Once on scene, look at the location, building, facility, and so on. Consider the *seven sides* as you would a structure fire: four sides, roof, basement, inside. This will help you determine potential location and flow of hazmat, and location of victims, exposures, and so on.
- *Identify* the hazmat ASAP. If possible, ascertain this en route from dispatch. Chances are, you will be dispatched to an unknown product, vapor, or release. If you cannot get more information from dispatch, then find the site manager or RP on scene ASAP. If possible, obtain any SDS.
- Refer to your DOT Emergency Response Guidebook *(ERG)* for details regarding guide number and title of the hazmat, potential hazards (fire, explosion, and health effects), public safety, protective clothing and evacuation, and emergency response (fire, spill or leak, and first aid). If in doubt or dealing with an unknown substance, or mixed load, refer to ERG Guide Section 111.
- If possible, get a *lap* of the scene, without going downwind or downhill, to get the big picture and identify the direction of any leak or plume.
- Your *initial size-up* will conclude with an arrival report to incoming units; however, remember that size-up is continuous. As you talk to the RP, bystanders, occupants, employees, and attempt a lap, you will garner more critical information. The *IOCAN* system is designed to give a quick arrival report. Your arrival report is likely dictated by your SOGs. Consider any local hazmat SOGs when giving your arrival report.

Apparatus Placement

Apparatus placement should be considered early, as you typically do not get a second chance. Plan ahead based upon your type of apparatus, mission, SOGs, and order of arrival.

- Approach from *upwind and uphill* to prevent exposure to your crew.
- A first-arriving engine will likely attempt to see *three sides* of the scene, allowing the officer to have a tactical advantage before stepping off the rig. A hazmat incident could involve fire, explosions, or leaks, so be cautious and consider long-term impact of your staging area.
- The *hot zone* is any area that will cause damage to apparatus, including but not limited to collapse, thermal insult, excessive smoke, hazardous materials, power lines, and so on. Ensure that apparatus is staged out of the hot zone, including potential changes in conditions and wind shifts.
- Consider a *staging* area and potential route of travel for incoming units like hazmat teams (fig. 14–5).

Water Supply

Water supply will likely be for decontamination during a hazmat incident. Water supply must be considered as soon as possible to take advantage of any hydrants while responding into the scene.

- *Who* is securing the water supply? *Where* is the water supply? Is it reachable by one engine, or is this an extensive lay? Will water tenders be required?
- *What* is the supply? A dry or wet forward lay by the first-due engine from the hydrant to the fire? A wet forward lay by the second-due engine? A driveway/split lay? A reverse lay by the second engine from the first, out to the hydrant? A tank transfer? A water shuttle from a distant hydrant, or a drafting operation?
- Once a patent water supply is established, it should be announced on the tactical channel.

Command

Command must be established and announced. *There can only be one IC at a time, and the IC must be on scene to assume command.* While your SOGs will determine specific parameters, some fundamental best practices exist.

- When the first-arriving officer is a chief or staff officer with no tactical capability, assuming command is the most likely option. Developing an IAP and calling additional

Fig. 14–5. Special operation units like a hazmat team and others will need staging.
Source. Photo courtesy of Sacramento (CA) Metropolitan Fire District

resources will typically be the best decision. That said, attempt to contact the RP and get SDS or identify the product.

- *Name* the incident and identify the *ICP.*
- You may find yourself working with *law* enforcement early and enter into *unified command. See chapter 16 for more on unified command.*
- Your *county hazmat team or agency* should be notified ASAP. Even if you have a response team, the county may want to be notified.
- Call *additional resources/alarms* early and confirm the staging location. This would include hazmat teams, medic units, air/light and even companies to assist with decontamination.
- Set up your *ICS* early. A hazmat group is adequate for most incidents; however, a hazmat branch would be more appropriate for larger incidents with multiple teams operating.
- Your county *Office of Emergency Management/Services (OES)* may also require notification.
- Do you have adequate channels for *communications?* Consider a separate tactical channel for the hazmat group/branch.

Strategy

A hazmat response is going to have a lifesaving priority initially until you get a VP. That said, you may have to determine if the strategy is offensive, defensive, or combination based upon any potential threat from the hazardous material. This may require a combination strategy as victims are brought from the defensive space/exclusion zone to the offensive space/support zone. Quickly de-escalating to an incident stabilization priority will prevent responders from taking too much risk, so clarify that ASAP.

Safety

As stated previously, *isolate and deny entry* ASAP to stop the damage. That said, a hazmat incident will be the result of another event, such as a structure fire, vehicle accident, building collapse, active threat, and so on. Each of these has separate and distinct safety concerns in addition to the three primary listed above. Keep those in mind and set up your divisions/groups and safety officer accordingly.

- An ASO is required on all hazmat responses per 29 CFR 1910.120. This is in addition to the incident safety officer. The ASO should have the technical level of training to properly support entry team operations.

Rescue

A hazmat incident poses additional unknown risks to the average firefighter in their full PPE and SCBA. This means that running into an area with an unknown leak or vapor could be fatal, despite structural PPE and SCBA.

- That said, *immediate victims* who are within line of sight and can be quickly retrieved require further information, rapidly obtained, to determine if a save is possible. Refer to SDS, DOT ERG, your policy, and your SOGs. In any case, a crew should be tethered with a safety line and remain in sight.
- *Gross decontamination* takes place when large numbers of people who are contaminated are sprayed with water in an early attempt to minimize health risk and injury. This can be done early by a first arriving company, provided the hazmat is not water reactive. Refer to the DOT ERG.
- *Evacuation* is the most likely form or lifesaving activity you will perform to remove people from the area and may require a separate group supervisor, depending upon the size and scope of the area to be evacuated.

- *Medical treatment and transport* must be considered early. Even if there are no civilian victims, your crews will require medical monitoring and always have at least one transporting medic unit on scene for the "what if?"

Exposures

For a hazmat, exposures could be other areas where victims and materials may be located away from the incident. This could be *interior* exposures within the building due to HVAC spreading materials, or *exterior*. Consider *downwind* exposures due to hazardous materials releases and plumes which may necessitate evacuations of communities down wind.

Confinement

Locate the exact area of the release/leak and prevent spreading the problem.

- If you have hazardous materials-contaminated patients, it is critical that the walking wounded not spread the hazardous materials to hospitals.
- An *entry team* will be supported by a *back up team* and *technical reference* specialist. Each of these components work under the hazmat group supervisor (fig. 14–6).
- *Technical decontamination* will be required for any entry and back up teams that enter the exclusion zone.

Extinguishment

You may require multiple areas for decontamination, both gross and technical. Ensure water supply is adequate. This may require *multiple lines*, engine companies, or a *decontamination trailer*. Consider *water reactivity* and look at SDS or DOT ERG for alternative methods of decontamination.

Fig. 14–6. A hazmat entry team containing the spill at the task level.

Source. Photo courtesy of Sacramento (CA) Metropolitan Fire District

Overhaul

Ensure you have *all* victims, and all responders are accounted for. This includes knowing *who* was transported and *where* they were transported. Family will want to know. Decontaminate all equipment as well.

Ventilation

For a hazmat incident, ventilation refers to extreme *weather* exposure to victims while they are being treated and awaiting transport. Consider *shelter* like large tents or buses.

Salvage

Talk to the owner/occupant/*RP* or manager to transfer care of the scene to the appropriate party.

Support

Here, we consider what is needed to support the IAP. Several agency-specific resources are certainly available. As a minimum, consider the following:

- *hazmat team*
- *Law enforcement (LE)*
- *Investigator*
- *Public information officer (PIO)*
- *Safety officer (SO)*
- *Red Cross*
- *Utility companies (gas, electricity, and water)*
- *Rehab*
- *Air/light unit*
- *Gross decontamination/decon trailer*
- *Buses*
- *OES*

Tactical Worksheet—Hazardous Materials Incident

S	Size-up	☐ FPODP ☐ VP ☐ Weather	☐ 7-Sides ☐ Identify ☐ ERG	☐ Lap ☐ IOCAN
A	Apparatus	☐ Upwind/Uphill ☐ 3- Sides	☐ Hot Zone ☐ Staging	
W	Water Supply	☐ Who	☐ Where	☐ What
C	Command	☐ Name ☐ ICP ☐ Law	☐ UC ☐ County hazmat ☐ Add Res	☐ OES ☐ ICS ☐ Comms
S	Strategy	☐ O	☐ C	☐ D
S	Safety	☐ Isolate/Deny entry	☐ ASO	
R	Rescue	☐ Immediate victims ☐ Gross Decon	☐ Evac ☐ Medical T/T	
E	Exposures	☐ Interior	☐ Exterior	☐ Downwind
C	Confinement	☐ Location ☐ Entry	☐ Back up ☐ Tech Ref	☐ Tech Decon

E	Extinguishment	☐ Decon Trailer	☐ Number of lines	☐ Water reactivity
O	Overhaul	☐ Who	☐ Where	☐ Equipment Decon
V	Ventilation	☐ Weather	☐ Shelter	
S	Salvage	☐ Who	☐ Where	☐ RP
S	Support	☐ hazmat Team ☐ Law ☐ Invest ☐ PIO	☐ Safety ☐ Red Cross ☐ Util ☐ Rehab	☐ Air/Light ☐ Gross Decon ☐ Buses ☐ OES

ICS and Communications for Hazardous Materials Incidents

Like any incident, the use of ICS on hazmat incident will vary based upon the size and complexity of the event. You may only require a hazmat group supervisor if you have a small spill (fig. 14–7); however, you may require multiple branches if you have hazardous materials and/or significant fire and rescue challenges with dozens of patients (fig. 14–8).

A separate tactical channel should be considered for hazmat operations. Consider secondary devices and if your frequencies would activate a device. If the incident warrants and you

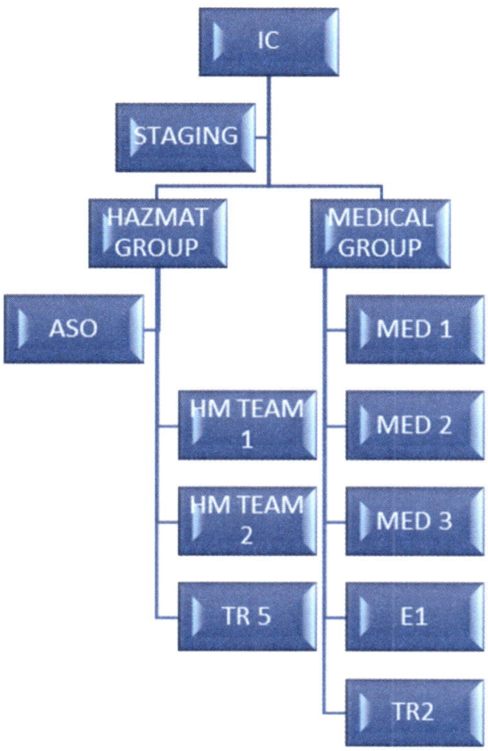

Fig. 14–7. A simple hazmat organizational chart. Companies and other resources assigned to the hazmat and medical groups would report to their respective supervisors.

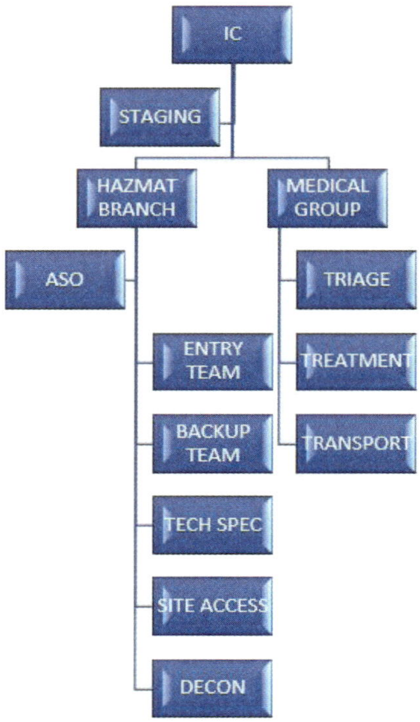

Fig. 14–8. Example of a moderate-sized hazmat organizational chart. Each component under the hazmat and medical groups would represent several resources assigned to each.

have branches established, multiple tactical channels may be required for hazmat, medical, fire/rescue branches, and so on. Specific positions may be utilized to break down the division of labor to a more manageable span of control. Each position can be added as needed and as the incident escalates (fig. 14–9).

Summary

Hazardous materials response and mitigation is consistently evolving, with new threats emerging. From the "suspicious white powder" calls post-9/11, to the proliferation of fixed and mobile meth labs; biological challenges of bird flu, Ebola and COVID-19, to now lithium-ion batteries—the challenges are infinite. If you are an incident commander in an agency that has taken on the responsibility of the AHJ for hazmat response and management, you must stay current and engaged. As a formal leader in one of these organizations, it is also critical that you have an awareness of the higher degrees of regulation associated with hazardous materials incidents and mitigation. With all that is demanded in the discipline of hazmat, it is also important to remain humble and accept that often times we may not be the best or only agency to handle these responses, and that the potential for management through a unified command (chapter 16) is much higher.

Chapter Review

Review Questions

1. Describe a contamination reduction zone.
2. What are the responsibilities of the technical reference specialist?

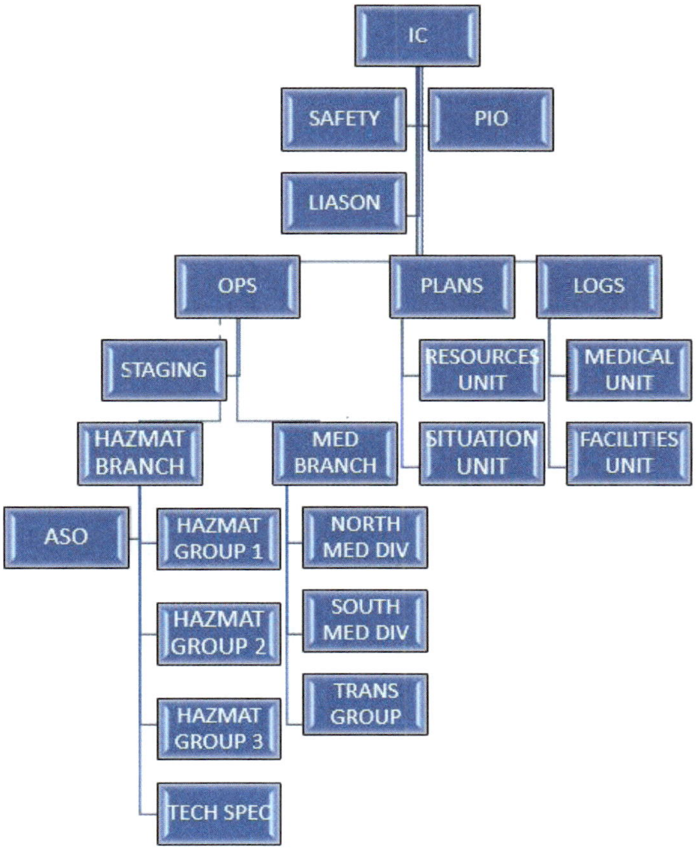

Fig. 14–9. Example of a complex hazmat organizational chart. Each hazmat group will have entry, backup, decon, and any other positions/functions necessary. Each medical division will have triage and treatment units. This would likely be a Type 3 incident.

3. Describe victim profiling for hazardous materials incidents.
4. What are the medical considerations for a hazardous materials incident with regard to victims and technicians?
5. Discuss the distinct safety concerns, roles, and requirements for hazardous materials incidents.
6. Explain some of the expanded communication demands of a hazardous materials incident.

FESHE Strategy and Tactics (C0279) Related Content

The content contained in chapter 14 provides detailed information specific to hazardous materials incidents. While the FESHE C0279 course does not specifically reference hazardous materials incidents, they involve the same processes and principles of structural firefighting requiring size-up, determination of incident priorities, selection of appropriate strategy, and coordination of tactics and resources. The chapter format supports the C0279 course outline by detailing information on hazardous materials incident operations, communications, and coordination of these events and ICS/NIMS roles and responsibilities to support and manage them.

Access the Tactical Review Guide Appendix C#7: Spill on Freeway and C#8: Sick Building for objectives-based scenario exercises.

NFPA 1021 Job Performance Requirements

The information in this chapter can be utilized to support training and educational programs associated with the Emergency Services Delivery Fire Officer I JPR 4.6, 4.6.1, 4.6.2, Fire Officer II JPR 5.6, 5.6.1, and Fire Officer III 6.6 and 6.6.1. For larger-scale incidents and pre-planning, the content in this chapter also supports training and education for Fire Officer III 6.6.3 Develop a plan for the organization given an unmet need for resources that exceed what is available in the organization.

Access the Tactical Review Guide Appendix C#7: Spill on Freeway and C#8: Sick Building for 1021 Job Performance Requirement-based scenario exercises for Fire Officer I, II, and III.

Endnotes

1. "List of Codes and Standards," National Fire Protection Association, https://www.nfpa.org/for-professionals/codes-and-standards.
2. FIRESCOPE, *Field Operations Guide—ICS 420-1*, ICS Publication, 2022.

15

Multi-Casualty Incidents

Fig. 15–0. View from shooter's window at Mandalay Bay.
Source. Photo courtesy of Clark County Sheriff (NV)

Introduction

The fireground IC must be proficient at MCI command because any response could become an MCI, and unfortunately, mass shootings are on the rise, such as the Route 91 Music Festival Shooting in Las Vegas, NV (fig. 15–0). Multi-casualty incidents have been around for decades. Structure fires, motor vehicle accidents (MVA), hazardous materials incidents, industrial accidents, and other incidents have grown into MCIs since firefighters have been responding to fires (fig. 15–1).

The difference now is two-fold. First, MCI protocols at the national and local levels have become more sophisticated, requiring the astute officer to understand how to apply them to save the most savable lives. Second, manmade disasters (active shooter, terrorist acts, etc.) have created a greater potential of a large MCI that will involve unified command and other challenges, including public information management and transition to a Type 3 incident or above. The unified command component is presented in much greater detail in chapter 16.

Time is the most critical factor in an MCI. Patients must be rapidly triaged, treated, and transported to definitive care. The IC must be proficient in commanding an MCI just as effectively as any fire. Like any incident, the components and principles of incident command that you have learned in this book will apply to the MCI. Perform an ongoing mental size-up with FPODP, bridge the tactical gap, stay ahead of the incident power curve, give clear objectives, decentralize command with incident command system (ICS), and so on. In addition, you must possess a working knowledge of your local MCI protocols, nomenclature, and resource levels.

Units responded to an MVA on the freeway. Upon arrival, the first-due engine reported multiple vehicles involved, including the need for extrication. I ordered an additional engine, truck, and several transporting medic units. Some occupants had no complaints, some had significant injuries and were entrapped. Crews utilized simple triage and rapid treatment (START) to quickly organize the patients.

Upon arrival and assumption of command, the incident was organized into a rescue group and a medical group. The rescue group objective was extrication of all entrapped victims. The medical group objectives included triage, treatment, and transport of all patients. Unified command was established with the highway patrol. Since the patient count was only eight, the MGS was able to contact the trauma center and coordinate the transportation destinations of all the patients. The incident was mitigated relatively quickly.

Fig. 15–1. Management of patient care alone at multi-casualty incidents is a challenge. Many times, they are associated with the simultaneous mitigation of at least one hazard.
Source. Photo courtesy of South Metro Fire Rescue (CO)

Evolution of MCI Response

Previous and current generations of responders are often dispatched to a violent crime and required to stage until law enforcement declares the scene secure. This erring on the side of caution and firefighter safety is commonplace. Fire dispatch centers communicate laterally with law enforcement dispatch centers to notify fire and EMS when they are cleared to enter the scene. This can cause myriad problems. One example is the lag time between cops on scene getting information to staged fire and EMS crews, creating a significant delay in treatment of patients.

Another example is that conditions often change and the scene that is declared secure deteriorates by the time fire and EMS arrive. This occurred on a shooting one night when we were staged awaiting the all clear from law. We were notified by dispatch that the scene was now all clear per law and went in. Once inside the scene, multiple gang members had guns drawn and threatened to shoot our crew if we didn't move fast enough to save their fallen colleague.

A major problem in the past was the abject lack of coordination between law and fire/EMS during active-threat incidents, such as an active shooter. Lack of communication, unified command, use of ICS, and competing objectives would create chaos. Cops and fire would set up different command posts that did not talk to each other or identify priorities, strategy, or joint tactical objectives. Resource requests would be conflicting or redundant, straining the entire system.

CASE STUDIES IN COMMAND

Century Theater Shooting

CAPTAIN (RET) RYAN FIELDS-SPACK, AURORA (CO) FIRE DEPARTMENT

On July 20, 2012, a shooter entered the Century Movie Theater in Aurora, Colorado, and opened fire (fig. 15–2). In just over two minutes, he shot 70 people—with 12 succumbing to their injuries.

As with many incidents, the initial dispatch did not convey the scope of the situation and came in as a "person shot." Thus, a single complement of engine, battalion chief and ambulance were dispatched. Upon arrival, the battalion chief noted over 1400 people streaming out of the theater complex, many of them covered in blood. Starting from this point, the battalion chief had the challenge of both setting up the scene and working to decipher what had happened—and how many resources were needed. In short, he was behind the power curve from the beginning due to diminished initial situational awareness.

Fig. 15–2. The Century Theater in Aurora Colorado. *Source.* Photo courtesy of Aurora (CO) Fire Department

The Challenges

Three specific challenges confronted responders during the Aurora theater shooting. After the incident, the Aurora city council contracted with the company Tri-Data Division, System Planning Corp. to conduct a formal after-action report. The lessons learned can similarly be applied to many of today's active shooter incidents. For Aurora specifically, Tri-Data identified three core communication challenges:[1]

Challenge 1

Police and fire commanders did not establish a unified command.

As responders approached the scene at the theater on July 20, 2012, they faced multiple choke points, a fire alarm activated in the background, and over 1400 patrons streaming into the parking lot. Per Tri-Data, "There was a failure to establish direct communications links between the police and fire incident commanders and, to a lesser extent, with responding mutual aid units." This was due in large part to the geographical distance between the police and fire commander on the scene.

Challenge 2

The police department was unable or did not know how to communicate directly with the fire department.

Dispatchers, in addition to tracking and assigning resources, were also receiving sequential 9-1-1 calls that needed to be answered. As a result, "multiple requests for rescue and EMS from police in the field were acknowledged by the police dispatcher but not received or not acknowledged by the fire department incident commander, and needed resources were not assigned." Further, ". . . requests [were] being made directly to the police dispatcher by multiple field units, which made it difficult to identify duplicate requests and keep track of EMS assignments. Any time information must pass through multiple hands it is possible there will be miscommunication, duplicate requests, and lost requests."

Challenge 3

Police and fire department personnel did not participate in regular communications interoperability drills.

While the sheer volume of injured on the scene exacerbated the communication difficulties, the following radio interoperability challenges were noted: "Police channels were not available on the Fire Department 'Suppression' Fleet channels. Accessing interoperable talk groups required switching away from primary police or fire channels, making it highly possible that critical transmissions would be missed."

How Can *Calm the Chaos* Help?

Implementing the strategies and tactics taught in this text can vastly improve upon the preparedness of the next commander at an incident like this. Chief Kastros has articulated that public safety agencies, whether law enforcement, fire, or EMS, are bound by three general rules as they respond to emergencies and apply to everyday calls for service, as well as those once-in-a-career active shooter events. The rules include:

Every first responder executes an emergency response from one of three points of view:

- Task saturation: A paramedic treating a wound or an officer providing cover will, justifiably, be focused on the task at hand—not on activity happening beyond six feet away.
- Tactical approach: Company officers are guiding the tactical activities of their own teams, and they may not be aware of other tactical activities taking place.
- Strategic focus: Incident commanders are charged to identify a strategic mode and goal for the incident and make tactical assignments to accomplish it.

Those points of view take place on emergencies with varying levels of complexity. They can be simple, complicated, or complex.

- Simple emergencies: First responders face simple emergencies daily, and they typically employ a step-by-step process to bring them under control.
- Complicated emergencies: A manhunt for a dangerous person or a cardiac arrest with challenging arrhythmia are complicated emergencies. They require a good deal of thought to bring about a positive resolution.
- Complex emergencies: An active shooter is a complex emergency—one with a malicious actor seeking to one-up the last big active shooter event and make a name for themself. These events morph with every iteration and challenge even the most robust responding agency. While public safety seeks to train for an active shooter, they can never quite prepare for the next scenario.

There are three basic types of incidents first responders face every day:

- Law Enforcement Only.
- Fire/EMS Only.
- Combined law enforcement, fire, and EMS event—where all agencies are called to the same location. These are complex emergencies with all agencies engaging

in their respective task-oriented, tactical, and strategic activities at the same time. These combined complex events require a level of coordination, communication, and trust that is increasingly difficult to achieve.

By implementing the strategies in this textbook, you will be better prepared to manage the scene of the next complex active shooter incident. You will calmly and confidently:

- Establish a strategic unified command.
- Identify tactical divisions and groups.
- Assign task-specific actions to effectively extricate, triage, treat, and transport the patients to lifesaving care.

The responders to the Aurora Theater shooting performed incredibly well during a once-in-a-career event. Today, with the lessons we learned and the guidance offered in this text, we can be exceptional and bring calm to chaos on the next one.

The modern response to MCIs has evolved out of the military and from necessity. The adage, "Slow is smooth; smooth is fast" made popular from the movie *The Shooter* is a great way to approach an MCI. As paramedics and advanced life support came of age in the early 1970s and America was coming out of Vietnam, it was the perfect merging of these two historic moments. Paramedics were a rare commodity in the days of Johnny and Roy on the hit show *Emergency!* Qualified paramedics on scene, like medics in battle, represented a limited echelon of skills and equipment that provide advanced life support capabilities. The need to perform triage in the street on larger incidents mimicked the need in the field of battle so the most good could be done for the most people, with limited resources.

Any significant MCI will have a shortage of resources, initially. The ability to quickly triage patients and call vast amounts of resources that can be expeditiously assigned and organized is vital to saving lives.

Sorting the patients into immediate, delayed, and minor categories will give those who need the advanced care the highest priority. *It's all about time, and the golden hour cannot be paused. We must act with the utmost efficiency.*

The three primary objectives in an MCI are triage, treatment, and transport of patients. Everything the incident commander does must support these three critical tactical objectives. Surrounding this nucleus of effort must lie an incident command system that remains ahead of the curve regarding resource requests, communications, apparatus, and equipment movement in and out of the scene and maximizes limited resources initially. This is all within a potentially larger incident like a high-rise fire, collapse, explosion, major vehicle accident, or active shooter. Each of these incidents will also tap resources and cause chaos around the incident with additional responders like law enforcement, media, site managers, school principals, and potentially countless other stakeholders.

Advancements in strategies like unified command and mutual aid, along with tactical advancement like rescue task forces, have bridged the gap between fire and law. A rescue task force (RTF) is comprised of fire/EMS personnel embedded with law enforcement to comprise a team with the objective to enter the space immediately adjacent to the threat, and remove patients to casualty collection points. A casualty collection point (CCP) is a location where victims are brought from the hot zone/defensive space, to the warm/cold zones/offensive space to be treated and transported. This allows EMS to follow up and provide medical treatment and transportation much sooner.

CASE STUDIES IN COMMAND

Twin Parks Apartments Incident

CHIEF FRANK LEEB, FIRE DEPARTMENT OF NEW YORK CITY (FDNY)

I was at the FDNY Fire Academy on Randall's Island on the morning of January 9, 2022. I was the *Chief of Training* for the FDNY's 17,000 employees and was in my fourth-floor office catching up from a busy week. On this Sunday, I also had the command chief duty, which meant that I was responsible for citywide duty as the command chief and would respond to large-scale or unusual incidents during my 24-hour tour.

The fire academy is a relatively short ride to the Bronx. The report of the fire was received for 333 East 181 Street (Twin Parks North West). While responding, the first-to-arrive units were receiving reports from the Bronx dispatchers of unconscious victims on the third floor. The working structural fire was soon confirmed and upgraded to a second alarm.

With the information I was receiving, I was quick to recognize this escalating fire as something out of the ordinary and began responding on transmission of the second alarm and arrived shortly after the transmission of the third alarm. Upon my arrival I would assume command. However, I would largely assist and support the command structure already in place and functioning well. While commanding this fire, we had a ring-side view from the front of the structure to the incredible and lifesaving work of our members.

The FDNY responded to what would become one of the worst fires in New York City's history. Yet, at the same time, the fire would result in a historic number of rescues and lives saved, testing the training, dedication, professionalism, and resolve of FDNY members. Tragically, 17 lives were lost at this horrific fire. We are all profoundly impacted when we are unable to save one victim, no less an unimaginable and heartbreaking 17.

While mourning the loss of so many lives, we must also acknowledge the FDNY members from fire and EMS who put forth a heroic effort to rescue and save nearly 100 occupants from the Twin Parks North West building. FDNY members play to win, saving lives at all costs. This FDNY "play to win" mindset on every run of every tour, every day, was evident on this day, at this fire. It is the winning mindset of the FDNY and our members' dedication to training that prepares and positions them to perform under even the most demanding conditions, such as those confronted at the Twin Parks North West fire.

The fire began in Apartment 3N, a duplex apartment on the second and third floors of the 19-story residential apartment building located at 333 East 181st Street. The entrance to 3N was on the third floor, and a stairway inside the apartment descended to the second floor ("duplex down"). The fire, which was caused by an electric space heater, began in a bedroom on the lower level—the second floor. The fire area at the Twin Parks operation was extensive, consisting of both levels of Apartment 3N (the apartment of origin), Apartment 3J (across the hall from 3N) and the third-floor public hallway. With the doors open to both 3N and 3J, both apartments and the public hallway quickly became part of the fire area. Given the science of fire dynamics (including flow path and the stack effect), it is clear that high pressure within Apartment 3N moved fire, heat, and smoke toward lower-pressure areas on the third floor and vertical stairwell shafts. However, an integral part of the story was a chain of open doors that allowed smoke to permeate the building.

The open door to the fire apartment, open stairwell doors on numerous floors and open doors to apartments on several floors above the fire floor allowed deadly, toxic smoke to quickly spread to apartments throughout the structure. Video evidence shows that as the fire department was arriving, smoke had already reached the 19th (top) floor. This left precious little time for rescues, requiring a fifth alarm assignment and herculean FDNY effort to save lives in imminent peril.

In total, there were more than 60 fire victims, and FDNY members removed more than 30 victims from the structure in cardiac arrest. Seventeen people succumbed to their injuries, including eight children. A majority of the victims who died were at least ten floors above the fire—nine of the fatalities were found in stairwells on the upper floors, far from the fire. All of the deceased died from smoke inhalation (fig. 15–3). Both fire and EMS members performed many courageous actions inside and outside the fire building.

While this incident took place in a 19-story structure, many of the lessons that can be taken from this fire are applicable for many other incidents that have the potential to produce a large number of victims. This can include anything from an overturned school bus to a carbon monoxide (CO) emergency during a crowded religious service to a fire in a nursing home or hospital. Simply stated, you don't need to have tall structures to respond to large-scale incidents.

My goal is that I can transfer several key takeaways to provide insight so you can better prepare, develop, and employ similar strategies and tactics before the next time you respond to large scale incident—wherever that may take place. It is about transferring my experience to you (fig. 15–4).

There are so many important lessons to reinforce from the Twin Parks Fire. Below I present 10 for your consideration. I chose these specific 10 to highlight for use in this book, as these are items that can easily be overlooked and not often discussed.

Fig. 15–3. A memorial for the victims at the Twin Parks Incident
Source. Photo courtesy of FDNY

Fig. 15–4. IC and team at the Twin Parks Incident
Source. Photo courtesy of FDNY

1. **Incident priorities, command and control, and leaders' intent**—These items are very much interrelated. It is vital for the incident commander to clearly communicate the incident priorities. Often times, the incident priorities may appear obvious; however, incident priorities change during the course of the incident and must not be left for others to guess or interpret. This is best accomplished when the leader's intent is clearly communicated and it is all tied together by an efferent and clear command and control structure.

2. **Resource management**—Tracking and calling for needed resources early on—especially when the incident is producing large numbers of victims. Simply stated, more victims require more resources.

3. **Time to task completion**—If you do not know how long it takes to stretch and operate a hoseline on the third floor or how long and how many firefighters it takes to remove an adult victim, it is difficult to calculate your resource needs. This must be drilled upon and times tracked. Data-driven decision-making makes for a more efficient and productive fireground.

4. **Street management plan**—Access and egress for ambulances is most often not considered early on at an incident and sometimes not until victims are produced. This can add unnecessary delays when removing victims from the scene. Think about the street management plan early on.

5. **Communications**—Clear, calm, and concise. Be mindful to not communicate unnecessary information, as critical information must be able to be transmitted. Equally important—listen to transmissions, as so much situational awareness can be gained. Remember, as the victim count increases, so does the number of priority transmissions. Locating a victim is a priority transmission. What does the radio traffic sound like when there are 60 victims?

6. **Radio frequencies**—Consider additional radio frequencies for use as command or secondary tactical channels as the incident expands. Non-critical communications taken off the primary tactical channel ensures important transmissions are not stepped on or missed.

7. **Surge capacity**—Have a written plan and know the number of patients you can bring to a hospital without overwhelming the emergency room. Can your emergency room properly care for several people in cardiac arrest?

8. **Reflex time**—Understand how long it will take to have adequate resources on scene. Consider the time needed for 5, 10, 30, or more ambulances to arrive.

9. **Maintaining a manageable span of contro**l—We all know when it comes to the ideal span of control that three to seven (with five being ideal) is the goal. Strive to stay within this goal and consider sectors (divisions/groups), both horizontal and vertical, to stay compliant with span of control.

10. **Stay current**—Modern incidents require commanders that have an in-depth knowledge of today's structures and what is burning. Understanding fire dynamics enables today's incident commander to proactively and appropriately implement strategies, tactics, and tools that demonstrate mastery of the modern fire environment. Training and education are a never-ending process—keep learning.

SAW-CSS-RECEO-VSS

The SAW-CSS-RECEO-VSS system has been used nationwide for years and is easily applied to MCIs. You will notice that *italicized words correspond with key points on the tactical worksheet* that follows.

Size-Up

Remember, size-up is a mental, ongoing process.

- Start with your *FPODP or FIRST* algorithm.
- *MCI type* refers to the nature of the MCI, or what caused it. Three types exist. First is a trauma MCI, like a MVA or a shooting. Second is a Hazmat MCI, like a leak or exposure to a hazardous material. Third is a medical MCI, like a sick building or several people with food poisoning. If this is an explosion or suspected intentional act, consider secondary devices.
- *Type* of patients refers to how many patients and what level triage they are currently. Based upon the START system, patients will be triaged in one of three categories: immediate, delayed, and minor.
- Your *initial size-up* will conclude with an arrival report to incoming units; however, remember that size-up is continuous. As you talk to the RP, bystanders, occupants, employees, and attempt a lap, you will garner more critical information. The *IOCAN* system is designed to give a quick arrival report. Your arrival report is likely dictated by your SOGs. Consider any local MCI protocols when giving your arrival report.

Apparatus Placement

Apparatus placement should be considered early, as you typically do not get a second chance. Plan ahead based upon your type of apparatus, mission, SOGs, and order of arrival.

- A first-arriving engine will likely attempt to see *three sides* of the scene, allowing the officer to have a tactical advantage before stepping off the rig. An MCI could involve fire or explosions, so be cautious and consider the long-term impact of your staging area.
- The *hot zone* is any area that will cause damage to apparatus, including but not limited to collapse, thermal insult, excessive smoke, hazardous materials, power lines, and so on. Ensure that apparatus is staged out of the hot zone, including potential changes in conditions.
- The *travel plan* is critical to identify early before more units arrive on scene. The more patients, the more apparatus and ambulances will be coming through the scene. Depending upon the size and scope of the incident, some ambulances may make more than one round trip to and from the hospital. A one-way travel route for ambulances to pick up patients from treatment areas is ideal and reduces congestion and potential for backing accidents.
- Consider a *staging* area in addition to the travel plan and a *helispot* for air ambulances to land and take off.

Water Supply

Water supply will likely not be a factor unless you have a working fire with MCI or need water for decontamination during a hazardous materials incident. Water supply must be considered as soon as possible to take advantage of any hydrants while responding into the scene.

- *Who* is securing the water supply? *Where* is the water supply? Is it reachable by one engine, or is this an extensive lay? Will water tenders be required?
- *What* is the supply? A dry or wet forward lay by the first-due engine from the hydrant to the fire? A wet forward lay by the second-due engine? A driveway/split lay? A reverse lay by the second engine from the first, out to the hydrant? A tank transfer? A water shuttle from a distant hydrant, or drafting operation?
- Once a patent water supply is established, it should be announced on the tactical channel.

Command

Command must be established and announced. *There can only be one IC at a time, and the IC must be on scene to assume command.* While your SOGs will determine specific parameters, some fundamental best practices exist.

- When the first-arriving officer is a chief or staff officer with no tactical capability, assuming command is the most likely option. Developing an IAP and calling additional resources will typically be the best decision. That said, during an MCI, you may begin START triage.
- *Name* the incident and identify the *ICP.*
- You may find yourself working with *law* enforcement early and enter into *unified command. See chapter 16 for more on unified command.*
- Call *additional resources alarms* early and confirm the staging location.
- Set up your *ICS* early. A medical group is adequate for most incidents; however, a medical branch would be more appropriate for larger incidents.
- Do you have adequate channels for *communications?* Consider a separate tactical channel for the medical group/branch.

Strategy

Obviously, an MCI is going to have a lifesaving priority. That said, you may have to determine if the strategy is offensive, defensive, or combination based upon any potential threat from an active shooter or hazardous material, for example. This may require a combination strategy as victims are brought from the defensive space to the offensive space.

Safety

The three primary safety concerns with MCIs are scene *security* from threats, *secondary devices* from terrorist acts, and *hazardous material* exposure. That said, an MCI will be the result of another event, such as a structure fire, vehicle accident, building collapse, active threat, and so on. Each of these has separate and distinct safety concerns in addition to the three primary listed above. Keep those in mind and set up your divisions/groups and safety officer accordingly.

Rescue

The three primary tactical objectives for an MCI are triage, treatment, and transport. Each will require supervision of some level. For most MCIs, a medical group supervisor will suffice and manage all three objectives.

- If the incident is large enough where a medical branch is established, the more supervision will be required. A triage unit leader may be required just to supervise the triage of vast number of patients.
- A treatment unit leader will potentially have an immediate, delayed, and minor treatment area for the patients with manager supervising each area.
- A patient transportation unit leader/group supervisor may be required for transportation of patients via ground and air ambulances.
- Consider RTFs working under a rescue group supervisor.
- On a larger incident, a medical communications coordinator (MCC) would be tasked with contacting the trauma center to determine where patients are to be transported.

Exposures

For an MCI, exposures could be other areas where victims may be *located away* from the incident. This could be due to stray bullets, or contaminated victims moving offsite. Consider *downwind* exposures due to hazardous materials releases and plumes.

Confinement

As with the shooting at Mandalay Bay in Las Vegas, victims may be spread over a large area and they may be mobile. Walking wounded are a common phenomenon with large-scale MCIs.

- If you have hazardous materials contamination, it is critical that the walking wounded not spread the hazardous materials to hospitals.

Extinguishment

Is whatever caused the MCI neutralized? It may be an *active threat*, or a *hazardous material leak*. Ensure that no additional injuries are sustained.

Overhaul

Ensure you have *all* victims, and all responders are accounted for. This includes knowing *who* was transported and *where* they were transported. Family will want to know.

Ventilation

For an MCI, ventilation refers to extreme *weather* exposure to victims while they are being treated and awaiting transport. Consider *shelter* like large tents or buses.

Salvage

Talk to the owner/occupant/*RP* or manager to transfer care of the scene to the appropriate party.

Support

Here, we consider what is needed to support the IAP. Several agency-specific resources are certainly available. As a minimum, consider the following:

- *Law enforcement (LE)*
- *Investigator*
- *Public information officer (PIO)*
- *Safety officer (SO)*
- *Red Cross*
- *Utility companies (gas, electricity, and water)*
- *Rehab*
- *Air/light unit*
- *Hazmat team*
- *Buses*
- *Air ambulance*

ICS and Communications for Multi-Casualty Incidents

Like any incident, the use of ICS on an MCI will vary based upon the size and complexity of the event (see figs. 15–5, 15–6, and 15–7). You may only require a medical group supervisor if you have a MVA with 5–10 patients; however, you may require multiple branches if you have hazardous materials and/or significant fire and rescue challenges with dozens of patients.

A separate tactical channel should be considered for medical operations. If the incident warrants and you have branches established with over 50 patients, for example, multiple tactical channels may be required for triage, treatment, and transportation so one channel does not become overwhelmed.

Specific positions may be utilized to break down the division of labor to a more manageable span of control. Each position can be added as needed and as the incident escalates. Below is a list of positions to consider for MCI operations of a larger scale. You will likely not use all positions on the day-to-day MCI responses. However, the ability to scale quickly and stay ahead of the incident power curve is vital in the once-in-a-career incident. You will undoubtedly have local protocols, policy, and SOGs. Below are definitions and positions to consider as the MCI expands. Consider this a toolbox from which to choose to command and coordinate an MCI.

Medical Group Supervisor

Like any tactical supervisor, a MGS bridges the gap between the units working at the task level and the IC at the strategic level. The MGS will likely have the objectives of triage, treatment, and transport of patients and medical communications with the trauma center that is responsible for coordinating patient transportation destinations. In a smaller, simpler MCI, this may be all that is required to manage the medical functions of the incident. If an incident was

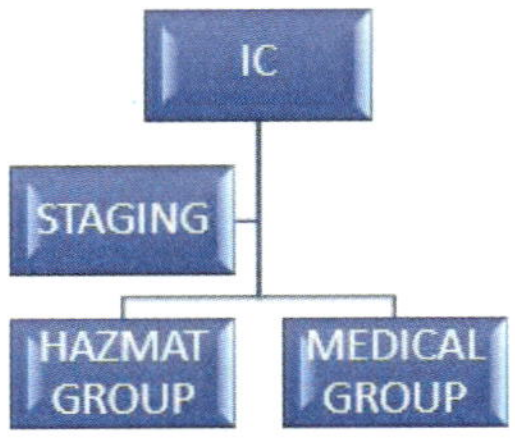

Fig. 15–5. A simple MCI organizational chart

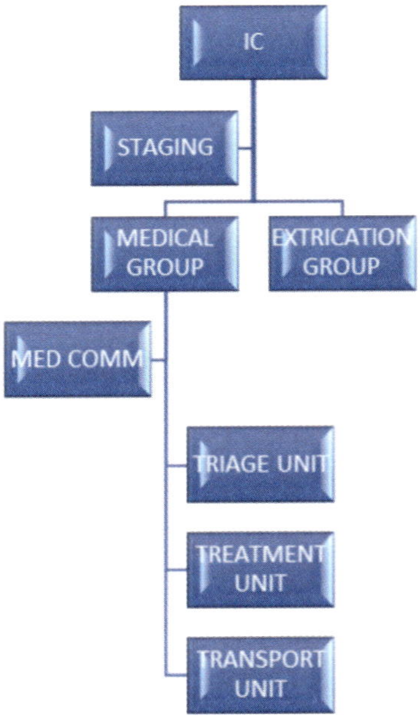

Fig. 15–6. Example of a moderate-sized MCI organizational chart

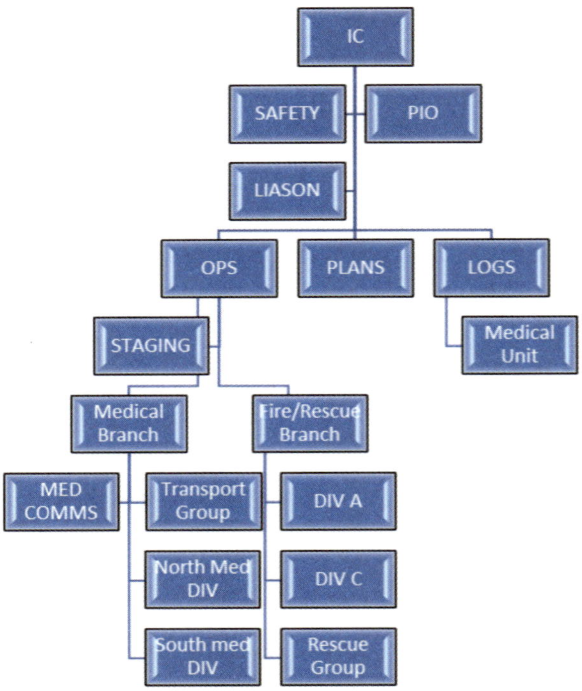

Fig. 15–7. Example of a complex MCI organizational chart. Each medical division will have triage and treatment units.

geographically spread out over a large area, multiple medical *division* supervisors may be required to function as a tactical supervisors to break down the incident to a manageable span of control. For example, Medical Division North, Medical Division South, and so on.

Medical Branch Director

The medical branch director (MBD) is a step above a MGS and would be utilized to manage larger MCI operations where multiple medical divisions are established as described previously.

Transportation Unit Leader/Group Supervisor

The Transportation tactical supervisor may be a *unit leader* or a *group supervisor*, depending upon the scope of the incident. The transportation *unit leader* would be utilized on small- to medium-sized incidents and report to the MGS. On a larger MCI when a medical branch is established, a transportation *group supervisor* would be utilized and report to the medical branch director. *There should be one transportation unit leader/group supervisor per incident to prevent duplicate ordering of ground and air ambulances that could over-tax the system. In either case, they coordinate with the medical branch/group/divisions.*

Triage Unit Leader

The triage unit leader will have single resources reporting to them with the objective of triaging all patients into immediate, delayed, minor, or deceased categories. The triage unit leader would coordinate laterally with the treatment unit leader and vertically with the medical group/division supervisor.

Treatment Unit Leader

A treatment unit leader has the primary tactical objective of supervising and supporting the three treatment areas: immediate, delayed, and minor. Each of these areas would have an area manager in charge of the respective area. The treatment unit leader would coordinate laterally with the triage unit leader and vertically with the medical group/division supervisor.

Medical Communications Coordinator

The MCC is responsible for coordinating patient transportation destinations. The MCC communicates and coordinates between the trauma center that assign patients to various hospitals, the treatment unit leader, and the patient transportation unit leader/group supervisor.

Additional Positions

Other positions that may be utilized on extremely large incidents where resources are available include:

- Immediate, Delayed, and Minor Treatment Area Managers
- Ground Ambulance Coordinator
- Air Ambulance Coordinator
- Patient Loading Coordinator
- Medical Supply Coordinator
- Transportation Recorder
- Morgue Manager
- Helispot Manager
- Staging Area Manager
- Medical Unit for First Responders

WISDOM FROM THE MASTERS

Management of the Intentional Mass Casualty Incident—A New Concept

CHIEF ERIC SAYLORS, EL CERRITO (CA) FIRE DEPARTMENT

The following section is a framework for effectively managing an intentional mass casualty incident (IMCI). The framework starts with a big-picture view of the events, followed by a small-picture view of operations. The big-picture view outlines the differences between an IMCI and an (unintentional) MCI, identifies the principle of the problem, creates a common operating picture, and orders the incident priorities into strategic and tactical objectives. The small-picture view describes the current practices supported by research at a tactical level when encountering a massive killing perpetrated by an intelligent actor.

Big Picture—MCI versus Fast-Moving IMCI

This section divides mass casualty incidents into two types of events: MCIs and IMCIs. IMCIs and MCIs consist of patients involved in a shared event resulting from an accident. Think of vehicle collisions, mass transit derailments, or hazmat exposures. IMCIs consist of victims resulting from an intentional event instigated by an *intelligent actor (IA)*. The incidents are deliberate and, therefore, may have secondary events in the future. Think of a school shooting, the Boston Marathon bombing, or a vehicle driving into a crowd. The presence of an *intelligent actor* is the distinguishing factor between an MCI and an IMCI. Intelligent actors create an unknowable threat to responders that resists timely incident stabilization. In other words, responders must take action in a threat zone before the scene is safe or cleared. Operating in a threat zone requires a different mindset than a traditional MCI. The following outlines this mindset by describing the principle of the problem, the common operating picture, and incident priorities.

Principle of the Problem

Time is everything. Victims of IMCIs are typically bleeding to death at a rapid rate. Most walking wounded flee the scene once the killing starts, leaving behind the critically injured. Evidence suggests roughly half of the victims bleed to death without intervention in the first ten minutes. Surgery is the only thing capable of saving most victims; therefore, the time from injury to surgery is critical. Any action that delays the victim getting to surgery costs lives. IMCIs are fast-moving MCIs that require immediate action and transport, skipping many activities prescribed in traditional MCIs, such as triage, patient counts, and triage tags. The goal is to control massive hemorrhages on scene and transport every victim to surgery within 45 mins.

Common Operating Picture

Most EMS doctrine addresses patients with specific complaints requiring prescribed care. Patients tend to have multiple caregivers per patient in a stable environment, as opposed to victims, who are people in a threat environment that require immediate removal. Think of victims as those encountered in a burning house, drowning, or trapped in confined spaces. The priority for victims is to remove them from the threat zone. People found in burning homes are not triaged before removal, nor are downing victims assessed in the water. Fast-moving IMCIs produce victims. Victims must be removed from the threat zone before transitioning to patients and receiving traditional care.

Threat zones can be secured or cleared. A *secured* threat zone has an element of force protection that can return effective fire but is not under fire. In other words, you have an LE officer with a gun to protect you. Areas with a stable law enforcement presence are secured. A *cleared* threat zone is an area that received a systematic search by law enforcement, eliminating any credible threat. Clearing areas takes enormous time and should happen after all victims and fire/EMS staff are removed. In an analogous way, the fire service can think of clearing as an overhaul after a fire; it happens once the threat is contained and the victims are gone. The goal of responders during an IMCI is to only operate in secured areas before the terrain is clear. Finally, there are no *safe* areas when encountering an intelligent actor.

Zones of Operation—Direct Threat, Indirect Threat, and Evac Zone

Events with hostile intelligent actors construct zones of operation called direct threat, indirect threat, and evac zone. Each has operational priorities and limitations. The direct threat (DT) zone typically refers to the line of fire or path of explosion. Without adequate training, equipment, and force protection, fire or EMS personnel should not operate in the DT zone. The indirect threat (ID) zone is outside of the line of fire or path of explosion. The ID zone is contiguous with the DT zone, where fire and EMS personnel control massive hemorrhage and extricate victims. Finally, the evac zone is the location victims are moved for transport. An evac zone is not a casualty collection point (CCP) but a loading site for victims from manual to vehicle transport. CCPs are used in rare occurrences when the manual movement of victims is impossible due to an engagement with a hostile force or bottleneck in the terrain. Managing a CCP is a specialized skill beyond most first responders' training and equipment and should only be utilized as a last resort.

Incident Priorities, Strategic Objectives, and Tactical Objectives

The progression of incident priorities on a fast-moving IMCI is life, scene security, and recovery. Meeting incident priority of life preservation requires law enforcement to stop the killing by neutralizing the threat, followed by fire and EMS services stopping the dying by providing bleeding control and movement of victims to definitive care. Law enforcement forms contact teams and moves toward the stimulus, constantly engaging the IA(s) until they are neutralized. Once the threat is neutralized, fire and EMS crews operate in the ID zone to control massive hemorrhage, extricate the victims to the evac zone, sort victims based on injury, and transport the victims to definitive care.

Meeting the incident priority of scene security includes the removal of fire personnel and victims from initial threat zones to prevent additional injuries, device triggering, or evidence destruction. Finally, a large number of transported victims can move the location of an IMCI to an MCI at the hospitals. Additional resources should be assigned to hospitals to assist with patient care if needed.

Meeting the incident priority of recovery includes providing victim notification/unification centers at the initial scene and hospitals. In addition, the responders need immediate mental health assessment and treatment before going home. Finally, resources and plans for long-term mental health needs for the responders, victims, victims' families, witnesses, and community must be addressed.

The proceeding incident priorities and objectives are listed in the ad hoc IAP that follows:

Incident Action Plan Framework Based on *NFPA 3000*,[2] Tactical Emergency Casualty Care (TECC), and the Hartford Consensus[3]

Incident Priorities

- Life
  - Scene security
    - Recovery

Strategic Objectives

- Law—Neutralize threat (stop the killing)
- Law—Clear building
- Fire—Provide bleeding control and movement of victims to definitive care (stop the dying)
  - Remove all unnecessary fire personnel from scene
  - Assist hospitals with impact
    - Provide for immediate recovery
      - Provide for early recovery

Tactical/Operational Objectives

- T—Threat suppression (law)
- H—Hemorrhage control
- RE—Rapid extrication
- A—Assess victims
- T—Transport victims
  - Full personal accountability report (PAR) of units assigned to incident
  - Dispatch engine/trucks to hospitals to assist with treatment/triage
    - Notify city office of emergency management
    - Set up victim notification and unification centers at scene, at hospitals, and at a remote location
    - Use PIO for social media and media messaging to victims' families
      - Provide mental health needs for responders, victims, victims' families, and witnesses

Small Picture

Empirical evidence suggests the decades-old rescue task force (RTF) is objectively ineffective. The diamond formation of the RTF is logistically heavy and slow, relying on a maneuver known as diplomatic protection run backward. In the tactic, firefighters are surrounded and escorted as diplomats into the threat area instead of out of it. In addition, the RFT typically attempts to triage victims in the threat zone while struggling to move them to the evac zone. As a result of struggles with an RTF, a team out of Sacramento, California, consisting of medical doctors, researchers, combat psychologists, fire/EMS personnel, and military special forces operators developed an alternative tactic known as a rescue strike team (RST). Over two hundred repetitions of RST in multiple environments yielded times 74% faster than an RFT tactic. The study measured the time of fire/EMS entering the threat zone to when all victims and rescuers arrived in the evac zone. The following describes the basics of the RST.

The RST assumes that law enforcement should devote all officers to suppressing the threat by killing the IA, locking them down, or driving them from the scene. Once the threat is suppressed, there is no need for a diamond formation with a one-to-one ratio of fire/EMS personnel to law officers. Instead, the fire/EMS personnel are escorted to the victims by one LE officer through a secured corridor. The LE escort is part of the initial contact team and knows the best path to the victims. The RST operates under a rescue unit leader (RUL), who is typically the first captain on the scene. The RUL makes contact with the LE escort and confirms the threat is suppressed. The RUL then leads the on-scene fire/EMS crews with the LE escort to the victims. The RUL is the eyes and ears of the IC, the level of accountability for all rescuers in the threat zone, and direct contact with LE. The RUL should always stay within arm's reach of the LE escort to provide constant communication with LE. The link-up of the RUL and LE escort links communications between fire and law early on in the incident before the establishment of unified command, regardless of radio differences.

An RST can consist of 4 to 20 rescuers. Evidence suggests 20 rescuers with a small wound kit and a moving device such as a drag sled or Fox Trot litter can control hemorrhaging and move 50 victims 200 feet in about twelve minutes within a secure corridor. Members of the RST team focus only on controlling massive hemorrhages and moving victims. Victims are treated and moved on a first-come, first-serve order. Triage is discouraged in the threat zone, as it is unreliable and timely. Victims are either dead and marked with a purple glow stick or removed. In addition, victim counts are discouraged and replaced with round number estimates. For example, a victim estimate is roughly 10, 20, 30, 60, and so on. Exact victim counts in the threat zone are timely and unreliable, as the counter cannot know the number of dead and alive. As the last live victim is removed from the threat zone, the RUL announces to the IC that the final victim and rescuer are coming out and moves to the evac one with the LE escort.

As victims arrive in the evac zone, they are sorted into three categories based on wound location: abdominal, chest, or extremity, a process known as the ACE acronym. Wound treatability drives the logic of the ACE acronym. Both abdominal and chest wounds are in non-compressible areas, requiring surgery ASAP, whereas tourniquets can control bleeding from extremities on scene. Therefore, given a choice, victims should leave the scene in the order of the ACE acronym; abdominal first, then chest, and finally extremities.

The RST is a simple and effective tactic based on an ink blot maneuver, as opposed to the diplomatic protection of the RTF. In an ink blot maneuver, force protection secures an area so that logical units have freedom of movement in a secured area. RST is not the only effective tactic in an IMCI, but it is the next evolution. As of this writing, it is yet unknown. The purpose of this is an introduction and awareness to the RST concept.

Summary

Fire department incident commanders must be proficient at managing a multi-causality incident, as nearly any response has the potential to become one. Management of patient triage, treatment, and transport alone is a challenge, and when this operation is occurring simultaneously with the mitigation of at least one hazard, it will test even the most experienced IC and organizations. The pace of the media coverage of significant events requires proactive planning for information management, as does the degree of coordination with patient prioritization and facility destinations. As the MCI is often an incident within an incident that requires management, these added external communication demands become an incident outside the incident that require management and should not be overlooked.

Time is the most critical factor in a multi-casualty incident, as patients must be rapidly triaged, treated, and transported to definitive care. As it has been repeated throughout this text, being aggressive is proactively seeking opportunities to address the time that you can control. Staying ahead of the incident power curve for the MCI starts ahead of the incident with an IC who has a working knowledge of EMS protocols, nomenclature, resource levels and destination capabilities. With the increased potential for hostile events to become an MCI, proactive planning and practice for managing through unified command (chapter 16) is a must. As it has been shared through the "wisdom from the master's" contributions, from apartment fires to active shootings, the next major MCI could be your next response. Maintain the "what if" mindset and be prepared.

Chapter Review

Review Questions

1. Describe START.
2. What are the pros and cons of models which require law enforcement securing or clearing a scene for fire department or EMS response?
3. What are the three primary objectives in a multi-causality incident?
4. Describe a casualty collection point.
5. What are the responsibilities of a transportation unit leader?
6. Explain some of the expanded communication demands of a multi-casualty incident.

FESHE Strategy and Tactics (CO279) Related Content

The content contained in chapter 15 provides detailed information specific to multi-causality incidents. While the FESHE CO279 course does not specifically reference multi-causality incidents, they involve the same processes and principles of structural firefighting requiring size-up, determination of incident priorities, selection of appropriate strategy, and coordination of tactics and resources. The chapter format supports the CO279 course outline by detailing information on multi-causality incident operations, communications, and coordination of these events, and ICS/NIMS roles and responsibilities to support and manage them.

NFPA 1021 Job Performance Requirements

The information in this chapter can be utilized to support training and educational programs associated with the Emergency Services Delivery Fire Officer I JPR 4.6, 4.6.1, 4.6.2, Fire Officer II JPR 5.6, 5.6.1, and Fire Officer III 6.6 and 6.6.1. For larger-scale incidents and pre-planning, the content in this chapter also supports training and education for Fire Officer III 6.6.3 "Develop a plan for the organization given an unmet need for resources that exceed what is available in the organization."

Endnotes

1. "Aurora Century 16 Theater Shooting: After Action Report for the City of Aurora," Tri-Data Division, System Planning Corp., April 2014, https://www.courts.state.co.us/Media/Opinion_Docs/14CV31595%20After%20Action%20Review%20Report%20Redacted.pdf.
2. National Fire Protection Association (*NFPA) 3000: Standard for an Active Shooter/Hostile Event Response (ASHER) Program* (NFPA, 2024), https://www.nfpa.org/codes-and-standards/3/0/0/3000.
3 Lenworth Jacobs et al. and the Joint Committee to Create a National Policy to Enhance Survivability from Mass Casualty Shooting Events, "Improving Survival from Active Shooter Events: The Hartford Consensus," Journal of Trauma and Acute Care Surgery 74, no. 6 (June 2013): 1399–1400, https://journals.lww.com/jtrauma/Fulltext/2013/06000/Improving_survival_from_active_shooter_events__The.3.aspx.

16

Unified Command

Fig. 16–0. Sacramento Metropolitan Fire Battalion Chief and County Sheriff representative initiate unified command.

Introduction

Whether a medical aid, house fire, high-rise fire, or major disaster, every incident starts at the local level. On the headline-grabbing big ones, some poor responder arrives on scene first and realized, "Oh man, this is way beyond me!" As legendary Los Angeles City Battalion Chief John Mittendorf would say, "Why did I come to work today?"

When you arrive on scene and realize that this is a massive incident, take a deep breath, remain calm, and size up your situation using the same FPODP method you would use on a room-and-contents fire in a small house. Set the tone for the incident with the first arrival report. Remember the old saying, "as the first hoseline goes, so goes the rest of the incident." Another way to think is to realize that you have a tremendous opportunity to set the incident on the right foot and keep everyone calm.

CASE STUDIES IN COMMAND

Madison Incident

CHIEF KILEY KEELEY AND CAPTAIN SHAWN LEMON, SACRAMENTO (CA) METROPOLITAN FIRE DISTRICT

(Lemon) As firefighters, we spend much of our time training for the unexpected, as impossible as that may be. We pride ourselves on being flexible problem-solvers, often called upon to assist the communities we serve. As a newer captain within my organization, I made it a point to develop outside-of-the-box thinking and solid decision-making. To make this possible, we *must* have a strong understanding of our priorities and our capabilities as a unit as well as a department. As leaders within the organization, understanding the importance of gaining situational awareness and developing an action plan that accomplishes desired goals creates the foundation for the rest of the call. A bad foundation can doom an incident before it ever has a chance to succeed. Contrarily, a strong foundation can overcome small mistakes made throughout an incident and still support a successful outcome.

To give some background, Sacramento Metro Fire operates a three-man engine company with very clear responsibilities. E42 is one of Sac Metro Fire's RIC Engines and is equipped with specialty equipment. This equipment ranges from air monitoring to forcible entry tools that aren't part of the normal compliment. The apartment complex where this call took place is a two-story complex with units on the front and back. This complex has a one-way-in, one-way-out-style parking lot shaped in a U around the structure.

In late June, 2022, E42 was given an unexpected opportunity to change the lives of some members of the community, as well as their own. E42 was dispatched early in the morning for a domestic disturbance and were requested, by Sacramento Sheriff's Officers on-scene, to stage near the incident. The crew of E42 responded to a location across the street from a mid-size garden-style apartment complex. An update was given through dispatch that there was a male adult inside an apartment holding his one-year-old child hostage with a knife and was now attempting to start a fire. From that update, the complexity of the call significantly increased.

E42 dressed for fire and began reviewing the complex map locating the exact apartment involved. E42 was shortly updated with information that there was a well-developed fire and that crews were now cleared to enter. E42 responded and balanced the call from a domestic disturbance to a full commercial assignment, which adds four engines, three trucks, two battalion chiefs, and one medic.

Upon arriving at the entrance of the apartment complex, we were met with our first unexpected situation. SAC Sheriffs had been on scene of this call for over an hour and a heavy police presence blocked the access to the complex, with police cars parked side-by-side running the length of about 100 yards. With the entrance we had committed to blocked, we made the decision to gain situational awareness on foot, get eyes on the scene and update responding crews as needed.

Upon reaching the involved building I found a well-involved first- and second-floor apartment with extension reaching into a common attic. This was located mid-span within the apartment block, providing additional exposures to the Bravo, Charlie, and Delta side. I made contact with officers on scene and updated responding crews of the access issues, development of the fire and requested a second alarm due to the large increase of incident complexity. I was informed by the sheriff officer in charge of the location of the male assailant and the child. They were known to be in the bathroom attached to the master bedroom. This room was part of a shared wall with the apartment on the Delta side. Officers had attempted to make access through the wall, but were unsuccessful.

I tasked my firefighter with obtaining a chainsaw from the engine. I updated responding crews that we had a known location of victims and we were going to be breaching the wall of an adjoining apartment to attempt access. At this point of the incident, the second-arriving engine had made it to the scene, made access through the exit, and was in position for fire attack. With the cover of the sheriffs, we were able to open the wall and use our TICs (thermal imaging cameras) to locate the armed assailant, who was bent over in the bathtub. I again updated command that we had made access and requested additional resources to assist us.

Accepting the risk involved, both my firefighter and I made access to the smoked-out bathroom, where we were able locate the armed assailant and the child, both unconscious in the bathtub with the water running. My firefighter reacted quickly and removed the unconscious child from the tub. The child was quickly passed from fire personnel to the officers in the adjoining apartment providing us with cover. My firefighter and I grabbed the unconscious male and quickly developed the plan of moving him, first to the toilet seat next to the tub, and from there, a second move from the toilet to the bottom of the cut hole that was located just above the toilet. At this point, we were met by a second engine company that assisted us in moving the unconscious male to the adjoining apartment. I informed command that all occupants had been removed from the IDLH environment.

The focus was now from lifesaving priority to scene stabilization, due to how involved the building still was in active fire. We swapped out our air tanks and returned to command for an additional assignment.

It wasn't until a few weeks later that we received official word that both individuals involved had made a full recovery. This was a very dynamic call; it could have had a very different outcome. I feel our success lay in the freedom to make the decisions that would make a difference, and in no small way was due to the actions and support of the sheriff's department.

Emergency scenes often contain some sense of chaos at first, but our training and experience quickly bring a systematic plan and a sense of calm for mitigating practically any incident. Most of the time, we all begin on the same page with a common dispatch, relevant updates, and there is a common sequence of events that we rely on to help bring that sense of calm and organization. When things start out differently, especially on significantly complex incidents, then our training, experience, communication, and abilities are pushed to the limits.

(Keeley) Upon starting the vehicle and exiting the apparatus bay, I heard traffic on the tac channel. This was unusual to hear something so early in the call. It was E42 giving an update that they had a well involved first-floor apartment fire lapping to the second floor. They reported they had two victims inside the fire unit and they were entering a neighboring unit, breaching a wall to rescue the victims. This call was different and would require our members to perform flawlessly due to the significant safety hazards and life risks to the fire department members and to the public. The access, building location, and the significant law enforcement presence would all add complexity.

This call was in Battalion 5 (BC5), but close to my battalion, Battalion 7. I knew that I would be arriving simultaneously to BC5. BC5 arrived and established command, while I went forward to take a division. As I approached the fire unit, heavy fire was rolling out of the lower unit and into the unit above, with additional extension to the attic. Multiple Sacramento Sheriff officers with assault rifles and body armor were all over the scene. They gave me an update that we had an adult male with a knife, holding his infant child. He had barricaded himself in the master bedroom of the apartment and lit items in the room on fire. He then retreated to the bathroom and they had not seen him since. I heard a saw running in the neighboring unit. I sent additional crews from my position into the neighboring unit to assist the rescue while other engines were used to attack the fire on the first and second floors. Searches and evacuations were also in progress in the exposure units.

The baby was removed first, unconscious. I provided an update to command, who had already requested multiple ambulances and another alarm. Command was able to remain strategic by allowing tactical division supervisors to run certain aspects of the scene. Next, the adult male was brought out, also unresponsive. Both victims were transported to our burn center in critical condition. The fire was contained to two units, ventilation was completed, and the building was evacuated. No other injuries occurred.

While forward task-level and tactical functions were coordinated, we needed a unified command with law enforcement. They had so much more information based on their experience and knowledge of the scene. They had been on scene and in the apartment, and therefore had good intel on items we would only hear of much later. Law sets up their ICP at a remote location, which in this case was about half a mile away. Fire sets the ICP close to the incident to aid in gathering situational awareness and to allow for face-to-face updates when possible. This difference in our practices did not provide for an opportunity to get into true unified command.

Members of the fire department and sheriff's department were heroic that morning. Both lives were saved, and only two apartments were destroyed by fire. Many conversations with our agencies have taken place and we are working on plans to overcome our unified command challenges.

Post-incident thoughts:

- The importance of flexibility. Just like all firefighters, we arrive to fires with the mindset that we are there to put the fire out. Is that always the job of the first-in engine? Are there objectives that can make a more significant difference? For example, search, command, rescue?
- Chiefs trusting and supporting the crews. What can a department do to develop the trust between company-level crews and command? On this call there were a lot of decisions and requests made prior to the arrival of a command officer. I was humbled by the trust placed in me (a company officer) when I made the decision to immediately add a second alarm and divert from the normal expectation of pulling hose and putting out fire. Trust must be developed between the company and the command level.
- Unified command. After this call it was clear that we had gaps in communication and decision-making between our two agencies. Sacramento Metro and Sacramento sheriffs both recognized the importance of communication and alignment of command. There are already steps being taken to improve communication and streamline responses.

Unified Command

Unified Command (UC) incidents still have the potential for the NIOSH 5 to come into alignment. In fact, there is greater potential, since multiple agencies must now work together and overcome boundaries and differences in accepted risk, nomenclature, accountability, SOGs, and command systems.

The NIOSH 5 can have an even greater potential in UC. The challenges become exponentially increased the more agencies are involved. Just between fire and law, the NIOSH 5 manifest in ways that include, but are not limited to the following:

- *Risk assessment*—Fire and law enforcement look at risk very differently. A cop's job is to go into an active shooter situation, whereas fire may stage. Conversely, we go into burning buildings that cops stay out of, and for good reason. We stay in our lanes based upon the job, training, equipment, and inherent risk respective to our missions.
- *Communications*—We have different terminology, protocol, frequencies, dispatch centers, and norms. Cops predominantly still use 10-codes, while fire uses clear text.
- *Accountability*—We have unique means of accounting for each other, including utilizing different names and labels for varying resources.
- *Incident command*—Law does not utilize ICS as commonly as fire. This creates further issues. However, they will follow our lead. As seen in the above call, there were two separate ICPs.

- *SOGs*—Obviously, the difference in policy, protocol and SOGs is significant, which means varying priorities, resource allocations, and even the potential for conflicting operations.

We responded to a large multi-vehicle crash on a major freeway. There was a report of dozens of vehicles involved before anyone was on scene, including hazardous materials, semi-trucks, and fire (fig. 16–1). We were miles away and could see the plume of smoke.

The first-arriving acting captain arrived on scene and gave an amazing arrival report. It was calm, clear, confident, and concise: "Engine 71 on scene of a large multi-vehicle accident with fire and entrapments. Unknown number of vehicles. Engine 71 is assuming Highway 5 Command. Command Post is mobile. Requesting five additional ambulances, five additional engines, two additional trucks, and two water tenders. . ."

It was amazing. He was as calm as can be. We were all impressed while still responding and knew that he was going to set the incident up for success. It made us step up our focus and professionalism because he set the tone. He would call additional resources, direct them to approach from different on-ramps, and when the battalion chief arrived and assumed command, the command transfer was smooth. It was clear that this acting captain had trained a lot. He was unphased, as if to say, "Okay, seen this before. Let's get to work." That kind of professionalism can only come from dedication and a lot of training.

The accident would involve over 30 vehicles, including multiple semi-trucks, hazardous materials spills, extrications, and fire. Five people were killed. Dozens more were transported using the regional MCI protocols. The incident was one of the largest in California and was ultimately divided into a fire/rescue branch, hazmat branch, and medical branch. Safety, information, and liaison officers were all part of the command staff. Unified command was established between the fire department and California Highway Patrol. It took several days (operational periods) before the freeway was opened again. Countless other agencies responded, from Fish and Game, to Coroner, to County and State OES. This grew to a Type 3 incident requiring significant logistical support and debriefings.

It all started with that first-arriving acting captain. He used the principles you have learned in this book to effectively calm the chaos. He performed a mental size-up before giving an arrival report. He stayed ahead of the incident power curve when he called the world to come and help. He gave clear objectives to incoming units, and he established command, all while remaining calm.

You never know what the next shift or call will bring. You may be the first to arrive on the next 9/11. Utilizing the methods learned throughout this book will make you successful.

Fig. 16–1. Smoke could be seen from miles away after this incident on I-5.
Source. Photo courtesy of California Highway Patrol

WISDOM FROM THE MASTERS

Murrah Federal Building Bombing—Oklahoma City

MARK S. GHILARDUCCI, DIRECTOR OF THE CALIFORNIA GOVERNOR'S OFFICE OF EMERGENCY SERVICES

Throughout my three-plus decades working in public safety, starting as a first responder in the field and working my way up through the ranks, ultimately as the director of the California Governor's Office of Emergency Services, I have utilized ICS exclusively, at every level, to manage a diverse number of incidents, both small and large, one-dimensional, and complex. During my time, I have responded to countless incidents in California and across the country, from fires to earthquakes, search-and-rescue operations to hazardous material incidents, transportation accidents to multi-casualty incidents. By implementing ICS immediately and building an objective-driven IAP, I have been able to effectively coordinate resources, manage incidents, and ensure that everyone involved in an incident remained on the same page, met designated objectives, and operated safely, with minimal confusion or duplication of effort.

Many of the events I responded to started with an initial response of local jurisdictional resources, where we initially approached the scene by establishing a designated IC, developed a quick action plan, and implemented operations. At times, this was all that was needed to effectively manage, coordinate, and mitigate the situation. However, on other occasions, events grew into incidents requiring the coordination and integration of multiple agencies and resources, often from different organizations or disciplines, but all requiring the need to be on the same page. One such incident that I managed as the IC alongside the chief of the Oklahoma City Fire Department was the US&R operations at the 1995 bombing of the Murrah Federal Building in Oklahoma City (fig. 16–2).

Fig. 16–2. Oklahoma City Bombing aftermath
Source. Photo courtesy of Mark S. Ghilarducci

This was a non-traditional, high-profile, and extremely complicated event, where the incident command system played a crucial role in the success of the incident. Early on, an ICS structure was established to help coordinate and manage the rescue and recovery efforts and the multiple agencies and organizations involved. It provided a standardized structure for communications, establishment of measurable objectives by operational period, effective decision-making, and resource allocation, which ensured a more efficient and effective use of resources, enhanced victim accountability, and supported personnel safety throughout the multiple-week event. The ICS helped manage the enormity of the event, which included more than 3000 responders over 25 operational periods, through the establishment of key operational objectives that addressed the continued rescue and recovery of victims; appropriate number and type of personnel assigned; allocation of personnel working both outside and inside the building; evacuation procedures in the event of secondary collapse; establishment of divisions and branches and/or assignment of single specialized resources; establishment of a security perimeter and traffic plan; identification and confirmation of the number of people missing and their work location; search zones and building stabilization procedures; incident communications plan; a medical and safety plan; and specialized situational awareness such as weather and building schematics.

Managing a rapidly evolving incident can be a daunting task, especially when unexpected events occur, or the situation requires immediate lifesaving actions (fig. 16–3). The incident command system provides a clear framework for managing resources and is an essential tool that offers structure, flexibility, and organization for any type of incident. This systematic tool can be utilized by a first-arriving engine company or EMS unit to manage a single vehicle accident, as well as for the management of a multi-alarm, multi-agency event, or large-scale multi-jurisdictional/regional disaster.

Whether at the incident scene or at the Emergency Operations Center, ICS ensures that everything is well-coordinated and runs smoothly by providing a standardized all-risk incident management concept, built around incident action planning, defined measurable objectives, logistics support, and operational tactics. It allows responders to rapidly integrate into an organizational baseline structure that matches the complexities and demands of the incident or multiple incidents, without being hindered by jurisdictional boundaries or differences in terminology.

Fig. 16–3. Los Angeles County—Federal Emergency Management Agency US&R Task Force performing rescue *Source.* Photo courtesy of Mark S. Ghilarducci

A key element of ICS is its considerable internal flexibility. It can expand or contract to meet unique needs or types of incidents. This flexibility makes it a very cost-effective and efficient management tool for both small and large, traditional, and non-traditional situations.

I have also utilized ICS to manage multiple non-traditional events, such as two Super Bowls and other similar large-venue events, humanitarian support for migrants crossing along California's southern border, swiftwater rescues, post-fire debris removal operations, security events and civil unrest, and of course, the response to COVID-19.

Organized into five functional areas, ICS sets in place a standard organization that includes five functional elements: command, operations, planning, logistics, and administration/finance, and the establishment of a unified command structure which allows all personnel, agencies, and organizations involved in the response effort to work together under a single command structure with a designated IC. This baseline structure ensures that any response effort, no matter the cause or type of event, is organized in a consistent manner, is effectively coordinated, and is scalable, meaning that the organization can expand or contract as dictated by the incident's needs. Whether responding to a complicated natural disaster, a structure fire, or a single-patient EMS event, the ICS consistently ensured that I managed the incident effectively, safely, and efficiently (fig. 16–4).

Incidents today are often unpredictable and can rapidly evolve into extreme events. Incident managers need to gain the experience to make the right calls and be willing to consider all aspects of information that can determine the coordination and direction of the incident. Utilizing ICS as your baseline incident management tool, built around situational awareness, event intelligence, input from technical experts, development of an action plan, and actionable objectives will minimize tunnel vision and maximize the safe and effective mitigation of an incident, no matter how large and how many agencies respond.

Fig. 16–4. Director Ghilarducci at an incident briefing
Source. Photo courtesy of Mark S. Ghilarducci

The Impact of 9/11 on Unified Command

UC has become much more prevalent across America since September 11, 2001. On 9/11, America was attacked on a scale and magnitude that had not been seen since Pearl Harbor on December 7, 1941. In addition, FDNY and the city of New York, by sheer size and scope, were not accustomed to calling mutual aid, as they had ample resources to handle day-to-day incidents, and several significant ones. The result was several key lessons learned regarding the national-level of response and need for unified command and coordination of resources on an unprecedented scale.

FEMA US&R Sacramento, California Task Force 7 (CA TF-7) responded on September 11, 2001 to New York City, and we spent 11 days at Ground Zero (fig. 16–5). Virtually all the FEMA US&R Task Forces would respond in the weeks following the attacks.

Since we responded on the day of the attacks, we had a first-hand experience of how the lack of a NIMS slowed our response and ability to coordinate with FDNY. The FEMA system launched the US&R teams; however, there was not a means to coordinate the response with FDNY chiefs and city leaders, so the process of getting to the pile was slow and frustrating.

In the days that followed, there would be countless citizens, fire departments, construction workers, and other resources that would self-dispatch into the scene, creating even more chaos. These well-intentioned individuals and groups looked like ants on a hill while they unintentionally created a greater burden to the ability to organize the incident and ensure the IAP was effective.

As you can imagine, the scene was massive and required unprecedented command and coordination efforts. In the face of such a disaster, especially not knowing how many of their own were lost, the FDNY members were awesome. They worked tirelessly and many would later succumb to the effects of exposure during their heroic efforts in the years that followed. We had lost brothers as well. The national US&R community is a tight-knit group and we had built friendships through training together with the FDNY Task Force members and through the Fire Department Instructors' Conference (FDIC).

As we attempted to engage in the operations, the size and scope of the destruction made our ability to communicate and coordinate with the appropriate leaders extremely challenging, to say the least. Barriers to communication, lack of defined processes, freelancing, and field personnel not knowing our mission and capabilities all added to the myriad of challenges. Many of our counterparts were buried in the pile. The Special Operations Command (SOC) units responded, and many had died. FDNY is a massive organization and SOC/US&R

Fig. 16–5. Sacramento Task Force 7 responded to New York City following the 9/11 attacks. *Source.* Photo courtesy of CA US&R Task Force 7

members comprise a very small percentage of the total personnel. So, in short, no one knew who we were.

One defining moment was when I spotted an FDNY Task Force 1 tent in the sea of humanity, equipment, and carnage. I stepped into the tent and thought I saw a ghost. It was my friend, Bobby Athanas from FDNY Rescue 3. I was sure he was dead. We embraced tightly. He then told me that another dear brother, Dennis Mojica of Rescue 1, had died. This would be the roller coaster of emotions that defined our 11-day deployment.

Unified command is all about relationships before the incident. Firefighters and all emergency responders and military personnel know that trust is the key to success. Relationships and trust built *before* an incident will pay huge dividends. In the midst of the chaos, seeing a familiar face in a different uniform brings relief and can be an anchor point from which to build a unified IAP.

Homeland Security Presidential Directive-5

On February 28, 2003, Homeland Security Presidential Directive-5 (HSPD-5) was launched. This created the NIMS to galvanize the lessons learned from 9/11 into a plan to move forward and be better trained and equipped in the future.

The following key components are directly out of HSPD-5:[1]

Purpose

To enhance the ability of the United States to manage domestic incidents by establishing a single, comprehensive national incident management system.

Policy

To prevent, prepare for, respond to, and recover from terrorist attacks, major disasters, and other emergencies, the United States Government shall establish a single, comprehensive approach to domestic incident management. The objective of the United States Government is to ensure that all levels of government across the Nation have the capability to work efficiently and effectively together, using a national approach to domestic incident management. In these efforts, with regard to domestic incidents, the United States Government treats crisis management and consequence management as a single, integrated function, rather than as two separate functions.

Tasking

The Secretary (of Homeland Security) shall develop, submit for review to the Homeland Security Council, and administer a National Incident Management System (NIMS). This system will provide a consistent nationwide approach for Federal, State, and local governments to work effectively and efficiently together to prepare for, respond to, and recover from domestic incidents, regardless of cause, size, or complexity. To provide for interoperability and compatibility among Federal, State, and local capabilities, the NIMS will include a core set of concepts, principles, terminology, and technologies covering the incident command system; multi-agency coordination systems; unified command; training; identification and management of resources (including systems for classifying types of resources); qualifications and certification; and the collection, tracking, and reporting of incident information and incident resources.

Other regions of the country, like California, have used unified command on a regular basis for decades. Each fire season, the California Master Mutual Aid System is utilized, which is coordinated by the California Office of Emergency Services (CAL OES). Through CAL OES, fire departments throughout the state respond in mutual aid to major wildfires. Strike teams of engines and other resources commonly will travel hundreds of miles up and down California to respond and fight catastrophic wildfires that threaten and destroy entire communities (fig. 16–6).

As a member of a Type 1 IMT, I was fortunate to see the full scope of ICS in action on large-scale wildfires. The United States Forest Service (USFS) had utilized unified command and ICS across America. The key to the success of the interoperability of so many different agencies working together was mutual respect and building trust.

UC does not have to be reserved for large-scale incidents like 9/11 or a major wildfire. UC is extremely effective on daily Type 4 and 5 incidents, from vehicle accidents to hazardous material responses. The scope of this chapter is on such incidents; however, key points of the larger components of NIMS are included.

You may have a Type 5 incident that quickly scales, through unified command, and becomes a Type 3 incident or greater. All incidents begin locally with that first-due officer. So, understanding the nuances of UC and how it can quickly grow and how you can set it up with a strong foundation as the initial IC will help to ensure you are successful. The key to successful UC is relationships and training before the event. Trust and relationships must be built between responders of varying agencies well in advance if you expect to have an effective and efficient unified response. That trust can only come from training and time.

Fig. 16–6. A CAL OES engine is staffed with local firefighters on a mutual aid fire
Source. Photo courtesy of Sacramento (CA) Metropolitan Fire District

A Local Example

We responded to a report of a hazardous material in the playground of an elementary school. Units arrived, reporting a plastic soda bottle with an unknown substance and report of several children feeling sick. This quickly became a hazardous materials response and multi-casualty incident (MCI), simultaneously (fig. 16–7).

Fig. 16–7. Pre-planning schools are now a vital part of incident preparation.

Size-up was as follows:

- Facts:
  - Elementary school with at least five children complaining of illness.
  - Possible hazardous material in the C side playground.
  - Afternoon, clear weather, and 70°F. No wind.
- Probabilities:
  - Confirmed VP that may grow in number.
  - Substance may be a hazardous material, or not.
  - Parents will quickly descend upon the school for their kids and create an accountability problem.
- Own situation:
  - Initial response was one engine, one fire medic ambulance, one battalion chief, and a sheriff.
- Decision:
  - Lifesaving priority.
  - Combination strategy.
- Plan of operations:
  - Isolate and deny entry to the C side playground.
  - Triage, treat, and transport patients with appropriate decontamination.
  - Call hazmat team, additional engine, truck, and five medic ambulances.

I balanced the response to include the resources above, and more were added. Sheriffs arrived and I suggested to the senior officer that we enter into UC with the rear of my vehicle as the ICP. He readily agreed. I requested that the school principal join us at the ICP. Our IAP consisted of a life priority, combination strategy, with the following unified incident objectives:

1. Establish an interior perimeter/exclusion zone around the hazardous material (defensive space until arrival of hazmat team) (law/fire/EMS).
2. Declare an MCI. Remove, treat, and transport any sick or injured children and/or adults (fire/EMS).
3. Establish an exterior perimeter to prevent any parents or outside people from accessing or leaving the school and compromising student accountability (law).
4. Identify, isolate, and mitigate the hazardous material (fire).
5. Provide appropriate levels of decontamination (fire/EMS).
6. Account for all students who are transported and/or remaining on site (law/fire/EMS).
7. Shelter all uninjured/non-contaminated adults and children in place in the gym by class (law).

8. Give accurate and timely information via news and social media to allay concerns of parents (law/fire).
9. Reunify children and parents (law).
10. Inform and coordinate with county OES for hazmat removal and documentation (fire).

By the end of the incident, we transported 29 students to local hospitals. Several ambulances made more than one round trip. Reunification between parents and children was challenging yet performed very efficiently between the school staff and law enforcement. All of the patients were children, and most did not have a complaint other than "feeling funny" which, seemed to be a psychogenic response. The amount and quantity of hazardous material was negligible. This turned out to essentially be a great UC drill. Lessons learned:

- Law enforcement will follow your lead. They know that the fire department typically has a better working knowledge of ICS. If you suggest UC, they will most likely oblige.
- Depending upon the nature of the incident, interior and exterior perimeters are crucial to controlling the scene. This is most necessary in an MCI and hazardous materials incident, but could apply elsewhere.
- Sharing objectives at the ICP between law and fire is vital to getting on the same page. Each IC should have an aid to assist with respective radio traffic.
- Divisions and groups work especially well in unified command. Here we had the following:
    - Medical group with objectives of triage, treat, and transport.
    - Hazmat group with objectives of identify, mitigate, decontaminate.
    - Law group with the objectives of inner/outer perimeter control and reunification.
    - PIO with objective of timely and repetitive information.
    - Rather than a liaison, the school principal was identified as an additional IC and remained at the ICP. This proved vital to clear communications with school faculty and staff, understanding the layout, attendance, and support services available. It also proved vital to accurate parental information dissemination (fig. 16–8).

UNIFIED COMMAND

The acronym UNIFIED COMMAND is for the specific nuances of the unified command incident. Once the UC is established, consider the appropriate SAW-CSS-RECEO-VSS worksheet (high-rise fire, vegetation fire, MCI, hazmat) to establish your incident tactical and control objectives. Structure fires will likely not be in unified command; however, a large vegetation fire may utilize UC between jurisdictions or between law and fire for evacuation. At the very least, an evacuation group may be supervised and staffed by law enforcement. I have used this several times with much success.

We responded to a law enforcement assist that involved a barricaded suspect with his child held hostage. This became a Type 3 incident and lasted several days. The UNIFIED COMMAND system worked well. We had unified briefings with sheriffs and the provided them with ladders, lighting, fire standby, and ultimately EMS (fig. 16–9).

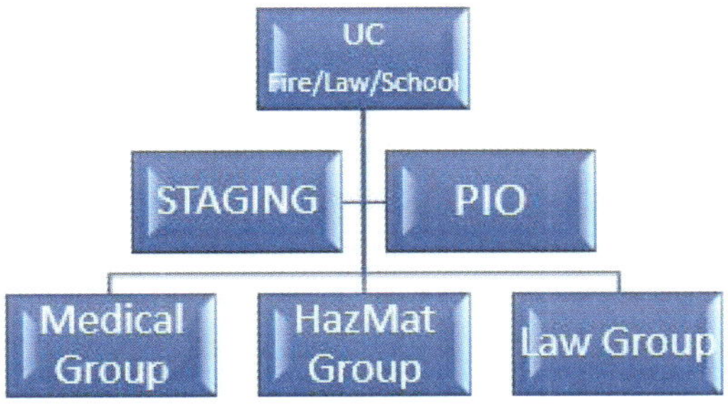

Fig. 16–8. ICS for the hazmat MCI

UNIFIED COMMAND

Understand the situation: Size up and look beyond your scope/discipline. Unlike the unifocal nature of structure fires, a unified command situation requires you to think of other disciplines, objectives, and jurisdictions. Look at the total incident from all aspects: fire, EMS, law, highway patrol, hazardous materials, OES, city/county, public works, state, and federal agencies (Coast Guard, Fish and Game, Air Resources, County Health, etc.).

Notify appropriate agencies: The AHJ is the lead agency, by decree, and has the formal/legal power and responsibility to control and enforce rules, regulations, standards, and other guidelines affecting the boundaries of the area under its charge. The AHJ will often determine the priorities and IAP for a particular operation, or at least have representatives on scene to do so. For example, the AHJ for a highway is the state highway patrol. The AHJ for crime within city limits is the police department. The AHJ for structure fires is your FD. The AHJ for hazardous materials incident may be the FD, PD, or local OES.

Notify the AHJ and any agencies that may have operational/jurisdictional authority besides yours. If in doubt, give them your size-up and ask if they intend to respond. *Informing* another agency is not the same as *involving* another agency. Just because you call and they respond does not necessarily mean that UC is automatic. Consider the following incident types and agencies:

- Vegetation fire: Law enforcement for traffic control and evacuation. Again, they may or may not enter into UC with you and simply fill a tactical role (evacuation group); however, if you have a large fire along a freeway, chances are highway patrol will be in UC with you, for example. Also consider another fire agency like a state fire department or federal agency like the U.S. Forest Service for fires on state/federal lands and forests.

Fig. 16–9. Sheriffs' officers handed the child to Sacramento Metro crews after the threat was neutralized. He was fine.
Source. Photo courtesy of Sacramento (CA) Metropolitan Fire District

- MCI: An active shooter, for example, would obviously be a UC incident with law enforcement. As previously discussed in the case study, if the incident is at a school, Heaven forbid, consider the principal or their designee to join you in UC. The principal is ideal as they have the final say and this will reduce reflex time for decision-making.
- Hazardous material incident: Law enforcement or the county OES may have authority, depending upon your city charter, even if they do not have response capabilities. A hazardous material release on private property could necessitate a liaison or UC with the site manager.

Identify the need for UC: Does the incident go beyond just calling additional agencies and escalate to unified command? When in doubt, call for help and ask your boss or counterpart if UC is appropriate. It cannot hurt and will keep you ahead of the incident power curve. Triggers for UC include, but are not limited to:

- Vegetation fire that crosses jurisdictional boundaries into another city or agency response area.
- Hazardous materials incident in a waterway, on a highway, or at school or other government facility.
- MCI involving a shooting at a government building or on a highway.
- Fire or other significant incident in a school.

Find your counterpart: Find the AHJ representative and the highest-ranking officer/authority figure in the other discipline. This will require effort and you may need a runner. Using dispatch is not an effective means, unless attempting to find their ICP.

Information exchange: Give and gather info. Do so in context of their needs. The more you empathize and anticipate their needs, the more they will reciprocate. CAN reports are an excellent tool.

Establish unified command: Make it official and notify your respective resources/people and dispatch centers. Identify and communicate the ICP.

Develop the plan: Identify joint priority, strategy, and tactical objectives.

Communicate the plan: Ensure that the priority, strategy, and tactical objectives of the incident are communicated to everyone and set up a communications plan for strategic and tactical operations. If you do not have common communications, then identify a single PIO and briefing area for media.

Objectives: Establish groups and/or divisions to manage your objectives. The tactical objectives of the incident must be jointly communicated and in terms that all disciplines understand (i.e., set up rescue task forces for an active shooter incident with the objective of rapid triage and removal of victims in the warm zone).

Map it: Obtain and share a map/plot plan/site plan/Google Earth image of the incident location, including roadways.

Mark it: Utilize ICS to mark the incident divisions, branches, staging areas, access points, medical treatment areas, helispots, traffic patterns, and so on. Identify inner/outer perimeters, hot/warm/cold zones, and so on.

Accountability: Ensure all personnel know their job and their boss. Use ICS groups, divisions, and branches for accountability and safety. Consider a safety officer as well.

Needs assessment: Conduct an operational resource needs assessment and anticipate logistical resource needs such as lighting, feeding, shelter, and so on.

Deploy the plan: Briefings and CAN reports should be appropriately established through groups/divisions/branches.

CASE STUDIES IN COMMAND

Champlain Towers Condominium Collapse

CHIEF SCOTT DEAN, CITY OF MIAMI (FL) FIRE-RESCUE DEPARTMENT

In the early morning of June 24, 2021, the Champlain Towers, a multi-story beachfront condominium located in Surfside, Florida collapsed. The collapse was a devastating tragedy, resulting in the loss of 98 lives and injuring many others. The incident shocked the local community, as well as people across the United States and around the world, who followed the ongoing search-and-rescue efforts and mourned the loss of those affected by the disaster (fig. 16–10).

During this timeframe I was the assistant fire chief for the Special Operations Division for the City of Miami Fire-Rescue Department. One of the areas of responsibility as the Chief of Special Operations is the US&R, Florida Task Force 2 (FL-TF2). FL-TF2 is one of eight State of Florida nd one of twenty-eight Federal (FEMA) task forces. This team responds to natural (i.e., hurricanes, earthquakes) and man-made (i.e., terrorist events) disasters throughout the country when needed. I've been on the team since early 2001 and I've been a task force leader since 2011. During my tenure I've been deployed to over 30 disasters to include the 9/11, Hurricane Katrina 2005, Haiti Earthquake 2010, Hurricane Florence 2018, Hurricane Michael 2018, Puerto Rico Earthquake 2020, Surfside Building Collapse 2021, Kentucky Flooding 2022, and Hurricane Ian 2022, to name a few.

For the Champlain Towers Collapse, I was deployed as one of the task force leaders for FL-TF2. During the first operational period I was asked to be the division/group supervisor for the 12:00 p.m. to 12:00 a.m., Alpha shift. As the division/group supervisor I focused on two objectives: (1) the safety of all the responders working on the building, and (2) move the building as quickly as possible to try to rescue anyone that could be alive (fig. 16–11).

There were several concerns that needed to be addressed during the deployment of resources to this disaster. These concerns consisted of the unknown amount of missing people, tremendous amount of smoke/fire/water, questions of how did the building fall and will the remaining building fall on top of the rescuers, COVID-19 was in full throttle, religious concerns, politics, weather, Category 1 Hurricane approaching, relationships with those missing in the building, body recovery, tremendous number of personal belongings, and so forth. So, where does one begin?

Early on, we created divisions to have accountability of all the teams that were working the building, but this ended up not being ideal because the footprint was a little

Fig. 16–10. A condo collapses June 2021 in Surfside, Florida
Source. Photo courtesy of Miami (FL) Fire-Rescue

Fig. 16–11. Firefighters focused on the safety of first responders and quickly trying to rescue people.
Source. Photo courtesy of Miami (FL) Fire-Rescue

over 1.25 acres. Due to the limited geographical footprint, we ended up deciding to take all the task force squads working on the Alpha shift and create one team. This allowed us to keep communication confusion down and improved coordination of the rescue efforts. For example, FL2 squad 1, FL3 squad 1, and FL4 squad 1 were modified to squads 1 through 15. Each squad would just fill the gap for whichever team was up for rehab. No squad had a set geographical location assigned to them and this allowed for a much higher work rate of speed, and it maximized workflow. The fact that the teams were able to move 22 million pounds of debris in just 21 days is a testament to their dedication, expertise, and hard work (fig. 16–12).

Accountability was critical for the safety of all the responders working at the site, and keeping detailed logs of which squads were working and in which location of the building collapse was essential. My biggest worry that constantly gave me tremendous strain was the concern that what was left standing of the 12-story building would fall on top of all the rescuers, especially since the cause of the original collapse was unknown. We worked alongside some of the best structural engineers and safety officers in the business to come up with what would be a suitable area of operation while trying to be as safe as possible in an unsafe environment (fig. 16–13).

As a leader of a major disaster response like the Surfside Building Collapse, it's essential to consider not only the safety of the responders but also the emotional toll that such an experience can have on them. Responders are exposed to traumatic and often heartbreaking situations, and they may struggle with the aftermath of their work long after the event has ended. Providing support and resources to help responders process their experiences and cope with the emotional impact of their work is critical to ensuring their long-term well-being (fig. 16–14).

Fig. 16–12. Maximizing work flow enabled crews to move 22 million pounds of debris in 21 days.
Source. Photo courtesy of Miami (FL) Fire-Rescue

Fig. 16–13. It was critical to keep logs tracking where responders had worked at the site of the building collapse.
Source. Photo courtesy of Miami (FL) Fire-Rescue

Fig. 16–14. Responders exposed to trauma may struggle long after an event ends.
Source. Photo courtesy of Miami (FL) Fire-Rescue

These are many of the significant takeaways from the Surfside Building Collapse response:

1. Allowing the family members to come to the site and witness the destruction firsthand can be a way for them to find closure and process the grief.

2. One team, one mission! Put all the squads in numerical order, squads 1 through 15. Consolidating all the squads into one team can increase efficiency and speed up progress.

3. Emotional scars. . . Responders are also human beings, and it's important to recognize and address the emotional toll that such an experience can have on them. It's okay not to be okay!!!! I recall the first opportunity that I got to take a nap; I woke from a nightmare that I had that took me back to the 2010 Haiti Earthquake response. I never thought about that response until that moment.

4. Communicating with the family members of responders and providing resources for them to understand and cope with potential symptoms of trauma can be helpful in facilitating their loved ones' transition back to normalcy. We set up a teams meeting with all the family members so that we could explain to them the potential symptoms that their loved ones might show once they returned home.

5. Always be inclusive, because you don't know what you don't know. Being inclusive and open to help from other organizations or individuals can lead to more efficient data collection and a more effective response overall. A prime example of this is the Israel Defense Force (IDF) that showed up to the scene to provide assistance. They were an incredible resource for data collection, which was essential so that we could locate everyone that was missing.

6. While not everyone can be saved, making a positive impact and trying to improve the situation as much as possible is crucial.

7. Deal with one issue at a time. Dealing with issues as they arise and addressing them one at a time can help manage the overall response and prevent becoming overwhelmed.

8. Try to find some good in everything bad. Finding positives or silver linings, such as the recovery of all missing individuals in this case, can bring some sense of closure or comfort to those affected.

9. Make a difference in everything you do. Striving to make a difference and improve the situation is an important mindset to maintain during any disaster response.

This building collapse response took all eight State of Florida Task Forces and five Federal Task Forces being deployed for 29-days to locate 100% of the missing people. This response was a coordinated effort by all the police departments, first responders, fire departments, search-and-rescue teams, and volunteers to complete the mission.

Scaling to a Larger Incident Type

You will initially be in a Type 5 or Type 4 incident. The scope of this book is primarily the Type 5 and 4 incident; however, scaling to a Type 3 incident is more likely than in years past due to the advent of Type 3 IMTs that respond to operational area (county-level) incidents. The IC must know what will happen to set up the incident to transition to the Type 3 level.

Most likely, you will be in unified command with another agency, likely law, before the transition to a Type 3 team occurs. As you are commanding the incident, your counterparts and/or bosses will be listening, and once they trigger a Type 3 IMT response, you will be notified. This will likely happen by phone. SOGs and policy differ. You may also find yourself as one of several incidents that are happening concurrently. This could be the case with multiple fires due to red-flag weather conditions, arsonists and/or civil unrest, a natural disaster, or a series of manmade disasters. This may trigger a complex or area command format with an emergency operations center. Figure 16–15 shows a possible ICS for a unified command with law enforcement for an MCI.

A *complex* is two or more individual incidents located in the same general proximity that is assigned to a single incident commander or unified command to facilitate management. An *area command* is an expansion of the incident command function primarily designed to manage a very large incident that has multiple management teams assigned (fig. 16–16).

However, an area command can be established at any time that incidents are close enough that oversight direction is required among incident management teams to ensure conflicts do not arise.[2]

An *emergency operations center* (EOC) is the physical location where the coordination of information and resources to support incident management (on-scene operations) activities normally takes place (fig. 16–17). An EOC may be a temporary facility or located in a more central or permanently established facility, perhaps at higher level of organization within a jurisdiction.[3]

If multiple incidents/locations are being managed, an Area Command team may be established to coordinate and prioritize resource requests from the various locations/incidents. This could be a Type 3 IMT. The EOC supports the IMT.

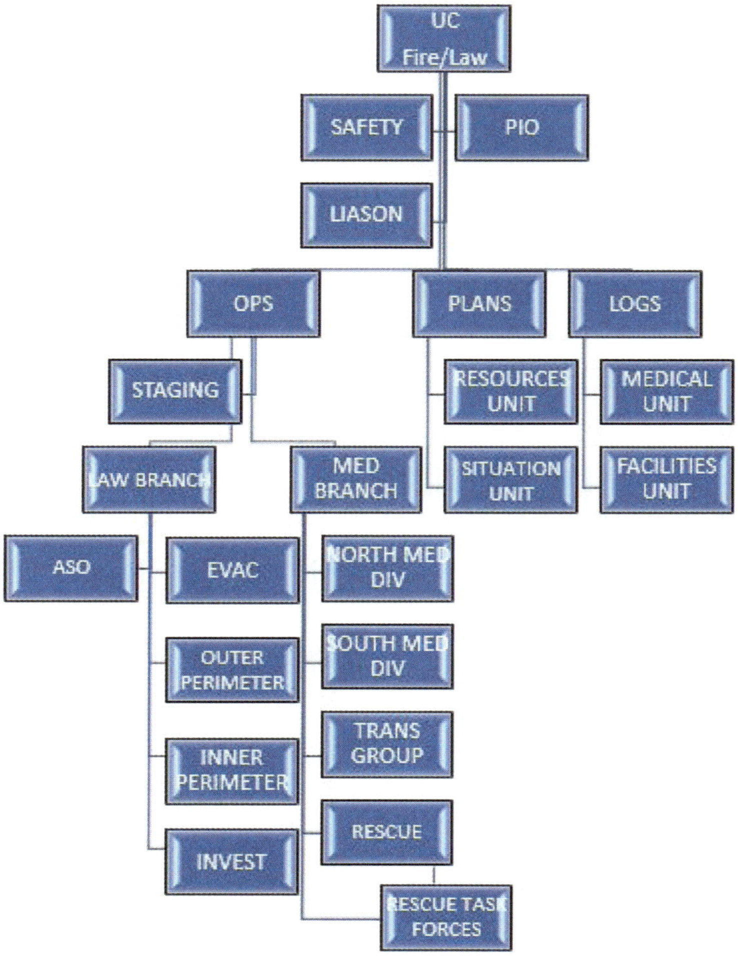

Fig. 16–15. An example ICS for unified command

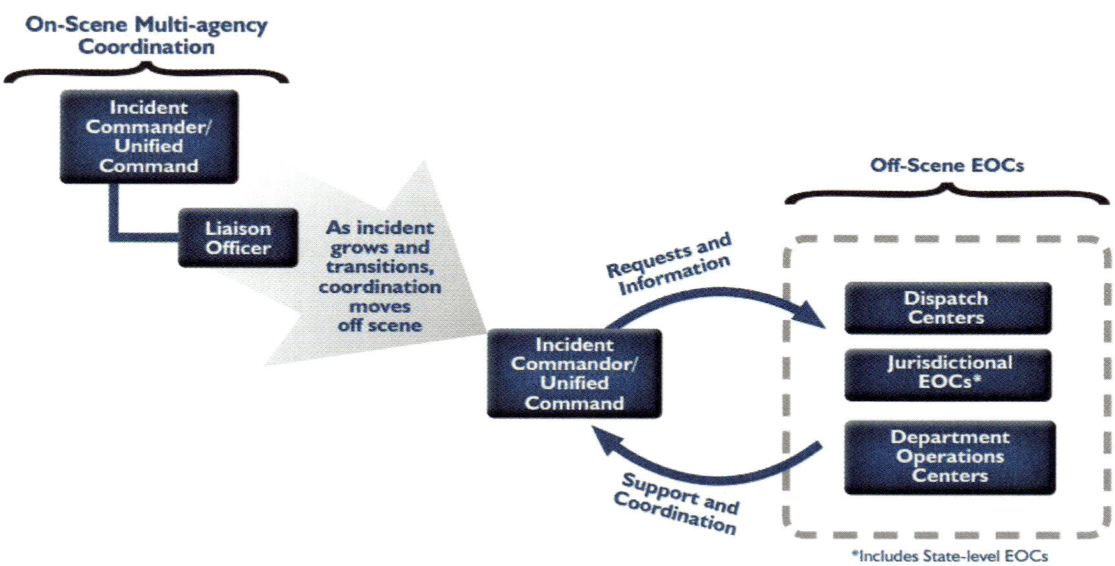

Fig. 16–16. Coordination of an incident moves off scene as it grows and transitions.
Source. Photo courtesy of FEMA

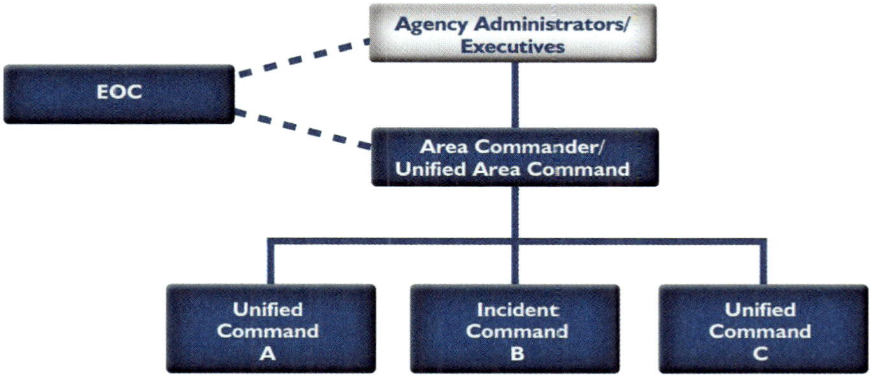

Fig. 16–17. The Emergency Operations Center supports the incident management team.
Source. Photo courtesy of FEMA

CASE STUDIES IN COMMAND

Route 91 Music Festival Shooting

CHIEF GREG CASSELL, CLARK COUNTY (NV) FIRE DEPARTMENT

The Route 91 Harvest Festival was a large, outdoor concert festival featuring some of the largest names and acts in Country music. Beginning in 2014, the event took place for three consecutive nights on a 17-acre lot located on the south end of the Las Vegas Strip. Growing in popularity each year, the festival hosted as many as 25,000 attendees each night. The location for this event is in the jurisdiction of the Las Vegas Metropolitan Police Department (LVMPD) and the Clark County Fire Department (CCFD).

The third and final night for the 2017 festival was Sunday, Oct. 1. With more than 22,000 attendees and hundreds of workers on site, the last act was on stage and performing the closeout set. The weekend event had been a huge success. Unfortunately, that reality quickly changed.

On the 32nd floor of a resort hotel across the street from the festival, a lone gunman opened fire on the concert venue with multiple high-powered rifles. In a timeframe of about 10 minutes, more than 1,000 rounds rained down on the venue, resulting in a horrific scene. More than 850 people were injured, more than 420 of those had at least one gunshot wound and ultimately 60 people lost their lives.

This attack led to a large response from multiple emergency service entities located within the Las Vegas Valley. Within 10 minutes of the initial call alerting police and fire/EMS to the attack, the process to establish a UC between LVMPD and CCFD had begun. A short time later UC was set up at the police station just down the street from the venue. This transition was not perfect but was functional and only happened because of the mutual training and understanding brought about by years of hard work and relationship-building between the LVMPD and CCFD leadership and members.

On the morning of October 2nd, the initial response period ended, and the next operational period began with LVMPD, CCFD, and the Federal Bureau of Investigation (FBI) listed as the unified incident commanders. This format lasted for two days. But how did we get to this point, law enforcement and fire service in a functioning UC on a major incident?

In November 2008, Mumbai, India suffered a coordinated, multi-point terrorist attack that lasted for four days and resulted in 160 deaths. The scale of this attack caught the attention of law enforcement agencies across the world, including the LVMPD and the Los Angeles Police Department (LAPD). Not long after this attack, these two departments joined up and sent representatives to Mumbai to learn firsthand about the details of the attack and most importantly, lessons learned from Mumbai's law enforcement representatives. The information that came back from that trip ultimately led to the transformation of law enforcement and fire service relationships across southern Nevada.

At the time of the Route 91 Harvest Festival attack, LVMPD and the CCFD were several years into a collaborative effort to prepare our respective departments and regional counterparts for a large-scale incident of some type. This was no easy task, as resistance from both sides was a significant obstacle to developing, testing, and evaluating potential response procedures and policies. However, given the state of the world and, for that part, our country at the time, having a hostile mass casualty incident (HMCI) take place in our jurisdiction was a matter of when, not if, and everyone knew it. Unfortunately, that reality exists in every town, large or small, urban or rural, across the United States.

From 2010 to 2013 there were many meetings, policy drafts, and several dozen training drills at venues ranging from schools, government buildings, shopping malls, and other soft target locations. As expected, there were several mistakes, bad ideas, and concepts that were exposed during these drills. However, those failures provided opportunities for mutual discussion and operational improvement by both law enforcement and fire. Ultimately, a policy was solidified by all involved to include support from the labor unions representing responders on both sides. At this point all law enforcement entities and fire departments began training their personnel to respond together under a UC. This also led to LVMPD and CCFD integrating many training aspects from new recruits to supervisory level promotional training, resulting in not only familiarity of procedures, but person-to-person recognition between law enforcement and fire personnel when the attack happened. This fact was noted by many of those who responded to this incident.

Fast forward back to the night of October 1st, 2017. As chaos was unfolding on the Las Vegas Strip, senior leaders from LVMPD and CCFD were descending on the area from all directions. The 17-acre concert venue quickly became a 3.5 square mile crime scene as victims fled in all directions, including to the perimeter of the international airport. Within minutes of the call coming in the first CCFD battalion chief (B2) arrived, established command, requested a multiple alarm response, and quickly located a sergeant from LVMPD who was at a major intersection just north of the scene. The sergeant advised B2 that his lieutenant was enroute to the South-Central Area Command police station, six blocks south of their current location, and that location was going to serve as the ICP. With that information, B2 directed Battalion 6 (B6), who was arriving from that direction, to go to that location, meet up with the LVMPD lieutenant, and at that point he would transfer command to B6. As previously mentioned, this was not as smooth as anticipated; however, with the arrival of an LVMPD captain and a CCFD deputy chief, who had previously trained together on large-scale events, UC was transferred to them for the remainder of the initial response period.

One of the biggest challenges for the UC in handling this incident is what is known as "echo calls." For more than an hour, 9-1-1 calls were coming in from nearly a half dozen resort properties on the Strip and the airport saying someone was shot at that location. Fearing that we were experiencing a multi-point attack, similar to Mumbai, UIC had to send resources to those locations to handle what was reported as "another active shooter." These efforts took almost two hours and dozens of personnel to handle. Fortunately, there were no other shooters and the gunshot victims at these locations had actually been shot at the concert and fled to these other locations seeking shelter and help.

Lessons learned from an incident at a night club about seven weeks prior to the attack were key to the UIC's decision-making process during the attack. In that event, a fight in a night club led to numerous 9-1-1 calls reporting an active shooting situation at various locations, all of which were unfounded. The debrief provided by LVMPD to both law enforcement and fire service on that incident was fresh in everyone's mind as the attack on the concert unfolded.

Many lessons were learned as a result of this incident. From incident command to emergency management functions supporting more than 4,200 people in need of some form of assistance, to the massive undertaking of providing for the mental health needs for all involved, and the amazing work done by the medical examiner's office, we all learned a lot. We also realized that our regional, multi-disciplinary, multi-year efforts to prepare for some kind of large incident paid off (fig. 16–18).

Lessons Learned

1. Sudden need to debrief a large percentage of your organization. We had over 100 personnel from CCFD on-scene. Our employee assistance program and peer support team were nowhere big enough to handle this impact. The IAFF was a tremendous help, as they sent two teams of personnel to help us. Building

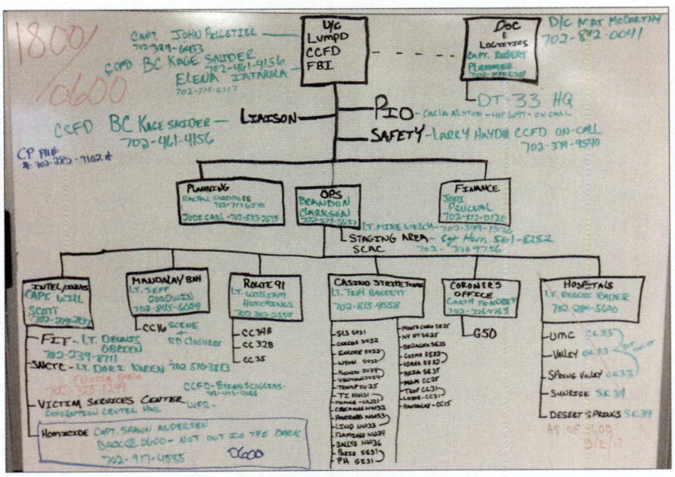

Fig. 16–18. The ICS for the Route 91 Music Festival shooting
Source. Photo courtesy of Clark County (NV) Fire

relationships with local groups that provide counseling, trauma yoga, meditation, spiritual needs, or other similar services is strongly advised.

2. The size and magnitude of the family assistance center/reunification center was underestimated. Along with that, we never considered the need for a long-term resiliency center to provide ongoing assistance to survivors before this event. It's been over five years, and our resiliency center is still operating.

3. Support hospitals in times of patient surge from a hostile MCI. We developed and implemented a response model called "Hospital Area Command."[4] Even though our event produced hundreds of patients, one hospital seeing 200 patients arrive in an hour, the number of patients that can collapse a hospital's emergency resources is scalable. For example, a small town with a 10-bed emergency room and a staff of 10 will be completely overwhelmed if 50 people are injured during an attack on a local sporting event or other gathering.

4. Continue to build and foster relationships with other response agencies. Use the "One Team, One Fight" mentality. No matter what your job is, law enforcement, fire, EMS, hospital RN, doctor, or respiratory tech, your goal when you go to work is to do your part to save lives. That is our fight, and all those mentioned are on the same team.

5. Continue to work on communications. Hold exercises that involve line-level responders, supervisors, and communication centers. In those exercises, practice using different radio channels in the ICP and at branch or division/group level operations. Identify and practice ways to disseminate this information during the initial response phase and ongoing operational periods. We found some success by taking pictures on smart phones that were then texted to staff in the field.

Summary

Unified Command (UC) is a situation that you are more likely to encounter than ever before. The attacks on September 11, 2001 changed the landscape of incident command and coordination between agencies and jurisdictions. We must train and build relationships with our allied agencies well before an incident occurs. Trust is crucial. This includes law enforcement, EMS, public works, and countless other agencies that you may never have expected. In addition, state and federal resources may be involved if/when your incident escalates.

The acronym UNIFIED COMMAND can be utilized to train and engrain key components of setting up successful UC operations. Communications, like any incident, will make or break success. The UC arena is exponentially more challenging for the NIOSH 5, especially in accountability and communications.

The principles you have learned throughout this book definitely apply to UC: Perform a thorough, ongoing size-up with FPODP, stay ahead of the incident power curve by calling resources early, bridge the tactical gap with ICS, and so on.

If the incident scales to a Type 3 level, understanding the moving parts around you will make you more successful. This includes the use of a Type 3 IMT, the EOC, and area command for multiple incidents. Remember, it all starts with the first-arriving responder, who must remain calm and understand that the tone is set with the first arrival report. Take a deep breath, size up your situation, and trust your training.

Chapter Review

Review Questions

1. Explain why there is greater potential for the presence of the NIOSH 5 at incidents involving unified command.
2. What are the key components of Homeland Security Presidential Directive-5?
3. Define the acronym UNIFIED COMMAND.
4. What is the difference between a complex and area command?

FESHE Strategy and Tactics (C0279) Related Content

The content contained in chapter 16 provides detailed information specific to incidents involving a unified command structure. While the FESHE C0279 course does not specifically reference unified command incidents, they involve development of a command structure and NIMS compliance for appropriate coordination of resources. The chapter format supports the C0279 course outline by detailing information on unified command incident operations, communications, and coordination with regard to ICS/NIMS roles and responsibilities.

NFPA 1021 Job Performance Requirements

The information in this chapter can be utilized to support training and educational programs associated with the Emergency Services Delivery Fire Officer III 6.6 and 6.6.1 and Fire Officer IV 7.6.1 and 7.6.2. For larger-scale incidents and pre-planning, the content in this chapter also supports training and education for Fire Officer III 6.6.3 "Develop a plan for the organization given an unmet need for resources that exceed what is available in the organization." The chapter content also supports the education and training associated with the Emergency Management professional qualifications for Fire Officer III 6.8.

Endnotes

1. "Homeland Security Presidential Directive-5," U.S. Department of Homeland Security, February 28, 2003, https://www.dhs.gov/sites/default/files/publications/Homeland%20Security%20Presidential%20Directive%205.pdf.
2. FIRESCOPE, *Field Operations Guide—ICS 420-1*, ICS Publication, 2022.
3. FIRESCOPE, *Field Operations Guide*.
4. Greg Cassell et al., "The Clark County (NV) Hospital Area Command Collaboration," *Fire Engineering*, May 1, 2023, https://www.fireengineering.com/fire-ems/the-clark-county-nv-hospital-area-command-collaboration/#gref.

Section IV

Appendixes

A

Los Angeles City Fire Department Informational Summary—Green Sheet

Los Angeles City Fire Department
Informational Summary Report

GREEN SHEET

Injuries/Commercial Structure Fire

May 16, 2020
Boyd Incident
CA-LFD-001073

This report is intended as a safety/training tool, aids in preventing future occurrences, informs interested parties, and identifies lessons learned. This Informational Summary Report is prepared and distributed within thirty (30) days following the initiation of a Serious Incident Review Team to investigate a serious incident. Information contained herein is subject to revision as further investigation is conducted and additional information is developed.

Commercial Structure Fire
May 16, 2020
Boyd Incident
Page 2

SUMMARY

On Saturday, May 16, 2020, at approximately 18:27:00 hours, the Los Angeles City Fire Department (LAFD) responded to a structure fire at 327 East Boyd Street in the Skid Row area of downtown Los Angeles. The first arriving resources found a one-story commercial occupancy with light, intermittent brown smoke showing from the top of two closed rolling steel doors. The subject occupancy was doing business as a wholesale distributor to dispensaries and smoke shops, and it had closed for the day prior to the time of alarm.

Resources from the first alarm assignment secured a water supply, forced entry, deployed hand lines, laddered the building, engaged in fire attack, and initiated timely vertical ventilation operations while in the offensive mode. Due to the excessive storage encountered at the entry point, there was a delay in gaining access; nonetheless, fire attack teams quickly developed a course of action to begin their advance into the structure.

Conditions noticeably worsened over the next several minutes. Initially, members described a notable change in smoke emitting from the front of the occupancy, and then a rapidly increasing "freight train" or "jet-like" sound, accompanied by rapid and escalating "popping-like" noises. Subsequently, companies initiated a rapid retreat—both off the roof and out of the building. As firefighters began initiating their escape, a large volume of smoke ignited, and then immediately following, fire "blowtorched" out of the front of the occupancy, across the street, and out of the rear portion of the roof.

As a result of the catastrophic fire event, just eleven (11) minutes into the incident, twelve (12) firefighters were injured. The injuries were sustained when the firefighters were engulfed by a large volume of fire, some while retreating from interior firefighting positions, and others as they were descending the aerial ladder. All told, eleven (11) firefighters sustained minor to critical burns, and one (1) additional firefighter suffered a lower extremity injury.

Fireground emergency communications, which included both, "Emergency Traffic" and "Mayday" transmissions, were declared on several occasions—moments before the evacuation, during the retreat, and after the fire phenomenon. Immediately after the "blowtorching" fire receded, Company Accountability Reports (CAR) together with Personnel Accountability Reports (PAR) were requested by the IC.

Commercial Structure Fire
May 16, 2020
Boyd Incident
Page 3

In the moments that followed, injured firefighters along with firefighters working in the immediate area began to effect rescues, account for personnel, and removed Personal Protective Equipment (PPE) for the most critically burned. The burned firefighters were assisted by additional resources arriving on scene into awaiting ambulances. Additionally, a Medical Group was established to triage and transport a total of eleven (11) firefighters to the closest Medical Trauma/Burn Center.

Remaining resources continued to fight the fire defensively, utilizing large diameter hand lines and master stream appliances, until a knockdown was finally determined at approximately 20:08:00 hours.

The cab of the first arriving Truck, with its deployed aerial ladder sustained significant damage, as both were in the immediate path of the flames. The first arriving Engine sustained minor heat damage to the rear of the apparatus due to the fact that the driver had positioned the apparatus past the occupancy. Lastly, multiple tools and equipment were irrevocably damaged or deemed missing as a result of this incident. The total loss is still being tabulated.

NARRATIVE

On Saturday, May 16, 2020 at 18:26:38 hours, Metro Fire Communications (MFC) received multiple 911 calls with reports of smoke coming from a building on Boyd Street in downtown Los Angeles. The first caller provided the specific address of 327 Boyd Street and identified the building as "Smoke Tokes." The dispatcher created the incident and resources from the LAFD were dispatched at 18:27:47 hours (call processing time, one (1) minute and nine (9) seconds). The incident time clock was initiated at the time of alarm and the following resources were dispatched:

(2) Battalion Command Teams
(1) Truck Company
(1) Light Force (Truck Company and 200 Series Engine Company)
(3) Engine Companies
(1) Hazardous Material Squad
(1) Paramedic Rescue Ambulance
(1) 800 Series Rescue Ambulance

An additional Paramedic Rescue Ambulance (RA) assigned to the district, but not yet attached to the incident, was a few blocks away when they were alerted of the structure

Commercial Structure Fire
May 16, 2020
Boyd Incident
Page 4

EMS/Administrative Channel 4. MFC instructed the RA to communicate on Administrative/Channel 9 and attached them to the incident at 18:29:50 hours. No other communication was made by the RA.

Resources from the first-in district arrived on scene at 18:30:26 hours (operational response time, two (2) minutes, thirty-nine (39) seconds). The first arriving truck reported a one-story commercial occupancy with moderate smoke coming from the Alpha side. The Truck Captain declared that the resources on scene would be operating in the Offensive Mode and proceeded to assume command as "Boyd IC."

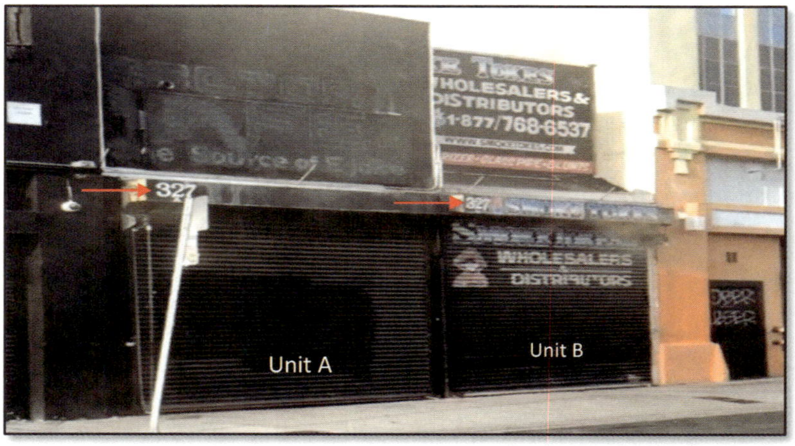

The building was approximately 25' wide by 110' long and located in a row of commercial occupancies. The one-story fire building was flanked by a two-story commercial building to the West (Bravo Exposure) and a four-story commercial building to the East (Delta Exposure). The involved structure was constructed of cinder block walls and was divided into two separate occupancies by a partition wall running from the front to the back (North to South), with a mezzanine in the middle. The store front was secured with two rolling steel doors, identified as Unit A (left side) and Unit B (right side). Closed at the time of alarm, the store was doing business as a wholesale and distribution center for supplies associated with dispensaries and cannabis smoke shops. There was excessive storage present in both of the units, and the building was not sprinklered. (See complete building description and photos under Conditions on page 16)

At 18:30:43 hours, the first aerial ladder truck pulled up to the front of the occupancy,

Commercial Structure Fire
May 16, 2020
Boyd Incident
Page 5

line in a two piece configuration from the hydrant to the fire, utilizing a hydrant located at the Southwest corner of Boyd Street and San Pedro Street

The IC (1st Truck) placed the following resources as follows:

1st Truck – Roof (CII, A/O, TOP, TILL, INS)
1st Engine – Fire Attack (C1, ENG1, HYD1, NOZ1)
2nd Engine – Back up Fire Attack (C2, ENG2, HYD2, NOZ2)
3rd Engine – Delta Exposure
1st 800 Series Rescue Ambulance – Attached with Roof (DRVR8, ATTN8)

The first Engine on scene pulled past the structure, spotted the apparatus in a defensive position, and pulled a running line for supply. Multiple handlines were deployed off the transverse bed, including a 1¾" and 1½" for fire attack, and eventually a 1¾" protection line to the roof. A 35' extension ladder was thrown by the truck company personnel to Unit A and the aerial ladder was placed to the roof, over Unit B.

At 18:31:18 hours, the first arriving Battalion Command Team transitioned command with the first truck on scene and continued with offensive operations, placing the balance of the assignment as follows:

1st Paramedic Rescue – Attached with Fire Attack (DRVR1, ATTN1)
2nd Paramedic Rescue – Medical Group (DRVR2, ATTN2) *Reported to Fire Attack
2nd Truck – Bravo Exposure
Hazardous Material Squad – Standby
2nd Battalion Command Team – Asked to provide a survey lap and assess the exposure on the Delta side

The Inside Member (INS) assigned to the first truck company on scene initiated forcible entry by cutting the bottom bar slide locks on two rolling steel doors on the Alpha side (Unit B first, followed by Unit A), and then rolled up the door to

Commercial Structure Fire
May 16, 2020
Boyd Incident
Page 6

In the order prescribed below, the remaining truck company personnel, consisting of the Tiller Member (TILL), Top Member (TOP), and Apparatus Operator (A/O) ascended the aerial ladder to the roof of the involved occupancy, carrying the roof kits and chainsaws aloft.

Brown smoke was seen filling the street as it pushed out of Unit A. In front of the structure fire attack teams donned their facepieces and prepared to deploy their hose lines.

At 18:33:04 hours, Fire Attack made access to Unit A and stated, "… it looks like it's going to be coming from the door on the right." Excessive storage in Unit A forced Fire Attack resources to look for another means of access.

A decision was made to wait for Unit B to be opened in order to determine if there was better access.

At 18:34:00 hours, difficulties in opening the Unit B slide locks caused INS to decide to cut the rolling steel door to make entry. INS and the Hydrant Member (HYD3) from the third Engine Company began making a three-sided cut on

Commercial Structure Fire
May 16, 2020
Boyd Incident
Page 7

Resources assigned to exposures described the smoke as acrid and gag inducing, resulting in multiple firefighters donning their facepieces and going on air, in front of, and adjacent to the involved structure.

As the roof team transitioned from the aerial ladder to the roof of the involved occupancy, moderate, lazy smoke was observed on two fronts—emanating primarily from the alley on the Charlie side as well as from the small, lengthwise gaps that ran between the fire building and the taller Bravo and Delta exposures. Additionally, a roof vent adjacent to the parapet, towards the rear on the Delta side, also showed smoke ventilating with pressure.

At 18:33:40 hours, A/O made an inspection cut and small hole towards the Alpha/Bravo corner, out of the way of the path of travel. Once completed, pressurized, brown smoke exited out of the hole.

The roof structure was comprised of pre-fabricated wooden "I" beam joists, spaced approximately 12 inches on center, double stemmed every 4 feet, and resting on galvanized joist hangers. The roof decking consisted of plywood with rolled composition and was supported by nailing blocks spanning every 4 feet. The joists were also supported by the partition and pony wall that separated the two occupancies.

At 18:34:08 hours, the roof team sounded out toward the center of the roof, making their way toward the back of the structure. The A/O observed smoke emitting from a roof vent and decided to begin cutting the ventilation hole approximately 65' to 70' in from the Alpha side.

The initial heat hole was cut approximately 8' long by 2' wide, capturing the roof over both Unit A and Unit B. As the roof team pulled the material, pressurized brownish-grey, and sometimes what was described as, "clear" smoke, billowed from the hole and pushed toward the North (Charlie side) away from the roof team.

Commercial Structure Fire
May 16, 2020
Boyd Incident
Page 8

Firefighters from the 800 Series RA (DRVR8 and ATTN8) followed by the Truck Captain (CII) ascended to the roof and joined up with the ventilation team. DRVR8 threw a drop bag from the roof to pull up a protection hose line. The Engineer (ENG2) attached a 1¾" hose line to the drop bag and ensured the line went aloft.

At 18:34:42 hours, the IC asked Fire Attack what type of fire load was in the occupancy and what type of business it was.

At 18:34:50 hours, describing Unit B, Fire Attack responded with the following: "We've made access. We have floor to ceiling of product. We're trying to find an entrance. It's going to be real hard for us to get in. We have a moderate amount of smoke right now coming out of the Alpha side, both entrances."

After forcible entry was completed on Unit B, Fire Attack was confronted with an extensive amount of shelving and merchandise stacked up against the rolling steel door. With no viable access, Fire Attack decided to initiate their attack through Unit A.

As a result of the excessive storage and challenges with forcible entry, Fire Attack was delayed approximately six (6) minutes from the time the first resource arrived on scene.

At 18:35:10 hours, resources assigned to back up Fire Attack, which consisted of five (5) members (NOZ2, DRVR2, C2, HYD2, and ATTN1), progressed into the involved structure ahead of the assigned Fire Attack company (NOZ1, C1, HYD1) through Unit A. As they advanced into the occupancy, they experienced heavy brown smoke at the ceiling, with moderate visibility and low heat.

The aisleway to Unit A was approximately 42" wide and was strewn with boxes and copious amounts of melted candle wax. The cluttered aisle was lined with commercial storage racks on either side. In the front, they were loaded with boxes, stacked to the ceiling, comprised of glass jars, candles, vapes, pre-rolled cones and various smoking products. In the rear portion of the

Commercial Structure Fire
May 16, 2020
Boyd Incident
Page 9

At approximately 10' to 15' into the structure, the fire attack team encountered a metal 6 step rolling ladder, which took up most of the aisleway. They shuffled around the moveable step ladder, and C2 asked for the ladder to be removed to the exterior. Multiple firefighters attempted to transfer the rolling ladder behind them; however, the ladder never made it outside. A minimum amount of water was flown from the 1¾" hose line, as the seat of the fire was never found. A long pike pole was used to check for extension and it was determined that a mezzanine existed over the middle of the unit, leading Fire Attack to believe the fire was above the mezzanine.

At 18:35:23 hours, a firefighter (EIT) from the second arriving Battalion Command Team communicated to the IC to order more resources.

At 18:35:28 hours, the IC ordered two additional tasks forces.

As the roof team extended the initial heat hole, Roof Division communicated with Fire Attack, "Fire Attack from roof... open up your streams. Aim them to the ceiling on the Charlie side... aim toward the Charlie side." The message was not answered and as a result, the IC clarified the message with Roof Division and then relayed the message to the fire attack team.

At 18:36:58 hours, the back-up Fire Attack team and remaining firefighters (DRVR1, ATTN2, and INS) advanced an 1½" hose line behind Fire Attack, pulled hose, and removed boxes and single sheets of oriented strand board (OSB) that was blocking the entrance.

At 18:37:00 hours Fire Attack stated, "Fire Attack copies, we're working our way back there."

At 18:37:30 hours, Roof Division communicated with the IC and stated, "IC from Roof Division... you have exposures to the Bravo and Delta side. We have a hole opened up. We're getting heavy fire out of our hole toward the Charlie side." The original ventilation hole, after being extended twice, was cut to approximately 8' long by 6' wide.

Commercial Structure Fire
May 16, 2020
Boyd Incident
Page 10

Heavy fire pushed out of the ventilation hole, forcing the roof team to back away from the operation. Deciding to don structural gloves at this point, A/O gave the chainsaw to TILL and provided direction for TILL to start a new ventilation hole approximately 6' back from the initial heat hole and to work against construction. TILL initiated a new ventilation hole and the roof team began to pull the material. Pressurized smoke surged from both ventilation holes and the flames that followed were approximately 20' to 30' high.

At approximately 18:38:00 hours, while extending the second hole, the roof team heard sounds coming from inside the structure. Initially, they described hearing loud, "popping-like" noises that escalated rapidly in volume and frequency. In a matter of seconds, the "popping" noises were accompanied by a "jet-like" rumble. The members described physically feeling the roof vibrate as the sound grew louder. The A/O decided it was time to retreat and backed the roof team away from the hole. DRVR8 opened up the 1¾" hose line on a straight stream, directly into the smoke and fire coming from the hole. The roof team — at the direction of A/O and CII — retreated to the Alpha side of the roof back toward the aerial ladder and 35' extension ladder.

At 18:38:08 hours, Fire Attack stated, "We're going to be delayed getting to the seat of the fire."

18:38:15 hours, MFC communicated to the IC that 10 minutes had elapsed in the incident.

At 1838:24 hours, the Engineer (ENG1) ran up to C1, grabbed him and stated, "We got fire, heavy, heavy black smoke coming from the back." C1 backed up into the street to visualize the conditions being described.

Commercial Structure Fire
May 16, 2020
Boyd Incident
Page 11

At 18:38:30 hours, CII walked back towards the Alpha side from the middle of the roof and stated, "IC from Roof." There was no response from the IC.

At 18:38:40 hours, CII looked over the parapet near the 35' extension ladder and communicated, "Emergency Traffic, Emergency Traffic."

As conditions on the roof worsened, the roof team began hurriedly, grabbing their tools and equipment. CII yelled at the roof team to leave the tools and get off the roof. The firefighters, one by one, pulled themselves onto the aerial ladder in the following order: ATTN8, DRVR8, TILL TOP, A/O, CII. ATTN8 climbed on first and then DRVR8 climbed on next and then slid the hand rail backwards to make room for TILL. TOP used the opposite side of the aerial so that A/O could climb on at the same time. CII eventually transitioned to the aerial ladder and was the last one off the roof.

On the inside of the structure, about 5' to 10' beyond the rolling step ladder, Fire Attack heard "popcorn-like" noises that escalated quickly in both volume and frequency. The environment darkened and shortly thereafter, a "jet-like" sound began to crescendo within the structure. At that moment, multiple firefighters started to yell, "Get out!" As the C2, NOZ2, DRVR2, HYD2, ATTN1 and NOZ1 began to turn around, fire began to encapsulate them and blowtorched past them, out through the portal. Firefighters inside ran toward the exterior openings on the Alpha side.

The firefighters who had advanced the deepest into the structure now found themselves trapped by the rolling step ladder. As they fought to escape, they crawled through boxes and shelving. Eventually, each member beyond the step ladder was able to force their way around the step ladder and out to the curb, as fire continued to race past them.

Commercial Structure Fire
May 16, 2020
Boyd Incident
Page 12

At 18:38:42 hours, in the front of the building, "dirty" black smoke pushed out from Unit A across the street, enveloping the firefighters operating in front of and adjacent to the involved occupancy. As they backed up from the entrance, the smoke ignited, sending approximately 30' flames across the street and into the air.

As the smoke ignited and began to roll out of the Alpha side, fire began to appear from a rolling steel door on the Charlie side of Unit A, and rapidly expanded into an evolving fire ball. Small containers of nitrous oxide chargers ruptured, shot through the roof as well as out of the Charlie portal.

DRVR8, TILL and ATTN8 rapidly descended the aerial ahead of the deteriorating conditions and found temporary refuge on the trailer of the truck.

At 18:38:50 hours, EIT2 positioned on the street just to the west of the involved structure communicated, "Mayday, Mayday, Mayday."

At 18:38:52 hours a Battalion Emergency Medical Services (EMS) Captain requested to be attached to the incident with MFC on Administrative/Channel 9.

On the Alpha side, flames disappeared momentarily and the smoke appeared to temporarily diminish. Then without warning, an enormous fire ball blowtorched out of the structure reaching across the street. The fire ball emanated from the front of Unit A, followed by Unit B. Fire was seen reaching the buildings across the street and rising into the atmosphere approximately 60' high above the roof line. It scorched everything in its path, including the Truck and the rear portion of the Engine.

At 18:39:00 hours, the IC initiated the Emergency Tones for approximately 7 seconds followed by, "Mayday! Mayday! Mayday! All companies off, out of the building! Mayday! Mayday! Mayday! All companies we have a Mayday situation!"

Commercial Structure Fire
May 16, 2020
Boyd Incident
Page 13

As the fire engulfed the interior fire attack teams and blowtorched out of the units, C1 was outside the occupancy after being warned by ENG1. HYD1 and ATTN2 were close to the exit and were forced out across the street. ATTN1 and HYD2 had fire blow past them as they crawled to escape the intense heat. Deeper into the interior, DRVR2 maneuvered around the rolling step ladder, crawled through the flames and was pulled from the Alpha side portal into the street. NOZ1 negotiated the same step ladder and found the cinder block wall to reference for escape. NOZ2 crawled over boxes, struggled with the ladder and crawled to the portal to be pulled out by other firefighters. C2 was the last one out and managed to squeeze past the rolling step ladder and crawl through the excess storage until an awaiting firefighter pulled him out from inside the structure.

As the remaining roof team mounted the aerial ladder, flames completely engulfed the rungs, cutting off their escape. TOP descended the aerial ladder with the chainsaw. As he attempted to step on each rung, the concussive force that followed the fire ball caused his foot to slip under the rung and halted his progress. At that moment, he was enveloped by the fireball. At the same time, A/O and CII transitioned from the roof and moved quickly down the aerial ladder behind him. CII and A/O caught up with TOP who was struggling to get loose from being trapped in the rungs. CII ran into A/O and propelled him into TOP, who had broken free and resumed his descent on the rungs. TOP fell once again as A/O unintentionally toppled over him from behind.

As TOP and A/O attempted to get free from being entangled with each other, their axes became wedged in the aerial ladder. One of the axes splintered from the force and one scabbard separated from the leather belt, leaving the axes wedged on the aerial ladder. Entangled with each other, TOP and A/O forcefully pushed themselves down the rungs until they both ended up face down in the aerial ladder. At the same time, CII fell into the ladder face first after missing a step and all three were engulfed in intense fire for

Commercial Structure Fire
May 16, 2020
Boyd Incident
Page 14

way down the aerial, each of them falling out onto the turntable deck. Eventually, CII was able to stand up and walk the remaining portion of the aerial ladder until he paused and sat down on the rungs at the base of the aerial.

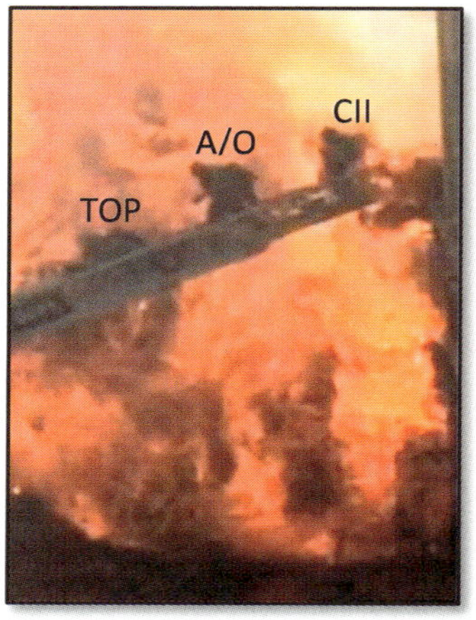

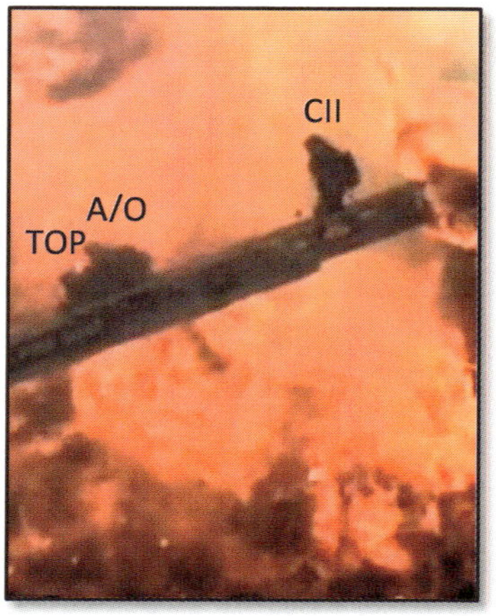

The extreme fire phenomenon lasted for approximately 45 seconds before it receded back into the structure. Intense fire conditions continued to show from the roof and Charlie side for the next minute.

At 18:39:17 hours, the Battalion Chief (BC2) from the second arriving Battalion Command Team stated on Command/Channel 11, "We're going to need a lot of companies. We're going to need some rescues."

Commercial Structure Fire
May 16, 2020
Boyd Incident
Page 15

Firefighters from the interior came out one by one. At the turntable of the Truck, TOP was pulled by TILL and jumped from the turntable deck to the ground. TILL reached up and pulled A/O off the pedestal onto ATTN8 and then assisted CII down the rest of the aerial ladder. CII descended from the turntable to the ground.

The A/O from Squad (A/O2) and ENG1 utilized available hand lines to carefully spray down and cool burned firefighters prior to their transition to awaiting ambulances.

At 18:40:00 hours, the IC stated, "All companies on the Boyd incident we need to conduct an immediate PAR and CAR."

C1 communicated a "Mayday" and reported multiple firefighters down. The number increased from two (2) firefighters and then, in a subsequent radio message, to ten (10) downed firefighters. The IC communicated that a Medical Group was being established and help was on the way.

At 18:40:33 MFC dispatched multiple resources to support the incident.

Continued requests from the IC for a PAR for the Truck Company, went unanswered.

Injured firefighters made their way to the intersection of Boyd Street and San Pedro Street where a Medical Group was established. Some firefighters were assisted by arriving resources and were carried down the street. Others were assisted in walking down the street, with the most critical firefighters being placed on gurneys.

A police escort was established and EMS facilitated communication with the receiving hospital. All eleven (11) burned firefighters were loaded into three (3) RA's, including one in the driver and passenger seat of the front cab and driven to the closest Medical Trauma/Burn Center.

Remaining resources continued to fight the fire and a knockdown was established at 20:08:00 hours.

Commercial Structure Fire
May 16, 2020
Boyd Incident
Page 16

CONDITIONS

At the time of the incident, the temperature was approximately 75 degrees with a wind out of the West at 6 mph. The business was closed at the time of the fire.

Call processing times and operational response times were better than average. Crew continuity was strong. Most of the members assigned to the first alarm were working at their regular assignment or had years of experience at the assignment previously, or in the surrounding Battalion. Drivers were familiar with their routes and responded expeditiously. Company discipline was evident and Standard Operating Guidelines were followed. Supervisors were knowledgeable of the district and effectively communicated the conditions of the incident. The IC was the normally assigned Battalion Commander.

According to the Los Angeles Department of Building and Safety (LADBS), records and permits indicate that the structure was erected as new construction, on or about July 28, 1998. Building permits showed a plot plan with an original footprint of 25' wide by 94' deep and approximately 24' high.

The building was divided into two separate occupancies, separated by a partition wall that ran North to South (front to back), comprised of 2'x6' members and sheeted with dry wall that extended to the bottom of the wooden "I" beam joists. The spaces above the wall and between the joists were not blocked off, causing horizontal openings between the two occupancies. Both occupancies were outfitted with commercial grade storage racks that were being used to hold excessive storage.

An unpermitted addition was added to the rear of the structure that was approximately 18' in length and approximately 25' in width. The addition had a roof line that was 7' lower than the original building roofline. The roof supporting members were comprised of eight (8) open web bar joist trusses, approximately 6" - 8" in height. The rear wall of the addition appeared to be constructed with steel beams and an "I" beam header as the main structural component, protected by a stucco exterior.

The subject address was not found in the LAFD Fire Prevention database and records of inspection were never completed.

The fire phenomena event that trapped and simultaneously injured 11 Firefighters is attributable to the involved fire units containing large, "live" fire loads of pallets, cardboard boxes, smoking and dispensary products and supplies, butane (C_4H_{10}), nitrous oxide (N_2O), and the lack of a sprinkler system.

Commercial Structure Fire
May 16, 2020
Boyd Incident
Page 17

The cause of the fire – which originated in ordinary combustibles – is under active investigation by the LAFD Arson Section and the United States National Response Team (NRT).

Although this fire could initially be classified as a routine ventilation limited structure fire, the excessive quantity of N_2O along with smoking and dispensary supplies and products changed the normal fire progression and made this a unique and extraordinary event.

Since the two units were separated by an unpermitted partition wall that lacked fire blocking at the roofline, the fire was able to quickly communicate between joists to the neighboring unit. The fire continued to preheat and consume all available fuels in both units, including stored butane and ordinary combustibles, causing the large storage of N_2O chargers to rupture.

The release of the oxidizer (N_2O) into the fuel rich atmosphere created a volatile chain of events that caused the spontaneous release of N_2O from their containers, supplying the necessary amount of oxygen to support the type of event witnessed at the Boyd Street incident. The incredibly large, violent release of energy was the result of rapidly expanding water vapor, carbon dioxide and extreme heat.

The ventilation team executed several large ventilation openings over the seat of the fire. Nevertheless, no amount of vertical ventilation on the 110' x 25' roof would have been able to safely release the amount of energy and fire that occurred, nor is it likely to have changed the outcome of this fuel and oxidizer driven event.

The Bravo and Delta walls of the occupancy were constructed of cinder block, with no openings for the blast energy to escape or be absorbed. Although the rear of the structure had suffered some damage from exploding butane and direct fire impingement prior to the event, forcible entry had not occurred.

In addition to the ventilation holes, the only other openings for which the large, rapidly expanding heat energy could escape were the occupancy openings at the front of the structure. The rolling steel doors were completely opened; however, excessive storage conditions created small, restricted aisles and narrow openings.

The evidence suggests that this configuration created a "nozzle-like" effect that funneled the intensified energy out of the front of the structure—while firefighters were attempting to retreat from the interior and off the roof.

An extensive search of the structure determined that at the time of the fire, the

Commercial Structure Fire
May 16, 2020
Boyd Incident
Page 18

Based on the investigation, it is estimated that the involved units contained hundreds of butane cartridges and thousands of 8 gram N_2O chargers.

The Boyd event was an extremely violent, hot, and rapidly expanding fire due to the abundant fuel supply in the form of smoke (carbon monoxide/CO), plastics, butane, and N_2O. Of particular concern was the extremely large quantity of the N_2O.

Although N_2O is a non-combustible gas, N_2O, when heated, will decompose creating an exothermic reaction. During the decomposition phase, its compounds split releasing oxygen (O_2), and nitrogen (N_2). If N_2O is mixed with a fuel source it will also create a rapidly expanding pressure wave of carbon dioxide (CO_2) and water vapor (H_2O).

The small N_2O cylinders found on scene of the Boyd incident contained approximately 8 grams each of net product and have been known to release their contents in temperatures as low as 106 degrees Fahrenheit. As the fire intensity and pre-heating increased, the release of the N_2O increased. The temperature around the N_2O storage was approximately 400 – 900+ degrees Fahrenheit early into the event.

According to industry experts, when large quantities of product contained in the cylinders are stacked in quantity and exposed to heat, they can release or explode from the outer edges of the storage configuration, working inward until a critical temperature is reached. This can cause a simultaneous release of a large portion of the product to occur with violent results.

The rapidly occurring chain reaction of the heated, fuel rich environment and rapidly increasing supply of N_2O were likely the source of the "popping-like" noises and rumbling energy that many firefighters on scene of the incident reported hearing, deep within the structure, just prior to the fire event.

In contrast, a fire in this type of occupancy of ordinary combustibles would progress and free burn - utilizing all available fuel and oxygen until it begins to run out of either fuel or oxygen. If the occupancy had high amounts of storage (fuel) and limited airflow, it would become a ventilation limited fire, where the fire begins to slow until a source of oxygen is introduced. In extreme cases, a back-draft can occur if an opening is created at street level prior to vertical ventilation operations being completed. Other well-documented phenomena of these types of fires are flashover, flame-over or smoke explosion.

At this incident, the fuel rich environment's demand for oxygen was supplied by way of the oxidizer (N_2O) thus minimizing the necessary flow-path dependent supply to allow

Commercial Structure Fire
May 16, 2020
Boyd Incident
Page 19

building event that did not respond to conventional firefighting flow-path control and ventilation methods.

Alpha side looking toward the Delta exposure

Alpha side looking toward the Bravo exposure

Alpha side looking into Unit A

Alpha side looking into Unit B

Commercial Structure Fire
May 16, 2020
Boyd Incident
Page 20

Charlie side looking towards the Bravo exposure

Charlie side looking towards the Delta exposure

Charlie side looking into Unit B

Charlie side looking into Unit A

Spent Nitrous Oxide Chargers in Unit B

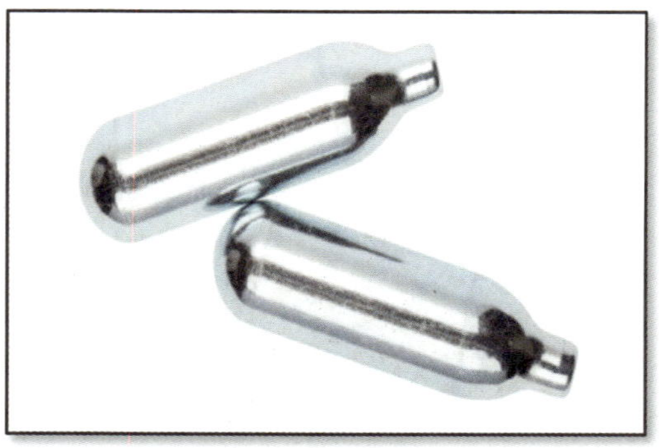

Nitrous Oxide Chargers 8oz.

Commercial Structure Fire
May 16, 2020
Boyd Incident
Page 21

Butane Canisters

Spent Nitrous Oxide Chargers

Butane Canisters in Unit B

Spent Nitrous Oxide Chargers in Unit B

Contents in Unit A

Nitrous Oxide Chargers on Bravo exposure roof

Commercial Structure Fire
May 16, 2020
Boyd Incident
Page 22

Perspective of roof from Bravo exposure looking back towards the Alpha side

Perspective of roof from Bravo exposure looking back towards the Charlie side

Commercial Structure Fire
May 16, 2020
Boyd Incident
Page 23

INJURIES/DAMAGES

A total of 12 firefighters were injured. Injuries ranged from partial thickness to full thickness burns, respiratory involvement, and one member suffered a lower extremity injury. Significant heat damage to one Aerial Ladder Truck and minor heat damage to the rear of one Engine Company was sustained. Multiple tools were either unrecoverable or significantly damaged. Personal Protective Equipment (PPE) including: Turnout Ensemble, Self-Contained Breathing Apparatus, Portable Radios and Thermal Imaging Cameras are currently being tested and findings will be provided in a later report.

All PPE from the injured members were collected, cataloged and inspected. The PPE were displayed on a mannequin and photographed to capture normal positioning on a body. A majority of the injured members showed signs of direct thermal impingement on multiple components of the PPE ensemble.

The following is a list of the members with associated burns:

CII	Burns to: Head, shoulder, both hands, both calves
AO	Burns to: Right ear, right arm, right hand, both calves
TILL	Burns to: Left ear, left hand
TOP	Burns to: Lower back, both arms, right hand, left foot
CI2	Burns to: Head, both ears, shoulders, back, both hands
NOZ2	Burns to: Head, back, both arms, right leg
HYD2	Burns to: Left hand
NOZ	Burns to: Head, back, both arms, both hands, both legs
ATTN2	Burns to: Head, back, both arms, both hands, both legs and buttocks
ATTN1	Burns to: Ears and back
DRVR1	Burns to: Right hand

Commercial Structure Fire
May 16, 2020
Boyd Incident
Page 24

Aerial Ladder placement

Aerial Ladder Truck placement

Aerial Ladder Truck Tractor and Trailer damage

Aerial Ladder Truck Exterior Cab damage

Aerial Ladder Truck Interior Cab damage

Aerial Ladder Truck Interior Cab damage

Commercial Structure Fire
May 16, 2020
Boyd Incident
Page 25

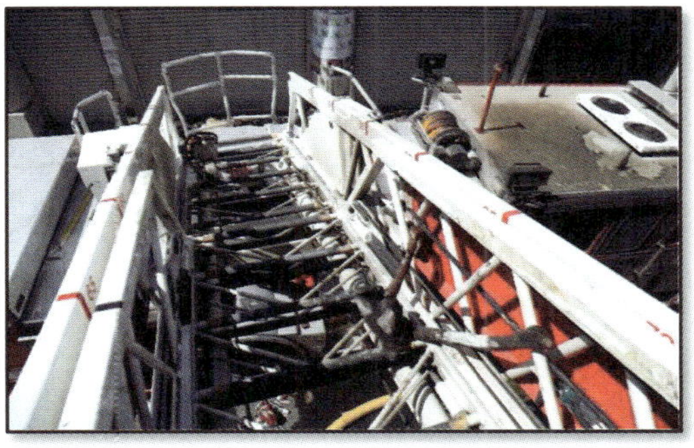

Aerial Ladder and Equipment damage

Aerial Ladder damage

Engine placement

Engine hose cover damage

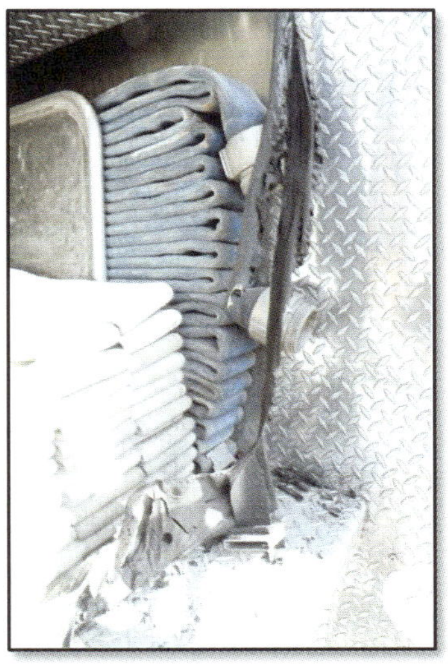

Engine hose cover damage

1 ¾" hose line damage

Drop bag damage

35' Extension Ladder damage

Commercial Structure Fire
May 16, 2020
Boyd Incident
Page 26

SAFETY ISSUES FOR REVIEW

- Review the appropriate donning of Personal Protective Equipment during emergency operations, which includes all issued turnout-coats, turnout-pants, turnout-boots, firefighting helmets, firefighting hoods, firefighting gloves and SCBA facepieces with voice amplification systems.

  - Toxic smoke exposure continues to be the main cause of injuries and death and in the fire service. It is the long-term effects of these exposures that have a significant consequence on firefighter health. Materials have changed over time and the variability of products of combustion have become more toxic. (A. Purser, Robert L. Maynard, James C. Wakefield, 2016 *Toxicology, Survival and Health Hazards of Combustion Products*) It is the unexpected and the potential consequence that we must prepare for. These environments are identified as an *Immediately Dangerous to Life or Health* (IDLH) environment. National Fire Protection Association (NFPA), a standard in the fire service,

    1. NFPA 1-2018, 3.3.51 **Immediately Dangerous to Life or Health (IDLH).** Any condition that would pose an immediate or delayed threat to life, cause irreversible adverse health effects, or interfere with an individual's ability to escape unaided from a hazardous environment. [1670, 2009]

    2. NFPA 1500, 7.10.7 When engaged in any operation where they **could** encounter atmospheres that are IDLH or potentially IDLH, or where the atmosphere is **unknown**, the fire department shall provide and require all members to use SCBA that has been certified as compliant with NFPA 1981.

- Review the LAFD Risk Management Policy
- Review "Mayday" procedures
- Review radio communication procedures when a "Mayday" is in progress
- Review reporting to Staging or Base procedures during emergency operations
- Review Incident Organization responsibilities
- Review the fourteen structural "watch outs"
- Review fire ground survival procedures
- Review firefighter accountability procedures

Commercial Structure Fire
May 16, 2020
Boyd Incident
Page 27

LESSONS LEARNED

1. **Personal Protective Equipment (PPE)** – Wearing all appropriate Department approved PPE (turnout-coats, turnout-pants, turnout-boots, firefighting helmets, firefighting hoods, firefighting gloves and SCBA facepieces with voice amplification systems) as intended — based on the ***potential*** of an incident — will save lives and reduce risk of injury. Emphasize the practice of going "on air" when working in, around, or above an IDLH. On scene members stated that their lives were saved or injuries reduced because of their decision to don their appropriate PPE's and SCBA.

2. **Accountability** - Detailed accountability of all members before, during, and after the incident is critical to determining the scope of a firefighter emergency and ensuring firefighter safety. During a catastrophic event, the smallest details become extremely important, such as the accuracy of F-11 staffing rosters, apparatus personnel accountability magnets, company identifying helmet passports, database inputted radio identifiers, command situation status/resource status, F-666, MCI procedures, and Personal Accountability Report (PAR) procedures.

3. **Occupancy Identification** – Recognize, analyze, and base strategic decisions on the ***potential*** hazards associated with drug industry related occupancies. The hazards of this relatively new industry continue to evolve and change rapidly. Coordination between the FPB, Joint Hazard Assessment Team (JHAT), Joint Regional Intelligence Center (JRIC), and Emergency Operations field resources is essential to stay current on trends in the industry that pose operational risks to LAFD members. Continually study Blue Sheets, Green Sheets, After Action Review Reports (AARR), and Line-of-Duty Death (LODD) Reports to gain knowledge on associated risks.

4. **Incident Organization** – "Organize early" During developing incidents, it is essential to assign Division/Group Supervisors for span of control, accountability, safety, and "ground truth" situational awareness. Division/Group supervision assists in defining roles and clearing up communication, especially during "Mayday" incidents. Consideration should be given to clearly defining who is in charge when multiple companies are entering a single point of entry. According to Book 30, the use of Divisions/Groups in the Incident Command organization provides a standard system to divide the incident scene into smaller manageable components. Whenever two resources are assigned to the same geographic area or function at an incident, one of the officers shall be placed in command of that geographic area or function

Commercial Structure Fire
May 16, 2020
Boyd Incident
Page 28

5. **Radio Discipline** – Maintain radio discipline during all emergency incidents, with particular attention to maintaining a clear tactical channel during all phases of a firefighter emergency. Companies must remain committed and disciplined to their company's task and continue firefighting operations during a "Mayday." Utilize alternative radio channels or face to face interaction to communicate with the IC, Division/Group Supervisors, or other companies during "Mayday" events. Additionally, do not call MFC to inquire about the incident or request a move-up during chaotic incidents. Members of MFC require all efforts to be focused on supporting the incident with minimal distraction.

6. **Emergency Evacuation** – Due to the inherent dynamic nature of firefighting, a continually evaluated and communicated plan for rapid egress (evacuation) is essential when working at an incident. Recognize, communicate, and mitigate potential dangers of narrow or blocked aisleways, stacked contents, or excessive storage. Limiting entry and assigning companies to clear the egress route should be considered. Develop company SOG's for emergency evacuation from the interior as well as the exterior or roof.

7. **Back to the Basics** – LAFD operations are based on fundamental fireground tactics designed to be proactive to incident potential. Utilizing the aerial ladder instead of solely ground ladders was a significant factor in this incident. Continue to emphasize that firefighting basics are being accomplished at every incident which includes, but is not limited to, methodical hose lays, developing supply lines, additional ground ladders, organized ventilation operations, forcible entry, and establishing a medical component early. Build on the basic firefighting and command fundamentals by constantly challenging your process with increased scope and complexity of training scenarios or simulations.

8. **Training Culture** – Continue to promote and emphasize the LFD proactive training culture abiding by the philosophy of "Train as if your life depends on it…because it does!" Many of the members interviewed stated that they relied on skills and knowledge learned during fundamental company level training, After Action Reviews (AAR), and organizational training such as the LFD Live Fire/Smoke Recognition and Fireground Survival Programs to perform during a life-threatening emergency. Emphasize the priority of such training and use repetition to instill deep learning.

9. **Crew Resource Management (CRM)** – "If you see something, say something!"

Commercial Structure Fire
May 16, 2020
Boyd Incident
Page 29

toward the rear of the building. According to accounts, the communicated awareness assisted in reducing the number of injuries or even prevented death. Emphasize that all members operating at an incident have the ability and responsibility to serve as a lookout for others.

10. **Post Mayday Events** – Recognize when to reassign or remove affected members of all ranks from operational and decision making positions after mayday, near-miss, or extremely stressful experiences. Cognitive ability decreases significantly during and after times of emotional duress and replacing affected positions should be considered for members' safety and mental health.

REFERENCE MATERIAL

- Departmental Bulletin 15-11, Risk Management Policy and Firefighter Emergency/Mayday Procedures
- Departmental Bulletin 16-05, Emergency Operations – Efficiency and Safety
- Volume 1 2/1-42.20 Firefighter Emergency
- Volume 1 2/3-20.30 Reporting to Base or Staging
- Volume 2 3/7-40.25 Wearing of Personal Protective Equipment (PPE)
- Volume 3 6/3-00.00 Self Contained Breathing Apparatus – General (SCBA)
- Volume 3 6/6-03.01 Module I – Laying a 4" Supply Line
- Volume 3 6/3-06.00 Lessons Learned
- Book 6 Lesson #20 Aerial Ladder Operations Spotting and Raising
- Book 6 Lesson #21 Aerial Ladder Operations Use of Aerial Ladder
- Book 8 Module #8 Apparatus Spotting and Placement
- Book 30, Incident Command Structure - Divisions/Groups, Basic Operational Approach
- Book 70 Multi-Casualty Incident Command System
- Book 101 Rapid Intervention Company- Firefighter Emergency
- Book 101 Rapid Intervention – Radio Assignments and Apparatus Roster Systems
- Book 101 Rapid Intervention – Escape Routes
- Book 101 Rapid Intervention – Look Outs/Conditions
- Training Bulletin 36, Identification of Phencyclidine and Clandestine Laboratories
- Training Bulletin 76, Company Operations
- Training Bulletin 90, Fire Burns – Structural Collapse and Fire Ground Survival
- Training Bulletin 106, LCES and "Watch Outs" for Structure Fires – Interior Firefighting "watch outs"
- Training Bulletin 152, Butane Honey Oil Recognition and Hazards
- Training Bulletin 165, Illegal Marijuana Cultivation
- NFPA 1561 – Standard on Emergency Services Incident Management System and

Commercial Structure Fire
May 16, 2020
Boyd Incident
Page 30

Street view from Google Maps

Aerial imagery of 327 East Boyd Street (indicated by arrow)

Commercial Structure Fire
May 16, 2020
Boyd Incident
Page 31

Aerial drone imagery from overhead perspective

Aerial photography from South (Alpha) perspective

Commercial Structure Fire
May 16, 2020
Boyd Incident
Page 32

Aerial drone photography from the West (Alpha/Bravo) perspective

Aerial photography from North (Bravo/Charlie) perspective

Commercial Structure Fire
May 16, 2020
Boyd Incident
Page 33

Aerial photography from North/East (Charlie/Delta) perspective

Commercial Structure Fire
May 16, 2020
Boyd Incident
Page 34

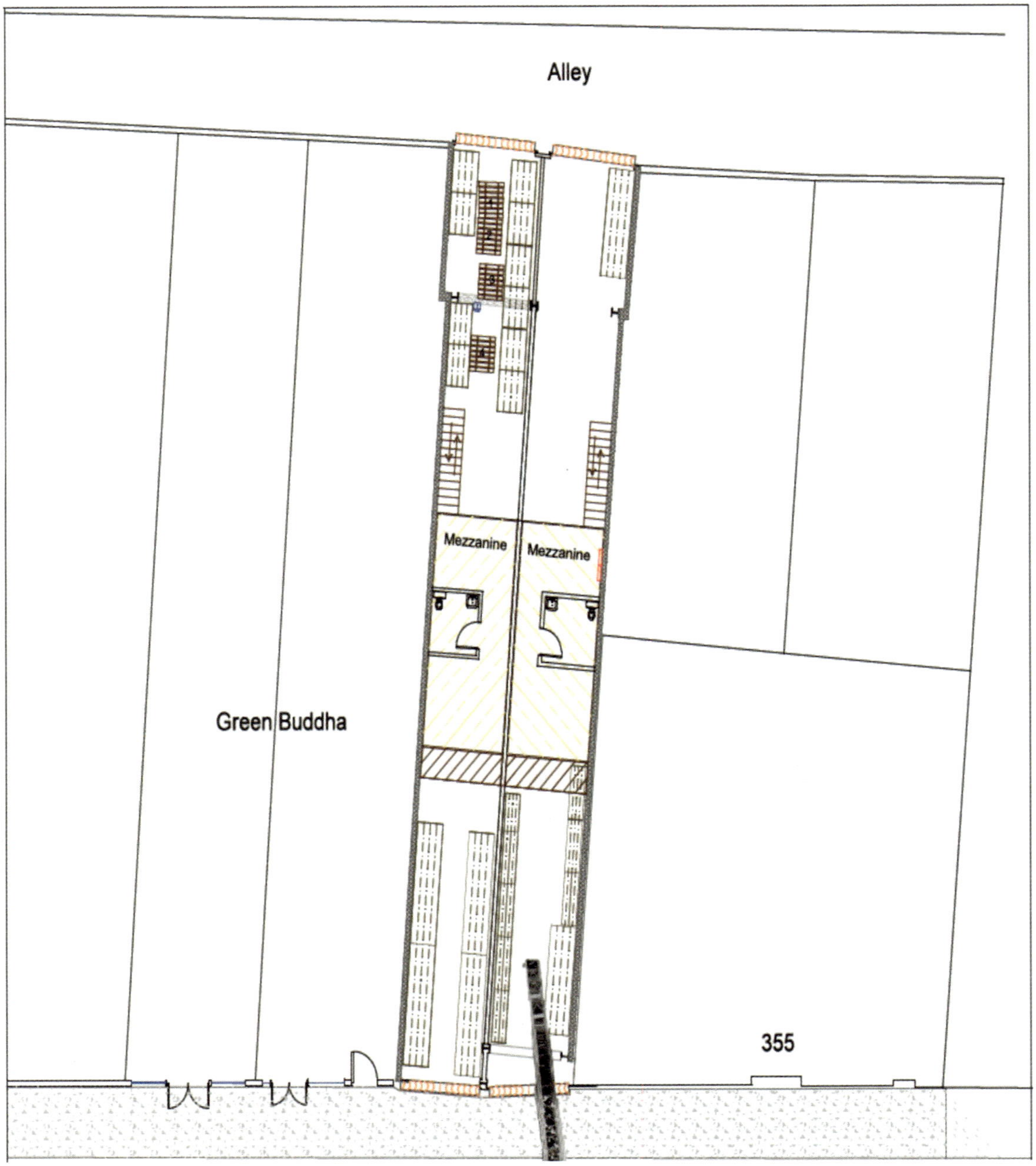

2-D image of 327 East Boyd Street (Courtesy of NRT)

B

Tactical Review Guides

This tactical review guide (TRG) appendix is added so you can apply what you have learned by combining several key elements in the book at once. This is a small sampling. For additional simulator training, 115 simulations are available online in the *Mastering Fireground Command—Calm the Chaos* online workshop. Scan the following QR code for more information or go to TrainFirefighters.com.

The TRG appendix has eight examples of incidents. Each one has the following components:

- **En route:** A listing of considerations while responding to the respective incident. This includes a company officer and a chief officer list.
- **Arrival:** An FPODP size-up for the incident based upon the first picture. Both *own situation* and *plan* of FPODP have company and chief officer considerations, as they may differ. At the end of FPODP will be additional pictures to consider.
- **Operations:** An abbreviated SAW-CSS-RECEO-VSS listing of operational considerations.
- **ICS:** An example organizational chart utilizing ICS for the incident. This will vary depending upon your response and staffing levels. This is simply an example.

Tactical Review Guide #1: Three-Story Victorian Home with Basement Fire, Extending

En route

Company Officer	Chief Officer
Time/Weather	Time/Weather
Location/Type of structure	Location/Type of structure
Route/Hydrants	Victim profile
Victim profile	Tac channel/Other units
Tac channel/Other units	Updates/Additional alarms, resources

Arrival

Facts: Three-story Victorian house with smoke from basement

Probabilities: Moderate victim profile with occupants possibly above fire; fire could rapidly extend above

Own situation:

- Company: First-arriving company with balance responding.
- Chief: First-arriving chief with one to two companies on scene, balance responding.

Decision: Life saving priority; offensive strategy

Plan:

- Company: Quick lap/evacuate. Firefighter forces entry at basement access doors on A/D corner. Engineer stretches wet line to be ready for fire attack. Second-in company complete water supply. (Note: if fire showing upon arrival, consider transitional attack prior to advancing into structure.)

- Chief: Set up command. Objectives: confinement, search, vent. Consider Division A for companies working in the main house/upper floors, and Division D for companies working in the basement. These are based upon sides by which companies are accessing the inside of the house. Ensure a transporting medic is on scene.

Operations

Safety: Victim profile is critical. While you want to get above to evacuate/search, be cautious over fire. See more safety considerations on overhaul section. Two-out initially, rapid intervention crew (RIC) ultimately. RIC or two-out can do utility control.

Rescue: Primary and secondaries starting above the fire on the first floor. Do not take for granted that the basement is clear. Search all floors. Search basement ASAP during attack. Separate crews for secondary.

Exposures: Floors above. Due to balloon frame construction (note windows aligned on C and D sides) consider extension in walls, leading to attic extension. This is common in basement fires with balloon frame construction.

Confinement: Confine to the involved basement, if possible. Check extension in interior basement stairwell, floors, walls, and attic above.

Extinguishment: Attack may start from outside basement access on A/D corner. Check interior stairs leading down to basement as well for extension. This is a two-line fire, plus a two-out line (likely all 1¾"). One in basement and one on first floor.

Overhaul: Ensure basement fire is totally out. Ensure no extension above, again in interior stairwell, floors, walls, and attic. When in doubt, open it up. Salvage first if you have discretionary time.

Vent: Hydraulic vent from inside basement is ideal. Positive pressure ventilate (PPV) in upper floors *as long as there is no fire*. If fire extends to attic, may have to consider vertical vent, based upon your standard operating guidelines (SOGs). Should not be necessary if you stay ahead of it in the walls.

Salvage: Likely, what is in the basement is lost. Salvage above. Pictures, pills, and pets.

Support: Air unit, investigator, utility companies. Main house may or may not be livable based upon smoke damage and structural integrity. Also, utilities may be totally compromised or may be isolated. When in doubt, keep them out.

ICS

- Division A
- Division D
- RIC group
- Medical

Tactical Review Guide #2: Two-Story Taxpayer, Fire from a Side First-Floor Restaurant

En route

Company Officer	Chief Officer
Time/Weather	Time/Weather
Location/Type of structure	Location/Type of structure
Route/Hydrants	Victim profile
Victim profile	Tac channel/Other units
Tac channel/Other units	Updates
	Medic unit, additional companies

Arrival

Facts: Two-story taxpayer. Fire from first floor, A side restaurant

Probabilities: High victim profile. Exposed residences above the fire restaurant.

Own situation:

- Company: First-arriving company with balance responding.
- Chief: First-arriving chief with one to two companies on scene, balance responding. Called additional medic unit.

Decision: Life saving priority; offensive strategy

Plan:

- Company: Enter A side. Transitional attack. Notify incoming companies. Passing command, however you are still in command until command assumed by second-due or chief. Firefighter stretches line, engineer completes water supply. Get victim profile of commercial unit and initiate evacuation of the first-floor commercial occupancies.
- Chief: Set up incident command post (ICP) across street, A side. Objectives: confinement, search, vent, medical, evacuation, and check exposures above ASAP. Ensure at least one transporting medic on scene, and available at all times. Set up Division A initially, as you will have multiple companies working/entering from the A side. This includes crews working on the second floor that may have entered the A side. Also, Division C is appropriate from the rear alley for access and search. Also, if you have multiple victims, set up a medical group for treatment and transport. Have a company report to the rear and give a CAN report as a minimum.

Operations

Safety: RIC or two out should be a high priority, based upon your SOGs.

Rescue: First, search fire unit and second floor units. Separate crews for secondary in all units. Highest priority is above, as these are living units. Note the door to the right of the restaurant door. This will access the upstairs units. Evacuate the ground-floor commercial units.

Exposures: Living units above are highest priority. Check extension in B and D units adjacent to the fire unit.

Confinement: Contain fire to unit of origin, then the floor. Expect extension into upper living units and even attic.

Extinguishment: Quick transitional may work to keep fire confined to first floor. *Get upstairs ASAP!* This is minimum a two-line fire on the first floor, plus a line upstairs to check extension.

Overhaul: Check above fire on both floors and in attic.

Vent: Hydraulic vent is a great option here, especially if you have minimal staffing. PPV, *as long as there is no fire.* Ensure you have lines in place and an exhaust identified (likely out fire unit).

Salvage: Work with manager to salvage priorities in store. Smoke removal above.

Support: Law for traffic control, air unit, investigator, utility companies, Red Cross if you have displaced residents above, PIO. Coordinate with management of whole building, if possible.

ICS

- Division A
- Division C
- RIC group
- Medical

Tactical Review Guide #3:
Strip Mall—Large Regional, Fire Running Mansard

En route

Company Officer	Chief Officer
Time/Weather	Time/Weather
Route/Hydrants/FDC	Location/Type of structure
Victim profile	Victim profile
Tac channel/Other units	Tac channel/Other units
	Updates
	Medic unit, additional companies, second alarm

Arrival

Facts: One-story strip mall. Heavy smoke from center unit.

Probabilities: Moderate to high victim profile rapid fire progression. Attic extension.

Own situation:

- Company: First-arriving company with balance responding.
- Chief: First-arriving chief with one to two companies on scene, balance responding. Called additional medic unit and alarm.

Decision: Life saving priority; combination strategy (transitional or defensive for involved unit, offensive to exposures)

Plan:

- Company: Ensure evacuation while firefighter stretches line. 2½" attack line. Quick transitional attack. Passing command, however you are still in command until command is assumed by second-due or chief. Engineer completes water supply. Get victim profile of occupancies.
- Chief: Set up command. Objectives: confinement, search, vent, medical, evacuation. Ensure at least one transporting medic on scene, and available at all times. Set up Division A initially, as you will have multiple companies working/entering from the A side. Consider Division C for eyes on the rear. Also, Division D for span of control as needed if you have extension. Expanded communications plan for extended attack/escalating incident.

Operations

Safety: RIC or two-out should be a high priority if evacuated, based upon your SOGs. Victim profile may drive deferring RIC. Consider fire exposure to panelized roof construction. HVAC on roof. Limited access C side.

Rescue: Search fire unit and B1/D1 exposures ASAP. Separate crews for secondary in all units.

Exposures: B1/D1 & D2 units and common attic of fire building.

Confinement: Contain fire to unit of origin. Check above drop ceilings, attic, and facade storefront.

Extinguishment: Quick transitional, however be cognizant of water supply. Expect more fire after initial knockdown. This is 4-line fire, 2½" on main body of fire. 1¾" in exposures minimal. If you have an extended attack, ensure two-out line is in place and staffed. Keep in mind, 2½" lines are labor-intensive. Ensure adequate resources and long-term water supply.

Overhaul: Ensure overhaul of attic and facade storefront.

Vent: If your FD performs vertical vent, consider inspection, heat cuts, and strip/trench cuts if you have extended fire attack and potential attic run. Roof division would be appropriate, as this will likely have multiple companies on the roof. PPV, *as long as there is no fire*. Ensure you have lines in place and an exhaust identified (likely out fire unit or C sides of each involved occupancy).

Salvage: Securing the building is a high priority. Contact building manager and business owners.

Support: Law, investigator, utility companies, air unit, rehab. Coordinate with mall management.

ICS

- Division A
- Division C
- Roof division (If our FD performs vertical ventilation)
- RIC group
- Medical

Tactical Review Guide #4:
Mayday—Walmart. Fire in B Loading Dock

En route

Company Officer	Chief Officer
Time/Weather	Time/Weather
Location/Type of structure	Location/Type of structure
Route/Hydrants/FDC	Victim profile
Victim profile	Tac channel/Other units
Tac channel/Other units	Updates
	Medic unit, additional companies, second alarm
	Multiple long-term water supplies and FDC

Arrival

Facts: One-story big-box Walmart. Moderate-heavy smoke showing from A/B corner near garden supply and B loading dock.

Probabilities: High victim profile, evacuation in place with occupants not leaving. Rapid fire progression.

Own situation:

- Company: First-arriving company with balance responding.
- Chief: First-arriving chief with one to two companies on scene, balance responding. Called additional medic unit and alarm.

Decision: Life saving priority; offensive strategy; be ready to switch to combination or defensive strategies.

Plan:

- Company: Stretch 2½" line to B entrance at loading dock (person-door to the right of the loading dock. Update alarm and gather information from occupants as to the location of fire and begin evacuations of all occupants from A side. Notify incoming companies (second-due to stretch a 2½" to A side for confinement/search) water supply and fire systems to have separate supplies. Passing command, however you are still in command until command assumed by second-due or chief. Firefighter stretches line from best location nearest the fire near loading dock, engineer completes water supply.
- Chief: Set up command. Objectives: confinement, search, vent, medical, evacuation. Ensure at least one transporting medic on scene, and available at all times. Set up Division B first and then Division A, as you will have multiple companies working multiple locations. Ensure fire stream coordination by divisions at entry points with NO opposing lines. Also, if you have multiple victims, set up a medical group for treatment and transport. Direct divisions of priorities of rescue and limit hose line advancement based on trigger points of smoke/fire conditions and building degradation.

Operations

Safety: RIC or two-out may be deferred initially due to higher victim profile. Definitely want RIC in place when you have adequate resources. Follow your SOGs for this type of occupancy. Moderate smoke from both sides of a large warehouse can lead to heat buildup and flashover. Concrete or masonry walls limit breaching for access. High-piled stock throughout store. HVAC on roof. The fire load in this store is immense. With minimal resources and growing fire, it may be all about evacuation and that's it.

Rescue: Determine victim profile ASAP from manager regarding employees. Evacuation will likely be in process, but don't be surprised if a lot of customers are still lingering in the store. Support evacuation. Target known victims *and* search near seat of fire first. Separate crews for secondary searches.

Exposures: Interior is open, heat and smoke can spread quickly.

Confinement: Contain fire to area of origin. Advance with risk versus reward in mind. High probability of extension into all areas of the big-box building. First line should be stretched closest to the fire. In this case, person-door on B side next to loading dock. If fire doors between rear stock area and front retail area, then close them

Extinguishment: Fire attack likely most efficient/fastest from the entrance nearest the seat of the fire. This is a multiple-line fire (2½" with back-up lines at each entrance), each with a separate water supply. Line placement must be coordinated by division supervisors to eliminate opposing hoselines. If you are losing, let it go and evacuate all members.

Overhaul: High-piled stock can create a hazard.

Vent: If your FD does vertical vent, then it would be appropriate. Inspection, heat and strip cuts as needed. Limit time on panelized roof based on fire activity. Coordinate with attack crews. PPV, *as long as there is no fire*. Ensure you have lines in place and an exhaust identified (likely out entrance nearest C side door).

Salvage: Main office, computers. You are not trying to save inventory.

Support: Air unit, investigator, utility companies, PIO. City or county health department food damage. Coordinate with management.

MANAGING THE MAYDAY: Once the Mayday is declared, stop all radio traffic and get a clear report from the crews declaring the Mayday, so the situation is clearly understood. Get a location, unit, name, assignment/air, resources needed (LUNAR) or use your agency's designated policies to gather information. Determine if assistance is necessary. Ensure crews working near the Mayday are aware. Remember that nearly 94% of all Maydays are addressed by crews already adjacent to the person/crew needing assistance. This will require Division B and RIC to coordinate face-to-face for resource sharing. If RIT/RIC is deployed, RIT/RIC supervisor will manage the Mayday. Division B will ensure that the fire is contained. Assign additional resources to the RIC/RIT group. Appropriate resource ordering ahead of the Mayday will help support a successful deployment. We always want a transporting medic unit on-scene and ready to receive a patient. For a commercial fire, ideally having 2 engines and 1 truck forward staged and ready for work is key. *Do not forget additional overhead*. Once the Mayday is declared those companies get assigned to RIT/RIC and additional alarms and medics are called for backfill. *Do not switch channels too soon.* You may be able to manage the Mayday *without* adding a channel. If your policy calls for a transition in your communications plan, initiate that very carefully and ensure that all crews move to the newly designated channel. PAR companies off the primary tactical channel onto a secondary tac channel. Leave the RIC Group and downed firefighter on the primary tactical channel. Remember it can take up to 12 personnel to extract one trapped firefighter safely and efficiently. This will be a critical event and critical incident stress management (CISM) will be necessary.

ICS

- Division B
- Division A
- Roof division (If your FD performs vertical ventilation)
- RIC group
- Medical

Tactical Review Guide #5: High-Rise Fire—Upper Floors

En route

Company Officer	Chief Officer
Time/Weather	Time/Weather
Location/Type of structure	Location/Type of structure
Route/Hydrants	Victim profile
Victim profile	Tac channel/Other units
Tac channel/Other units	Updates
	Second alarm or greater and transporting medic units if not already assigned

Arrival

Facts: 12-story high-rise apartment building with working fire, 9th floor. Report of multiple victims trapped in multiple units on 9th and 10th floors.

Probabilities: Confirmed multiple victim profile. Exposed units adjacent, and floors above.

Own situation:

- Company: First-arriving company with balance responding.
- Chief: First-arriving chief with one to two companies on scene, balance responding. Called second alarm and five additional medics, enroute. Calling 3rd and 4th alarm on arrival with base identified.

Decision: Life saving priority; offensive strategy

Plan:

- Company: Depends upon your SOGs. Likely enter A lobby for recon. Notify incoming companies. Passing command, however you are still in command until command assumed by second-due or chief. Firefighter brings high-rise pack. Engineer pumps FDC and coordinates water supply with next in engine.
- Chief: Set up command. Objectives: confinement, search, vent, medical, evacuation. Ensure multiple transporting medics on scene, and available at all times. Set up Division 9 and 10 initially, as you will have multiple companies working multiple floors. May need more. Also, set up a medical group for treatment and transport of multiple victims. Think MCI.

Operations

Safety: RIC or two-out may be deferred initially due to confirmed victim profile. Definitely want RIC in place on upper floors when you have adequate resources, possibly not until third alarm or later based upon victims.

Rescue: Victim profile: Attempt to get specific apartment number and/or floor of confirmed victims trapped. Target known victims and search involved unit/floor, then above. Start with the area directly near/above fire. Primary and secondaries in the main unit and check adjacent units also. Separate crews for secondary in all adjacent/upper units. Consider evacuations versus shelter-in-place for units further away from fire, based upon amount of smoke and fire extension.

One effective tactic is rapid ascent teams utilizing a rescue stairwell that is separate from the attack stairwell. This removes victims away from the fire , while fire attack from separate stairwell. Prevents congestion in one stairwell for everything. Be aware of stack effect of smoke on upper floors and evacuate accordingly.

Exposures: Floors above and units adjacent to the fire unit. Stairwells.

Confinement: Keep fire out of hallway and stairwell. Fast water is critical. Fire can extend through stairwells, chases, ducts, elevator shafts, and exterior windows lapping. By the time you make the fire floor, it will be likely too late and fire will be in the hallway as is evidenced by conditions. Confine fire to involved unit, if possible, then hall, then floor. FDC should be in the stairwell, check pre-plans. While quick attack from stairwell is important, consider defending the common-floor area and units toward the **unburned area** as well. This may require lines from the rescue stairwell. Always be wary of opposing hoselines.

Extinguishment: This is a four-line fire (likely all 2½"). Two lines 9th floor and two lines on 10th floor. This requires a lot of companies to get in place and maintain with rotation of crews. Could be more if goes above 10th floor.

Overhaul: Check between floors, walls, chases, and shafts. Coordinate with building engineer.

Vent: PPV *as long as there is no fire*. Ensure you have lines in place and an exhaust identified (likely out fire unit). You may have PPV on both floors 9 and 10. They should coordinate through Divisions 9 and 10. May also use HVAC system to isolate. Hydraulic out windows of involved units will also be fast and effective.

Salvage: Pictures, pills, and pets. Adjacent floors may be tenable. Water on floors below. Depends upon utilities and smoke/fire damage.

Support: Light/Air unit, investigator, utility companies, Red Cross, PIO, safety, rehab unit, and law. Coordinate heavily with management and building engineers. Medical will be significant, either from burns, smoke inhalation, and so on. A medical group should be set up early with a supervisor and multiple medic units. Also, systems group should be set up to coordinate elevator control, HVAC, sprinklers/pumps, public address and be tied at the hip with building engineers.

ICS

Sample Organizational Chart with respective objectives:

Division 9	Division 10	Rescue Group	Systems Group	Medical Group	RIC Group	Staging	Rehab	Base

In addition, safety, PIO, and liaison officers should be considered.

- **Divisions 9 and 10:** Fire attack and search from the fire attack stairwells. Coordinate water supply to the FDC.
- **Rescue group:** Target known victims trapped locations, if known through victim profile. Support Divisions 9 and 10 with searches from the rescue stairwell. *And*, ascend to the stairwell to evacuate occupants from smoke due to stack effect on upper floors.
- **Systems Group:** Coordinate elevator control, HVAC, sprinklers/pumps, PA, and be tied at the hip with building engineers.
- **Medical group:** Triage, treat, and transport all patients, whether civilian or firefighters.
- **RIC group:** Objective is RIC on fire floors. Could be staged on rescue stairwell or staging to prevent congestion. Should be established after adequate number of companies are attacking the fire and searching. Follow your SOG. Multiple companies required.
- **Staging:** Two floors below fire floor on floor 7 with extra crews, bottles, tools, and so on. Rehab location as well and RIC to prevent congestion.
- **Rehab:** On staging floor with fluids, bottles, and so on for crews. Need to account for crews.
- **Base:** Outside the building where apparatus is staged, and fresh crews prepare to enter building.

As you can see, the span of control for the IC can get blown out quickly if/when multiple floors become involved. Depending upon your resource levels, this could easily become a 5-plus alarm fire. One way to bring the span of control back into a manageable level is to establish branches or sections, based upon your SOGs.

- Operations section:
  - Division 9
  - Division 10
  - Rescue group
  - Medical group
  - RIC group
  - Staging
- Logistics section:
  - Systems group
  - Rehab group
  - Base

Even if you added a complete command staff (safety, information, liaison officers), the span of control for the IC would still only be 5:1 with the above format.

Tactical Review Guide #6:
Vegetation Fire—Terrain with Structures Threatened

En route

Company Officer	Chief Officer
Time	Time
Fuel/Weather/Topography	Fuel/Weather/Topography
Location/Access to incident	Location/Access to incident
Fire history	Fire history
Victim profile	Victim profile
Additional resources	Additional resources/Air support
Law enforcement	Law enforcement
Tac channel/Coordinate with law?	Tac channel/Unified command with law
	Values at risk/Structure defense

Arrival

Facts: Three-acre grass and brush with spotting up a slope, moderate-fast rate of spread, immediate structure threat at the head and left flank of the fire

Probabilities: Injured/non-ambulatory persons, rapid fire progression, structural involvement, extended attack incident

Own situation:

- Company: First-arriving company on a wildland assignment.
- Chief: First-arriving chief with one to two companies on scene, balance responding.

Decision: Life saving priority; offensive strategy

Plan:

- Company: The slope would cause a significant delay in stopping the forward progress of this fire. Go directly to residences above. Evacuate ahead of the fire, protect threatened structures and civilians. Order additional resources.
- Chief: Establish command (possibly unified with law), order additional resources, including air support, if available. Reinforce structure defense efforts. The tactical priority is getting ahead of the fire for structure defense. This will protect values at risk and put crews in best position for life saving strategic priority. Could go into extended attack. Initial ICS will likely include Division B (left flank), Division O (head of the fire) and X (right flank). Air Ops should be ordered ASAP if available.

Operations

Safety: Lookouts, communications, escape routes, safety zones, 10 Standard Firefighting Orders, 18 Situations that Shout Watch Out, power lines, snags/widow-makers. Use law to evacuate and traffic control. With this slope, rolling debris could be a hazard. Also, evacuating vehicles could also cause congestion down the mountain.

Rescue: Evacuate ahead of the fire, provide medical triage, treatment, and transport, as necessary. Use law to evacuate.

Exposures: Get ahead of the fire ASAP. Structure defense equals life saving ahead of the fire at houses/buildings. Check structures where the fire front has already passed to ensure no ignition/rekindle.

Confinement: Control objectives: Define containment lines (roads, rivers, handlines, dozers) and reinforce, as necessary. For this fire, you will likely need progressive hoselays on the left and right flanks.

Extinguishment: Control the perimeter (100 feet in or per department policy). Ensure 100% mop up around all structures. Fully check all structures (attic spaces, interior, basements, decks) for potential fire extension. Full extinguish all sheds, fences, and so on, on exterior of structure.

Overhaul: Mop up according to department policy (100 feet in, 300 feet or 100%). Ensure routine patrols of fire-affected areas. Make sure all fire lines are wind tested prior to demobilization of resources.

Vent: Consider weather changes (use spot weather forecasts, sling weather on sight, etc.), announce changes in wind/weather to all crews.

Salvage: Contact responsible party for property.

Support: Law enforcement (unified command), investigator, PIO, safety officer, Red Cross for displaced residents, OES, utility company, rehab, air operations, dozers, hand crews, logistical support for crews.

ICS

Divisions around the perimeter is the most effective means of organizing the incident. Rather than a structure protection group, divisions will likely all have the same baseline objectives: perimeter control *and* structure defense. The key is the have the correct type and number of resources on each division, based upon their conditions and structure potential. Division O (the head) would be the priority.

If air operations are involved, then an air attack group is appropriate.

- Division B
- Division O
- Division X
- Air Ops group
- Staging
- Evacuation group (law)
- Public information officer

Tactical Review Guide #7: Spill on Freeway

En route

Company Officer	Chief Officer
Time/Weather	Time/Weather
Location/Access to incident	Location/Access to incident
Victim profile	Victim profile
Additional resources/Hazmat team	Additional resources/Hazmat team
Law enforcement/Highway patrol	Law enforcement/Highway patrol
Tac channel/Coordinate with law?	Tac channel/Unified command with law

Arrival

Facts: 6,000-gallon DOT407 tanker rolled over, poison placard visible, blocking all lanes of traffic

Probabilities: Injured persons, leaking product.

Own situation:

- Company: First-arriving company on a traffic collision response with a medic unit.
- Chief: First-arriving chief with one to two companies on scene, balance responding (hopefully you added resources).

Decision: Incident stabilization; combination strategy

Plan:

- Company: Isolate and deny entry, locate the driver, attempt to identify the product.
- Chief: Establish command, establish hot, warm, cold zones. Notify appropriate agencies. Unified command with law enforcement (highway patrol) and possibly the highway department/Department of Transportation (DOT) is ideal. Initial isolation distances of a half mile are recommended by North American Emergency Response Guide (NAERG).

Operations

Safety: Uphill/upwind approach, reference NAERG for appropriate isolation and evacuation distances, assign a safety officer, ensure traffic safety plan is implemented.

Rescue: Rescue those in immediate danger, evacuate the area (reference NAERG), provide medical triage, treatment, transport as necessary, consider gross decontamination for exposed patients.

Exposures: Ensure hot, warm, cold zones are established, ensure downwind shelter-in-place and/or evacuation is in place.

Confinement: Locate source of the leak, entry and back-up teams, technical reference, decontamination.

Extinguishment: Threat of ignition, size of potential fire event, deployment of appropriate-size lines, consider potential water reactivity of substance.

Overhaul: Clean up contractor (through the shipper/product owner), appropriate disposal of product.

Vent: Ensure atmospheric monitoring is completed in downwind evacuation/shelter-in-place areas.

Salvage: What needs to be decontaminated, responsible party/owner of product is on site.

Support: Law enforcement (unified command), PIO, safety officer, county health department, OES, utility company, rehab, breathing support, gross decontamination of personnel, PPE, and equipment.

ICS

- Hazmat group
- Medical group
- Safety (including Hazmat ISO)
- Public information officer
- Unified command with highway patrol

Tactical Review Guide #8: Sick Building

En route

Company Officer	Chief Officer
Time/Weather	Time/Weather
Location/Access to incident	Location/Access to incident
Victim profile—Number of potential patients	Victim profile—Number of potential patients
Additional medics/Other resources	Additional medics/Other resources/Hazmat
Law enforcement/Postal inspector response	Law enforcement/Postal inspector response
Tac channel/Coordinate with law	Tac channel/Unified command with law
Approach from upwind/uphill	

Arrival

Facts: One-story main branch post office, chemical odor inside, multiple patients complaining of eye irritation, nausea, vomiting, and difficulty breathing

Probabilities: Persons needing rescue, hazmat incident, criminal act

Own situation:

- Company: First-arriving company with medic and battalion chief responding (hopefully you added resources).
- Chief: First-arriving chief with one to two companies on scene, balance responding (hopefully you added resources).

Decision: Life saving priority; offensive strategy

Plan:

- Company: Declare hazmat MCI, rescue those in immediate danger, triage. Isolate and deny entry.
- Chief: Assume command. ICP upwind, A side. Confirm hazmat MCI , establish unified command with law and post office. Objectives: isolate, deny entry, triage, treat, transport, hazmat. Call hazmat team.

Operations

Safety: Establish hot, warm, cold zones, ensure site access control, decontamination prior to transport, Safety Officer assigned and site safety plan. Call hazmat response team. Assign hazmat safety officer.

Rescue: Rescue/extract patients, evacuate, triage, treat, transport (remember to decontaminate first). Set up treatment areas.

Exposures: Check interior and exterior structure, ensure downwind evaluation/evacuation, reference NAERG for evacuation distances.

Confinement: Locate source of the leak, entry and back-up teams, technical reference, decontamination.

Extinguishment: Threat of ignition, size of potential fire event, deployment of appropriate-size lines, consider potential water reactivity of substance.

Overhaul: Clean up contractor (through the shipper/product owner), appropriate disposal of product.

Vent: Horizontal ventilation, consider downwind impacts, atmospheric monitoring.

Salvage: Consider impacts to packages and parcels in the building, what needs to be decontaminated, ensure postmaster and/or postal inspector is on site.

Support: Law enforcement (unified command), PIO, safety officer, utility company, rehab, breathing support, gross decontamination of personnel, PPE, and equipment.

UNIFIED COMMAND

U—Understand the situation (size-up)
N—Notify appropriate agencies
I—Identify the need for unified command
F—Find your counterpart
I—Information exchange
E—Establish unified command
D—Develop a plan
C—Communications plan
O—Objectives
M—Map it
M—Mark it
A—Accountability
N—Needs assessment
D—Deploy the plan

ICS

- Unified command with law and postal service
- Medical group
- Hazmat group
- Safety officer
- Public information officer

C

Fire and Emergency Services Higher Education Model Curriculum: Strategy and Tactics (CO279)

Course Description

This course provides the principles of fireground control through utilization of personnel, equipment, and extinguishing agents

Prerequisites

Principles of Emergency Services (CO272)

Course Outcomes

Upon completions of this course students will be able to:

1. Discuss fire behavior as it relates to strategies and tactics.
2. Explain the main components of pre-fire planning, and identify steps needed for a pre-fire plan review.
3. Identify the basics of building construction and how they interrelate to pre-fire planning, strategy, and tactics.
4. Describe steps taken during size-up.
5. Examine the significance of fireground communications.
6. Identify the roles of the National Incident Management System (NIMS) and Incident Command System (ICS) as it relates to strategy and tactics.
7. Demonstrate the various roles and responsibilities in ICS/NIMS.

Course Objectives

Upon completion of this course, the student shall be able to:

1. Create a strategy and implement appropriate tactics.
2. Possess a working knowledge and execution of ICS/NIMS at the incident.

Course Outline

I. Fire chemistry and concepts (see chapter 6)

   a. Heat transfer
   b. Principal fire characteristics of materials
   c. Fire classifications

II. Extinguishing equipment (see chapter 6)

a. Extinguishing equipment
b. Fire apparatus
c. Personnel requirement

III. Visual perception (see chapters 4, 5, 7)

a. Pre-planning
b. Size-up

IV. Pre-Fire planning (see chapters 4, 5, 7, 8, 9, 10, 11)

a. Concept
b. Phases
c. Methods
d. Format
e. Occupancy classifications
f. Building types

V. Basic division of tactics (see chapters 5, 6, 7)

a. Size-up (chapter 5)

i. Facts
ii. Probabilities
iii. Own situation
iv. Decision
v. Plan of operation

VI. Rescue (see chapters 5, 6, 7)

a. Life safety problems of fire (chapter 6)
b. Determination of life hazard (chapter 5)
c. Rescue resources and operations (chapters 5, 6, 7, 8)

VII. Exposures (see chapters 6, 7, 8, 9, 10)

a. Principle contributing factors
b. Exposure protection operations

VIII. Confinement (see chapters 6, 7, 8, 9, 10)

a. Fire separations
b. Fire loading
c. Built-in protection
d. Operations

IX. Ventilation (see chapters 6, 7, 8, 9, 10)

a. Relationship to objectives
b. Equipment
c. Roof types
d. Methods

X. Salvage (see chapters 7, 8, 9, 10)

a. Relationship to objectives
b. Equipment
c. Operations during fire
d. Operations after fire

D

NFPA 1021: Fire Officer Professional Qualifications for Emergency Services Delivery and Emergency Management

NFPA 1021: Standard for Fire Officer Professional Qualifications for Emergency Services Delivery

Fire Officer I

4.6 Supervising emergency operations and deploying assigned resources in accordance with local emergency plan.

4.6.1 Develop initial action plan, given size-up information for an incident and assigned emergency response resources, so that resources are deployed to control the emergency.

4.6.2 Implement an action plan at an emergency operation given assigned resources, type of incident, and preliminary plan, so that resources are deployed to mitigate the situation.

4.6.3 Develop and conduct a post-incident analysis given a single unit incident and post-incident analysis policies, procedures, and forms, so that all required critical elements are identified and communicated, and the approved forms are completed and processed in accordance with policies and procedures.

Fire Officer II

5.6 This duty involves supervising multi-unit emergency operations, conducting pre-incident planning, and deploying assigned resources.

5.6.1 Produce operational plans, given an emergency incident requiring multi-unit operations; the current editions of *NFPA 1600*, *1700*, *1710* and *1720*; and AHJ-approved safety procedures so that resources and their assignments are obtained, and plans carried out in compliance with *NFPA 1600* and approved safety procedures resulting in the mitigation of the incident.

5.6.2 Develop and conduct a post-incident analysis given a multi-unit incident and post-incident analysis policies, procedures, and forms, so that all required critical elements are identified and communicated, and the approved forms are completed and processed in accordance with policies and procedures.

5.6.3 Prepare a written report, given incident reporting data from the jurisdiction, so that major causes for fire service demands are identified for various planning areas within the service area of the organization.

Fire Officer III

6.6 This duty involves managing multi-agency planning, deployment, and operations.

6.6.1 Prepare an action plan given an emergency incident requiring multiple agency operations, so that the required resources are determined and the resources are assigned and placed to mitigate the incident.

6.6.2 Develop and conduct a post-incident analysis given a multi-agency incident and post-incident analysis policies, procedures, and forms, so that all required critical elements are identified and communicated, and the approved forms are completed and processed in accordance with policies and procedures.

6.6.3 Develop a plan for the organization given an unmet need for resources that exceed what is available in the organization, so that the mission of the organization is capable of being performed in times of extraordinary need.

Fire Officer IV

7.6 This duty involves developing plans for major disasters.

7.6.1 Develop a comprehensive disaster plan that integrates other agencies' resources given risk, vulnerability, and capability data, so that the organization can mitigate impact to the community.

7.6.2 Develop a comprehensive plan given data, so that the agency operates at a hostile event, integrates with other agencies' actions, and provides safety and protection of members.

NFPA 1021: Standard for Fire Officer Professional Qualifications for Emergency Management

Fire Officer III

6.8 This duty involves policies, procedures, and programs for the role of the fire service in the community's emergency management plan and the roles of local, state/provincial, and national emergency management agencies.

6.8.1 Develop a plan for the integration of fire services resources in the community's emergency management plan, given the requirements of the community and the resources available in the fire department, so that the role of the fire service is in compliance with local, state/provincial, and national requirements.

E

Quick Action Plans—Leveraging High-Rise Building Intelligence for Incident Commanders

By Jack J. Murphy, former Fire and Life Safety Director, New York Hospital/Cornell Medical Center and coauthor of *High-Rise Buildings: Understanding the Vertical Challenges* (Fire Engineering Books)

The fireground is a poor environment in which to make decisions grounded on guessed-based knowledge as an incident is rapidly unfolding. In a high-rise building environment, the need for building intelligence is more prevalent due to the complexity of the structure, including the building systems and various floor layouts, among other factors. A building intelligence (BI) plan implemented within the National Incident Management System (NIMS) will also assist the incident commander (IC) with a well-thought-out and prepared incident action plan (IAP) to better guide the overall fireground strategy, tactics, risk management, and firefighter safety. BI can be the foundation to formulate a battle plan with decision-based options for safe fireground strategies and tactics.

Fire companies need to play a more active role to gather building information for future fire and all-hazard (non-fire) emergencies. Today, the words *pre-incident* plan and *pre-plan* should be replaced with more related words beholding our enemy, the building, with terms resembling a battle plan readiness by executing a structure reconnaissance (recon) to gather battlefield intelligence. The distribution of this relevant building data will support fireground operations by providing manageable intelligence in the form of quick action plans (QAPs) for an initial engagement and further support fireground operations with decision-based knowledge as the incident unfolds.

Leveraging Manageable Response QAPs

In the National Fire Academy course "Command and Control of Incident Operations" pre-fire planning is clarified as a method of gathering and analyzing building information to develop an IAP. As an IC, sound building intelligence formulates the battle plan strategy, recommending choice of tactics and detailing the challenge of life safety. It also assists in resource evaluation. This QAP application in an electronic database is a mind-jogger or data stimulation for rapid decisions, incident scene safety data, and building plans that can be comprehended effortlessly in 30 to 60 seconds.

An e-building intelligence card (eBICard) can be broken down into three manageable levels and made available during the various QAP response phases:

- QAP 1/Basic Level: Provides e-building intelligence while en-route for the initial first-due units to begin fire or emergency operations and alerts of real-time temporary fire protection system impairments and construction alteration projects, as well as safety and precautions to firefighters. It also has an initial size-up application embedded.
- QAP 2/Intermediate Level: Supplementary e-building intelligence to quickly deploy personnel with more details intended for a strategic and tactical response to an incident that includes a building (main level) map and vertical riser illustrations of stairways, elevators, indoor (floor) mapping, location of people with disabilities, and mechanical equipment rooms (see fig. E–1).
- QAP 3/Comprehensive Level: This will have the most detailed building intelligence intended to support command as the incident develops, with specific field operation guidelines, including critical structure data that are building specific.

A sound QAP e-building intelligence program will be able to enhance firefighters' safety by adopting a "Know Before You Go" (KBYG) state of mind for a battle-ready fire unit.

Lessons-Learned: Hospital Newborn Wing Mass Evacuation

During the early morning hours of December 11, 1992, a vicious nor'easter storm pounded New York City with 50–75 mph winds and heavy rains. The high winds created an East River surge flooding and pushed it up and over the FDR 6-lane highway with a water deluge back pressure destroying the gaskets on the hospital's waste piping connections.

After the hospital's monthly emergency generator test while on-duty as the fire/life safety director (FLSD), a fire alarm activation for a strong chemical odor was emitting from the sub-basement area of a 10-story newborn baby wing. This building is one of 18 interconnecting structures at a major hospital complex which run three blocks in midtown Manhattan.

On the way to the incident the FLSD was able to ascertain the situation and did a flow-up call to Manhattan Fire Dispatch requesting a hazmat unit response. While awaiting the arrival of the initial fire units, the hospital crash cart was deployed. Absorbent padding was applied to redirect the water away from a ramp area that led to another attached hospital's electrical transformer room. Upon arriving, the 10th battalion chief was provided with the facility's pre-incident plans (i.e., floor plans, building system riser diagrams, etc.) concerning the waste piping system, and apprised about the initial diking actions and elevator cars being recalled to the lobby. The chief also mentioned hazmat was still approximately 20 minutes out due to the weather. With the strong odor starting to permeate the upper floors, a horizontal inpatient and staff evacuation was ordered (see fig. E–2). Having previously gathered BI about inpatient bed counts and evacuation interior routes played a key role for this mass evacuation. Firefighters and hospital staff were able to safely relocate all occupants to attached buildings.

During the evacuation, many hospital staff were trying to converse with the IC, but the FLSD was able to ward off any who had no role to assist with the response actions but were just being inquisitive. The liaison role of a competent building intelligence representative (BIR/e.g., FLSD, building engineer) within a NIMS structure can be invaluable asset to an IC.

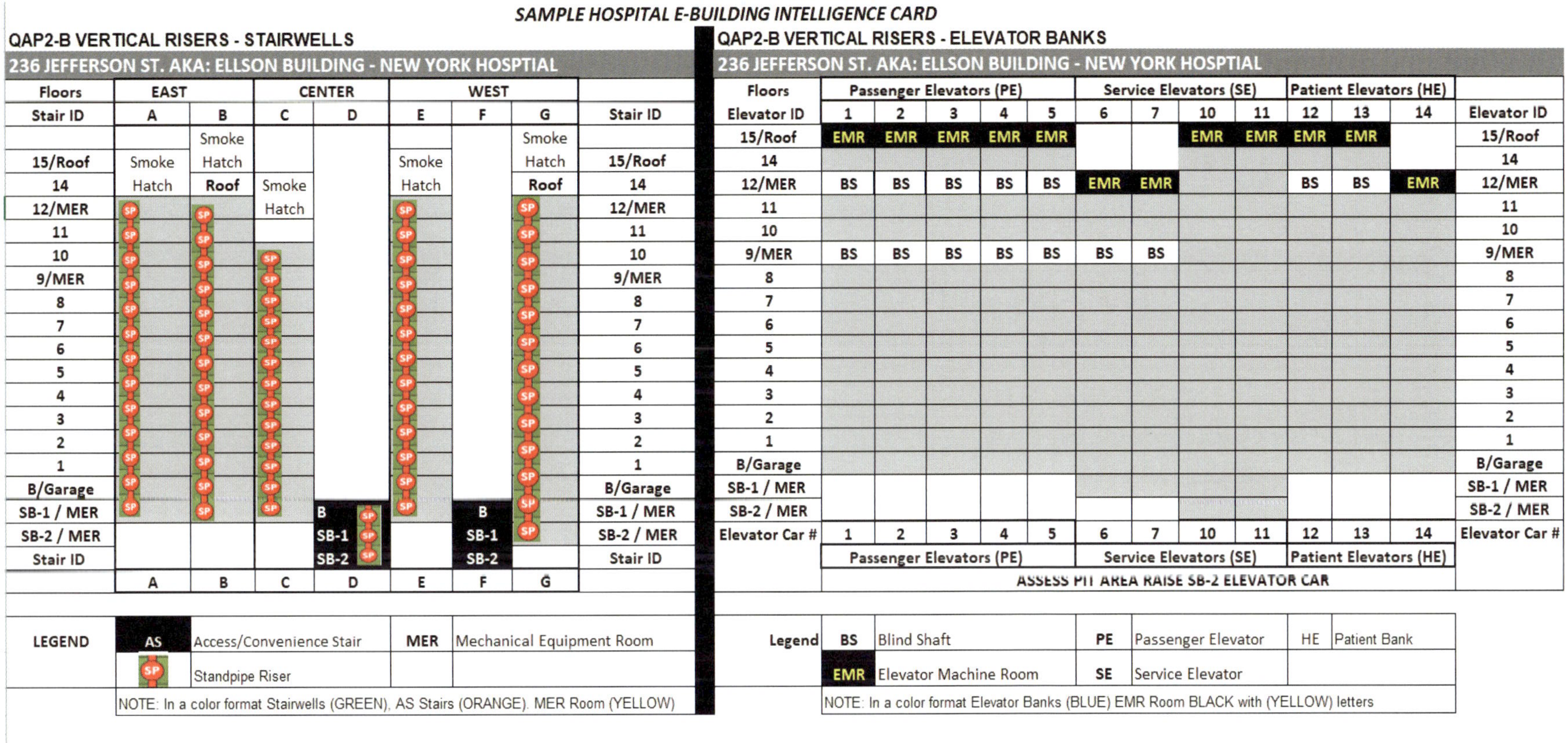

Fig. E–1. QAP-2 vertical risers—stairwells and elevator banks chart

236 JEFFERSON ST. NEW YORK HOSPTIAL		
IN-PATIENT FLOORS AND BED COUNTS		
FLOOR	**OCCUPANCY**	**BEDS**
14	IN-PATIENT	**48**
12	MER & KITCHEN	
11	PSYCHIATRY	**64**
10	IN-PATIENT	**84**
9	MER ROOM	
8	IN-PATIENT / ***ICU (8W)***	**112**
7	IN-PATIENT	**68**
6	IN-PATIENT / ***ICU (6S/6W)***	**130**
5	IN-PATIENT / ***ICU (5S)***	**116**
4	IN-PATIENT / ***ICU (4W)***	**111**
3	OPERATION ROOMS	**24**
2	IN-PATIENT / ***ICU (2S/2W)***	**66**
1	EMERGENCY ROOM	**48**
B/Garage	Garage	
SB-1	MER ROOM	
SB-2	MER ROOM	

236 JEFFERSON ST. NEW YORK HOSPTIAL					
HORIZONTAL EGRESS PATHWAYS EVACUATION BUILDING					
Floors					
14	ELLSON				
12/MER					
11	ELLSON	AUCH	ILL	KARROS	HALTON
10	ELLSON	AUCH	ILL	KARROS	HALTON
9/MER					
8	ELLSON	AUCH	ILL	KARROS	HALTON
7	ELLSON	AUCH	ILL	KARROS	HALTON
6	ELLSON	AUCH	ILL	KARROS	HALTON
5	ELLSON	AUCH	ILL	KARROS	HALTON
4	ELLSON	AUCH	ILL	KARROS	HALTON
3	ELLSON	AUCH	ILL	KARROS	HALTON
2	ELLSON	AUCH	ILL	KARROS	HALTON
1	ELLSON	AUCH			
B/Garage					
SB-1 Tunnel	TO: COLLEGE BUILDINGS, RESIDENTIAL AND 2 OTHER HOSPITALS				

Fig. E–2. Hospital in-patient floors bed counts (left) and horizontal egress pathways for an evacuation (right)

Note: The foundation for better building intelligence was in position with the first building information card (hardcopy) enacted in New York City soon after September 11, 2001. Since then, it has been sanctioned both at the national and the state levels by the International Fire Code (2012), Section-508 Fire Command Center; *NFPA-1620 Standard for Pre-Incident Planning* (2010) with the current 2020 issue in Section 4.5.2 Electronic Version; Insurance Service Office (ISO/2013), Fire Suppression Rating System, Section-H; and in *the NFPA 1521 Fire Dept. Safety Officer Professional Qualifications*: Section 5.33 Fire Suppression—Identify and estimate building/structural collapse hazards, given a building fire incident, a building collapse incident, reconnaissance opportunity, and establish AHJ pre-incident building plan information. . . and appropriate adjustments are made to the IAP by the IC to improve member safety.

F

Report Writing

Report writing is a vital skill for the incident commander, chief officer, and company officer. Whether you are first-due, assume a division/group, deploy a single hoseline, or assume command for the duration, you must be able to produce a well-written report. These are public documents that may be used in litigation, both criminal and civil. Families, insurance companies, lawyers, judges, other fire departments, and a host of other parties may read the report of your fire. Do you want to present yourself as a skilled professional or a third grader? Unfortunately, many fire officers have not been taught how to write. We hope several examples in this appendix will help.

Regardless of the report-writing program your agency uses, you will likely be required to produce a narrative of your assignment. As the IC, you will be required to provide a narrative of the entire incident, from dispatch to the last unit leaving the scene. This is a freestyle description of the events, conditions, assignments, ICS structure, accountability, IAP, and resolution of the incident. We will provide several examples for several different types of incidents.

Your agency will likely have a policy for report writing, along with a training guide on how to complete reports (see fig. F–1). This should include how to fill out the various fields in the report-writing program, and possibly sample narratives. The following narratives and generic examples will give you the end goal for which to aim. Again, as with all lessons in this textbook, adapt for your agency-specific policies, practices, and SOGs.

Incident #1—House Fire

House Fire (First-Due Engine 1)

E1, E2, TR3, Medic 1 and Chief 1 were dispatched to a house fire at 123 Main Street. Engine 1 and Medic 1 arrived to find a single-story home, approximately 1500 square feet, with heavy fire from the rear and smoke from multiple windows on the front. This was a low victim profile, as we had a credible RP/owner outside stating everyone was out of the house and accounted for.

Engine 1 had a water supply across the street. E1 established Main Street Command and initiated fire attack with a 1¾" line through the front door. Medic 1 was assigned IRIC/2-out and deployed a second line for 2-out/door control and protection of egress. Command was passed to E2 Officer, who assumed command upon arrival and his crew performed primary

MIDWEST CITY FIRE DEPARTMENT TRAINING DIVISION

ISO CLASS 1

SOAP Narrative Template:

The SOAP template is a simple guide to help format a report narrative.

S – Subjective information: This is what you were dispatched to or told.

O – Objective information: This is what you observed first hand.

A – Assessment: This is what you determined the situation was.

P – Plan: This is what you did or directed.

Fire Example:

S – E6 dispatched to business alarm at 123 main. Dispatch reports Zone 6 Smoke.

O – Upon arrival found nothing showing from the store with contractors present.

A – It was determined that the alarm activation was due to construction dust.

P – Unable to reset panel due to dust. Foreman Smith advised to take system off line while performing work. 205 was advised of the situation by phone since this was our second response of the day. E6 cleared, scene left in care of contractor.

EMS Example;

S – E2 dispatched to accident on the highway. Updated by dispatch this was possibly on the off ramp to Sooner.

O – Two vehicles involved in the accident were located at Casey's Gas Station on Sooner, not on the highway. Damage to both vehicles was minor and parties were exchanging information.

A – Minor auto accident

P – Cancelled highway accident response and returned all fire units to service. Continued PD for report and EMS for patients. Checked patients for injury and EMS completed refusal. Checked vehicles for leaking fluids and found none. E2 completed assignment and left the scene in care of PD.

Fig. F–1. Example report writing and training guide policy

search. Command was then finally transitioned to Chief 1. E2 was assigned Division A. E1 then reported to Division A.

The fire was contained to the kitchen. E1 reported knockdown to Division A and then transitioned to salvage and overhaul. E1 and Medic 1 were released by Main Street Command after the after-action review and Investigator 1 arrived on scene.

Same House Fire (Second-Due Engine 2)

E1, E2, TR3, Medic 1 and Chief 1 were dispatched to a house fire at 123 Main Street. E2 arrived and assumed Main Street Command from E1 who was initiating fire attack through the front door with a 1¾" attack line. E1 had a patent water supply. Medic 1 had deployed a secondary line for 2-out/IRIC. E2 Crew initiated primary search. E2 Officer performed a lap of the structure. The fire was located in the kitchen in the rear. There were no outside hazards noted and no basement or cellar identified.

The owner on scene reported everyone out of the structure (low VP). Chief 1 arrived and assumed Main Street Command from E2. E2 was then assigned to assume Division A with the tactical objectives of fire attack, search, and horizontal ventilation. E1, Medic 1, E2 Crew, and TR3 were assigned to Division A. E2 Crew initiated primary search, and TR3 performed horizontal ventilation and secondary search. The fire was quickly knocked down. Division A reported primary and secondary searches clear to Command. Division A crews performed salvage and overhaul until released after the after-action review.

Same House Fire (Chief 1)

E1, E2, TR3, Medic 1, and Chief 1 were dispatched to a house fire at 123 Main Street. E1 and Medic 1 arrived reporting a single-story home, approximately 1500 square feet, with heavy fire from the rear and smoke from multiple windows on the front. This was a low victim profile as they had a credible RP outside stating everyone was out of the house and accounted for. Chief 1 balanced this to a working fire assignment, which added Chief 2 and E3.

E1 initiated fire attack with a 1¾" line through the front door with a patent water supply established. Medic 1 established IRIC with a second attack line ready. E1 passed command to E2, who initiated primary search.

Chief 1 arrived with TR3 and Chief 1 assumed Main Street Command from E2. The ICP was located directly across the street from the front of the house. This was a life saving priority, offensive strategy and low VP was confirmed on the radio.

E2 was then assigned Division A with the tactical objectives of fire attack, search, and horizontal ventilation, with E1, Medic 1, E2 Crew, and TR3 working for him. Upon the arrival of E3, they were assigned to assume RIC, with the tactical objectives of softening the structure, securing utilities, and preparing for RIC operations. RIC and Division A reported directly to Command. Medic 1 was in a position to transfer to medical if needed. Chief 2 was cancelled.

Division A reported a primary search complete at 10:12 hours. Fire was contained at 10:16 hours. Division A reported secondary search complete/all clear of the structure at 10:20 hours.

Investigator 1 arrived and determined that food on the stove was the cause of the fire in the kitchen. See Investigator 1 report for more details. There was minor extension into the attic space above the hood system in the kitchen. Cellulose insulation was removed from the area to prevent rekindling.

American Gas and Electric was notified and secured utilities upon arrival. The owner, Mr. John Q. Public, was briefed on the conditions and actions taken. He stated that he needed his medications, which were retrieved by Division A/TR3. He had no medical complaints and he was going to tay with family. He was given report information prior to our departure from the scene.

A brief after-action review was conducted with all companies on scene and the scene was cleared at 11:55 hrs. Mr. Public stated that he would coordinate a board-up crew for the windows and sliding door on the rear of the house with his insurance company. Mr. Public was given Chief 1 cell and email for any needs.

Incident #2—Commercial Structure Fire

First-Due Engine 101

E101 and M101 were dispatched to a commercial structure fire at 1234 Auburn Blvd. Upon arrival, E101 reported a large warehouse, approximately 150 × 150 feet, with heavy smoke coming from the rear loading dock. A hydrant was located in the rear parking lot, approximately 200 feet away. E101 directed the next-due engine to lay out from Engine 101, pump back from the hydrant in the rear, and assume command from E101 upon arrival.

The manager on scene stated that there was a small gasoline spill inside the rear stock area and was ignited. He reported that employees were in the process of evacuating the building. A high VP was declared due to employees in the process of evacuating. The fire had increased and was involving rack storage and pallets of materials. E101 initiated fire attack with a 2½" line. M101 was directed to assist with the stretch, but stage away and be available for possible medical needs with their gurney and EMS gear at E101.

E103 arrived and command was passed to E103, who completed a water supply and assisted with fire attack. E101 worked under Division C for the entire incident. We changed SCBA bottles twice, notifying Division C, and continued fire attack until the fire was extinguished. We assisted with salvage and overhaul. Command conducted an after-action review and E101 was released.

Second-Due Engine 103

E103 was dispatched to a commercial structure fire in a warehouse. E101 arrived first, reporting a large warehouse with heavy smoke visible from the rear and high VP. E101 requested a reverse lay out to a hydrant in the rear. Upon arrival, E103 stopped at E101, E103 Officer and Firefighter disembarked and E103 Officer met face-to-face with E101 Officer. Command was transferred to E103 Officer. Engineer 103 laid out to the hydrant in the rear parking lot and pumped back to E101. I assigned my firefighter to join E101 crew with the stretch of their 2½" attack line. M101 was retrieving medical equipment to be ready in the rear for any possible victims.

Command was mobile and a second alarm was called, as fire had grown significantly. Staging for the second alarm was located at Auburn and Fulton. TR106 arrived and was assigned to assume Roof Division, with the tactical objective of a roof report and vertical ventilation. E106 also arrived and was assigned to deploy a 2-out line to protect E101 in the rear. Battalion 7 arrived with E/TR106 and assumed command after a brief transfer of command. E103 was then assigned Division C, with E101, E103 Crew, and E106 working for me. Division C tactical objectives were fire attack and search. TR109 was also assigned to Division C to assist with search. Battalion 13 assumed Division C from E103. We worked under Division C for the remainder of the incident until released.

First-Due Chief Battalion 7

Battalion 7, Battalion 13, E101, E103, E109, E106, TR106, TR109, and Medic 101 responded to a report of heavy smoke from a commercial building at 1234 Auburn Blvd. E101 arrived first and reported heavy smoke from the rear of a large 150 × 150-foot commercial warehouse. E101 went to the rear and initiated fire attack with a 2½" line and requested the second-due engine to perform a reverse lay to a hydrant in the rear. He upgraded the fire to a high VP, as employees were in process of evacuation. E103 arrived, assumed command from E101 and established a water supply and a secondary attack line. Command requested a second alarm with staging for the second alarm at Auburn and Fulton Avenue. Battalion 7 requested another transporting medic unit enroute for a total of three, due to the high VP.

Battalion 7 arrived and assumed command from E103. The ICP was on the C side of the structure in the rear parking lot. This was a life saving priority, offensive strategy, high VP. E103 was assigned to assume Division C, with E101, E103 Crew, and E106 working for him, and the objectives of fire attack and search. He requested another transporting medic unit, as M101 was also assigned to him, assisting with fire attack. I notified him that we had three medic units assigned to the incident. Two 2½" attack lines were in the building.

TR106 was assigned Roof Division, with the tactical objectives of a roof report and vertical ventilation. TR109 arrived and was assigned the tactical objective of primary search under Division C. The manager was directed to coordinate with TR109 Captain to assist with confirmation of employees out of the structure.

E109 arrived and was assigned to enter the A side exposure unit of the building and give a CAN report. Law enforcement was requested for traffic control on Auburn Blvd. M106 arrived and was assigned to initiate medical with the objective of standing by for treatment and transport on the C side. EMS 1 arrived and assumed Medical Group with M106 and M109 working for him.

Battalion 13 arrived and was assigned to assume Division C from E103. Division C requested two additional engine companies for crew rotation. E105 and E110 from the second alarm were assigned to Division C.

E109 gave the following CAN from the A side exposure unit: Conditions were moderate smoke, a fire wall was holding, and all employees had evacuated. He deployed a 2½" line and requested an engine and truck to assist and prepare if conditions worsened. Battalion 5 on the second alarm was assigned to assume Division A, with E109, E20, E19, and TR20 working for him. Division A tactical objectives were to hold the fire wall to protect the A side exposure unit and coordinate with Roof Division and Division C accordingly.

A third alarm was requested, with gas and electrical utility companies, rehab unit, light and air unit, and investigator. Roof Division requested another truck to assist with a trench cut at the fire wall between the C side involved unit and the A side exposure unit. TR23 was assigned to Roof Division for that objective. RIC was established with Engine and Truck 21, with Battalion 3 as RIC Group Supervisor.

Division C reported a primary search complete at 15:30 hours. The fire was knocked down and contained to the C side fire unit at 15:35 hours. A secondary search was completed with all clear declared at 15:43 hours.

The business manager, Mr. John Smith, was given report information. Salvage and overhaul ensued. Mr. Smith coordinated with the building owner and Command gave the building owner, Mr. Frank Jones, additional information, including incident number for a report. Mr. Jones coordinated a fencing company for the involved unit on the C side of the building. The A side exposure unit had minor smoke damage and the manager, Mrs. Simpson, was also given report information. All parties were given Battalion 7 contact information for support.

Investigator 1 arrived and confirmed that the fire started due to a spill of gasoline near an open flame. Rehab was called and after a brief after-action review on scene, the incident was reduced to two engines and two trucks for salvage and overhaul.

Battalion 7 terminated command and the incident was closed at 17:45 hours.

Incident #3—Vegetation Fire

First-Due Engine 311

E311, E312, E313, WT1, and Chief 1 were dispatched to a report of a grass fire at the corner of Highway 16 and American Blvd. E311 arrived to find a 3-acre grass fire in light flashy fuels, moving at a moderate rate of spread to the northeast. Potential was for 25 acres with residential structures at risk sparsely located in the fire's path to the east.

E311 added three additional pump and roll engines and two additional water tenders. E311 established Highway 16 command and initiated direct attack, pump and roll on the right/hot flank. E312 arrived on scene and was directed to come behind E311 on the hot flank for a tandem attack with the objective of getting to the head of the fire before it reached any structures.

Command was passed via radio to Chief 1 upon his arrival. E311 was assigned by Command to assume Division X Supervisor with the objectives of perimeter control and structure defense. I had E312, WT1, E321, E322, and WT2 working in Division X. The fire damaged fencing, outbuildings, tractors, and a trailer. There were no injuries reported.

First-Due Chief 1

E311, E312, E313, WT 1, and Chief 1 were dispatched to a report of a grass fire at the corner of Highway 16 and American BL. E311 arrived, reporting a 3-acre grass fire in light flashy fuels, moving at a moderate rate of spread to the northeast. Potential was for 25 acres with residential structures at risk sparsely located in the fire's path to the east.

E311 added three additional pump and roll engines and two additional water tenders. E311 established Highway 16 command and initiated direct attack, pump and roll on the right/hot flank. E312 arrived and assisted with tandem attack on the hot flank behind E311.

Chief 1 arrived and assumed Highway 16 Command from E311. This was a life saving priority due to homes in the path of the fire, and a combination strategy was declared with the ICP at the corner of Highway 16 and American Blvd. A structure alarm assignment was ordered to respond ahead of the fire at the cluster of homes in the path of the fire at the end of Saddle Road.

E311 was assigned to assume Division X, with E312, WT 1, E321, E322, and WT2 working for him. His standing tactical objectives were perimeter control and structure defense. E313 initiated direct attack on the left/cold flank. He was assigned to assume Division B with WT3 and E323 working for him. His standing tactical objectives were perimeter control and structure defense.

Upon the arrival of Chief 2 at the end of Saddle Road, he was assigned to assume Division O with the objective of structure defense. He was given the entire structure alarm assignment consisting of E1, E2, E3, TR1, and Medic 1. Medic 1 was directed to stand by for medical.

The fire was contained to 19 acres with fencing burned along the back of 51 Saddle Road, 53 Saddle Road, and two outbuildings were lost behind 51 Saddle Road, along with two tractors and a semi-trailer.

Division O coordinated support and report information for the owners of 51 and 53 Saddle Road. Highway Patrol was requested for traffic control. Crews continued to mop up the fire for 3 hours with zero smokers within 200 feet of the perimeter. Rehab 1 was called. Winds were 2–3 mph at the time of the last unit leaving the scene at 21:30 hours. All snags and fence posts were completely extinguished, as were any identifiable hot spots within the burn.

Incident #4—Multi-Casualty Incident

Chief 1 Report

Battalion 1, E1, and M1 were dispatched to a shooting with suspect still on scene with law enforcement at Jackson High School. Units arrived and staged across the street, awaiting the clear-to-enter order from law enforcement. Battalion 1 went to the parking lot of the school while Engine and Medic 1 staged. Sheriff Deputy Sgt. Smith was on scene, reporting that the threat was neutralized and 5 victims were shot.

Battalion 1 notified Engine and Medic 1 to come into the scene immediately. Battalion 1 established Jackson Command with unified command established with law enforcement (Sgt. Smith) and called 3 additional engines, 5 additional transporting medic units, and 2 air ambulances code 3. This was a life saving priority, offensive strategy. The ICP was in the parking lot, student drop-off area.

Engine 1 and Medic 1 initiated triage and treatment of 5 confirmed patients. Two were shot in the torso (immediate), and three were shot in the extremities (delayed). The helispot was located on the sports field on the south side of the campus. E2 and E3 arrived and assisted with patient care. E4 went to set up the helispot and assumed Helispot Manager.

EMS Officer 1 arrived and was assigned to assume Medical Group Supervisor. He had E1, E2, and E3 performing patient care. M1 moved the first critical patient to the helispot. M2 moved the second critical patient to the helispot. Each patient was transported to Trinity Trauma Center by air ambulance. Medics 3, 4, and 5 transported the delayed patients to American Hospital Code 3.

Law enforcement confirmed that one threat was neutralized. All patients were transported, and no additional patients were identified. Law enforcement remained on scene for investigation. Fire remained on scene to assist law with reunification of parents and identification of patients and transportation destinations for all parents of victims.

See EMS reports for additional information.

Glossary

active accountability. The constant awareness of crew locations and assignments by division and group supervisors through visual, hands-on, face-to-face, and radio communication.

aggressive. Anticipation in planning, assertive decisions, and proactive actions.

ALS-BASE. Attack, lobby, staging, and base; used for high-rise buildings.

anchor and hold. A tactic utilizing control lines and large water streams when attempting to stop fire spread in a wildland arena.

anchor point. The location from which a vegetation fire attack is initiated.

apartment complex. A series of multifamily dwelling (MFD) buildings located near each other.

area command. An expansion of the incident command function primarily designed to manage a very large incident with multiple management teams assigned.

arrival report. After a mental size-up, the first-due company or chief officer's initial radio report to incoming units to communicate arrival, conditions, actions, and needs, including the assumption or passing of command for subsequent responding resources.

assistant safety officer (ASO). Officer who, separate from the incident safety officer, must have the same training as the entry team and who focuses solely on hazmat operations safety.

auditory exclusion. The condition in which an individual is focused on a particular sound or voice and does not hear other background sound.

big box occupancies. Extremely large retail outlets of between 100,000 and 200,000 square feet or more. Examples include Home Depot, Walmart, and Target.

branch. A geographic or functional unit composed of multiple groups or divisions.

breach-enter-search. Accessing a compartment or area for search operations through a wall and not a traditional opening such as a door or window.

BRO-HAM. Boundaries, resources, objectives—hostile, agile, mobile; used when working as a division or group supervisor.

bump and run. A tactic in which resources typically move ahead of the fire front in the spotting zone to extinguish spot fires and hot spots, defending as many structures as possible.

casualty collection point (CCP). A location where victims are brought from the hot zone/defensive space to a warm/cold zone/offensive space to be treated and transported.

center hall apartments. Apartments designed in such a way that door to each individual unit opens to an interior enclosed hallway.

centralized command. A command structure in which all resources report directly to, and all tactical decisions are made by, the incident commander (IC).

check and go. A rapid evaluation to check for occupants requiring removal or rescue.

combination strategy. A strategy indicating that the incident has an area where an offensive strategy is being employed while a defensive strategy is being employed simultaneously.

command channel. A frequency dedicated to two-way communication between dispatch and responding units, or between units transmitting to each other.

command lag. The time difference between an incident within an incident (IWI) occurring and the IC responding to it.

command maturity. The ability to detach from the emotions of the situations and be calm, rational, focused, thoughtful, systematic, and professional.

command presence. The ability to exude confidence that inspires your troops to follow you.

commander's intent. Also called leader's intent, the incident priority/strategy declaration upon the assumption of command. For fire officers, it is the clear priority/strategy declaration that tells units what the overarching plan is. The three components are task, purpose, and end state. By understanding these three factors, subordinates at all levels can act on behalf of the IC by whatever means available, within the parameters of the commander/leader's intent.

complex. Two or more individual incidents located in the same general proximity that are assigned to a single incident commander or unified command to facilitate management.

concrete tilt-up. Buildings characterized by pre-poured concrete walls that are set and connected to the roof assembly for structural stability.

contamination reduction zone (CRZ). The area between the exclusion zone and the support zone in which decontamination occurs.

control objectives. Objectives based on operational or geographic boundaries of an incident. A structure fire may have a control objective of containing the fire to the room or unit of origin. For larger fires, it may be to the wing, floor, or building of origin. A vegetation fire may have roads, rivers, or control lines.

decentralized command. When the IC has divided the incident into manageable parts and given tactical divisions or group supervisors the authority and resources to complete objectives, account for crews, communicate the strategy, watch out for safety issues, manage the risk, and keep the span of control manageable.

defensive strategy. A strategy indicating that personnel are engaged in a backward-moving posture to engage in less aggressive and lower-risk operations to accomplish incident objectives and prevent further loss or damage.

direct attack. Also known as an offensive strategy, this occurs when firefighters remove the flames from the fuel by directly extinguishing the fire with water, usually with a mobile attack or progressive hose lay.

disciplined initiative. The characteristic of taking action consistent with the commander's intent.

discretionary time. When there is enough time to make a decision and weigh various options.

dispatch channel. A frequency dedicated to outgoing incident dispatches.

division supervisors. Supervisors who focus on a geographic area, such as the front of a building, third floor, roof, or flank of a fire.

eave attack. To remove the soffit and flow water along the eave line.

entry team. The technical hazmat team that enters the exclusion zone to mitigate a spill or leak.

escape routes. The path that firefighters use to reach a safety zone or get away from the fire.

exclusion zone. The area immediately around a hazardous material leak or spill in which specialized personal protective equipment (PPE) must be used by technical-level responders.

extended attack. The phase of wildland fire management when initial attack resources are no longer capable of achieving the objectives (usually control) set for the incident and additional resources and overhead are committed, generally, extending incident activities into another operational period.

fractional effective dose (FED). Describes the percentage of the population for which conditions have become untenable.

FIRESCOPE. Firefighting Resources of California Organized for Potential Emergencies.

firing operation. A defensive tactic whereby fire is placed on the ground by qualified firefighters to remove fuel ahead of the fire.

FIRST. Facts, intuition, resources, strategy, and tactics; used for mental size-up.

flanks. The outer left and right edges of the fire line.

fog of war. A phenomenon whereby an individual is confused or uncertain about the situation, including the status of the enemy (fire), troop strength, and positioning on the battlefield (building or area affected).

forward staging. When personnel who are fully equipped with PPE and self-contained breathing apparatus (SCBA) come with tools and equipment to a forward area to be imminently assigned.

FPODP. Facts, probabilities, own situation, decision, plan; an order of operations for size up.

fuel. Any material that burns, such as grass, brush, trees, or manufactured materials.

garden apartments. Apartments designed in such a way that the door to each individual unit opens to the outside.

golden hour. Getting patients medical care within this 60-minute time frame is critical to their survival.

green. Fuel not yet burned.

gross decontamination. The act of decontaminating large numbers of people who may be exposed to hazardous materials. This may be done with water alone by firefighters and hoselines.

group supervisors. Supervisors who focus on the functional components of an incident, such as fire attack, search, ventilation, medical, or rapid intervention crew (RIC).

head. The leading fire edge.

heat release rate (HRR). The rate at which heat energy is generated by burning.

heat transfer. The exchange of thermal energy to a fuel by the mechanisms of conduction, convection, and radiation.

heel. The area of fire origin; where the base of the fire is located.

hotel. A center-hall type of structure typically composed of multiple stories with enclosed stairwells and interior hallways that access guest rooms.

hydraulic ventilation. The use of water through fog streams to evacuate smoke, heat, and gases from a room.

hydroxocobalamin. A hydrogen cyanide poisoning antidote, more commonly known by the trade name cyanokit. Hydroxocobalamin detoxifies the cyanide and converts it into a form of vitamin B12, removing it from the hemoglobin and allowing for oxygen to attach again.

Incident Command System (ICS) formula. B + C + R = ICS, where B = building (or situation), C = conditions, and R = resources. It determines your ICS template.

incident stabilization priority. The mode when all civilian life loss risk has been mitigated and the operations are focused on preventing further escalation, property loss, and bringing the situation under control.

incident action plan (IAP). The amalgamation of the priority, strategy, tactical objectives, communication, resource allocation, and organization of the incident.

incident chiropractor. The concept that alignment by the IC is maintained between the strategic, tactical, and task levels. Division and group supervisors are critical to help maintain this alignment by carrying out the IC's strategy.

incident commander (IC). The person who determines the incident priority and strategy, develops the incident action plan, and orders resources to mitigate the problem.

incident power curve. The common sequence of events that occurs in most emergency incidents and fires in which conditions appear to improve while the demand upon the IC and tactical supervisors is increased and the potential to become overwhelmed is greater.
incident priority. The most overarching urgency (life saving, property, or environment) that determines the incident strategy, objectives, resources needs, and level of gain or risk to be considered.
incident strategy. The determination of the position (offensive, combination, or defensive) from which resources will engage the problem.
incident within an incident (IWI). A secondary event that happens during an emergency operation that escalates demand significantly enough to warrant additional resources and incident action planning
indirect attack. Also known as a defensive strategy, this occurs when the fuel is removed from the fire. Examples would be cutting lines or firing operations during a wildfire.
initial attack. The resources initially committed to an incident.
investigation mode. The term used to indicate that the first company has no indication of fire or smoke (other than a possible audible alarm) and has chosen to engage in further investigation.
IOCAN. Identify, object, conditions, actions, and needs; used for arrival reports.
kind. A classification of resources in the incident command system that refers to function (e.g., hand crew, helicopter, engine, and dozer).
large estate dwellings. Houses that exceed 3,000 square feet, per National Fire Protection Association (NFPA) 1700. Also known as McMansions.
law of conservation of energy. The principle that energy can neither be created nor destroyed, it can only be transformed from one form to another.
LCES. Lookouts, communications, escape routes, safety zones; used for vegetation/wildland fires.
leader's intent. Also called commander's intent, a clear, concise statement about what people must do to succeed in their assignments. It delineates three essential components: task, purpose, and end state.
Level 1 staging. Staging that occurs when resources stand by in the direction of travel, within a block of the incident.
Level 2 staging. Staging that occurs when a separate and specific location has been established away from the main incident, usually when greater alarms are called.
lifesaving priority. Indicates that there is civilian human life at risk and savable. Therefore, a higher amount of risk may be taken by responders.
multi-casualty incident (MCI). An incident with more than one casualty.
medical communications coordinator (MCC). Person responsible for coordinating patient transportation destinations. The MCC communicates and coordinates between the trauma center that assigns patients to various hospitals, the treatment unit leader, and the patient transportation unit leader/group supervisor.
mission command. The military approach to command and control that empowers subordinate decision making and decentralized execution appropriate to the situation. Mission command requires an environment of trust and shared understanding among commanders, staffs, and subordinates.
mission command mindset. Command through controlling standards, expectations, and performance. This is a 24-hour-a-day enterprise consisting of training, drilling, pre-planning, coaching, empowering, communicating, analyzing, operating, debriefing, reviewing, and revising.
mission orders. A military term for tactical/control objectives. For fire ICs, these are tactical objectives such as rescue, exposure control, confinement, extinguishment, overhaul,

ventilation, salvage (RECEO-VS), and control objectives such as confining the fire to the room or unit of origin.

mobile attack. Also known as pump-and-roll, this occurs when apparatus is capable of driving and pumping water simultaneously.

motel. A garden-style building with doors opening to the outside atmosphere, often next to parking.

multi-family dwelling (MFD). A structure designed to be occupied by more than one family as the primary living space.

offensive strategy. A strategy in which personnel are engaged in a forward moving posture to engage in more aggressive operations to accomplish incident objectives.

operations triangle. The elements critical to successful operations: strategy and tactics, ICS, and standard operating guidelines. All three are required and work together in a symbiotic, interwoven fashion.

passing command. The act of notifying subsequent responding units that the initial company officer plans to relinquish command as soon as the next officer arrives and assumes command.

personnel accountability report (PAR). A unit roll call, historically conducted at time intervals (e.g., 10 or 15 minutes), per the standard operating guidelines (SOGs) of the department, or when a strategic shift or sentinel event occurs, such as switching to a defensive strategy or a roof collapse.

pincer search. Multiple search operations or multiple search points where crews are entering multiple locations for known victims in unknown locations.

positive pressure ventilation (PPV). The tactic of using fans to pressurize a space with outside air to exhaust smoke, gases, and products of combustion out of the building.

prep and defend. A tactic used when both a safety zone and temporary refuge area (TRA) are present and adequate time exists to safely prepare a structure for defense prior to the arrival of the fire front.

prep and go. A term implying that some preparation of the structure may be safely completed prior to resources leaving the area.

progressive hose lay. A wildland firefighting tactic whereby hose is deployed and extended continuously in successive sections, from the nozzle, for as long as necessary to contain the fire property/environment priority. It indicates that all civilian life is free of risk, the incident is controlled and stabilized, and now the focus of the response is on preventing further damage to property and the environment. Minimal risk is allowed in this situation.

rapid ascent team. One or more crews utilized to access upper floors of a high rise for reconnaissance or to conduct search and rescue operations as soon as possible.

rate of spread (ROS). The speed at which a fire grows away from the area of origin. The rate is usually measured in chains per hour, with one chain equaling 66 feet.

reading smoke. Assessment of the volume, velocity, density, and color of smoke on the fireground.

reconnaissance. Also known as recon, the act of further surveying the battlefield (commercial building, apartment complex, industrial facility, area, etc.) to better ascertain the enemy (fire, hazmat, threat, etc.), strength, location, routes of travel, and obstacles to access. The objective is to gain more information than a size-up from the street alone would allow.

red flag conditions. Weather conditions in which there is a greater potential for vegetation fires igniting and growing rapidly due to high heat, low humidity, and high winds.

reinforced response. Those resources requested in addition to the initial response.

roof report. The statement to the IC from the roof division regarding the construction, conditions, and load (CCL) on the roof.

rescue task force (RTF). In the context of an MCI, it is composed of fire/emergency medical services (EMS) personnel embedded with law enforcement to create a team with the objective to enter the space immediately adjacent to the threat and remove patients to casualty collection points.

safety zone. A location to which firefighters can escape if conditions worsen rapidly.

SIGNAL. Species, incapacity, gender, number, age, location; used with a confirmed victim profile.

single family dwelling. A stand-alone residential structure built for the occupancy of one family.

site access control. The act of preventing persons from going into and out of the exclusion zone to prevent cross contamination.

size-up. An ongoing mental process, from pre-planning, SOGs, and training to after-action review, more training, and SOG revision.

span of control. The number of resources a commander can effectively manage. Historically, the optimal span of control is 5:1. Span of control is calculated by the number of things that require the attention of the commander. This is not limited to assigned resources. Responding to the dispatch channel, driving in traffic, and mobile data terminals are additional examples.

spotting. Embers casting out ahead of the fire, creating new fires and causing rapid fire growth.

stack effect. The movement of air in a high-rise building due to external and internal temperature fluctuations whereby lethal smoke and gases can settle far above the fire floor and other locations, putting occupants in peril.

START. Simple triage and rapid treatment system; used to organize and effectively provide medical care to patients.

strike team. Five of the same kind and type of resource, with a leader.

support zone. The area farthest away from the exclusion zone in which support functions such as command and medical are staged and working.

survivability profiling (SP). The educated art of examining a situation and making an informed and intelligent decision based on known events or likely circumstances to determine if civilians can survive the existing smoke and fire conditions.

TACOS. Tactical decision making, active accountability, communications, ongoing size-up, safety officer; the job of the tactical supervisor.

tactical channel. A frequency dedicated to operations for a particular incident or region.

tactical discipline. The act of restraining oneself in the moment to consider alternative options that could make conditions more favorable, rather than reacting in the moment with adrenaline or emotion.

tactical gap. The space between the strategic-level IC and the task-level crews working where risk assessment, communications, accountability, and situational awareness become distorted.

tactical level. The level of the incident between the strategic-level IC and the task-level crews working.

tactical objectives. The incident's operational and functional goals to be accomplished. Examples for structure fires are rescue, exposure control, confinement, extinguishment, ventilation, salvage, overhaul, RIC, rehab, medical treatment/transport, and support.

tactical supervisors/officers. Officers who make operational decisions, assign crews to specific tasks, account for their personnel, and assess safety issues to accomplish incident tactical objectives.

tactical patrol. An operation where the key element is mobility and continuous monitoring of an assigned area.

tactical points of opportunity (TPOs). Building construction, access, layout, and other physical features that can be leveraged to increase the impact of resources to accomplish tactical objectives.

tanker. A fixed-wing aircraft designed to drop water on wildland fires.

task force. A combination of kinds and types of resources with a specific task or objective.

task saturation. The phenomenon whereby an individual performing tasks is overwhelmed, unable to think tactically, hear radio traffic, multitask, or see the bigger picture.

taxpayer. Also known as residential-over-commercial buildings, this is a term for a building with one or more retail-type businesses on the ground floor and residences on the upper floors. Sometimes the upper floors may have offices. The challenges with these buildings include fire load, occupancy, and access.

technical decontamination. The act of decontaminating the entry and backup team members with multiple steps and solutions appropriate to a hazardous material. If an entry is required, a backup team must be ready. This also requires technical decontamination for the team as they exit the area.

temporary refuge area (TRA). A preplanned area where firefighters can immediately take temporary refuge or relief without using a fire shelter in the event that emergency egress to a safety zone is compromised.

time to task (TTT). Also known as time to completion, the time it takes to accomplish a task once it is identified as necessary to the incident tactical objectives.

tunnel vision. Describes the limitation of only seeing what is immediately ahead and missing critical visual signs of changing conditions of smoke and fire, construction features, access and egress points, victims, bystanders, overhead lines, hazardous materials, and so on.

Type 1 incident. A large, federal-level event such as 9/11, large wildfires, and massive hurricanes.

Type 2 incident. Typically state-level incidents such as smaller fires, floods, and events that are located or impacting resources within that state.

Type 3 incident. Typically a county or operational area incident that lasts more than a day and involves multiple agencies.

Type 4 incident. Incidents that can be considered multi-alarm/multi-agency events (e.g., a multi-alarm structure fire) that last up to 24 hours.

Type 5 incident. A single-alarm, initial-response event of the type that happens every day, from medical aids to house fires, which can be mitigated in a few hours or less.

Type. A classification of resources in the incident command system that refers to capability. Type 1 is generally considered to be more capable than Types 2, 3, or 4, respectively, due to size, power, capacity, or in the case of incident management teams, experience and qualifications.

unreinforced masonry (URM). This describes buildings of brick, block, or concrete that is not reinforced with metal such as rebar.

values at risk. Items of value that could be threatened by a fire, including but not limited to structures, outbuildings, vehicles, utility infrastructure, fencing, stored materials, livestock, and so on.

vegetation fire. A fire that involves dry grasses, shrubs, and light and flashy organic fuels. These can range from roadside grass fires to large vegetation fires that engulf thousands of acres and structures with lives threatened.

vent point ignition. Ignition that can occur in any place where the combination of heat and fuel in the smoke meets an available oxygen concentration to support flaming combustion.

victim profiling (VP). The practice of determining the presence, status, and location of civilian victims.

water tender. A mobile ground apparatus composed of a large tank of water with the objective of providing remote water supply to other resources.

wildfires. Fires that are typically in larger, forested areas that can involve national or state parks, national forests, private land, and logging areas, and which may also encroach on populated areas.

wildland urban interface (WUI). The type of area where communities, neighborhoods, homes, and other buildings exist in regions of heavy vegetation growth and more mountainous topography.

Review Questions Key

Chapter 1

1. Describe the origin story of command and control.
   Page 4
2. Explain the Military Decision-Making Process (MDMP).
   Page 6
3. Describe the influence of Chief Llyod Layman in regard to fire officer development.
   Page 7
4. What does FIRESCOPE stand for?
   Page 9
5. Explain the generations of warfare and their influence.
   Page 13

Chapter 2

1. Define span of control.
   Page 22
2. Define and compare centralized and decentralized command.
   Pages 22 and 24
3. Explain what the commander's intent would include on the fireground.
   Page 25
4. What are the tenets of mission command?
   Page 27
5. Describe the operational core values of the Operations Triangle.
   Page 32

Chapter 3

1. Explain the importance of the 2009 revision to *NFPA 1021: Standard for Fire Officer Professional Qualifications.*
   Page 42
2. What does FESHE stand for and what is the purpose?
   Page 44
3. Describe some of the foundational characteristics of the National Incident Management System (NIMS).
   Page 46
4. Explain NFPA 1561.
   Pages 47 and 48
5. According to NFPA 1710, for what occupancies is it recommended that more than one member be assigned the duty of managing command?
   Page 49

Chapter 4

1. Describe the five incident types.
   Page 54
2. Explain the difference between size-up and an arrival report.
   Page 57
3. Explain a combination strategy.
   Page 60
4. Describe tactical objectives and control objectives.
   Page 61
5. Explain the tactical gap.
   Page 61
6. What are the three levels of staging?
   Page 64

Chapter 5

1. Explain tunnel vison and auditory exclusion and how they can impact situational awareness.
   Pages 74 and 75
2. Describe the importance of pre-planning and walkthroughs, including some of the components.
   Page 78
3. What are the NIOSH 5?
   Pages 81 and 82
4. Explain the difference between victim profiling and survivability profiling.
   Page 90
5. What are the four critical aspects of reading smoke?
   Page 98
6. Explain leaders intent.
   Page 101

7. Describe the command transfer process.
   Pages 106 and 107
8. What are some differences, benefits, and challenges associated with different command post locations?
   Page 111

Chapter 6

1. Explain heat transfer and heat release rates.
   Page 121
2. What are some of the complications of vent-limited fires?
   Pages 121 and 122
3. Describe sequential and simultaneous operations.
   Page 124
4. Explain the differences in statistical trends associated with firefighter and civilian fire fatalities.
   Pages 128 and 129
5. Explain how the term fire control differs from extinguishment.
   Page 133
6. What percentage of fire fatalities are associated with smoke inhalation?
   Page 141
7. What span of control considerations should be applied to apparatus pump operators?
   Page 137
8. Explain some of the NFPA 1710 recommendation differences between the four occupancy types with regard to tactical assignments.
   Pages 135, 140, 142, and 144

Chapter 7

1. Describe the "then/what if" mindset.
   Page 152
2. Explain the difference between command maturity and command presence.
   Page 154
3. Provide two examples of a combination strategy.
   Pages 155 and 156
4. Describe the challenges and preventative means associated with managing an incident within an incident (IWI).
   Page 158
5. Explain the command options/modes and situations for each.
   Page 164
6. What is the primary purpose of the tactical supervisor?
   Page 169
7. Define the acronym BRO-HAM.
   Page 171
8. Explain incident templating and the ICS formula.
   Page 180

Chapter 8

1. Explain some of the factors that contribute to complacency with regard to single-family dwelling fires.
   Page 197
2. Describe the differences between fire attack as a radio designator and a group supervisor position.
   Page 201
3. Provide three tactical considerations for garage fires.
   Pages 204, 205, and 206
4. Provide three tactical considerations for attic fires.
   Pages 209, 210, and 211
5. Describe the application of exterior control/transitional attack.
   Page 214
6. Provide an example of how you would utilize three division/group supervisors given a first-alarm single-family dwelling fire.
   Pages 216, 217, 218, and 219

Chapter 9

1. Explain some of the challenges and tactical considerations associated with an apartment complex versus a stand-alone apartment building.
   Page 225
2. Describe water supply considerations for responses to multi-family dwellings.
   Pages 229 and 230
3. The high life hazard of multi-family dwellings presents with a potential challenge to the medical capabilities of a fire response. Explain how to proactively address this.
   Pages 232, 234, 235, 241, and 242
4. Explain the unique challenges associated with row house construction.
   Page 236
5. What are some ways to address the needs of expanded communication associated with working incidents at multi-family dwellings?
   Page 249
6. Provide an example of how you would utilize four or more division/group supervisors given a first-floor fire in a three-story multi-family dwelling fire.
   Page 249

Chapter 10

1. Define a commercial structure and provide some examples.
   Page 256
2. What are some of the LAFD structure fire watch out situations?
   Pages 262, 263, 264, and 265

3. Explain some of the differences in tactical considerations between residential and commercial occupancies.
   Page 265
4. What is a roof report?
   Page 268
5. Describe cold smoke.
   Page 269
6. Provide tactical considerations for ventilation methods with regard to commercial occupancies.
   Pages 275, 276, and 277
7. What are some of the strategic and tactical considerations for strip mall fires?
   Page 307
8. What is a taxpayer?
   Page 290
9. Provide an example of how you would utilize four or more division/group supervisors given a center-unit fire in a five-unit strip mall fire.
   Page 307

Chapter 11

1. Explain the differences between staging and base with regard to a high-rise incident.
   Pages 317 and 318
2. Define a rapid ascent team.
   Page 321
3. What are some tactical considerations for ventilation of high-rise fires?
   Page 323
4. Describe the role and responsibilities of a lobby/systems group.
   Page 326
5. Explain the use of branches for ICS at a high-rise fire incident.
   Pages 327 and 328
6. Create an ICS chart for a high-rise fire with smoke visible from the street for a single apartment on the 12th floor of a 15-story residential high-rise building.
   Page 327

Chapter 12

1. Explain how setting up division/groups early can help prevent a Mayday.
   Page 335
2. What is the number one NIOSH 5 contributing factor?
   Page 336
3. Describe the foundation of two-in, two-out.
   Page 338
4. What does LUNAR stand for?
   Page 345
5. Explain the functions of a RIT/RIC crew and some of the tactical considerations.
   Page 338

Chapter 13

1. Describe vegetation fires, wildland fires, and the wildland urban interface.
   Pages 352 and 353
2. What are the three factors that drive the size and complexity of a vegetation fire?
   Page 359
3. Describe LCES.
   Page 362
4. What are the 10 Standard Fire Orders?
   Pages 362 and 363
5. Explain three of the tactical options for structure defense.
   Page 364
6. Explain why initial establishment of divisions on a vegetation fire would not follow traditional alphabetical order.
   Page 374

Chapter 14

1. Describe a contamination reduction zone.
   Page 381
2. What are the responsibilities of the technical reference specialist?
   Page 382
3. Describe victim profiling for hazardous materials incidents.
   Page 383
4. What are the medical considerations for a hazardous materials incident with regard to victims and technicians?
   Page 383
5. Discuss the distinct safety concerns, roles, and requirements for hazardous materials incidents.
   Page 390
6. Explain some of the expanded communication demands of a hazardous materials incident.
   Page 389

Chapter 15

1. Describe START.
   Page 398
2. What are the pros and cons of models which require law enforcement securing or clearing a scene for fire department or EMS response?
   Page 399
3. What are the three primary objectives in a multi-casualty incident?
   Page 402
4. Describe a casualty collection point.
   Page 402

5. What are the responsibilities of a transportation unit leader?
   Page 410
6. Explain some of the expanded communication demands of a multi-casualty incident.
   Page 410

Chapter 16

1. Explain why there is greater potential for the presence of the NIOSH 5 at incidents involving unified command.
   Page 425
2. What are the key components of Homeland Security Presidential Directive–5?
   Page 431
3. Define the acronym UNIFIED COMMAND.
   Pages 435, 436, and 437
4. What is the difference between a complex and area command?
   Page 441

Index

A

D

E

F

G

H

I

J

K

L

M

N

T

Z

About the Authors

Anthony (Andoni) Kastros began his fire service career in 1987 and retired as a battalion chief with Sacramento Metro Fire District after 32 years of active service. He is the founder of TrainFirefighters.com. and is the author of the Fire Engineering book and video series *Mastering the Fire Service Assessment Center*, 2nd edition. The text is used widely throughout the United States for officer development.

Anthony also authored the 3-part series *Mastering Fireground Command—Calm the Chaos.* This Fire Engineering book and video series focuses on fireground strategy, tactics, and ICS using actual incidents, video, and simulations.

In addition, Kastros authored the video *Mastering Unified Command—From Hometown to Homeland.* The video is focused on bridging the gap between fire and law for a true unified response and command platform to face the new normal that is prevalent in America today. The video features many experts from around the United States, including former Secretary of Defense and CIA Director Leon Panetta.

Kastros is an FDIC-International instructor, was the keynote speaker at FDIC 2013, and is the recipient of the 2019 George D. Post Instructor of the Year Award from the International Society of Fire Service Instructors and Fire Engineering.

Anthony hosts *The Command Show,* a monthly blog radio available from Fire Engineering. As a USAR Team Manager with Sacramento Task Force 7, he deployed to Ground Zero on 9/11. Anthony spent 4 years with a Type 1 Incident Management Team. He and his team at TrainFirefighters.com teach command, tactics, leadership, mergers, strategic planning, and labor/management relations throughout the United States. His faith and family are the most important things in his life.

Brian Brush got his start in the fire service in 1996 as a high school volunteer in Forestville, California. While attending Oklahoma State University School of Fire Protection Engineering, he worked as a training technician at Oklahoma State Fire Service Training and returned to California to work as a wildland firefighter as a seasonal in summers of 1998–2000 with Cal Fire.

He began a career at West Metro Fire Rescue in Lakewood, Colorado, as a firefighter paramedic in 2002. In his 13 years at West Metro, he was promoted to company officer, worked in special operations companies, and served as a rescue specialist with FEMA US&R CO-TF.

He returned to Oklahoma to be closer to family in 2015. He is currently the training chief at Midwest City Fire Department.

Brian holds a master's degree in Fire and Emergency Management from Oklahoma State University, is a designated chief training officer through the Center for Public Safety Excellence, and is a graduate of the National Fire Academy Executive Fire Officer Program.

An author for *Fire Engineering* since 2009, an instructor at FDIC since 2011, and FDIC keynote speaker in 2023, Brian is a *Fire Engineering* editorial advisor and serves on the FDIC advisory board.